Table of Contents

	Page
FOREWORD	xiii
ACKNOWLEDGMENTS	xv

SUSTAINABLE AGRICULTURE: RESOURCE MANAGEMENT STRATEGIES

Planning Agricultural Production Systems as Ecosystems for Environmental, Social, and Economic Sustainability
T. A. Dumper and J. M. Safley 1

Real Estate Tools for the Farmer: Habitat Preservation, Tax Benefits and Capitalization through Limited Development
G. Phillips 9

Virginia's Integrated Agricultural Nonpoint Source Pollution Control Strategies
M. B. Croghan 17

Involving Agricultural Producers in the Development of Localized Best Management Practices
R. M. Waskom and L. R. Walker 22

Risk Assessments: Management Tools for Preventing Water Pollution on Farms
E. Nevers, G. Jackson, R. Castelnuovo, D. Knox 30

SUSTAINABLE AGRICULTURE: MANAGEMENT SYSTEM TECHNOLOGIES

Economic and Environmental Analysis of Conventional and Alternative Cropping Systems for a Northeast Kansas Representative Farm
S. Koo and P. L. Diebel 38

Economic Evaluation of Variable Resource Management Based on Soil Productivity Variability of Selected Land Areas, Sioux County, Iowa
D. R. Speidel and R. E. Nelson 45

Integrated Water-fertilizer-pest Management for Environmentally Sound Crop Production
J. L. Fouss and G. H. Willis 53

Feasibility of Site-specific Nutrient and Pesticide Applications
J. C. Hayes, A. Overton, J. W. Price 62

Sustainable, Environmentally Sound Potato Production in Northern Maine
J. C. McBurnie 69

SUSTAINABLE AGRICULTURE: INFORMATION SYSTEMS AND EDUCATION

The Agriculture/Environment Interface: Locating the Relevant Literature
S. C. Haas 77

Cooperative Extension Service National Water Quality Database
C. E. Burwell and E. Fredericks 83

USDA's Water Quality Program – Environmentally Sound Agriculture
F. N. Swader, L. D. Adams, J. W. Meek 86

AGRICULTURAL MANAGEMENT SYSTEMS: NUTRIENTS

Analysis of On-Farm Best Management Practices in the
Everglades Agricultural Area
L. M. Willis, S. B. Forrest, J. A. Nissen,
J. G. Hiscock, P. V. Kirby.. 93

Save Energy, Resources and Money with IFAS Bahiagrass
Pasture Fertilization Recommendations
S. Sumner, W. Wade, J. Selph, P. Hogue, E. Jennings, P. Miller,
T. Seawright, M. Kistler, G. Weaver, G. Kidder, F. Pate 100

Movement of Fall-Applied Nitrogen During
Winter in Western Minnesota
J. A. Staricka, G. R. Benoit, A. E. Olness,
J. A. Daniel, D. R. Huggins 105

Nitrogen Management, Irrigation Method and
Nitrate Leaching in the Arid West:
An Economic Analysis Using Simulation
G. D. Miller and J. C. Andersen 113

Prediction of Nitrogen Losses via Drainage
Water with *DRAINMOD-N*
M. A. Brevé, R. W. Skaggs, J. E. Parsons, J. W. Gilliam 120

AGRICULTURAL MANAGEMENT SYSTEMS: EROSION CONTROL

Modeling Erosion from Furrow Irrigation in the WEPP Model
E. R. Kottwitz and J. E. Gilley 131

Is Conservation Tillage a Sustainable Agricultural Practice?
P. G. Lakshminarayan, A. Bouzaher, S. R. Johnson 139

Nutrient and Sediment Removal by
Grass and Riparian Buffers
J. E. Parsons, J. W. Gilliam, R. Muñoz-Carpena,
R. B. Daniels, T. A. Dillaha 147

AGRICULTURAL MANAGEMENT SYSTEMS: SOIL AND WATER

Terraces Technologically Obsolete or
Will Terraces Still Have a Future
D. R. Speidel 155

An Economic and Environmental Evaluation of
Conservation Compliance on a West Tennessee
Watershed Representative Farm
R. G. Bowling and B. C. English 160

Conjunctive Use of Surface and Ground Water for Crop
Irrigation in the Western United States
B. M. Crowder 168

Irrigation Storage Reservoirs as a Water Supply
Solution for Upper Telogia Creek
W. R. Reck 176

AGRICULTURAL MANAGEMENT SYSTEMS: POINT SOURCE MANAGEMENT

Chemical Mixing Centers for Loading and Mixing Chemicals
J. T. Wilson 184

State of the Art Agrichemical Handling Facility:
Butler's Orchard
J. G. Warfield, C. M. Gross, T. N. Davenport 191

Pesticide Containment: An On-farm System
R. V. Carter, Jr. 199

Pesticide Contamination of Mixing/Loading Sites:
Proposals for Streamlined Assessment and
Cleanup and Pollution Prevention
M. V. Thomas 203

Agricultural Chemical Spill Cleanup in Minnesota:
A Site-specific Approach
S. R. Grow..208

WASTE MANAGEMENT AND REGULATORY ISSUES

"Planned Intervention": Merging Voluntary CAFO Pollution
Abatement with Command-and-Control Regulation
L. C. Frarey and R. Jones..215

Policy and Science Addressing CAFO Odor in Texas
L. C. Frarey..223

Farmer Decision Alternatives
Under Environmental Regulations
M. Rudstrom, R. A. Pfeifer, S. N. Mitchell, O. C. Doering, III........231

Composting of Yard Trimmings – Processes and Products
R. A. Nordstedt and W. H. Smith...239

Beneficial Uses of Composts in Florida
W. H. Smith..247

AGRICULTURAL MANAGEMENT SYSTEMS: NUTRIENTS

Agricultural Surface Water Management (ASWM)
in Southwest Florida
S. L. Means and J. T. Mustion..254

Screening Tool to Predict the Potential for Ground Water or
Surface Water Contamination from Agricultural Nutrients
W. F. Kuenstler, D. Ernstrom, E. Seely..260

Swine Wastewater Treatment in Constructed Wetlands
P. G. Hunt, F. J. Humenik, A. A. Szögi, J. M. Rice,
K. C. Stone, E. J. Sadler 268

Nutrient and Water Management Practices for Sustainable
Vegetable Production in the Lake Apopka Basin
C. A. Neal, E. A. Hanlon, J. M. White, S. Cox, A. Ferrer 276

Impact of BMP's on Stream and Ground Water Quality in a
USDA Demonstration Watershed in the Eastern Coastal Plain
K. C. Stone, P. G. Hunt, J. M. Novak, T. A. Matheny 280

ANIMAL WASTE MANAGEMENT: DAIRIES

Dairy Park: An Environmentally Sound Alternative to
Current Dairy Production Practices
A. K. Butcher and R. Neal 287

Dairy Protects Environment Through
Implementation of Best Management Practices
E. O. Cooper...295

Design of a Rotationally Grazed Dairy in North Florida
M. P. Holloway, A. B. Bottcher, R. St. John 300

Dairy Loafing Areas as Sources of Nitrate in Wells
D. E. Radcliffe, D. E. Brune,
D. J. Drommerhausen, H. D. Gunther 307

Measurement of Seepage from Waste
Holding Ponds and Lagoons
G. G. Demmy and R. A. Nordstedt 314

ANIMAL WASTE MANAGEMENT: WASTE UTILIZATION

The AgSTAR Program: Energy for Pollution Prevention
K. R. Roos 324

Potential Options for Poultry Waste Utilization:
A Focus on the Delmarva Peninsula
C. A. Narrod, R. Reynnells, H. Wells..334

Opportunities Available to the Vertical Integrator or Food Processor to Achieve Water Quality Benefits in the Dairy and Poultry Industries
A. Manale, C. Narrod, E. Trachtenberg 344

Evaluation of Manure Water Irrigation
D. Morse, L. Schwankl, T. Prichard, A. Van Eenenaam...................353

ANIMAL WASTE MANAGEMENT: GENERAL

GLEAMS Modeling of BMPs to Reduce Nitrate Leaching in Middle Suwannee River Area
W. R. Reck..361

Quality of Well Water on Tennessee Poultry Farms
H. C. Goan, P. H. Denton, F. A. Draughon 368

Alligator Production in Swine Farm Lagoons as a Means of Economical and Environmentally Safe Disposal of Dead Pigs
W. R. Walker, T. J. Lane, E. W. Jennings,
R. O. Myer, J. H. Brendemuhl 373

Dead Bird Composting
N. N. Watts...379

URBAN/AGRICULTURAL INTERFACE: GENERAL

Preliminary Water Quality Results for A Rock Reed Filter Home Treatment System
G. Dougherty..383

Tree-Crop Polyculture: Small Farming in the Suburbs
J. A. Rogers..387

Hydraulic Properties of Re-Cycled Shredded Rubber Tires as Conductive Fill in Open Ditches
S. J. Langlinais..391

Petroleum Storage Tanks in the Agriculture Industry
C. G. Ward, J. R. Harriman, A. R. Fontaine 399

URBAN/AGRICULTURAL INTERFACE: WASTE UTILIZATION

Industrial Waste Streams as Nitrogen
Sources for Crop Production
S. P. France, B. C. Joern, R. F. Turco 407

Environmentally Sound Agriculture Through Reuse and
Reclamation of Municipal Wastewater
A. Roberts and W. Vidak 415

Diagnostic Evaluation of Wastewater Utilization in
Agriculture, Morelos State, Mexico
C. Rodriguez Zavaleta, L. Oyer, X. Cisneros 423

Newsprint and Nitrogen Source Interaction on
Corn Growth and Grain Yield
N. Lu, J. H. Edwards, R. H. Walker, J. S. Bannon 431

WETLAND TREATMENT SYSTEMS

Preliminary Effectiveness of Constructed
Wetlands for Dairy Waste Treatment
C. M. Cooper, S. Testa, III, S. S. Knight 439

Restoration of Wetlands Through the Wetland Reserve
Program: A Mississippi Perspective
R. L. Callahan and L. P. Heard 447

Habitat Protection and Land Development:
Exploring a Synergetic Approach
L. M. Zippay and J. W. H. Cates 453

SUSTAINABLE AGRICULTURE: MANAGEMENT SYSTEM TECHNOLOGIES

The Impact of the Citrus Conversion Process on
Ground and Surface Water Quality: A Case Study
A. N. Shahane 461

Spray Deposition and Drift in Citrus Applications
M. Salyani 471

Perennial Peanut in Citrus Groves - An Environmentally
Sustainable Agricultural System
J. J. Mullahey, R. E. Rouse, E. C. French 479

Biological Control of Silverleaf Whitefly: An Evolving
Sustainable Technology
P. A. Stansly, D. J. Schuster, H. J. McAuslane 484

Use of Fire as a Management Tool in
Alfalfa Production Ecosystems
W. C. Stringer and D. R. Alverson 492

SUSTAINABLE AGRICULTURE: GENERAL

Soil and Water Management for Sustainability
C. A. Bower 497

Bromide and FD&C #1 Dye Movement Through
Undisturbed and Packed Soil Columns
K. K. Hatfield, G. S. Warner, K. Guillard 499

Using GLEAMS to Select Environmental Windows for
Herbicide Application in Forests
M. C. Smith, J. L. Michael, W. G. Knisel, D. G. Neary 506

Sustainable Vegetable Culture on Erodible Lands
M. E. Byers, G. F. Antonious, D. Hilborne, K. Bishop 513

Nematode Management in Sustainable Agriculture
R. McSorley 517

Soil Contamination Caused by Discharge of a Mine Water
W. Wojcik and M. Wojcik 523

AGRICULTURAL MANAGEMENT SYSTEMS: SOIL AND WATER

Potential Impact of Proper Soil Water Management on
Environmentally Sound and Antitoxic Food Production
B. Maticic 533

Energy and Irrigation in Southeast Agriculture
Under Climate Change
R. M. Peart, R. B. Curry, J. W. Jones,
K. J. Boote, L. H. Allen, Jr. 543

Effects of Companion Seeded Berseem Clover on Oat Grain
Yield, Biomass Production, and the Succeeding Corn Crop
M. Ghaffarzadeh and R. M. Cruse 551

Role of Mycorrhizae in Sustainable Agriculture
D. M. Sylvia 559

Designs for Windbreaks and Vegetative Filterstrips that
Increase Wildlife Habitat and Provide Income
B. K. Miller, B. C. Moser, K. D. Johnson, R. K. Swihart 567

Application of MSF Program for
Simulation of Ecological Systems
W. Wojcik and M. Wojcik 575

Foreword

This national conference was planned to highlight state-of-the-art technology for sustaining an environmentally sound and productive agricultural industry in the urbanizing United States. Awareness of and sensitivity to the world's environmental problems are forcing society, including the agricultural sector, to re-evaluate its current modes of operation. Growers as well as environmentalists and government agencies need to implement and promote management practices which enhance environmental preservation while maintaining an adequate food supply and the economic vitality of agriculture. Much of the necessary information is available and needs to be integrated into a useable form in order to solve many of the environmental problems facing agriculture today.

This conference builds upon the successful approach used in the first Environmentally Sound Agriculture Conference held in Orlando, Florida on 16-18 April 1991 which was promoted only in the southeastern United States. With a concerted effort to obtain nationwide coverage, the content of this second conference addresses a broad range of issues and technology related to management implications. It is hoped that this conference will be instrumental in creating a more open environment for a more knowledgeable dialogue by all concerned regarding the issues of a sustainable and environmentally sound agricultural production system in the United States. The goal of the conference and this resulting proceedings is to facilitate communication and understanding among all concerned regarding techniques for achieving sustainable and environmentally sound agricultural production systems. Hopefully this conference will broaden the perspective of all of us and enhance the implementation of environmentally sound production practices on a broader scale.

Kenneth L. Campbell

Committee and Sponsors

Sponsored by:

    University of Florida
    Institute of Food and Agricultural Sciences

Co-Sponsored by:

    Soil and Water Division
    American Society of Agricultural Engineers

    Florida Section
    American Society of Agricultural Engineers

Conference Planning Committee:

    K. L. Campbell, Conference Chair
    W. D. Graham, Conference Co-Chair
    A. B. Bottcher

Acknowledgments

As Conference Chair, I want to thank the members of the conference planning committee, Drs. Wendy Graham and Del Bottcher, for their willing and dedicated efforts in helping to make this conference a success. As Conference Co-Chair, Dr. Graham has provided valuable input and support at every step in the development of the conference plans. My job would have been much more difficult and time-consuming without her input and help with the planning of many aspects of the conference. Dr. Bottcher provided support with his previous experience as Conference Chair for the first ESA conference held in 1991. His help was especially appreciated in making the contacts and arrangements for the keynote invited speakers for the conference. Additional thanks go to the staff of the UF/IFAS Office of Conferences and Institutes for handling publicity, facilities and physical arrangements details, and to ASAE Headquarters staff who were extremely cooperative in working with us to develop a timely and high-quality proceedings of the conference. And finally, the conference planning committee thanks all the authors who contributed so much valuable knowledge and information through their presentations and written papers recorded in this volume.

Kenneth L. Campbell

PLANNING AGRICULTURAL PRODUCTION SYSTEMS AS ECOSYSTEMS FOR ENVIRONMENTAL, SOCIAL, AND ECONOMIC SUSTAINABILITY

T. A. Dumper, J. M. Safley*

ABSTRACT

Successful agriculture is based upon an ability to manage ecosystems to meet social and economic needs. Measures of success include obtaining sustained production and profitability while meeting food or shelter requirements, maintaining the resource base, and protecting the environment.

Resource management must use integrated systems analysis to achieve sustainable agricultural success. An agricultural production system must be analyzed relative to the condition of its resource base, context within larger and smaller environmental levels, and with full consideration of human quality of life. Management of the agricultural production system must respond to objectives that directly influence the environment in which that agricultural ecosystem occurs, the sustained economic viability of the enterprise, and the acceptance of that enterprise within its social and economic framework. Planning for total resource management must reflect the agricultural ecosystem and functions of the agricultural production system as the product of the management systems that comprise it.

Planning for an improved agricultural production system can be characterized by its resource interactions and this can be shown by a matrix which relates environmental, social, and economic effects. An example of this matrix, using policies and concepts of the Soil Conservation Service, is included.

KEYWORDS: Agricultural production systems, Sustainability, Planning, Levels of systems, Resource interactions.

OVERVIEW

Introduction

Successful agriculture is the ability to manage ecosystems that vary in complexity while balancing social and economic concerns. This agroecosystem is formed from interactive living and nonliving entities, such as natural resources, that are managed toward specific objectives. These resources may be broadly classed in terms of land, water, plants, air, and animals. As people attempt to control an ecosystem for economic and social return, the perceived bounds of resource effect must be extended and management must interact at many levels to provide sustainability.

---

* T. A. Dumper, Freestone Farms Consultants, Tannersville, VA., formerly MSEA Coordinator, Midwest National Technical Center, Soil Conservation Service, USDA, Lincoln, NE, and J. M. Safley, Assistant Director, Ecological Sciences Division, Soil Conservation Service, Washington, DC.

Concepts of the Agricultural Production System as an Ecosystem.

Agricultural ecosystems have been described (Odum, 1984) as domesticated ecosystems that exercise the life processes of primary production, that consume and decompose within an abiotic world, and that result in energy flow and nutrient cycling. Systems are constrained by economic and social factors that control sustainability. The agricultural ecosystem occupies an intermediate position between natural systems, such as grasslands or woodlands, and fabricated systems, such as an urban community.

Agricultural ecosystems, as any other ecosystem, are solar powered, but differ from natural systems in that:

1. Energy sources, such as labor and processed fuel, are used to enhance their productivity.

2. Species diversity is frequently reduced through management to maximize yields of specific food and fiber species.

3. Dominant plant and animal species are limited by artificial, rather than natural selection.

4. System control is achieved through external goals and management, rather than through feedback of the components, even though the natural characteristics of the systems may oppose those management goals.

It is clear that agricultural ecosystems are managed as agricultural production systems through human intervention. Intervention is applied to meet economic and social goals, such as production or conservation. Even so, agricultural production systems maintain the functional processes of natural ecosystems; such as nutrient conservation mechanisms, energy storage and use patterns, and regulation of diversity. Our modern landscape is dominated by these systems; about three - fourths of the United States is occupied by agricultural production systems.

Agricultural production systems are any grouping of environmental, economic, and social resources that are used to produce food or fiber. The system includes all of the physical, chemical, biological, ecological, social, political, and economic functions and processes which define it. Extensive human skills are required to manage it to meet social and economic objectives of success and sustainability. Conceptually, these resources are a continuum, connected horizontally among the resources, and scaled vertically in size from the minute to the immense (Holt, 1987 and Safley et al., in press). The agricultural production system is, in fact and in function, a managed agricultural ecosystem.

Attributes and Functions of an Agricultural Production System

Agricultural production exists to meet the basic human needs of food and shelter, to secure additional social benefits, and to obtain a desired lifestyle. Actions that meet social and economic needs and support the welfare of the producer and their communities include participation in market systems. These market systems may affect the operation of the agricultural production systems.

An agricultural production system, like any ecosystem, is interactive. When one factor of the system is altered through management, an effect will likely occur elsewhere in the system.

These effects may occur in the quality of natural resources or in the system's economic and sociological linkages. The producer's ability to manage resources successfully depends on an ability to perceive the structure and linkages of the system.

Management goals for an agricultural production system must consider all economic, environmental, and social effects to optimize long - term yield and sustain its resource base. The producer must also maintain or improve the quality of economic, environmental, or social resources to levels that are defined as desirable by the community of the producer. These sustainability goals are generally met through management of finite soil and water resources. Additional resource goals are also required by society for air quality and for plant or animal resources.

Hierarchies of an Agricultural Production System

Variability is present at all levels of interaction among natural resources, from the scale of a producer's environment to the intricacies of global ecosystems. Physical and economic scales vary from small, supplemental gardens to regional or international corporations. Social and economic scales range similarly from a share cropper's market, or barter system, to global trade.

If the scale of an agricultural production system is confined to the activities of an individual producer, a slice of the vertical extent of the system is smaller and relatively less complex. Even at that level, social and economic concerns predicate the well being and sustainability of the producer. At every scale the system is characterized by interactions among basic resources such as land, geologic materials, water, air, and biological forms that live at the interfaces of these resources.

Social and economic demands on an agricultural production system, reflected in terms of profit goals, make it difficult to consider actions that occur only at an individual's level. Increases in technical, transportation, and communication needs require producers to consider broader and narrower levels of interaction that affect their production systems. The manager of an agricultural production system must consider its function at levels ranging from a basic biological environment to a global market. The extent of the consideration represents the scope of the significant impact on the agricultural production system at each level. The following environmental levels must all be considered in successful management.

o   Plant Level: The vigor and productivity of the plant must be considered with other natural resources in its environment. Considerations include nutritional balance, function of the soil and other geologic materials as a growth medium, availability of air and water to feed and sustain the plant, its interaction with other plants as a competitor, sustainer or in symbiosis, and its functional relations with microscopic and megascopic animals.

o   Field Level: The combined primary and secondary management of plants and animals within a field is evaluated in terms of the availability of soil, water, and air resources. There are also social characteristics of this management which may involve choices of tillage, planting and harvest methods, and the overall quantity and quality conditions of natural resources.

o   Enterprise Level: In an enterprise, one or more fields are managed to produce food and fiber in order to bring economic and social returns. This return may be in terms of sustenance for the operator's family or their neighbors, or in coinage of the realm obtained in a formal or informal market. Typical enterprises, composing agricultural production systems, include crop production, grass utilization, livestock production, or the production of grass or trees for economic return.

o   Farm Level: This level may consist of one or more enterprises that are operated as a unit for social and economic return. Each enterprise may affect the operation of other enterprises within the farm unit and the natural and human resources that compose the agricultural ecosystem. Successful management of the natural and human resources must merge the goals and objectives of each enterprise and their demands on natural resources. Human resources are used to manage a farm unit that has multiple enterprises. They must consider each enterprise as a subsystem of the total farm operation and forecast the interactions and impacts of each environmental level.

The farm level has a management parallel in natural ecosystems, such as grasslands or forests. The plants and animals of a forest or grassland community may be managed for multiple uses; such as timber or forage production, wildlife or livestock habitats, and for human recreation or aesthetic enjoyment.

o   Community Level: Interaction of human and natural resources of all individual farms must be considered to meet the objectives of the community. The definition of the community, however, reflects the impacts of the farm operation on any larger world. Community boundaries may be drawn from physical or geopolitical areas, social and economic ties to market, or the relationships of a communication network.

The watershed is becoming a popular choice to represent a community during planning and management because its conditions are bounded finitely by land and water resources. Air, plant, and animal resources, however, easily cross watershed bounds and market concepts may be regional, national, or even global. Community influences on an enterprise create market effects, social regulations, and environmental concerns that affect management of natural or human resources. Effects on an agricultural production system from human environmental needs may occur from threatened and endangered species laws, cultural preservation regulations, or national and international trade agreements. The interactions affecting an agricultural production system occur at all levels, from plant environment to global community, and interface with each part of the natural resources base.

PLANNING FOR A SUSTAINING AGRICULTURAL PRODUCTION SYSTEM

The planning of an agricultural production system must consider all resources and at all of the environmental levels. The planning framework has been well defined and usually consists of a sequence of actions that respond to a resource problem or a manager's need for social or economic change. A change from present conditions is defined as goals and objectives in environmental, economic, and social terms. Steps of this framework generally exhibit this flow:

1. Identification of a resource problem or a need to make a change in the management or operation of an enterprise,

and the goals and objectives desired to remedy the
problem or implement the change are defined.

2. Quantification, identification and analysis of the
environmental, economic, and social resources available
occurs to meet the goals and objectives.

3. Formulation and evaluation of alternative solutions to
the defined problems or changes are based on the
environmental, economic, and social characteristics of
the problem area.

Following these planning steps, the most advantageous social,
environmental, and economic remedy to the problem or need is
selected and implemented that will meet the goals and objectives.

The nomenclature of this process has varied over the past decades
although there has been a growing consensus on its characteristics. The planning process has been described within the
federal establishment in multiple documents and regulations, such
as the principles and standards and principles and guidelines of
the U. S. Water Resources Council (1983); individual policies and
procedures of resource agencies and groups, such as the SCS
National Planning Procedures Handbook (1994); and analogous
documents of state and local governments and private sectors.
These procedures are expected to comply with environmental,
social, and economic laws and constraints that are promulgated by
society at all community levels.

An agricultural production system is usually a farm or a ranch
that has one or more enterprises. These enterprises are systems
that are managed to produce food and fiber, such as crops, forage
or livestock. They use environmental, economic, and social
resources, are managed as resource systems, and each acts as an
ecosystem. The systems are interactive and include operations
such as a crop, forage, forest, nutrient recycling, feeding,
milking, or grazing management systems. They also include,
actually as subsystems, water, wildlife, nutrient, and pest
management systems.

Each enterprise is made up of sets of resource management systems
that support production. If the farm or ranch has multiple
enterprises, such as the production of both forage and livestock,
these production systems must themselves interact to obtain a
desired level of efficiency. When properly formulated, the
product is a sustainable, long - term yield. The resource system
interactions occur inside and outside the boundaries of the
enterprise and their effects extend to the communities.

Goals and objectives are set at the farm level to respond to
specific problems and needs encountered in managing resources at
the enterprise level. Problems or needed changes are defined as
any failure to meet desired production goals or resource
conditions. The inventory and analysis of the production or
resource condition of the interactive fields and enterprises are
carried out to address the management goals and objectives.

Effects of Change in Agricultural Production Systems

Planning of agricultural production systems focuses on management
and physical changes at the field level. Conservation practices,
best management practices, or changes in farming procedure or
method may be applied at the field level, but the effects occur

at all levels from the plant environment to the community. Even when a change is made to increase plant growth and vigor, it is still applied as a field practice. These field changes are typified by tillage, water delivery, drainage, or management methods for residue, plant protection, fertilization, harvest, machinery usage, or livestock management.

The effects of the application of a change at the field level will not be limited to that domain, but potentially may affect a system of which it is a member. The concept of "everything is affected by everything" applies. Consequently, interaction among the resources of the systems must be considered among all fields within an enterprise; among all enterprises within a farm; among all farms within a community; and among all communities within a region or greater area. A large community may be a watershed, county, state, nation, or a global entity depending on the significance of the effect and the management perspective needed for success and sustainability of the enterprise.

Concepts of success in management vary with the level and breadth of the effect and reflect the goals set for managing the environmental, social and cultural resources. Some of the goals and objectives are internal to the farm or enterprise and its resource base. Other goals are imposed on the farm because of social, environmental and economic desires of the larger communities in which the unit is located. Goals and objectives range by scale and level, from a desired crop yield to a regulation promulgated by a nation or by the world.

All natural, social and economic resources should be considered in forecasting effects of a plan for an agricultural production system. Multiple level forecasting is difficult and the complexity of defining and evaluating these resource effects is demonstrated by a resource interaction matrix, Fig. 1. This matrix uses some of the planning concepts of the Soil Conservation Service (1994), who define the environmental resource base as consisting of soil, water, air, animals, and plants and social and economic resources as human considerations. Levels of resource effects, shown in the matrix, range from the plant environment to a large community, as defined by this paper. The interactions, shown by example, are all potential, general resource effects, rather than a quantified impact that would occur in a specific location. In evaluating a change for a specific agricultural production system, the effect would be quantified.

CONCLUSIONS

Complexities of interactive effects of changing agricultural production systems and the need to plan them based on ecosystem analyses is well demonstrated by the matrix. Effects occur at all levels of all systems where there is a resource membership. Unfortunately, although this paper describes the complexity of the systems, it does not present a simple model to evaluate system planning.

A successful plan for an enterprise and the farm depends on the ability of its designers and managers to consider the total set of multiple level resource effects. Planning for the total system is necessary to approach sustainability of an agricultural production system. Our present ability to make this analysis and evaluation depends on the planner's skill to perceive the resource effects at all environmental levels. The complexity of these interactions and their effects makes this a challenging task.

Resource and Environmental Level Effects

Resource	Earth Resources (Soil)					Water Res	
	Plant	Field	Enterprise	Farm	Community	Plant	Field
Water Quantity	Water holding capacity	Adequacy for field crop	Adequacy to produce economical crop	Adequacy domestic ground water	Sufficient resources to support market	N/A	
Water Quality	Materials dissolved in soil water	Chemical movement to water	Nutrient balance among fields	Soil loss effect on quality	Effects on water quality standards	N/A	
Air	Gas exchange	Wind erosion damage	Yield reduction from plant damage	Effect on farm headquarters	Blowing soil effects on public safety	ET effects	
Plant	Rooting medium	Soil loss tolerance	Loss to yield	Management among fields	Crop yield for marketing	Plant viability or survival	
Animal	Microbial habitat	Pest management	Able to support activity	Balance among enterprises	Resource supports market participation	Meets habitat requirements	
Social	Choice of tillage equipment	Choice of management system	Management systems among fields	Able to participate in USDA programs	Able to participate in market	Plant water management	
Economic	Suitable economic base	Resource quality for profit	Long term profit	Management for sustainability	Market effects of area	Water availability economics	

(Matrix continues)

Figure 1. Typical Effects of Agricultural Production Systems Among Environmental Levels.

Agricultural production system relationships are so complex that it seems likely that a computer assisted evaluation and decision will be required (Safley et al., In press). A mental planning and evaluation model is limited to the personal perceptions and experiences of the planner and that generally provides little documentation. A mental evaluation model has limited ability to define optimum production or full environmental compatibility. Resource managers are still the decision makers and they are limited by knowledge availability and processing capability.

Decision support systems are being developed in research communities, but they are not yet available to solve problems at this level of complexity. During the period when our tools do not yet meet our needs, a significant technical challenge remains for the planner. The planner must attempt to inventory, analyze, evaluate, and implement agricultural production systems as ecosystems with unassisted human ability to provide the best product for their customers. We are all a planner's customers and have a stake in plan success because they are considering the large community effects of an agricultural production system.

REFERENCES

1. Holt, D. A. 1987. Agricultural Production Systems Research. Proceedings of the Annual Meeting of the Agricultural Research Institute, October 7-9, 1987, Washington, DC

2. Odum, Eugene P. 1984. Properties of Agroecosystems, In: Agricultural Ecosystems; R. Lowrance, B. R. Stinner, and G. J. House (eds.), John Wiley and Sons, New York, pages 8-11.

3. Safley, John M., T. A. Dumper, D. A. Holt, R. I. Papaendick, F. Thicke, and D. A. Bucks. In Press. Systems Approach to Soil and Water Conservation and Water Quality. In: Protection of Soil and Water Resources. (A joint project of USDA and the Ministry of Agriculture, CIS), USDA, US Government, Washington, DC

4. U. S. Water Resources Council. 1983. Economic and Environmental Principles and Guidelines for Water and Related Land Resources Implementation Studies.

REAL ESTATE TOOLS FOR THE FARMER: HABITAT PRESERVATION, TAX BENEFITS AND CAPITALIZATION THROUGH LIMITED DEVELOPMENT

Gary Phillips

ABSTRACT

Real estate is the basis of farmers' wealth and future benefit, but few farmers have more than a cloudy perception of how real property law operates "in the marketplace" or the numerous tools available to us for estate planning, capitalization and tax relief. Access to good land at an affordable price and capitalization for equipment and improvements are two of the gravest problems facing today's farmers, while burdening debt service has forced many out of business entirely. At the same time there is a cry throughout the country for responsible land management, preservation of farm lands, protection of waterways and diverse habitats, an impulse toward greater environmental responsibility than we have exhibited in our checkered history. Farmers need the tools to use their real property in nontraditional ways which also promote community and environmental goals.

This paper will evaluate mechanisms such as conservation and preservation easements, transferable development rights, and limited development of farm properties. Case studies will be explored, including the author's own projects and experiences as an "environmental developer" working with land conservancies, landowners, farmers and farming organizations.
Keywords: Conservation easements, Transferable development rights, Habitat preservation.

CONSERVATION EASEMENTS

"Land is the only thing that matters. Land is the only thing that lasts."
Scarlett O'Hara, in Gone With The Wind

" Can attorneys reach back in the history of the common law to find an estate in land that is the 'cutting edge' of the environmental movement - a technique that satisfies the most conservative of estate tax planners and provides land owners federal income tax deductions, state income tax credits and local property tax relief as well - and that meets a public policy need to find an alternative to yearly appropriations for public land aquisition budgets?" (Camilla Herlevich, 1993, writing in The North Carolina Bar Association Newsletter)

The answer is an emphatic *yes*. Ms. Herlevich, a Wilmington, North Carolina attorney and one of the founders of the North Carolina Coastal Land Trust, is one of a small number of attorneys who specialize in the preparation of conservation easements, which are agreements between landowners and non-profit conservation groups or public agencies which restrict, *in perpetuity*, the rights of landowners over a specific piece of property. In return, landowners are usually given a charitable contribution deduction to offset income, a decreased tax basis, possible state tax credits and grounds to have their property taxes reduced. Conservation easements are fast becoming one of the tools of choice to protect fragile and/or ecologically significant areas, to preserve open space, and to keep prime farmland from intensive subdivision.

 Imagine real estate not as a commodity like corn or cars, but as a "bundle of rights," or the ability to use land in certain ways. The importance of the conservation easement is that it is a very flexible and powerful technique which allows us to remove and/or change certain elements of this "bundle of rights," reaping benefit for the landowner, the community and the environment.

In most cases, the landowner gives away or restricts the *future development rights* of any property placed under a conservation easement. The goal of a conservation easement is to protect some important conservation quality of the land, such as habitat, open space, scenic views, water quality or agricultural value. Other restrictions may be added depending upon the nature of the property and the reasons it is in need of protection.

A gift of a conservation easement works to the benefit of the farmer or landowner in several ways. First and foremost, the federal government allows the loss in value resulting from the restrictions on the property to be taken as a charitable deduction, thereby reducing net taxable income, capital gains income and/or estate tax value.

Obviously, if the farmer is holding his or her land for intense development this has a limited application (though still some value, and conservation easement-based developments have been very attractive to the market). But if the farmer wants to preserve the land for farming and wants to see it passed to their children without being sold for the estate tax, conservation easements have immense implications. What is necessary is the commitment to protect what is special, unique and valuable *for conservation purposes* about the specific piece of property.

The process is simple enough: an appraisal is made of the value of the property before the conservation easement was given and of the value after the restrictions were placed on the property. The difference of the "before" and "after" appraisal is taken as a charitable deduction. This can be quite substantial, as we shall see.

Considerations in planning a conservation easement:

> 1) The tax code requires that the gift be "for conservation purposes" and accepted by a legitimate conservation organization or public agency. Therefore it is important that the donor is truly contributing something of value to the environmental well-being of the community. There are substantial penalties for overvaluing the easement.

> 2) Any rights not specifically given away by the easement are still held by the landowner, including the "quiet enjoyment" of the property and the ability to sell the land, subject to its restrictions, for the highest possible price.

> 3) The specific restrictions within any easement are negotiable, as long as they serve the conservation purposes specified. For example, easements designed to protect a fragile plant environment would probably not allow the taking of timber anywhere within the easement, while an easement aiming to save prime farmland from development might allow selective timbering as part of the normal course of agriculture.

> 4) Public access is not required under most conservation easements, though the organization which owns the easement must have access for enforcement purposes, usually with notice.

> 5) Certain restrictions apply to the deduction, including a 30% cap (of adjusted gross income) on the amount of deduction which can be taken in the year of the gift, though the remaining deduction may be carried over for five succeeding years. Any conservation easement strategy should be carefully reviewed by a tax planner or capable attorney.

> 6) Many states provide tax credits and other incentives to encourage the gift of conservation easements, and several states allow for cash payments to farmers who preserve significant farmland in this manner.

7) After the granting of a conservation easement, landowners are usually able to reduce their property taxes because of the reduction in value due to loss of development rights.

Case Study # 1

In the fall of 1989 my partner Jay Parker and I bought from an unsuccessful developer 179 acres at the confluence of the Haw River and Dry Creek in Chatham County, North Carolina. Motivated by fear of increased traffic through their sleepy neighborhood and the proposed changes to the century-old tree lined roadbed that gave access to their properties, the local community had successfully thwarted the developer for several years by challenging his legal access to the property. By the time we arrived on the scene things had gotten bitter indeed, and the developer was threatening to carve a gravel pit on the land.

This parcel had both botanical and historical significance, with features such as high hardwood bluffs above the river, a flourishing montane community of plants along Dry Creek and the remains of a 19th century bridge across the Haw. Jay and I negotiated for an easement with the adjoining landowners and chose a strategy so unusual that our potential investors disappeared overnight! First we divided the land into four parcels of 25-90 acres in size. Then we placed the entire property in a conservation easement which precluded any further subdivision and established large protected wildlife corridors along more than two miles of creek and river frontage.

We were able to take a charitable deduction of over $150,000 because of the loss of value attributed to the conservation easement. Perhaps more importantly, the easement drew buyers who wanted to live in a protected natural area. Contrary to our investors fears, the large tracts all sold in the first year,. We made a reasonable profit, habitat was preserved, and the local community was protected.

Case Study # 2

The Nature Conservancy has been working with local farmers along Virginia's Eastern Shore to develop conservation plans which are both environmentally and economically sound. Though each conservation plan is based upon the characteristics of an individual site, the plans have several aspects in common:

> 1) A primary goal is the protection of significant natural features and the land's relationship with water.
> 2) A conservation easement is used as the guiding document.
> 3) The easement prohibits inappropriate uses of the property but allows, within limits, agricultural, residential and conservation uses of the land.
> 4) The easement allows for limited development within the confines of its conservation goals.

Agricultural land as well as habitat is preserved by these agreements, and the twin benefits of possible tax relief and limited development insures that the landowner receives the economic value he or she needs from the property.

TRANSFER OF DEVELOPMENT RIGHTS

> "Property belongs to a family of words that, if we can free them from the denigrations that shallow politics and social fashion have imposed on them, are the words, the ideas, that govern our connections with the world and with one another: property, proper, appropriate, propriety."
>
> Wendell Berry, in <u>Meeting The Expectations Of The Land</u>

Everybody talks about TDRs but only a few communities have been successful in implementing them. Still, they have the potential for being one of the most powerful allies in support of farming and the farming lifestyle, particularly in the rapidly diminishing agricultural belts and communities outside our large population centers.

Briefly spoken, the concept works like this: A developer interested in doing a project (in a "receiving" area) pays a farmer or landowner (in a designated "sending" area) for their development rights, i.e., the ability to intensively subdivide their property, in return for some concessions from the planning jurisdiction, usually for a permit for higher density than zoning would allow. Remember our earlier comment that real estate is a "bundle of rights?" Well, the transfer of development rights merely means that a portion of the bundle has been taken from one property and allowed to be transferred to another. The farmer sells to the developer his or her right to intensively subdivide the property, and the developer uses that purchase to negotiate a concession with the planning jurisdiction. Thus farmland or habitat is preserved, the farmer is paid but able to continue farming, the developer has a more viable project, and community values are enhanced.

Montgomery County, Maryland, has a sophisticated TDR program designed to protect its prime agricultural land from encroaching development. In 1980 the county set aside 110,000 acres in an Agricultural Reserve and down-zoned the entire area to allow for only one dwelling per 25 acres. Farmers were allowed to sell one development right for each five acres of farmland owned. Developers, eager to increase the density of suburban subdivisions, had bought thousands of TDRs in the open marketplace by the mid-80s, at prices ranging from four to eight thousand dollars each. Thousands of acres have been permanently protected, and farmers have been compensated.

This is only one of the ways in which a TDR program can be established. Dozens of such programs exist across the country to varying degrees of success, and most can be contacted through local planning departments.

The American Planning Association (Roddewig and Inghram, 1987) has identified six essential steps in the formulation of an effective TDR system:

> * Identify the actors in the real estate marketplace effected by the TDR program and the economic motivation of each actor;
>
> * Identify potential receiving areas and thoroughly analyze the development opportunities and profits at various densities;
>
> * Identify and analyze potential sending sites, and balance environmental goals against economic realities;

> * Make a critical choice between a voluntary or mandatory program and between a totally private TDR marketplace or a quasi-public market assisted by a TDR bank;
>
> * <u>Make the program and the ordinances implementing it simple and flexible</u> (emphasis mine);
>
> * Insure adequate promotion and facilitation of the program once it is initiated, and insure that the program is designed to continue despite possible political changes.

Will TDRs evolve into a viable support structure for the preservation of farmland? These and other innovative stategies need to be implemented to insure that farmers share in the long-term economic value of the land.

LIMITED DEVELOPMENT

> "Good farming is farming that preserves the earth
> and its network of life."
> Donald Worster, in <u>Meeting The Expectations Of The Land</u>

Though my father grew up a subsistence farmer, we are no longer a nation of subsistence farmers. A farmer must farm and somewhere in the dirt he or she must turn up money. Sometimes that money must come from the sale of land. Most farmers I know think that the sale of any land is a personal failure, and all farmers want more land than they have, but limited development combined with a solid conservation easement for farmland preservation or habitat protection is a powerful tool in the landowner's favor, and one that can allow the farmer to keep the farm and keep farming.

Nearly 30% of all domestic farm production now occurs on the metropolitan fringe, according to Edward Thompson, Jr. (1992), Vice-President for Public Policy at the American Farmland Trust in Washington, D. C. There are many opportunities for farmers in these areas besides "the rock and a hard place" represented by the choices of selling out for development or watching real estate taxes escalate out of sight while public pressures threaten the viability of farming.

In my community I am known as an "environmental developer." I go into situations where, usually because of significant botanical and habitat areas, intensive development and the needs of the community are at odds. I try to bridge the gap between the economic needs of the landowner, the perceived needs of the community, and environmental protection considerations. Fortunately, wise land use can benefit all three.

Let me give you an example from my own experience of how this all works: Several years ago a handsome old lady came to see me, sporting a blue wool cape and an artist's beret. She looked to be over eighty, but very spry. After pleasantries were exchanged she made this speech: "I was told to come here because you are the man to handle my property. I have eighty acres on University Lake just a mile from town, which includes a place known as McCauley's Mountain. I have three goals for this property. 1) I want to give a portion of it away in memory of my dead husband, 2) I want to do this in a manner which will give me tax benefits, and 3) I want the sale of the remaining property to furnish me an income for life." She paused and then fixed me with a glare: "And don't make any assumptions, young man. I have a sister who is one hundred and two!"

That moment began a long, fruitful association with Mrs. Swalin, whose husband had founded the North Carolina Symphony, and who had lived on the land in question for over 20 years before moving into town.

We learned first that McCauley's Mountain was on the Inventory of Significant Botanical Areas prepared by The Triangle Land Conservancy, highly prized because of its relatively undisturbed summit and mature oak-hickory forest. Over 90 vertebrate species had been observed on the property over several years.

The Conservancy was enthusiastic when we approached them with the idea of a large gift. With the conservation gift (the summit of the mountain and all the land within the Inventory) as a touchstone we were able to sell three large surrounding tracts (each limited by deed to only one homesite) for a high premium. With the sale of her homeplace the property generated over $600,000 in cash, *and* a $300,000 charitable contribution deduction, used to offset her capital gains. This compared very favorably with selling to a developer for a housing subdivision, and all the sales occurred within a short period of time. The community benefitted, the environment benefitted, and the landowner's economic goals were met.

For other stories and strategies involving landowners and farmers in innovative partnerships with environmental organizations contact The Nature Conservancy (see Resources), which has identified Sustainable Development as one of its four important conservation goals for 1994.

Even the smallest farm can benefit from good land use planning. Reserving a small number of homesites in conjunction with the gift of a conservation or preservation easement or as a part of the transfer of development rights can extend the farmer's economic options. Whether homesites for family or a nestegg for troubled times, they will be useful. They can be sold to capitalize improvements or reduce debt, they can be left as a future resource, or they can be built on by family without having to re-enter the planning process.

SOME FINAL SUGGESTIONS FROM A DEVELOPER TO FARMERS

> "A farmer should live as though he were going to die tomorrow,
> but he should farm as though he were going to live forever."
> old Scottish wise saw

Avoid clearcutting. There is not room enough in this paper to enumerate the myriad of arguments against this destructive, earth wasting and economically unsound practice but follow this advice: Choose timber advisors carefully and demand selective cuts *at the appropriate time of year* with equipment that does not compact and destroy the soil. In our area, clear-cut properties lose up to 75% of their real estate value.

Keep buffers on creeks and rivers, preferably wide enough to be wildlife corridors (150-250 feet). These protect water resources, enhance wildlife and diversity, form the basis for important conservation areas and contribute strongly to the aesthetics of the farm.

Decrease the size of individual fields instead of clearing to increase them. Old-style boundaries such as hedgerows equal habitat and a protection from the elements, animals make better use of rotational grazing and, again, aesthetics (livability, human scale,) are enhanced.

Finally, create a careful site plan of the farm, *on paper*. The site plans I create include at least the following elements:

topography (your county planning dept. may be helpful in this)
water flow and water sources (streams, ponds, springs, drainage, etc.)
significaant trees and other botanicals, including fragile plant and animal communities
a botanical inventory, if possible
site/view, vista, etc.
tillable land
field/woods relationship
architectural and historical elements (old homesites, carriage roads, stone walls, barns, old cemetaries,etc.)

This is not a complete list, but I use these elements in every piece of land I evaluate and plan, including my own farm. If I plan a homesite or a capital improvement, I want to keep in mind the protection and nurturing of the resources listed above. This does not fit the preconceptions of industrial agriculture(an oxymoron, I believe). Wendell Berry, in What Are People For?, points the way to our responsibility when he writes:

> "But I am beginning to see what is needed, and everywhere the need is for diversity. This is the need of every American rural landscape that I am acquainted with. We need a greater range of species and varieties of plants and animals, of human skills and methods, so that the use may be fitted ever more sensitively and elegantly to the place. Our places, in short, are asking us questions, some of them urgent questions, and we do not have the answers."

It is time now to listen to the questions, however hard they may be. Those of us who love the land must listen, and then speak.

REFERENCES

Berry, Wendell. 1990. What Are People For? North Point Press, San Francisco. 210p

Herlevich, Camilla. 1993. "Conservation Easements in North Carolina: Private Property Rights Serving The Public Interest". The North Carolina Bar Association Newsletter, May, 1993. P. 6-8

Roddewig, Richard J. and Cheryl A. Inghram. 1987. "Transferable Development Rights Programs". American Planning Association, Planning Advisory Service Report Number 401. 39pp

Thompson, Edward, Jr. 1992. "Conservation Easements: Preserving American Farmland". Policy and Probate. Nov./Dec. 1992

RESOURCES

The Nature Conservancy, 1815 North Lynn Street, Arlington, Virginia 22209. Tel-703 841-5300

American Farmland Trust, 1920 N. Street NW, Suite 400, Washington, D.C. 20036 Tel-202 659-5170

Land Trust Alliance, 900 17th stree NW, Suite 410, Washington, D.C. 20006 Tel-202 785-1410

Trust For Public Land, 2100 Centerville road, Suite A, Tallahassee, Florida 32308 Tel-904 422-1404

Preserving Family Lands: Essential Tax Strategies For the Landowner, by Stephen J. Small. Published by Landowner Planning Center, P.O. Box 4508, Boston, Mass 02101-4508

Meeting the Expectations of the Land: Essays in Sustainable Agriculture and Stewardship, edited by Wes Jackson, Wendell Berry and Bruce Coleman. North Point Press, San Francisco

Camilla M. Herlevich, Attorney at Law (Conservation Easements), The Cotton Exchange, 321 North Front Street, Wilmington, NC 28401

Weaver Street Realty (conservation easement based development) 116 E. Main Street, Carrboro, North Carolina 27510

VIRGINIA'S INTEGRATED AGRICULTURAL NONPOINT SOURCE POLLUTION CONTROL STRATEGIES

Moira B. Croghan

ABSTRACT:

Much of Virginia drains into the Chesapeake Bay, an estuary that captured national attention and received funding for nonpoint source (NPS) pollution control several years before the rest of the nation. This caused Virginia to develop NPS abatement programs involving financial incentives, technical assistance and education, which are described. With the impending implementation of revised Coastal Zone Management (CZM) and Clean Water Acts (CWA), the state is examining more closely the patchwork of agricultural NPS abatement programs that has evolved. The objective is to overlay upon existing programs a predictable framework for treatment of agricultural NPS - - one that will allow agencies to direct limited resources more effectively, will communicate clearly to farmers what is expected of them, will allow the most appropriate control method to be applied, and will focus energies on watersheds and NPS sources that represent the greatest threats to water quality. A policy is being developed to help determine when a particular site warrants increased attention. It will use a matrix to consider NPS categories, sources, physiographic region, watershed conditions, site characteristics, program approaches, and delivery mechanisms. It aims to ensure that the most effective mix of strategies is applied in the locations whose treatment will result in the greatest NPS reductions.

Background

As research conducted under the estuary program of the Clean Water Act on the Chesapeake Bay demonstrated that nonpoint sources were responsible for at least half of the pollutant load to the system, funds became available to Virginia for NPS prevention.

A variety of programs evolved in response to this opportunity. Emphasis here is placed on state initiated measures, although it is important to note that many related federal programs co-exist and are critical to Virginia's total agricultural NPS reduction strategy, especially those conducted by the U. S. Department of Agriculture: Food Security Act plans for erosion control, demonstration watershed projects, the many educational efforts conducted by Cooperative Extension, and more.

Existing Programs

In the mid-eighties, initial funds through the Chesapeake Bay Program were used to hire one technical employee in each of the Bay watershed's Soil and Water Conservation Districts. These individuals perform a hybrid of services, capitalizing on Districts' familiarity with the agricultural sector. Their priorities have been technical assistance for the installation of BMP's, education of landowners about NPS, and more recently nutrient management and other farm planning services. During this same period, the state invested significant energies in helping districts become recognized leaders in local natural resources decisions. A drawback to districts is that they have such a broad mission and purpose that their resources are spread too thinly to assume many needed functions.

In the late eighties, additional funds were dedicated to employ eight state employees who develop detailed nutrient management recommendations for individual farm operators, captured in a "nutrient management plan". The plan has recently been automated to allow quick revisions and calculations as conditions change. Recipients first were targeted by operator willingness, approaching the perceived leaders in the agricultural community, and then by nutrient reduction potential, in particular operators with high animal waste production such as dairies and poultry operations. Nutrient management is the primary agricultural NPS strategy employed by Virginia to protect state waters and the Bay. There is an infinite need for this service, and field staff have been trained in sales techniques to enhance their ability to change producer behavior. The program is successful in part because it offers participants a cost savings. Since its inception, nutrient management plans have been written for nitrogen reductions totalling 5.5 million pounds, worth $1,100,000 and phosphate reductions totalling 4.8 million pounds, worth $11,100,000 to Virginia farmers.

Meanwhile, the state developed a geographic information system to assist in NPS reduction tracking and to categorize watersheds according to NPS pollution potential. Almost five hundred hydrologic units are ranked, based on land use, animal units, soil types and other characteristics, as having a high, medium, or low potential for NPS. This assessment provides an empirical, albeit imperfect, method to set program priority locations.

Cost-share funds for best management practice (BMP) installation in the Bay watershed have averaged about $1 million a year. The demand for funds outweighs the supply. Monies are allocated through Soil and Water Conservation Districts, with larger amounts being dedicated to districts with more animal units and cropland. Typically 30% of the Bay cost-share funds go to installation of animal waste control structures, and the remainder go to cropland and engineering practices such as buffers strips, no-till, and watering systems. In 1992, the fund allocation formula was changed to reflect the watershed ranking system described above, to direct more money to areas of greater water quality need. Each district, however, still has received baseline funding levels. This program has tended to support repeat customers. Some farmers forego the program for a variety of reasons, including reluctance to deal with the government. It has not been funded adequately to create significant NPS reductions, and its educational value is limited by the repetitiveness of the participants. Nevertheless, agricultural leaders feel that increased cost-share assistance is critical to agriculture coping with any new demands for NPS reductions that may be required through the revised CWA. Conservation and agribusiness organizations are lobbying for increased BMP dollars.

A tax credit for purchase of specialized, nutrient management farm equipment was approved by the legislature in 1990. It allows a state income tax credit of up to $3,750 annually for the purchase cost of manure applicators, sprayers for liquid pesticides and fertilizers, and related equipment. To be eligible, the farm must have a nutrient management plan approved by the Soil and Water Conservation District. About 80 plans a year have been prepared under this benefit program. Total statewide tax credits to farmers each year average $160,000. A drawback to this program is that most farmers do not pay high state income tax, so the incentive factor is limited.

In 1988 the Virginia General Assembly passed the Chesapeake Bay Preservation Act which requires local governments to enact ordinances to bring added protection to sensitive environmental features such as water courses. As the local programs have evolved, their primary impact upon farmers is the mandatory installation of vegetated buffers bordering sensitive environmental features like wetlands and streams. The Act also requires the preparation (but not necessarily the implementation) of a comprehensive "soil and water quality conservation plan" which includes a soil erosion, nutrient management and integrated pest management component. Over 22,000 tracts need such a plan; the law takes a "plan every acre" approach which presents a monumental task. State funds have hired 13 additional employees within Soil and Water Conservation Districts to prepare the plans. Buffer requirements can be reduced if a farmer demonstrates that their implemented farm plan provides the same element of protection as the required buffer. This program affects only the third of the state closest to the Chesapeake Bay, not the entire Bay watershed. It is Virginia's first foray into strict local land use controls for environmental protection, with the impact eased through its implementation by counties rather than the state. Enforcement of the law has not yet begun and as its regulations are revised, it may be amended to take a more targeted approach to farm planning.

In 1992, Virginia began a pilot project in total natural resource conservation planning (NRCP). Farm plans had become necessary for compliance with so many programs that farmers were confused and receiving conflicting advice from different agency advisors. The model format for NRCP has been developed which combines all the elements of a comprehensive farm plan. The key to the strategy is formation of a local team of technical experts from local, state and federal conservation agencies who communicate and coordinate with one another prior to translating a farmer's needs into a plan - - one which will help the operator comply with the myriad of federal, state and local recommendations and requirements. The team will function under the umbrella of the local Soil and Water Conservation District.

Farms in Virginia with high numbers of animal units for years have been subject to a permit and are regulated much the way point source dischargers are handled under the Clean Water Act. Sludge application sites are treated similarly. In 1990, through informal interagency means, it was required that these two kinds of sites also have a nutrient management plan approved by the state NPS agency to address quantities, location and timing of applications.

Since animal waste permitting requirements under the federal law do not affect dry wastes, several counties with intensive poultry productions have enacted local zoning requiring that a farmer have a nutrient management plan demonstrating 100% utilization of the waste prior to issuance of a building permit.

Complimenting the above techniques, a variety of educational opportunities has been available to farmers to learn about NPS prevention. In addition to that conducted by Cooperative Extension and the Soil Conservation Service, state funds have produced brochures about manure management, cost-share assistance and numerous other topics. Dozens of field days and demonstrations have been arranged by field employees, and one-on-one technical assistance is available for calibrating spreaders and explaining the results of soil, manure and tissue tests.

While several of the above initiatives began in response to Chesapeake Bay Program funds, all except the Chesapeake Bay Preservation Act now have been extended to cover the entire state, one-third of which does not drain into the Bay. Funding for some programs, however, is more abundant within the Bay watershed.

<u>The Future</u>

Despite the implementation of the programs above, public opinion and water quality conditions imply a need for intensified control of agricultural NPS. Virginia is wrestling with how to achieve increased protection within the existing framework or programs and statutes, and also refocus programs to assure resources are being used in the most water-quality efficient means. A piecemeal approach to agricultural NPS control, where services were largely dedicated according to farmer cooperation, is not likely to suffice under new Coastal Zone Management and Clean Water Act requirements. Resources are too limited too apply all the programs everywhere, and farm operators deserve to understand what will be expected of them.

A new phase in agricultural NPS control is approaching. It will rely less on voluntary approaches, particularly for sites whose NPS risk can be documented. The state is working to develop a way to determine the relative threat of any particular farm to water quality, and then to apply its programs commensurate to that risk.

A multi-dimensional matrix has been drafted to help make these decisions. As it is being refined, the state will seek support for it among farmers and the various agencies involved in agricultural NPS programs. The matrix considers NPS categories (nutrients, sediment, toxins and pathogens); agricultural source type (dairy, feedlot, cropland, grazing); physiographic region (mountain, valley, piedmont, coastal plain); watershed conditions (high, medium, low potential for NPS); site characteristics (slope, soil type, land use); program approaches (voluntary, incentive, regulatory) and delivery mechanisms (education, technical assistance, financial assistance, regulation). It aims to provide a tool for determining an effective mix of NPS strategies. While the matrix is not ready for adoption, it does highlight the issues pertinent to meaningful NPS control implementation.

Among NPS source categories, the matrix focuses attention first on nutrients and second on sediment, because research indicates these parameters correspond to the primary water quality impacts from agricultural land uses.

The matrix judges a farm location in three ways: by physiographic region, by watershed, and by site conditions.

The region helps to determine appropriate program delivery options - - coastal plain sites, with porous soils, will need different nutrient reduction approaches - - ones that especially protect groundwater resources.

The watershed demonstrates more localized potential for NPS problems within the physiographic region; hydrologic units densely populated with livestock need more attention than a low-animal-unit watershed, even if it contains a large dairy operation.

Within any given watershed, in particular those ranked as having a high or medium potential for NPS, high-risk site characteristics can elevate a farm so that it deserves attention. Proximity of a farm to a watercourse, water quality data from that stream, soil type, slope, and the existence/nonexistence of BMP's can demonstrate a more urgent need for NPS prevention strategies.

Once it can be concluded that a site warrants attention, there is consideration of program delivery options. These need to be meshed with the site's historical background with NPS programs - - has education and technical assistance been provided, has a nutrient management plan been prepared, is it being followed, have cost share funds been provided, and so forth.

The matrix helps field employees direct their time. Assigned to work within a certain area, a nutrient management specialist, for example, can employ the matrix to decide which farmers he or she next should approach.

The nature and degree of attention given any land user largely will depend on their voluntary willingness to apply "minimum treatment standards". Though not in regulation at this time, certain minimum standards have been drafted that correspond roughly to guidelines for management measures under the Coastal Zone Management Act. If a farmer with a high-risk site already is applying the erosion component of Resource Management System as defined in the Field Office Technical Guide of the U.S. Department of Agriculture - Soil Conservation Service, and is implementing a nutrient management plan, little attention is warranted.

When an operator is NOT found to be complying with minimum treatment standards, then program delivery options can be assessed, beginning with and progressing through education, technical/financial assistance and incentives, and regulation.

Potential regulatory approaches are being debated and are not settled. Virginia has the statutory authority to impose additional regulations under its water control law. However, agricultural land use issues have not received much attention and traditionally have not been managed by the agency implementing that law.

Refined legal authority and direction for prevention of agricultural NPS may be necessary. The matrix designed to help allocate technical assistance and financial incentive resources can be applied to determine the necessity of a particular farm needing enhanced regulatory attention. For example, some farms in high-risk locations, as determined by the matrix, may warrant some level of permitting.

There are several permit options, with the least intrusive preferred for operators willing to comply. Minimum treatment standards could be drafted into regulation providing a "general permit". Another route would be simply to require that each high-risk located farm maintain on premises certain records, which in the event of a complaint or other concern, could be analyzed by the state. Records maintained on crops grown, fertilizers, manures, and pesticides applied, BMP's installed, etc. would allow a determination whether or not increased NPS abatement measures need to be taken. Likewise, a plan for greater NPS control on such a farm could be captured in a permit or a consent decree.

The mechanics, implementation responsibilities, and authorities for such targeting of agricultural NPS strategic programs will be explored by Virginia over the next couple of years, as federal water quality mandates are clarified by Congress and the state determines ways to reshape those directives into programs acceptable to the farm community.

Involving Agricultural Producers in the Development of Localized Best Management Practices

Reagan M. Waskom and Lloyd R. Walker[1]

ABSTRACT

The Colorado Legislature passed the Agricultural Chemicals and Groundwater Protection Act in 1990 to prevent groundwater contamination due to the improper use of pesticides and commercial fertilizers. This law requires Colorado State University Cooperative Extension to provide education in Best Management Practices (BMPs) to give producers an opportunity to voluntarily adopt practices which protect the environment. If the voluntary approach is successful, further mandatory controls specified in the law may not be implemented. In order to achieve voluntary acceptance of BMPs by agricultural producers, our objective is to avoid a top-down, agency driven approach to the development of the required BMPs. Rather, development of BMPs in Colorado is being accomplished largely at the local level by producers, chemical applicators, and other experts within the affected watershed. Two high priority watersheds have been selected to initiate the development of localized BMPs: the South Platte River basin and the San Luis Valley. Small work groups of 10 to 15 agricultural chemical users have been formed to develop a comprehensive set of BMPs that are technically feasible, economically acceptable, and achieve state water quality goals. County Extension agents serve to facilitate discussions and reconcile outside reviews. Nine guidance principles for nutrient management, irrigation, wellhead protection, pest management, and pesticide handling are used to focus the local BMPs on the primary water quality goals. The producers participating in the work groups demonstrate the successful use of BMPs on their farms and foster grass-roots support for the voluntary approach. Initially, there was concern that practices developed by local producers would be mainly status quo with no real impact on groundwater protection. However, producers have taken an aggressive approach toward voluntary practices, realizing that they are preferable to mandatory regulations.

KEY WORDS   BPM, Pesticides, Fertilizer, Environment

INTRODUCTION

The Colorado legislature responded to public concern about pesticides and nitrates in drinking water by passing the Agricultural Chemicals and Groundwater Protection Act (SB 90-126) in 1990. This Act, co-sponsored by the state agricultural chemical industry, is designed to address activities which may result in agricultural chemicals entering the groundwater of Colorado. The implementation of the Act is funded by a 50 cents per ton tax on fertilizer and part of the pesticide registration application fees. The Act declares that the public policy of Colorado is to protect groundwater and the environment from impairment

---

[1] Reagan Waskom, Extension Water Quality Specialist, Colorado State University Cooperative Extension, Fort Collins, Colorado 80523 and Lloyd R. Walker, Extension Agricultural Engineer, Colorado State University Cooperative Extension, Fort Collins, Colorado 80523.

or degradation due to the improper use of agricultural chemicals, while allowing for their proper and correct use. It further calls for education and training of all agricultural chemical applicators to protect the state's water resources. Three state agencies are responsible for implementing the Act:

> The **Colorado Department of Agriculture** has overall responsibility for implementation of the Act.
>
> **Colorado State University Cooperative Extension** provides education and training designed to reduce groundwater contamination from agricultural chemicals.
>
> The **Colorado Department of Health** conducts a groundwater monitoring program to assist in identification of problem areas.

A three tiered response is specified for addressing potential and actual groundwater pollution due to agricultural chemicals. The first level of response is preventive, specifically the establishment and voluntary implementation of Best Management Practices (BMP's).

The second level of response to documented groundwater contamination is mandated practices within specified management areas. If voluntary prevention efforts fail to remedy a groundwater pollution problem, the Commissioner of Agriculture can adopt rules and regulations that become an Agricultural Management Plan (AMP).

If continued groundwater monitoring reveals that AMPs are not preventing or mitigating the presence of agricultural chemicals, the third level response will be employed. At this level, the Commissioner of Agriculture or the State Water Quality Control Commission may promulgate rules and regulations regarding the use of any agricultural chemical that has been identified through monitoring as creating or likely to create a pollution problem. For example, the use of Atrazine could be excluded from a designated area if water quality monitoring indicates continued loading in spite of attempts to voluntarily reduce its presence in groundwater.

Any regulatory approach is likely to use BMP's developed during the voluntary phase of the Act's implementation. Therefore, it is important that agricultural chemical users have an opportunity to participate in what may ultimately be a rule making process.

Role of Colorado State University Cooperative Extension (CSUCE)

The desired outcome of this Act is educationally induced change of practices resulting in more environmentally friendly agricultural chemical use. Thus, CSUCE as the outreach element of the state's land grant university is expected to provide technical expertise and educational methods to achieve this desired outcome. In a traditional approach, Extension specialists develop research based BMP's and county/area agents deliver this BMP message to agricultural chemical users. However CSUCE chose to modify this approach by involving agricultural chemical users in the process of developing and implementing BMP's which are sensitive to local conditions such as cropping practices, irrigation methods, soil, and climate. This is done for several reasons, including:

* The diverse nature of Colorado agriculture and agricultural chemical use requires crop and region specific suites of BMP's to effectively address statewide water quality concerns.

* A "one size fits all" set of BMP's would be vague, general and largely ineffective.

* The potential for BMP's to become part of a regulation suggests that those regulated be a party to the rule making.

* The experience and perspective of agricultural chemical users is acknowledged as a valuable and necessary resource to bring into the process.

* Adoption of BMP's will be more successful if a local constituency supports the BMP's.

Localizing Agricultural BMP's

Development of localized BMP's for agriculture requires an effective partnership of Cooperative Extension and local agricultural producers. Each must be committed to the process, fully participate in the process, and be respectful of the other partner. Cooperative Extension acts as the catalyst and driver of the process. Local agricultural producers act as reviewers, contributors and supporters of the process and end product.

Cooperative Extension specialists provide the research based BMP's. The premise is that adequate research has been conducted to identify BMP's which have a known beneficial impact on water quality. However, the localizing process acknowledges that while these practices are technically feasible, they may have shortcomings when applied in the real world of producers trying to make a living by farming or ranching. Thus the research based BMP's provide the body of knowledge for beginning the localizing process.

Cooperative Extension county/area agents initiate and guide the localizing process. Their first task is to assemble a group of innovative producers willing to participate in the process. This BMP work group should be limited to 15 members. It should target a specific crop and/or locale. The BMP work group is modelled after other organizational structures where producers are involved in a decision making process affecting their operations (i.e.,. grazing committees, irrigation companies or soil conservation boards). Several BMP work groups may be necessary in an area of diverse agricultural production to effectively address that diversity. Broad representation is sought, and should include members of local agricultural organizations (i.e.,. irrigation districts, soil conservation districts, etc.). The group might also include limited participation by crop consultants, the agricultural chemical industry and government agency staff (i.e.,. SCS, etc.) as appropriate. Another task for Extension agents is to facilitate discussions of the BMP work group and keep the group focused on the task of developing local BMP's. Finally, the agent will use the end product of the group, the localized BMP's, to work in partnership with group members to educate other producers in the area.

An important attribute of the local agricultural producers on the BMP work group is a commitment to change. They must acknowledge that a voluntary change of management practices is essential to addressing agricultural water quality issues. The most important assets they bring to the process is their knowledge based on experience and their ties to the agricultural community. This experience, combined with a commitment to the process of adopting BMP's, is the key to developing localized BMP's. Their task is to begin with the research based BMP's and identify appropriate practices for their local circumstances. The end product becomes a set of practices that tempers research based knowledge with the practical realities faced by producers. Work group members must then promote these practices within the organizations they represent.

Achieving consensus is the desired operating procedure of the BMP work group. Each research based BMP is reviewed by the group. Consensus is achieved through discussion and modification of the practice until it is deemed feasible, while accomplishing the water quality goal. While achieving consensus is generally more time consuming, it is the best method of localizing and developing support for BMP's. It is extremely important for work group members to develop a sense of ownership of the practices.

The reference document for the localizing BMP process is the Best Management Practices for Colorado Agriculture notebook developed by the Extension Water Quality Team. The notebook assembles research-based BMP's on topics including nutrient, pest, and irrigation management, wellhead protection and agricultural chemical handling. Each topic has a guidance principle and supporting research based BMP's. The guidance principle is the desired outcome to be achieved through local BMP's. The research based BMP's serve as a starting point for the local BMP work group deliberations. The product of the deliberations, the set of local BMP's are presented in a format similar to the Best Management Practices for Colorado Agriculture notebook. Desktop publishing is used to produce local BMP factsheets which are assembled as a local BMP reference notebook (see Appendix for examples of guidance principles and BMP's).

The local BMP's identified are demonstrated to other producers in the area to assist the voluntary adoption process. Cooperative Extension agents facilitate this process. However, producers serving on the local BMP work group must be willing to promote the practices developed. This may be accomplished through their participation in field demonstrations or public presentations in a variety of media.

Field Experience with the Localized BMP Process

Two high priority watersheds are sites for initiating the development of localized BMPs: the South Platte River basin and the San Luis Valley. These watersheds are extremely important to Colorado's agricultural economy, but also have been identified as having two of the most vulnerable aquifers in the state. BMP work groups have been formed in each of these watersheds. Only working agricultural producers or representatives of related organizations or agencies were solicited to serve on the work groups. Producers serving on the group were either nominated as representatives of local agricultural organizations or selected based upon agricultural expertise or leadership in a critical area. Local publicity on the purpose of the BMP work group also was used to make the agricultural community aware of the project and solicit members for the work group. Representation was sought from local irrigation districts, consultant groups, the agricultural chemical industry, and the USDA Soil Conservation Service. However, it was decided to limit such representation on the work group to one per agency or organization. This is to ensure a majority of working agricultural producers on the work group while acknowledging the important background support role of these agencies and organizations. It is extremely important to have the BMP work group producer driven. County Extension agents serve to facilitate discussions and document results. One producer, elected by the group, serves as chairperson/moderator.

The draft of localized BMP's was successfully developed through the consensus process in the San Luis Valley. However, it should be noted that two crops (potatoes and barley in rotation) predominate, thereby making the process more straightforward. Even in this situation, differences emerged; for example on manure handling, depending on the role livestock played in an individual's enterprise. In the South Platte valley, a more diverse agriculture has required a more careful focus of a BMP work group. At this point it appears that several work groups will have to be formed along crop enterprise lines in that area.

The draft localized BMP's for the San Luis Valley are in the hands of the work group members. Their current task is to present the BMP's to the groups they represent for review and comment. Publicizing the draft BMP's will be done at an upcoming annual potato/grain conference. The BMP work group chair will address the conference, explaining the project, describing the work group process and introducing the draft BMP's. Using such a popular, well attended event was determined to be the best means to publicize the project. The work group will meet after the conference to finalize the localized BMP's. Area Extension personnel will publish the BMP's, use them for workshops and other training opportunities and collaborate with work group members this growing season to promote the practices.

The USDA Soil Conservation Service in Colorado is playing a supporting but important role in the development of localized BMP's. The state SCS staff are committed to using the localized BMP's at the appropriate county level whenever feasible. Additionally, state nutrient and pest management standards and specifications are being linked to the localized BMP's with the desire to achieve agreement in all guiding principles, if not in all corresponding details. Local SCS personnel serve a vital linkage in helping to guide the BMP work group toward the adoption of practices that serve USDA goals as well as state water quality goals.

Guidance principles for nutrient management, irrigation management, wellhead protection, pest management, and pesticide handling are used to focus the local BMPs on the primary water quality goals. Initially, there was concern that practices developed by local producers would be mainly status quo, with no real impact on groundwater protection. However, producers have taken an aggressive approach toward voluntary practices, realizing that they are preferable to mandatory regulations.

CONCLUSIONS

The local, voluntary approach to solving water quality problems related to agriculture may progress more slowly than many in the environmental community deem acceptable. However, it should be understood that change occurs somewhat slowly in agriculture due to the extremely risky nature of farming and ranching. Producers in Colorado, concerned that state or federal regulations will seriously impair their economic situation are now more willing to make management changes, as long as their profitability is not compromised. Allowing the users of agricultural chemicals to (in a sense) self-regulate their activities, at worst maintains status quo; at best it provides an innovative and acceptable method of solving a problem that would be very difficult for the State to effectively regulate.

In Colorado, the voluntary adoption of BMPs by farmers, land managers, and other agricultural chemical users is being accomplished by allowing these groups to have significant input at the local level. If mandatory controls become necessary, the BMPs adopted at the local level will provide state regulatory agencies a head start on determining what controls are feasible, and perhaps increase producer compliance as well.

APPENDIX

Best Management Practices

Best Management Practices can be defined as recommended methods, structures, and/or practices designed to prevent or reduce water pollution while maintaining economic returns. Implicit within the BMP concept is a voluntary, site specific approach to water quality problems. Many of these methods are already standard practices for many farmers, and are known to be both environmentally and economically sustainable.

A set of general Guidance Principles is outlined to focus the local BMP work groups on the primary water quality problems. Specific BMPs and production alternatives fall under these guidance principles. The BMPs chosen for use at the local level ultimately must be selected by the actual chemical applicator because of the site-specific nature of groundwater protection. Site characteristics such as depth to water table, soil type, water holding capacity, and irrigation method determine the feasibility and risk of specific practices. For this reason, a basic premise is that agricultural chemical users must know the site specific variables at the application site, and have a good understanding of agricultural chemical properties and the influence of management practices in order to protect groundwater resources. The way Colorado's groundwater protection act is structured allows operators significant control of actual BMPs, as long as they meet the water quality goals outlined in the guidance principles.

The following guidance principles were developed to serve as general, goal oriented guidelines to protect water resources. Under each guidance principle, site specific BMPs must be selected which are tailored to specific crops and local management constraints. BMPs listed after each of the following guidance principles are examples selected from among the BMPs developed by the local working groups.

Guidance Principle 1.   **Protect wellheads from potential sources of contamination.**

- Periodically inspect and maintain well construction as needed.
- Install backflow prevention devices.
- Stay at least 100 feet from the well when mixing, loading, and storing of agricultural chemicals.
- Monitor well water quality periodically and know site-specific variables affecting aquifer vulnerability.

Guidance Principle 2.   **Manage irrigation to minimize transport of chemicals, nutrients, or sediments from the soil surface or immediate crop root zone.**

- Schedule irrigation according to crop needs and soil water depletion.
- Upgrade irrigation equipment to improve application efficiency.
- Time the leaching of soluble salts to coincide with periods of low residual soil nitrate.
- Reduce water application rates to ensure no runoff or deep percolation occurs during chemigation application.

Guidance Principle 3. **Manage nitrogen applications to maximize crop growth and economic return while protecting water quality.**

- Sample soil to a minimum depth of 2 feet., preferably to the effective rooting depth, to determine residual NO_3-N.
- Establish yield goals for each field based upon a documented previous 5-year average plus no more than 5 percent.
- Credit all sources of available N toward crop N requirements, i.e., organic matter and previous crop residues, irrigation water nitrate, soil nitrate, and manure.
- Use slow release N fertilizers and nitrification inhibitors as appropriate.
- Split N fertilizer into as many applications as economically and agronomically feasible.
- Avoid fall application of N fertilizers, especially on sandy soils and over vulnerable aquifers.
- A yearly N management plan should be developed for each field and crop.

Guidance Principle 4. **Animal wastes should be properly collected, stored, and applied to land at agronomic rates for crop production to ensure no discharge to surface or ground water.**

- Analyze manures for nutrient content and percent dry matter
- Reduce N fertilizer recommendations according to the amount of available N in the manure
- Document that the land base for manure application is sufficient for the size of the animal feeding operation
- Avoid manure applications on frozen or saturated soils and always incorporate after application.

Guidance Principle 5. **Manage phosphorus requirements for crop production to maximize crop growth and minimize degradation of water resources.**

- Implement standard SCS soil erosion practices and structures.
- Sample the tillage layer of soil to determine available soil levels and apply fertilizers according to soil test recommendations.
- Credit all available P from manures and other sources.
- Employ grass filter strips around erosive crop fields to catch and filter P in surface runoff.
- Incorporate surface applied P into the soil.

Guidance Principle 6. Utilize an Integrated Pest Management (IPM) approach in pest control decisions.

- Monitor pest and predator populations.
- Select crops and varieties which are resistant to pest pressures.
- Time planting and harvest dates to minimize pest damage.
- Rotate crop sequence to break up pest cycles.
- Spot treat or band pesticides instead of applying broadcast treatments.
- Employ beneficial insects and other biological controls.

Guidance Principle 7. Employ pesticides judiciously and use in a manner which will minimize off-target effects.

- All chemical applicators should receive thorough training and EPA certification prior to any use.
- Select pesticides based on site and management variables to minimize potential groundwater contamination.
- Chemical applicators should know the characteristics of the application site, including soil type, depth to groundwater, and erosion potential.
- Chemical leaching hazard, persistence, and toxicity should be compared to site specific conditions to determine suitability of the pesticide at that location.
- Application equipment should be inspected, calibrated, and maintained on a regular basis.
- Minimize pesticide waste and storage by purchasing and mixing only enough chemical to meet application needs. Utilize mini-bulk or refillable containers wherever possible to minimize container disposal problems.

Guidance Principle 8. Maintain records of all pesticides and fertilizers applied.

Records should be kept on:

- Irrigation water analysis.
- Soil test results.
- Projected crop yield goals.
- N fertilizer recommendations.
- Fertilizer and/or manure applied.
- Amount of irrigation water applied.
- Actual crop yields.
- All pesticides applied, including: brand name, formulation, EPA registration, amount and date applied, exact location of application, name, address, and certification number of applicator.
- Records should be maintained for at least three years.

RISK ASSESSMENTS:

MANAGEMENT TOOLS FOR PREVENTING WATER POLLUTION ON FARMS

E. Nevers, G. Jackson, R. Castelnuovo and D. Knox[*]

ABSTRACT

The Farmstead Assessment System (Farm*A*Syst) is a voluntary pollution risk assessment program that protects private drinking water wells. A unique self-assessment tool, it translates complex environmental, geophysical, and technical information into a useable format that allows farmers to evaluate a wide range of potential contaminants located in and around the farmstead.

Using a series of worksheets, the farmer evaluates sources of toxics, microorganisms, and nitrates. Specifically, activities and structure involving pesticide storage and handling, fertilizer storage and handling, animal waste management, hazardous waste management, household waste water, and petroleum storage and handling are analyzed. The worksheet information is further evaluated in terms of the soil, and geologic and hydrologic features unique to the site. The individual risks are ranked, and an action plan is formulated to bring high risks under control.

The assessment materials incorporate current state and federal regulations, and the follow-up action plan includes information on improved management practices, current technologies, and recommended facilities and structures. Evaluation results from Wisconsin, Minnesota, and Michigan indicate that farmers like and use this management tool. It is a farm-centered, systematic approach that increases knowledge, assesses potential risks, and generates a follow-up action plan.

Keywords: risk assessment, farmstead, groundwater, management practices, pollution sources

FARMSTEAD GROUNDWATER POLLUTION SOURCES

Overlooked Pollution Sources

Farmsteads includes farm buildings and the land around them. These sites are increasingly being recognized as potential sources of surface and groundwater contamination in rural areas. The concentration of potential contaminants and intensity of activities around farmsteads can generate significant amounts of nitrates, toxins, and microorganisms. Wells on the farmstead and local groundwater are often vulnerable to contamination from these contaminants.

While the current technology and regulations have focused on field practices as a source of both groundwater and surface water contamination, the farmstead has concentrated sources of toxics, microorganisms, and nitrates which act as quasi-point contaminant sources. Farmstead facilities can include petroleum tanks, pesticide storage and handling, fertilizer storage and handling, household waste water treatment systems, abandoned wells, livestock waste storage and management, hazardous waste management, milking center wastewater, and silage leachate. In survey sites where pesticide concentrations in water well samples exceeded standards, one Iowa study indicated point sources were responsible for 25% of atrazine detections above the health

---

[*] E.G. Nevers, Outreach Specialist, National Farmstead Assessment (Farm*A*Syst) Office, University of Wisconsin - Madison., G.W. Jackson, USDA Extension Coordinator, Farm*A*Syst, R. Castelnuovo, Attorney, Farm*A*Syst and D. Knox, USDA Soil Conservation Service Coordinator, Farm*A*Syst.

advisory level (Seigley and Halberg, 1993). A single event - a major spill or the back siphoning of pesticides down a drinking water well while mixing - can negate years worth of careful field management practices.

Types of Pollution Risks on the Farmstead

To characterize the nature of pollution risks associated with farmstead activities, a hypothetical farmstead has been analyzed. This farmstead is located on a farm that has one hundred animal units and two hundred acres of cropland in a corn/alfalfa rotation. The following table estimates the amount of potential pollutants that may be handled at this site in a typical year.

Table 1. Potential Pollutants at a Typical 100 Animal Unit Dairy Farmstead

Potential Pollution Source	Estimated Amount per year
Livestock Wastes	17,000 pounds of nitrogen
Fertilizer	15,000 pounds of nitrogen
Silage	30,000 pounds of nitrogen
Pesticides	700 pounds active ingredients (varies extensively)
Petroleum Products	3,300 gallons
Household Hazardous Wastes	10 pounds
Farm and Shop Hazardous Wastes	20 pounds (varies extensively)

Small amounts of these products can cause major groundwater contamination. Table 2 calculates the minimum amounts of nitrate, atrazine, gasoline and fecal coliform bacteria that would contaminate an aquifer that is 100 feet deep, with 30 percent porosity and 10 million gallons of water per acre of surface area.

Table 2. Amounts of Farmstead Pollutants Needed to Exceed Standards

Pollutant	Amount	HAL*
Nitrate	800 lbs	≥ 10 ppm
Atrazine	.25 lbs	≥ 3.5 ppb
Gasoline (1% benzene)	5 gal	≥ 5 ppb
Fecal Coliform Bacteria		detect violates HAL

* Health Advisory Limit

Potential for Groundwater Contamination

While farmstead activities can contribute to point source pollution, the design and depth of wells also significantly influences pollution potential with shallow wells, sand point and seepage wells being most vulnerable to contamination. Data from Illinois indicates, "Contamination of sampled wells was related to well construction and depth, with ... dug-type wells having about 25 times as many pesticides and 3 to 4 times as many detections for nitrate." (Church, 1993) The relationship between contamination and depth of the well can be seen in Table 3, a summary from the Iowa State-wide Rural Wellwater Survey (1990). Wells under fifty feet in depth had high rates of nitrate-n, coliform bacteria and pesticide detects than those over fifty feet.

Multi-state data from Wallrabenstein and Baker (1992) at Heidelberg College also show higher rates of contamination for wells less than fifty feet deep . In addition, their research looked at well type and the impact of the proximity of barnyards, mixing and loading sites and cropland to the wells. The water samples were tested for nitrogen and pesticides. Table 4 summarizes the relationship of well depth and well type for nitrates and atrazine. The proximity of various farm activities is summarized in Table 4. This data shows consistently that higher concentrations of nitrates and atrazine were found in wells within 20 feet of the list activities.

Table 3. Summary of Iowa State-Wide Rural Wellwater Survey Results

Water-Quality Parameter	All sites	Wells < 50 ft Deep	Wells 50-99 ft Deep	Wells ≥ 100 ft Deep
Wells: (686 wells)				
% wells of known depth (median well depth: 110 feet)		28%	21%	51%
Nitrate-N:				
% sites >10 ppm	18%	35%	32%	4%
Total Coliform Bacteria:				
% sites positive	45%	72%	52%	27%
Fecal Coliform Bacteria:				
% sites positive	7%	8%	12%	2%
Pesticides:				
% sites with detections	14%	18%	14%	9%
% sites with ≥2 detections	5%	9%	5%	4%
% sites > HAL*	1.2%	2.0%	1.4%	0.6%

* Health Advisory Level
Source: Iowa Department of Natural Resources

Table 4. Multi-State Summary of Well Characteristics

Factors	Nitrate Concentration			Atrazine Concentration, μg/L		
	Number of wells	<0.3	>10 ppm	Number of Wells	<0.05	≥3.0
Well Depth:						
<50 ft.	5145	46.7%	8.5%	1578	88.3%	0.4%
50-100 ft.	10,262	68.2%	3.5%	3119	92.3%	0.1%
>100 ft.	8406	70.5%	1.6%	2389	93.8%	<0.1%
Type of Well						
Drilled	13,054	64.4%	3.4%	6814	93.2%	<0.1%
Dug	1105	19.5%	12.3%	469	74.2%	0.6%
Driven	1342	54.5%	9.0%	992	95.3%	0.3%

Adapted from Wallrabenstein and Baker

RISK ASSESSMENTS

Using Environmental Audits for Pollution Prevention

To prevent pollution under these circumstances, a method must not only address facility design and management practices, but also respond to the problems of well design and proximity to potential pollution sources. Of the existing methods, a pollution risk assessment or audit has proven its utility and cost-effectiveness. Many industries are required to conduct environmental audits, or have initiated their own voluntary audits. The Tennessee Valley Authority's National Fertilizer and Environmental Research Center has developed a "Fertilizer and Agri chemical Dealer Environmental Evaluation." Several agricultural chemical companies are requiring their dealers to conduct environment audits. They have established minimum criteria which must be met if dealers wish to continue their dealership. Farmers and other chemical users have similar needs. They are liable for contamination on their properties and would benefit from participation in a program that identifies pollution risks and recommends practices that prevent pollution.

Table 5. Relationship between Nitrate and Triazine Contamination and Various Farm Activities

Characteristics	Number of Wells	Average Nitrate-N mg/L	Percent >10mg/L	Number of Wells	%Triazine >.05µg/L	%Triazine >3µg/L
Proximity To Cropland:						
within 20 ft.	1735	2.71	8.9%	919	14.41%	1.09%
within 200 ft	7002	2.05	5.9%	4668	9.98%	0.34%
within sight	3698	1.67	3.2%	2008	8.56%	0.10%
Proximity to Feedlot/ Barnyard:						
within 20 ft.	676	3.77	11.5%	359	16.43%	0.89%
within 200 ft	3145	2.25	6.3%	1802	10.15%	0.44%
within sight	3594	1.86	4.6%	2162	11.10%	0.23%
Mixing Chemicals:						
within 20ft.	1121	2.48	7.4%	672	13.84%	0.89%
not near well	15,791	1.81	4.5%	8370	8.92%	0.15%

Adapted from Wallrabenstein and Baker

As a result of pressures outside of government, environmental assessments or audits are also becoming a staple in business transactions. Lenders are increasingly requiring environmental assessments during property transfers and some are recommending assessments for approval of operating loans. These requirements are making it a necessity for owners and operators to modify some practices to reduce pollution risks. The Farm Credit Service has trained their loan officer in assessment requirements and the American Bankers Association has recommended that all banks require environmental audits during commercial property transfers. Increasingly, attorneys are also advising their agricultural lending clients to perform the same assessments to avoid expensive legal liabilities. Finally, the insurance industry has also increased pressure on property owners and operators. Environmental damage liability coverage has been removed from standard liability policies. In limited cases where it is available, coverage is very expensive. Thus, most farmers and rural residents have no insurance coverage for environmental contamination. Cleanup of environmental contamination problems can literally "cost them the farm."

The Farmstead Assessment Program

To protect his or her drinking water, a farmer needs to consider the following risk factors associated with the farmstead:
- facility design and location
- associated management practices
- proximity to wells
- soil and geology of the site

To help farmers, ranchers, and rural residents prevent pollution, a multi-agency coalition developed a voluntary Farmstead Assessment System, called Farm*A*Syst. With support from the Extension Service, Environmental Protection Agency and the Soil Conservation Service, the program has grown in a little over two years from two pilot programs in Wisconsin and Minnesota to a national network. Through an expanded partnership of farm organizations, private industry, and governmental agencies, the program has grown to more than forty states and one Canadian province.

Farm*A*Syst has been designed to help farmers, ranchers, and rural residents identify site-specific well water and groundwater pollution risks and develop voluntary action plans to reduce identified high risks (Jones and Jackson, 1990). It was developed in response to the grassroots concerns of farmers and rural residents about protecting well water quality. A recent Gallup survey shows that "farmers are more concerned about farm environmental issues today than they were five years ago." (Sandoz, 1993) With Farm*A*Syst, farmers have concrete and systematic methods to address water quality-related concerns.

More specifically, the Farm*A*Syst program aids farmers and ranchers in:
- understanding and identifying pollution risks associated with their farms and rural residences
- understanding how existing programs and policies can help prevent pollution
- identifying actions that will reduce pollution risks
- obtaining technical, financial and educational assistance to prevent pollution
- taking voluntary actions to reduce pollution risks

Farm*A*Syst translates these general objectives into practical results using a series of worksheets and factsheets. The worksheets provide a systematic framework for evaluating relative pollution risks at a specific site. The factsheets contain information on actions that reduce pollution risks and information on sources of educational, technical, and financial assistance. The potential pollution sources covered in the initial Farm*a*Syst materials include:

Wells	Pesticides	Livestock Waste
Hazardous Wastes	Fertilizers	Livestock Yards
Household Wastewater	Petroleum	Silage
Milking Center Wastewater		

Dividing risks into four categories ranging from high to low, the worksheets allow users to evaluate specific criteria on their property and then rank pollution risks associated with particular design and management factors. The worksheets are free-standing, so users select only those worksheets relevant to their needs. Applying the information from these worksheets, participants use a separate worksheet to evaluate groundwater pollution risks in terms of the soil, and geologic and hydrologic features unique to their property.

An overall evaluation sheet combines the findings from the site evaluation and the assessments to develop a relative risk ranking for that farmstead or rural residence. All high risk practices and structures that are identified are addressed in a voluntary pollution prevention action plan to reduce high risks. Educational materials also provide information on local sources of technical assistance and emphasize the benefits and cost-effectiveness of corrective or preventive measures. The program was initially designed as a self-assessment and can be effectively used in that capacity. However, pilot implementation experiences indicate the program is more effective when conducted with technical and educational assistance.

To facilitate the development of an action plan to correct problems and reduce potential high risk items, a Decision Support System (DSS) software has been developed by J. Anderson and P. Robert at the University of Minnesota. It is Windows-based and uses the assessment information to provide a site-specific list of recommendations for farmers. The risk ratings from each of the categories on a worksheet as well as the average worksheet and overall farmstead risk ratings are entered into the program. High and moderate-to-high risk ratings are displayed and generate an individual action plan. The action recommendations can be modified to reflect local or state regulations, and availability of local technical and financial assistance. In addition, this software will help local Farm*A*Syst project coordinators by providing electronic record keeping. As simple computations and queries can be performed, project data can be readily aggregated and analyzed.

Results from Assessment in Michigan, Minnesota, and Wisconsin

The Farm*A*Syst has been incorporated into USDA Demonstration projects and USDA Hydrologic Unit project. The first projects were located in Wisconsin and Minnesota. Farm*A*Syst has also been used in Wisconsin with state priority watershed projects, services offered by independent crop consultants and farm cooperatives, adult vocational and farm management programs, community wellhead protection, and drinking water education programs. Evaluations of the pilot test projects in Wisconsin, Minnesota, and Michigan have been performed and the results on farmer receptivity and high-risk frequencies follow. In Wisconsin and Michigan, post-assessment surveys were sent out to participants three to six months after the initial assessment. In the Minnesota Anoka Sand Plains Project, copies of the actual assessment were used for the evaluation. To maintain confidentiality, names and addresses were not used.

Data from farmers in Waupaca, Manitowoc, and Jefferson counties in Wisconsin is presented in Table 6. This table shows the percent of all farms on which the different potential sources of groundwater pollution were identified. It also shows the percentage of farms on which the farmers have planned or made changes to reduce the risk of groundwater pollution from these sources. As the table shows, petroleum is by far the most frequently identified source of risk on farms in this project. The next most frequent risks identified are pesticides and livestock yards. Fertilizer waste, silage, and milking center wastewater were identified infrequently on the participating farms. Also, as Table 6 shows, the percent of farms and changes made plus those with changes planned equals or exceeds the percentage of farms with risks identified. Thus, farmers appear to be following up in making changes to reduce risks identified by their assessments.

Table 6. Wisconsin: High Risks Identified and Changes Planned or Made

WORKSHEET N = 47	% of All Farms with High Risks Identified	% of All Farms with Changes Planned	% of All Farms with Changes Made
1 Wells	15	11	4
2 Pesticides	26	21	19
3 Fertilizers	9	6	6
4 Petroleum	60	49	21
5 Hazardous Waste	19	41	3
6 Household Wastewater	17	9	6
7 Livestock Waste	23	15	11
8 Livestock Yards	26	15	15
9 Silage	9	4	4
10 Milking Center Wastewater	6	4	2

Source: Lamm and Jackson

However, without the availability of the additional data, the specific nature of changes made or planned cannot be identified. Note that the percentage of changes planned plus the percentage of changes made usually exceeds the percentage of farms with high risks identified. This is because some farms with certain practices have been identified as moderate risks. The farmers are taking steps to reduce those risks. Also, farmers have made some but not all of the changes recommended to minimize the high risk identified on their farms.

Table 7 shows the reasons why changes in high risk situations were not made by participating farmers. The most frequent reason mentioned was that changes were too expensive. Other research would be useful in determining which particular changes are beyond the financial capability of the farmer and require funding assistance.

Table 7. Reasons Why Changes In a High Risk Situation Are Not Made

Reasons	Percent
Too busy	9%
Too expensive	19%
Don't believe it will make a difference	4%
Changes in management are too time-consuming	0%
Need more information and/or assistance	6%

This pilot project was delivered by crop consultants and farm cooperative staff and the survey results indicate that participating farmers felt good about the program. Sixty-four percent of the participants indicated that the program was useful to very useful and ninety-four percent rated the assessment process on a ten point scale of six or above (Lamm and Jackson, 1993).

Delivery of the Farm*A*Syst program through drinking water education activities also produced encouraging results. In another pilot project, forty-six percent of the drinking water program participants indicate they have taken action to reduce identified high pollution risks, and fifty-one percent indicated they intend to do so.

Minnesota versions of the Farm*A*Syst were completed in August 1991. Since then there have been several continuing projects have used the Minnesota Farm*A*Syst. The first efforts took place in USDA Demonstration and Hydrologic Unit project areas including the Anoka Sand Plains. One objective of this demonstration project was to conduct 100 interviews with individual project participants. To date, fifty-six assessments have been completed. Some preliminary results are presented in Table 8.

Table 8. Anoka Sand Plains Project: Participant Responses

Categories N = 56	% participants completing each worksheet	% moderate to high risk ratings and high risk responses
Household wastewater	100	0
Pesticides	100	18
Fertilizers	100	14
Petroleum	100	48
Hazardous waste	98	16
Household wastewater	100	20
Livestock waste	72	16
Livestock yard	85	30
Silage	72	21
Milking center	66	11

Source: Jackson and Anderson

An analysis of the completed worksheet summaries (Worksheet 12) shows that the overall rankings of 41 percent of completed assessments were in the moderate-to-high and high risk categories. This indicates that within the context of current accepted or recommended practices, there are a variety of improvements that can be made to practices around farmstead wells. A preliminary evaluation of those individuals completing assessments in the Anoka Sank Plains shows that every individual changed at least one practice. Changes range in expense and difficulty from installation of a pesticide rinse pad, to pumping septic tanks, moving pesticide storage areas and relocating petroleum tanks.

In Michigan, Cass County had the first Farm*A*Syst pilot. With a large number of pork producers, there was major concern about water quality problems resulting from confinement operations and manure storage facilities. Over 150 individuals participated in the program and 32 returned post-assessment surveys. The most frequently identified high risk items were: wells, septics and petroleum storage. A follow-up visit by the MSU-Extension specialist three months after the assessments were done, showed that twenty-seven out of forty-eight previously identified high risk items had been corrected to reduce or eliminate the risk potential. Many of the changes were low cost or management changes. Nine of the forty-eight would require major planning and expense.

Not Just for Farmers

Non-farm rural residents also have the potential to contaminate their drinking water well and the underlaying groundwater. Most rural homesteads use petroleum products and lawn and garden chemicals. They also generate household waste water and hazardous and solid waste. In many areas of the country, these non-farm homesteads outnumber the farmsteads. At least 20 states are developing a non-farm version of Farm*A*Syst known as Home*A*Syst which will specifically target these rural residents.

CONCLUSIONS

Farm*A*Syst is a unique program because it addresses a wide range of potential contaminants and remedies in a comprehensive, easy manner. It incorporates current regulations and the best available technologies and practices. For farmers, ranchers and rural residents, they now have at their disposal the means to accurately assess how their activities influence pollution risks. More importantly this pollution prevention tool assists individuals in taking decisive actions to preserve the quality of their drinking water; prevent groundwater pollution; reduce potential liability; and, protect their health.

REFERENCES

1. Church, J.A. 1993. Illinois agents instrumental in survey of state's groundwater quality. The County Agent. 54(3):23.

2. Iowa Dept. of Natural Resources. 1990. The Iowa state-wide rural well-water survey of water quality data: initial analysis. Technical Information Series 19. Iowa City, IA.

3. Jackson, G.W. and J.L. Anderson. 1993. Farmstead assessment for whole farm water quality protection. Proceedings of Agricultural Research to Protect Water Quality. February 21-24, 1993, Minneapolis, MN. Soil and Water Conservation Soc. 517-519.

4. Jackson, G.W., R. Castelnuovo, E.G. Nevers, and D. Knox. 1993. Preventing water pollution through farmstead management of pesticides and toxics substances. Presented at the American Chemical Society Annual Meeting and Conference. August 22-27, 1993, Chicago, IL.

5. Jones, S.A. and G.W. Jackson. 1990. Farmstead assessments: a strategy to prevent groundwater pollution. J. Soil and Water Conservation Soc. 45(2):236-238.

6. Lamm, T. and G.W. Jackson. 1993. Private sector delivery of farmstead assessments: a pilot project in three Wisconsin counties. Project Report (draft). National Farmstead Assessment Office. University of Wisconsin-Madison.

7. Sandoz News. 1993. Gallup says environmental issues prompting changes in attitudes, actions on the farm. Released 1-28-93. Des Plaines, IL.

8. Seigley, L. and G. Hallberg. 1993. Private well water in the Big Springs basin: 1981-1992. Iowa State Univ. Water Watch Newsletter 42:4.

9. Wallrabenstein, L.K. and D.B. Baker. 1992. Agrichemical contamination in private wells. Proceedings of the Focus Conference on Eastern Regional Ground Water Issues. October 13-15, Boston, MA. Ground Water Management, Book 13 of Series. Dublin, OH. 697-711.

ECONOMIC AND ENVIRONMENTAL ANALYSIS OF CONVENTIONAL AND ALTERNATIVE CROPPING SYSTEMS FOR A NORTHEAST KANSAS REPRESENTATIVE FARM

Seungmo Koo and Penelope L. Diebel[*]

ABSTRACT

The problem of surface-water contamination from atrazine, a widely used agricultural chemical, has recently arisen in northeast Kansas. Part of this area has been designated as a Pesticide Management Area by the U.S. Environmental Protection Agency (EPA). The objective of this study is to consider both the potential contaminant loadings caused by agricultural chemicals such as atrazine and the economic feasibility of several alternative cropping systems in northeast Kansas.

Final results indicate that three of four alternative systems have higher net returns than a conventional system and lower contamination levels of atrazine and sediment. However, nitrogen levels in runoff and leachate of the alternative systems were higher than those of the conventional system. The results imply that a policy targeting only a few chemicals may ignore changes in other contaminants levels. Although many agronomic benefits may be associated with the alternative systems in this study, potential pollution still exist.

KEYWORDS. Alternative cropping systems, Enterprise budgets, Non-point pollution simulation, Surface-water contamination.

INTRODUCTION

In recent years, concerns regarding agrichemicals, especially atrazine in northeast Kansas, have significantly increased because the area is a major source of water supply for Kansas City and nearby towns. One of the most commonly appearing chemical is atrazine, a herbicide that helps to selectively control broadleaf (dicot) weeds, such as pigweed, cocklebur, and velvetleaf, and certain grass weeds, especially in corn and sorghum. In 1959, atrazine was registered as an approved herbicide in Kansas. It is still widely used because it economically and effectively controls many weeds, reducing interferences (Regehr et al., 1992). Atrazine is also the most commonly used chemical among the agricultural practices in northeast Kansas (Diebel et al., 1993).

Atrazine contamination affects human health adversely. For instance, a 150-pound person would have a 50 percent probability of death from poisoning by ingesting about 0.75 pound of atrazine in a single dose. This dose is relatively non-toxic to other animals (Regehr et al., 1992). Due to the increased concerns, the U.S. Environmental Protection Agency (EPA) designated part of northeast Kansas as a Pesticide Management Area and reduced the acceptable atrazine residue levels in drinking water from 150 parts per billion (ppb) to 3 ppb in 1992. The Kansas State Board of Agriculture decided the maximum application of atrazine

---

[*] Graduate Research Assistant and Assistant Professor, Department of Agricultural Economics, Kansas State University.

should be at 2 kilograms per hectare in any given crop year in the Pesticide Management Area. The related area was the Delaware River Basin in northeast Kansas, including some parts of Nemaha, Brown, Jackson, Atchison, and Jefferson counties. This program was voluntary during the first year, but the regulations will become mandatory if no progress is made in reducing the atrazine levels in drinking water. The atrazine in runoff, in the PMA, has been monitored since January 1. of 1993 and will be monitored for at least two more years. However some of the rules for the PMA are mandatory now, such as the prohibition of atrazine use within 500 feet of any public water supply source. In addition, atrazine is banned from use on non-croplands, including highway and railroad rights of way and lawns.

In this voluntary situation, the farmer's production decisions are the most important factors in reducing contamination. If the farmer's primary goal is to maximize net return, he must now consider an additional objective that reduces the residue or the amount of atrazine use. The widespread adoption of alternative agriculture could be possible solution to the agricultural pollution problems if it can satisfy both goals. The alternative agriculture includes a spectrum of farming systems, ranging from organic systems that attempt to use no purchased synthetic chemical inputs to the conventional practice (National Research Council, 1987).

In this study, economic budgeting analysis and non-point pollution simulation are used to compare both the potential contaminant loadings caused by agricultural chemicals such as atrazine and the economic feasibility of the conventional and alternative cropping systems in northeast Kansas for the period, 1987 to 1991.

PROCEDURES

Diebel et al. (1993) collected producer survey data and established a representative farm scheme for northeast Kansas. These data were used to create a representative farm including (1)schedules of field operations from a survey of selected producers in northeast Kansas, ongoing research at the Corn Belt Experiment Station, and (2)average characteristics of the 332 farms in 14 county study areas for the period 1986-1990. These counties include Atchison, Brown, Doniphan, Douglas, Jackson, Jefferson, Johnson, Leavenworth, Marshall, Nemaha, Pottawatomie, Shawnee, Wabaunsee, and Wyandotte counties. The average dryland cropland is 259 hectares, with 37.4 percent owned and 62.6 percent rented.

Table 1 illustrates the structures of the conventional and four alternative cropping systems. Five distinct crop rotations make up the Conventional System. The corn-soybean rotation covers a total of 101.08 hectares, with 50.59 hectares in each crop. The sorghum-soybean rotation represents a total of 56.66 hectares. There are 45.12 hectares in a wheat-sorghum and wheat-soybean rotation. Continuous corn is grown on the remaining 12.14 hectares. Alternative System 1 is a 3 year wheat, sorghum and soybean rotation. This system has 86.66 hectares allocated to wheat interplanted with clover. Sorghum is planted on the other 86.33 hectares and soybeans cover the remaining 86.33 hectares. Alternative System 2 is a 2 year sorghum and wheat rotation. This system has equal hectares of each crop, and wheat is interplanted with vetch. Vetch is used as a nitrogen source and not harvested. Alternative System 3 is a 5 year rotation of alfalfa (3 years), wheat and soybeans. Each year, 51.8 hectares of new alfalfa is planted in the spring on soybean land. Alfalfa is harvested once in year one, three times in year two, and once in year three. The third year of alfalfa serves as a green manure for the subsequent wheat crop. Oats are interplanted with alfalfa in the first year. Alternative System 4 is a 7 year rotation of corn, soybean, corn, soybean, and alfalfa (3 years). This system has 73.66 hectares planted to corn and an additional 73.77 hectares is planted to soybeans with half of these hectares being planted to alfalfa in the spring. Alfalfa is not harvested in year one, but is harvested three times in year two and three. Oats are interplanted with alfalfa in year one. The final year of alfalfa provides nitrogen from fixation for following crop.

The number of chemicals including pesticides and fertilizers are indicated on Table 1. The Conventional System is the most chemical intensive among the systems. Although the Alternative Systems use several chemicals, the amounts applied are smaller than those of the Conventional System. The number of tillage operations under each system is one of the most important factors in the non-point pollution simulation model.

Enterprise budgets were developed by Diebel et al. (1993) to summarize the annual operating expenses and fixed costs of each system, making it possible to compare costs and 5 year average net returns of each cropping system. Table 2 contains costs under each rotational systems. The variable costs include the costs of labor, seed, herbicide, insecticide, fertilizer, fuel, oil, equipment repair, custom hire, and interest on variable cost. The fixed costs include real estate taxes, interest on land, share rent, depreciation and interest on machinery, and insurance and housing. Figure 1 compares the costs under each system. Alternative 2 has the highest total cost and Alternative 3 has the smallest total cost. The data regarding these costs were collected from Kansas Farm Management Association, Kansas Agricultural Statistics, northeast Kansas cooperatives, and a farmer's survey (Diebel et al., 1993). In calculating 5 year (1987-1991) average net returns, only the variation of crop yields was considered assuming constant costs over the period.

The non-point pollution simulation model, GLEAMS 2.0, was used to compare potential contaminant loadings under each system. The model required 5 sets of input data files, consisting of daily precipitation, hydrology, erosion, pesticide, and nutrient input data files.

Table 1. Structure of the Five Cropping Systems.

	Conv.	A 1	A 2	A 3	A 4
Number of :					
Rotations	5	1	1	1	1
Years a	2	3	2	5	7
Rotations (ha) :	C-Sb(101.08)	W/Cl-Sg-	W/V-Sg	Al/O-A2-	C-Sb-C-
	Sg-Sb (56.66)	Sb (259)	(259)	A3-W-Sb	Sb-Al/O-
	W-Sg (45.12)			(259)	A2-A3
	W-Sb (45.12)				(259)
	C-C (12.14)				
Number of Applications :					
Pesticides	10	4	5	0	0
Fertilizers	4	1	3	1	0
Tillage	C-Sb: 7	8	8	13	13
	Sg-Sb: 7				
	W-Sg: 6				
	W-Sb: 5				
	C-C: 4				

a All the rotations in the conventional system are 2 year rotations except continuous corn.

Table 2. Costs and 5-Yr a Average Net Returns, and 5-Yr Average Contaminant Loadings.

	Conv.	A 1	A 2	A 3	A 4
Variable Cost($/ha)	139.69	138.97	175.32	83.05	96.57
Fixed Cost($/ha)	285.73	271.04	272.37	264.20	264.20
Total Cost($/ha)	425.42	410.01	447.69	347.25	360.77
Net Return($/ha)	-10.26	41.56	-98.42	27.10	47.47
Atrazine Runoff(g/ha)	4.30	1.44	0.14	0.00	0.00
N runoff(kg/ha)	1.03	0.94	0.96	1.27	1.07
N leachate(kg/ha)	2.50	3.00	3.00	4.90	3.44
Sediment Yield(ton/ha)	7.88	5.20	3.02	3.15	4.60

a 1987-1991

Pesticide data files were not needed for Alternative system 3 and 4 because they did not use pesticides. Each cropping system (or rotation) was simulated once for every year in the rotation sequence. The starting dates were lagged each year so that each crop was simulated for every year. The annual contaminants levels were calculated by weighting the contribution of each crop by the hectares grown. The hydrology file contained data regarding temperature, radiation, and basic information of soil condition of the area. The erosion file contained information regarding soil condition and geological structure of the representative farm. The pesticide file contained chemical application schedules and the amounts of the chemicals applied. The nutrient file contained schedules and amounts of fertilizer applications, and types of the tillage operations were specified. The type of crops and simulation period are commonly specified in each file. The data for the simulation model were collected from soil survey of the study area (Soil Conservation Service, 1982), GLEAMS 2.0 User Manual, and expert opinion.

RESULTS AND DISCUSSION

Table 2 and Figures 1 and 2 contain 5 year (1987-1991) average net returns and contaminant loadings. All the contaminants in the area were simulated although the results of this study focus on atrazine in runoff, nitrogen in runoff and leachate, and soil erosion. Result of atrazine in leachate is not discussed because it was a very small amount or did not occur in most years.

Based on the assumption that other conditions such as variable and fixed costs, and crop prices are constant, variability of yields was examined. Yield data for 5 years was obtained from the Kansas Farm Management Association (1987-1991). Crop yield data of the alternative systems were the same data obtained from those of the Conventional System because the actual yield data were not available.

The Conventional System had a negative 5 year average net return. Alternative System 1 was the secondly highest ranked system based on net return. The relevant crops were the same as the Conventional System with the exception of corn. Alternative System 2 was the worst enterprise based on net return due to the composition of crops and higher variable costs. Alternative System 4 was the best production system based on net return. Alternative Systems 3 and 4 were comparable because both systems grow alfalfa. In addition to the selection and profitabilies of crops selected, it should be noted that Alternatives have fewer total costs than the other systems.

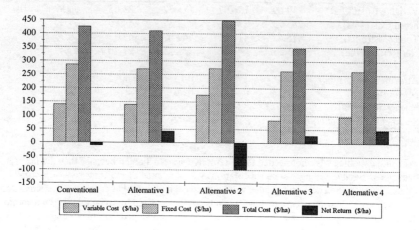

Figure 1. Costs and Five-Year Average Net Returns for Northeast Kansas from 1987 to 1991.

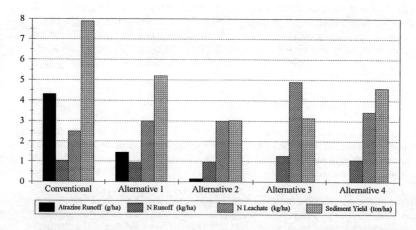

Figure 2. Five-Year Average Contaminant Loadings for Northeast Kansas from 1987 to 1991.

The Conventional System had the highest 5 year average of atrazine in runoff due to the intensive use. Alternative Systems 1 and 2 had lower levels of atrazine in runoff than the Conventional System. However, the atrazine levels in runoff under these systems cannot be directly compared because their application rates are different.

Alternative System 3 had the highest 5 year average level of nitrogen in runoff. Alternative System 4 was the second highest in nitrogen in runoff due to the different types and number of field operations, or nitrogen from crop fixation.

All the Alternative Systems had higher levels of nitrogen in leachate than the Conventional System. Alternative System 3 had the highest average level of nitrogen in leachate as well as nitrogen in runoff. Higher levels of nitrogen were found in runoff and leachate than other contaminants under Alternative Systems 3 and 4.

The Conventional System had the highest average level of soil erosion and Alternative System 2 had the lowest level. Among the alternative systems, Alternative System 1 had the highest average level of soil erosion.

CONCLUSION

One of the greatest limitations of this study revolves around data availability. Calculation of true five-year average net returns was impossible because of the lack of data regarding historical input prices. In addition, crop yields did not vary across systems because long-term average crop yields associated with the alternative systems were not available. Another data limitation arose in the non-point pollution simulation model. Because this study assumed a representative farm case, which considers the average of 14 counties in the study area, specific field observation could not be used.

Final results imply that a policy targeting only a few chemicals may ignore changes in the levels of other contaminants. For example, if farmers in the study area adopt one of the alternative systems, they could get benefits from of higher net returns and reduced levels of atrazine in runoff and soil erosion. However, the same system may result in producing more nitrogen in runoff and leachate than the Conventional System.

Converting the whole farm into a new cropping system may not be possible because of the high costs of transition and the need for production diversification under the Conventional System. Therefore, transitional systems are suggested to avoid those kinds of risks. Basic forms of transitional systems are introduced in Diebel et al's (1993) study, in which alternative systems are incorporated onto some portion of hectares in the Conventional System. Those systems can be introduced through educating farmers in the area about existing incentives, including economic and environmental benefits.

REFERENCES

1. Crowder, Bradley M. and Edwin C. Young. 1988. Managing Farm Nutrients: Tradeoffs for Surface- and Ground- Water Quality. Economic Research Service, U.S.D.A. Washington D.C. Agricultural Economic Report No. 583.

2. Diebel, Penelope L., Richard V. Llewelyn, and Jeffery R. Williams. 1993. An Economic Analysis of Conventional and Alternative Cropping Systems for Northeast Kansas. Report of Progress 687, Agricultural Experiment Station, Kansas State University. Manhattan, Kansas.

3. Hallberg, George R. 1987. Agricultural Chemicals in Groundwater: Extent and Implications. American Journal of Alternative Agriculture. 2(1): 3-15.

4. Hickman, John S. 1993. Basic Principles of Pesticide Runoff. Programs and Proceedings of Symposium on Agricultural Nonpoint Sources of Contaminants: A Focus of Herbicides. EPA and USGS. Lawrence, Kansas.

5. Kansas Agricultural Statistics. 1987-1991. Average Prices Received by Farmers. Selected Issues. Kansas State Board of Agriculture.

6. Kansas Farm Management Association. 1975-1991. The County Report. Department of Agricultural Economics and Cooperative Extension. Kansas State University. Manhattan, Kansas.

7. Knisel, W.G., F.M. Davis, and R.A. Leonard. 1992. GLEAMS Version 2.0 Part 3: User Manual. USDA.

8. Koo, Seungmo. 1993. A Comparative Analysis of Conventional and Alternative Systems in Northeast Kansas. Master's Thesis. Department of Agricultural Economics, Kansas State University. Manhattan, Kansas.

9. Leonard, R.A., W.G. Knisel, and D.A. Still. 1987. GLEAMS: Groundwater Loading Effects of Agricultural Management System. Transactions of ASAE. 30(5): 1403-1418.

10 National Research Council. 1989. Alternative Agriculture. Academy Press, Washington D.C.

11. Regehr, David L., Dallas E. Peterson, and John S. Hickman. 1992. Questions & Answers about Atrazine. Atrazine & Water. Cooperative Extension Service. Kansas State University. Manhattan, Kansas.

12. Soil Conservation Service. 1982. Soil Survey of Nemaha County, KS. Soil Conservation Service, USDA. Washington D.C.

13. Thomas, D.L., M.C. Smith, M.A. Breve, and R.A. Leonard. 1989. Alternative Production Systems. Proceedings of the CREAMS/GLEAMS Symposium. University of Georgia.

14. United States Department of Agriculture. 1978. Predicting Rainfall Erosion Losses. Agriculture Handbook No. 537. Washington D.C.

ECONOMIC EVALUATION OF VARIABLE RESOURCE MANAGEMENT BASED ON SOIL PRODUCTIVITY VARIABILITY OF SELECTED LAND AREAS, SIOUX COUNTY, IOWA

D.R. Speidel and R.E. Nelson*

ABSTRACT

Crop inputs in excess of the yield potential of a soil reduce profitability and increase pollution problems from leaching and runoff. Inaccurate fertilizer application and planting rates can also result in profit loss. Current farming practices plan only for the dominate soil in a field.

Differential soil management is a technique where application of fertilizer and planting rates are tailored to the capability of individual soil groups. Guided by digitized soil maps, recently developed equipment with on-board computers can adjust crop inputs while moving through the field.

A computer model was developed to determine the efficiency of this process. Nineteen farm operations were interviewed to establish the range of yield potential for predominate soils managed by each producer. Crop response for each soil was determined from previous research results. Predominate soil groups in the area evaluated include silty loams in upland positions and silty clay loams in upland, alluvial or bottomland positions.

The computer model showed that fields with the right combination of soils did have potential for additional profit when crop inputs were tailored to individual soil types. Therefore, benefit of differential soil management can be predicted for a given field with controlled crop inputs.
KEYWORDS. Differential soil management, digitized soil maps, computer-regulated crop inputs, soil productivity, variable resource management, income improvement.

BACKGROUND

The technological advances which have made agriculture highly productive rely on costly resources that can become pollutants in our environment. Procedures are needed that provide more efficient use of these resources to reduce pollution and to maintain profitable farming operations.

Farm operators normally apply fixed rates of crop inputs to an entire field based on the predominate soil in that field. Differential soil management, a new technique, makes it possible to match application rates to each soil mapping unit (SMU). Digitized soil maps are used in conjunction with computers to match fertilizer application rates to yield potential of each soil resource as the equipment crosses the field (Luellen, 1985).

The objective of this study was to determine if differential soil management is practical for use in Sioux County, Iowa. A survey was made of 19 farm operations to establish the range of yield potential for predominant soils in the county. A computer model was developed based on farm interviews, determination of corn yield (Zea mays L.) and crop response to management for each soil map unit. Comparison of single field crop budgets to differential managed crop budgets would provide a simple means to demonstrate which technique was more efficient for specific fields.

---

* D.R. Speidel, District Conservationist, USDA-SCS, Stanton, NE 68779 (402-439-2213). Graphics by R.E. Nelson, Professional Engineer (ret.), USDA-SCS, Norfolk, NE 68701. Original study completed Iowa State University as part of Master's thesis.

METHOD AND PROCEDURE

Sioux County is located in northwest Iowa. The principal soil is Galva (Typic Hapludolls), a silty clay loam derived from loess (SCS - Soil Survey of Sioux County, IA). Other important soils are Primghar, Radford, Ida and Moody. Farms in the Galva-Radford, Galva-Primghar and Galva-Ida soil associations were selected for this study. These soil associations represent 78 percent of the land area of the county.

Fifty-Six Primary Sample Unit sites (PSU's), generated randomly to sample 2 percent of the county for the National Resource Inventory (1982) were in the selected area. Nineteen representative farming operations were selected, using the following criteria:

(1) five representative PSU's for each soil association.
(2) continuous operation by the same operator from 1977-1986.
(3) field size 32.4 hectares (80 acres) or more contiguous within the PSU.

A questionnaire was developed and interviews made in 1987. Farm operators were asked to estimate the yield for each year. Questions included planting date, plant population, rotation, average yield per field, fertilizer management history and soil test levels. Areas within each field were identified that produced above or below average yields. In addition, the operator was asked how often the average yield was reached and to indicate any problem areas (wetness, drought, etc.). Areas well below or above the average were identified on an aerial photo. Every operator could remember select areas of the field with either high or low yields. These high or low producing areas were correlated with the soil map units (SMU's) of the field, and the acreage was determined from the soil map. A soil map unit was determined to be predominate if at least five producers identified it.

Table 1 lists the highest and lowest yielding field for each soil association and the number of years the operator believed these yields were attained. The range in yields for a field shows that specific areas within the field were variable. The frequency is the number of years out of 10 the operators predicted a field would reach the estimated yield.

Table 1
Field Yield Variability - Six Selected Farms

Soil Assoc. on PSU Site	High Yielding Field Frequency	Low Yielding Field Frequency	SMU's 1977-86 Yield Range
GP 1	150 bu. 9:10	130 bu. 8:10	120-175 bu.
GP 2	150 bu. 8:10	110 bu. 6:10	80-150 bu.
GR 1	170 bu. 7:10	100 bu. 8:10	100-170 bu.
GR 5	110 bu. 7:10	70 bu. 4:10	70-110 bu.
GI 1	120 bu. 7:10	90 bu. 7:10	80-125 bu.
GI 4	70 bu. 3:10	30 bu. 4:10	30-100 bu.

Ten predominant SMU's were identified, four in the Galva-Primghar (GP) association, four in the Galva-Radford (GR) association and two in the Galva-Ida (GI) association. Average yields, determined from the operator interviews from 1977-1986, were used to correlate these SMU's into categories with similar yield potentials and common soil characteristics. Four specific groups of SMU's with similar management requirements and yield potential were identified (Speidel, 1988). Each group has a yield potential variation of less than 1270 kg ha^{-1}. These groups were used in the crop budget analyses (Northwest Research Center, 1976-1986).

Two-thirds of the operators credit droughty soil conditions as a limiting factor in managing for optimum yields. Rainfall in the area averages 66-72 cm (26-28 in) annually, but may vary plus or minus 39 cm from the average (Northwest Research Center, 1976-1986). Wetness or flooding was a hindrance to half the operators in the management of each field. One-fourth of the operators identified hail or severely eroded soil as factors limiting their ability to improve production. One-half of the operators had applied improvements during 1977 to 1986 which included drainage, fertility management and soil conservation measures.

Soil tests indicated at least average management for all but one operator. Soil test levels were mainly medium to high for phosphorus and potassium. Only the Galva-Ida soil association had two PSU's that tested low to medium.

Crop response to management, by soil type, was determined from available research information (Northwest Research Center, 1976-1986). The records from two local research stations provided information on the yield response to fertilizer, population stress and soil moisture stress for the predominate soils in the study. ASCS and Iowa soil property and interpretative data records (ISPAID) were used to confirm yields.

The operator's experience and local research was correlated to determine the optimum nutrient and plant population levels. Crop response to variation of these inputs was determined from local research results. Plant population ranged from 40,640 to 66,040 seeds per hectare (16,000 to 26,700 seeds per acre). A nitrogen efficiency level of .02 kg of nitrogen per kg of corn (1.2 lbs per bushel) was determined from the survey. This compares well with the efficiency rate for nitrogen determined by the Northwest Research Center at Sutherland and Doon, IA (1976-86). Figure 1 shows the incremental response to management used in the model for each soil group.

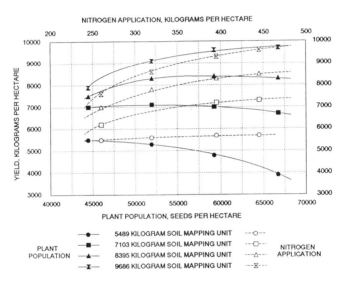

Figure 1. Yield response as affected by nitrogen application and plant population.

To develop the model, several economic assumptions were used. The economic principle used for the differential soil management model was profit maximization. Each SMU yield group had an optimum level of production it could reach and a unique incremental production relationship (Webb, 1984). To analyze the cost and return from this relationship, a crop budget was used. The Iowa State Extension crop budget format (Duffy, 1986) was expanded to include four different yield groups. The budget adjusted for each soil group's potential yield when managed for optimum production and for changes in production costs. The crop budget would compare returns of variable fertilizer and plant population to fixed management cost.

The crop budget shown in Table 2 was developed based on using inputs until the last marginal costs equals the marginal return. Parameters used to compare the systems were as follows:

Price   Value of corn.
Costs   Cost inputs of seed and fertilizer.
Returns   Yield for SMU managed field versus single field.

Table 2
Partial Crop Budget Summary
Soil Productivity Differential Management System
Corn Following Corn

	Yield Goal (kg ha^{-1})			
	5489	7103	8395	9686
Seed Cost ($)	37.74	44.04	50.32	56.62
Nitrogen Cost ($)	38.86	50.20	59.21	66.60
Phosphate Cost ($)	19.00	24.42	27.14	32.07
Potash Cost ($)	8.13	9.50	10.85	12.21
Machinery/Labor/Land ($)	496.48	513.26	526.58	540.52
Total Cost ($)	600.21	641.42	674.10	708.02

Nitrogen and plant population were two key variables built into the model. The crop removal rate for phosphorus and potassium determine the amount of these nutrients (Kipps, 1970). All other costs used current rates from Iowa State. The response of corn to nitrogen varied with each soil group at a given plant population (Fig. 1). Plant population rates were set to reach corn yield goals with average size ears of 180 grams (0.4 lbs). The response curves for both variables were plotted to determine the efficiency rate for each group of SMU's for four different fertilizer rates and four different plant populations. The results were used to build lookup tables for each SMU yield group.

Table 3 shows two lookup tables managed at the 7103 and 9686 kg ha^{-1} yield goals. SMU's with low optimum potential fail to respond to the additional nitrogen and stress from too high plant populations. High yield potential SMU's, if managed at the lower population or lower nitrogen levels, fail to achieve full potential.

Table 3
Variable Population
Yield response to variable population and fixed fertilizer management.

Lookup Table Summary (in bushels per acre)

Yield Goal 110 bu (7103 kg ha^{-1})					Yield Goal 150 bu (9686 kg ha^{-1})				
	SMU Yield Groups					SMU Yield Groups			
Plant Pop.	85	110	130	150	Plant. Pop.	85	110	130	150
18,000	86	109	112	115	18,000	88	112	118	120
21,000	84	110	123	127	21,000	85	113	131	135
24,000	80	109	124	131	24,000	81	112	132	149
27,000	65	104	120	132	27,000	67	108	130	150

Table 4 shows the results of applying the model to compare two PSU's.

Table 4
Net Income per Hectare

	PSU GI 1 7103 kg ha^{-1}		PSU GR 1 9686 kg ha^{-1}
Price Goal 37.2/metric ton			
Predominate SMU yield goal			
Fixed Management Cost ($/ha)	41.23	Fixed Management Cost ($/ha)	135.50
Variable Fertilizer ($)	58.62	Variable Fertilizer ($)	137.62
Variable Population ($)	51.93	Variable Population ($)	137.05
Variable Fertilizer & Plant Population ($)	65.60	Variable Fertilizer & Plant Population ($)	139.18

The two example PSU's in Table 4 have the following composition of soil yield groups. Eighty-eight percent of PSU GR 1 was composed of SMU yield group 9686 kg ha^{-1}. The balance of the SMU's in PSU GR 1 were in yield group 8395 kg hg^{-1}. In comparison, the PSU GI 1 had greater variability in soil yield groups. PSU GI 1 SMU yield group composition consisted of 70 percent 7103 kg hg^{-1}, 26 percent 9686 kg hg^{-1} and 4 percent in the 5489 kg hg^{-1} soil yield groups.

RESULTS AND DISCUSSION

The survey showed that farmers were aware of yield difference due to soil variation. Corn yield variations reached 5,166 kg ha^{-1} some years. The typical variance was plus 1270 kg and minus 1930 kg from the average. The operators' expectations for the field as a whole determined how each SMU was managed. Management in the Galva-Primghar association was directed at the somewhat poorly and poorly drained soils. SMU 310B2, a minor, eroded soil in the Galva-Primghar association, was one of the lowest producing soils, 1270 to 2540 kg ha^{-1} less than other SMUs. However, management of eroded soils was stressed in the Galva-Radford association. The soil 310B2 SMU and 310C2 SMU were predominant soils in the Galva-Radford association and yielded up to 2500 kg ha^{-1} higher than the same SMU's did in the Galva-Primghar association.

Fertilizer and farm chemical application, plant population and tillage practices were not adjusted for a specific soil. Non-predominant soils are then fertilized for 9686 kg ha^{-1} yield but yielded 5812 kg ha^{-1}. Analysis of the farmer survey suggests the following for Sioux County:

(1) Eastern section is managed at 9686 kg ha^{-1} (150 bu/ac) for corn.
(2) Central section is managed at 8395 kg ha^{-1} (130 bu/ac) for corn.
(3) Western section is managed at 7103 kg ha^{-1} (110 bu/ac) for corn.

The extent that other management practices influence yield is not clearly known (Miller and Shrader, 1972; SCS USDA Tech. Bulletin No. 883 and 888; and Western Research Center, 1969). According to farm managers interviewed in Sioux County, IA, in addition to adjusting fertilizer rates and plant population, consideration must be given to planting dates, drainage, soil and water conservation practices, timing of tillage and of application of manure and crop rotations. A marked effect on crop yield for a given SMU was observed in the western section of the county. The major SMU was eroded and use of manure application and contour farming was wide spread to improve soil tilth and conserve water. This was well illustrated by one producer where the topography was suitable for contouring, enabling farm operations of the moderately eroded soils to be separated from the somewhat poorly drained soils. The producer adjusted both plant population and nitrogen for two different yield goals. Yield on the moderately eroded soil was 1270 kg ha^{-1} higher than for similar soil in the eastern section of the county where differential soil management was not used.

Trends were observed in the three different sections of the county when analyzed. The advantage of using this management technique varied across the county. Reducing costs was the primary benefit in the Galva-Primghar sample. Improved production was more noticeable in the Galva-Ida sample.

The bar graph in Figure 2 shows the response of selected PSU's from the survey to differential soil management. The resources adjusted were varying the 1) fertilizer rate, 2) plant population and 3) both fertilizer and plant population to the economic optimum for each soil group. The results in Figure 2 show the potential advantage of differential soil management over fixed management. The assumed cost for the service is shown as the breakeven line. Moderate potential was found in the Galva Primghar and Galva Radford samples. The Galva Ida sample, however, showed twice the potential advantage over fixed management. This is due to the greater variability in productivity by at least 2540 kg ha^{-1} for a significant acreage of individual soils.

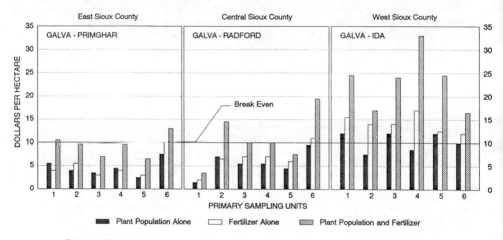

Figure 2: Predicted return, variable resource management vs. current management.

Differential soil management is profitable when returns exceed the service charge, provided all management inputs for optimum production for each soil group can be achieved. Current equipment service charges for variable rate fertilizer application range from $6 to $37 per hectare ($2.50 to $15.00 per acre) (Robert, et al., 1993). The impact of the management service cost is shown in Table 4 comparing the two PSU's from the Galva Radford and Galva Ida soil associations. The PSU GR 1 has only a $3.68 ha^{-1} profit margin to pay for the differential soil management service charge. However, PSU GI 1 has a $24.37 ha^{-1} potential economic benefit from variable resource management to pay for management charges.

When a single soil yield group represents a substantial part of the field, the field or weighted average method was a reasonable means of setting yield goals (Miller, 1986). However, when there are two or more soil yield groups representing a majority of the field, differential soil management is the superior method.

CONCLUSION

Farmers are aware of yield variations within a field and these yield differences do correlate with specific soil map units that can be delineated on standard soil maps.

Fields with wide soil productivity variation, if managed for the average potential yield of all soils present, may fail to reach the production potential for a third or more of the field.

Optimum resource utilization depends on the incremental relationship of the crop response to inputs on each of the different soil groups, the farmer's ability to manage these soils at the optimum response levels, current prices and costs and the proportion of different soil groups in a field.

Research is needed to better quantify the factors affecting the decision whether to select this technique. These include: 1) The crop response to changing commercially applied inputs on the different soil types. 2) The economic cost in time and effort required to manage two to four different soil groups scattered across a field for management with a commercial applicator or managed only by the individual producer. 3) Environmental benefits from the reduction in chemical runoff.

This study indicates that differential soil management of corn production in Sioux County, IA, is the superior method when the field meets certain parameters. The parameters found in the study indicate that the field has two or more distinct soil yield groups with neither group greater than 70 percent and a yield spread of 2540 kg ha^{-1} (40 bushel per acre). In circumstances where the environmental concerns are high, the field profile having these two distinct soil groups is less important.

BIBLIOGRAPHY

Duffy, M. 1986. *Estimated costs of crop production in Iowa 1987.* Pm-1712 Rev. Cooperative Extension Service, ISU, Ames, IA.

Kipps, M.S. 1970. *Production of field crops.* McGraw-Hill. New York, NY.

Luellen, W.R. November 1985. *Fine-tuned fertility: tomorrow's technology here today.* Crops and Soils Magazine. Publication American Society of Agronomy. Madison, WI.

Miller, E.L. and W.D. Shrader. 1972. *Effect of level terrace on soil moisture content and utilization by corn.* Agronomy Journal. Iowa Agriculture and Home Economics Experiment Station, Ames, IA.

Miller, G.A. 1986. *How to establish realistic yield goals.*, Pm-1268 Cooperative Extension Service. ISU, Ames, IA.

Northwest Research Center, Sutherland, IA, and Doon, IA . *Annual Progress Reports.* 1976-1986. Agriculture and Home Economics Experiment Station. ISU, Ames, IA

Speidel, D.R. 1988. Unpublished paper for Master of Agriculture thesis. *A soil productivity variability survey of selected land areas, Sioux County, IA.* ISU, Ames, IA.

U.S. Department of Agriculture, Soil Conservation Service. 1949. *Farm planning and management for soil conservation.* Inservice publication, Annual Reports. Erosion Experiment Station. USDA Tech. Bulletin No. 883 and 888 and Mimeographed publications, Iowa Agriculture Experiment Station, Clarinda, IA.

U.S. Department of Agriculture, Soil Conservation Service. 1982. *Basic Statistics 1977 National Resources Inventory,* Stat. Bulletin No 686. ISU Statistical Laboratory, Ames, IA.

U.S. Department of Agriculture, Soil Conservation Service. 1987. *Advance Report - Part 1 Soil Survey of Sioux County.* In cooperation with Iowa Agriculture and Home Economics Experiment Station, Cooperative Extension Service, Iowa State University and Division of Soil Conservation, Iowa Department of Agriculture and Land Stewardship, Des Moines, IA.

Webb, J.R. 1984. *Influence of nitrogen rates on corn yields.* Annual Progress Report Northwest Research Center Sutherland, and Doon, Iowa Agriculture and Home Economics Experiment Station. ISU, Ames, IA.

Western Research Center, Castana, IA. 1969. *Annual Progress Report.* Agriculture and Home Economics Experiment Station. ISU, Ames, IA.

Robert, P.C., R.H. Rust and W.E. Larson. 1993. *Soil specific crop management: a workshop in research and development issues.* American Society of Agronomy, Madison, WI.

INTEGRATED WATER-FERTILIZER-PEST MANAGEMENT FOR ENVIRONMENTALLY SOUND CROP PRODUCTION [1]

James L. Fouss   and   Guye H. Willis [2]
Member ASAE   Member ASAE

ABSTRACT

Farm management and operational decisions required in the future will be complex because soil-water management must be integrated with improved fertilizer and pesticide application methods to insure environmentally sound agricultural practices. Off-site environmental requirements may be more important and difficult to achieve than agricultural production requirements in the future. The "optimal" management of soil-water for agricultural cropland in humid areas of the U.S., e.g., control of water table depth in the soil profile via controlled-drainage and subirrigation, involves complex daily operational or management decisions because of the typical erratic spatial and temporal distribution of rainfall. Primary objectives of new research thrusts being undertaken include the development of integrated methodology needed to manage soil, water, ground cover, and agrochemical applications (i.e., fertilizers and pesticides) in such a way that agrochemicals are contained within their "action zones", thus reducing the potential or risk of surface and groundwater pollution. The new integrated management technology will also utilize rainfall probability data from the daily National Weather Service forecasts to assist in optimizing the operational control of soil-water and scheduling agrochemical applications to minimize potential losses in drainage flows or deep seepage. Improved soil-water management technology, including automated operation of water management systems, may make it possible to reduce the amount of fertilizers and pesticides used, thus increasing crop production efficiency and farmer profitability, while significantly reducing pollution potential. Computer simulation models are being developed through this research to embody the new integrated systems approach technology, and which will aid the engineer in designing integrated crop production systems and their operational methods. These computer models will need to be incorporated into decision support models (Expert Systems) which can be used by farmers and farm managers to operate water-fertilizer-pest management systems in accordance with the new technology and water quality standards. This paper outlines the needs for such integrated systems research, and describes a new comprehensive field experiment and model development project being conducted in the lower Mississippi Valley to develop this new technology for humid region climate conditions.

---

[1] Contribution from the Soil and Water Research Unit, ARS-USDA, in cooperation with the Louisiana Agricultural Experiment Station, LSU Agricultural Center, Baton Rouge, LA.

[2] The authors are Agricultural Engineer and Soil Chemist, USDA-ARS, Soil and Water Research Unit, P. O. Box 25071 - University Station, Baton Rouge, LA 70894.

BACKGROUND AND INTRODUCTION

About 25%, i.e., 40 million hectares, of the total U.S. cropland needs drainage (U.S. Dept. Agric., 1987). Much of this land is usually flat, highly fertile, and has no serious erosion problems. These potentially productive wet soils are primarily located in the prairie and level uplands of the Midwest, the bottom lands of the Mississippi Valley, the bottom lands in the Piedmont areas of the South, the coastal plains of the East and South, and irrigated areas of the West (Schwab et al., 1993). During most or part of the year these soils have shallow water tables that are potential sinks for agrochemicals that leach below the root zone. The frequently occurring shallow water table conditions in humid regions increase surface runoff with the potential of carrying residuals of applied agrochemicals (e.g., fertilizers and pesticides) to streams and lakes. In areas where water management facilities include subsurface drainage, agrochemicals that leach below the root zone may be carried off-site in subsurface drainage discharge.[3]

The discussions in this paper are concerned primarily with the conditions, practices, and research needs of the important humid region agricultural area in the lower Mississippi Valley. The lower Mississippi River Valley (LMRV) contains about 2.5 million hectares that have been assessed as highly vulnerable to groundwater pollution by leachable chemicals, and a slightly smaller amount of land evaluated as moderately vulnerable. There have been reports of nitrate and pesticide contamination in groundwaters in the LMRV (Williams et al., 1988; Calhoun, 1988; Cavalier and Lavy, 1987; Acrement et al., 1989; Whitfield, 1975). Water levels in the LMRV generally are less than 9 m below land surface, and are much closer to the surface (0 to 2 m) in southern areas. These shallow water tables fluctuate considerably and respond mainly to rainfall (1000-2000 mm annually), the major source of recharge for the LMRV groundwater (Whitfield, 1975; Morgan, 1961; Poole, 1961; Dial and Kilburn, 1980).

The widespread use of agricultural chemicals, including fertilizers and pesticides, has been brought about by economic factors and farmers' efforts to obtain a fair return on their investment in crop production systems. In the Lower Mississippi Valley (LMV), several conditions exist and practices are followed that may contribute to groundwater contamination potential via transport of nutrients and pesticides through the root and vadose zones of the soil profile: (1) A relatively shallow depth in the soil profile to the alluvial aquifer, (2) the current use of many water soluble chemicals, (3) permeable soils, such as silt and sandy loams, (4) cracks that develop in the soil upon drying, creating preferential flow paths deep into the soil profile, (5) large annual rainfall amounts, and (6) tillage, drainage, and irrigation practices. There are major economic reasons for the continued use of pesticides for the foreseeable future in U.S. agriculture, especially

---

[3] The reader is referred to a series of papers dealing with the "Effect of Agricultural Drainage on Water Quality in Humid Regions" that were published by ASCE in the conference proceedings, Management of Irrigation and Drainage Systems: Integrated Perspectives, R. G. Allen and C. M. U. Neale, (Ed.), July 21-23, 1993; see pp. 511-572.

in the LMV where hot and humid climate conditions enhance weed growth and insect infestation. Thus, there is the potential for extensive contamination of surface and groundwater resources in the LMV resulting from continued high fertilizer and pesticide use if current agrochemical application, soil and water management, and crop production practices are followed.

RESEARCH NEEDS

Integrated management of soil and water resources and agronomic cultural practices is necessary to insure environmentally sound crop production on shallow water table soils in humid areas (such as the LMV), and to reduce or eliminate potential pollution of water resources, including adjacent wetland areas. Integrated methodology is urgently needed to manage soil, water, ground cover, fertilizer and pesticide applications in such a way that fertilizers and pesticides are contained in their functional "action zones" of the soil profile. It should be possible to manage water table depth to control root-zone soil-water content such that plant fertilizer use efficiency is enhanced, thereby decreasing fertilizer needs and reducing nutrient pollution potential. Furthermore, recent research has shown that improved soil-water management increased early-season crop growth and provided canopy cover that decreased weed populations (Carter, 1990); this decrease could reduce herbicide needs and thereby lower pollution potential. There is also an opportunity to reduce pollution by overall pesticide management practices, particularly for those pesticides that degrade as a function of soil moisture and aeration conditions. The reduced need for fertilizers and pesticides would thus increase profits to farmers.

Variable rainfall amounts and uncertain times of occurrence in most humid regions, and especially in the LMV, create a need for water management systems which will compensate for both excess and deficit soil-water conditions. A dual purpose subsurface conduit system for subdrainage and subirrigation is becoming popular in many humid areas of the U.S. and eastern Canada (Skaggs, 1980; Fouss, et al., 1990). New technology is needed to properly design and operate these water table management systems; for example, when to subirrigate and when to control subdrainage effluent or permit "free" drainage of the soil profile to provide the best possible root-zone moisture conditions for crop production, while reducing runoff, erosion, and pollution. In many cases, automatic control of the water table management system may be merited (Fouss, et al., 1990). Figure 1 illustrates a pumped-sump water table management system in the controlled-drainage mode of operation. In some humid regions, the probability of predicted rainfall in National Weather Service forecasts, particularly the new *7-day Forecast* (SRCC, 1994), may soon become accurate enough to permit adjustment of day-to-day operation of water management systems (Fouss and Cooper, 1988) and/or timing of fertilizer and pesticide applications. Principal management objectives may be to: (a) increase the effectiveness of fertilizers and pesticides applied and reduce the potential of losses; (b) reduce the occurrences of severe excess soil-water events, and the duration of deficit soil-water conditions; and (c) improve the efficiency of utilizing rainfall received, thus minimizing the need for pumping irrigation water.

To quantify specific risks involved with various management practices, agricultural scientists are developing and using computer models and *expert systems* capable of predicting the effect of nutrient and water stress on crop yields and at the same time assessing the pollution potential to surface and groundwater resources. If nutrient use is reduced through improved management or facilities, the producer needs to know the risk and economic impact. Models can aid in evaluating alternative rates and timing of chemical applications, the use of alternative chemicals with different properties, and the optimum management of soil-water (controlled-drainage and irrigation), for soils with different physical and chemical properties, and wet or well-drained soils. To improve the predictions on a field or watershed scale, more details are necessary on preferential flow through soil macropores, on denitrification in the vadose zone and the groundwater, and on ways of incorporating spatial variability into computer simulation models. Methods to better predict the fate of nutrients and chemicals in the soil profile and groundwater, under different cultural and management practices, need to be developed and validated.

INTEGRATED SYSTEMS RESEARCH

A comprehensive research program has been initiated in the lower Mississippi Valley to develop the technology to design and operate an integrated system for water-fertilizer-pesticide management. The overall objective is to evaluate various soil and water management strategies in terms of reduction in agrochemical (fertilizer and pesticide) losses in surface runoff, subsurface drainage effluent, or deep seepage, and improvements in agrochemical use efficiency and crop yield potential. The research involves both modeling and field plot experimentation, plus laboratory investigations. Too many variables are involved to optimize system design and operational performance based on field tests alone. Model development phases were begun early, permitting preliminary simulation results of proposed water management systems operation to help in the design of the field experiment and identify treatments and required measurements.

Field Project Design

The field project is located on the LSU Ben Hur Research Farm near Baton Rouge, LA, on a Commerce silt loam soil which consists of layers of silt and clay mingled with sand lenses that were deposited by past Mississippi River overflows. Four water management treatments with four replications were installed in a randomized complete block design on 16 bordered 0.21-ha plots; the treatments include: (I) surface drainage only [the subsurface drainlines are plugged [4]], (II) conventional subsurface drainage [water table kept at the 1.25 m depth of the drainline or greater], (III) controlled shallow water table at 0.45 ± 0.05 m depth, and (IV) controlled deeper water table at 0.75 ± 0.05 m depth. Each plot, 35 m x

---

[4] These surface drained <u>only</u> plots may have the drainline plugs removed after 3-4 years to impose other types of subsurface drainage treatments. The plugs will be removed for 2-3 weeks each January (typically a very wet month) to allow the drainlines to be cleared of any sediment.

61 m, has a 0.15-m high earthen dike at the outer edge of each border, a 0.15 mm polyethylene subsurface barrier installed 0.3 m below the soil surface and extending down 2.0 m, three subsurface drainlines (102-mm diameter corrugated plastic tubing) installed at 15 m spacings and 1.25 m depth, a 300 mm diameter plastic pipe riser on the outlet of each drainline to control the outlet water level (these risers are housed in a 1.2-m square by 3.0-m deep steel sump), and an H-flume at the surface runoff outlet. Each plot is precision-graded to a 0.2% slope with about a 0.2% cross-slope. The drainlines next to the longitudinal borders control border or water table transition effects between adjacent plots. The area centered over the middle drainline (15 x 61 m) is assumed to be representative of an area in a larger field with the same drain spacing. The project area is equipped and instrumented for automatic measurement and control of water table depth, and microprocessor-controlled measurement and automatic sampling of surface runoff and subsurface drain outflow. Complete details of the experimental design, materials and equipment, operational procedures, instrumentation, and data acquisition for the project are presented by Willis et al. (1991, 1992).

Automated Water Table Control

For treatments (III) and (IV), the water table depths in the experimental plots are automatically maintained within the depth ranges specified by automated or *feedback* control of the water level at the subsurface drain outlet (i.e., the outlet riser pipe) to regulate subsurface-drainage and subirrigation flows (commonly referred to as *controlled-drainage and subirrigation*). For the controlled-drained mode of operation (during periods of potential excess soil-water), drainage water is pumped from the outlet riser pipe to a gravity flow channel to maintain the outlet water level within the desired range. Conversely for the subirrigation mode (during periods of potential deficit soil-water conditions), water is supplied from an external source (e.g., a well) into the riser pipe to maintain the desired outlet water level. In the project, all *on-and-off* cycles of drainage pumps and irrigation pumps (or valves) are activated, as needed, by a microprocessor data-logger/controller system which monitors outlet water levels via electronic water pressure sensors in each riser pipe. The data-logger/controller system also continuously monitors the water table depth in all plots (midway between drainlines) via electrical water level sensors installed in 50 mm diameter plastic pipes (water table "wells"). If a plot water table depth becomes too shallow or deep, the outlet water level control thresholds (i.e., *on-and-off* limits for drainage/irrigation pump operation) are automatically adjusted upward or downward (called *feedback* adjustment) as needed to compensate, and thus also control the plot water table depth within the desired range. A through discussion of *feedback* water table control is presented by Fouss et al., 1990.

Scheduling Agrochemical Applications and Subirrigation

Agrochemical management practices (application rate, mode, and timing) will be used on the 16 experimental plots to determine their effects on nutrient and pesticide losses in surface runoff, subsurface drain outflow, and deep seepage.

Application rates and modes will follow conventional recommended practices for the first few years of the project (corn will be grown), however, the timing of applications will be adjusted based on the probability of rainfall in the daily and 7-day weather forecasts issued by the National Weather Service (NWS). To reduce the risk or potential of agrochemical losses in runoff during rainfall events that occur soon after applications, scheduled applications may be advanced or delayed from the normal or desired timing based upon rainfall forecasts. An example 7-day forecast from the NWS is shown in Table 1 for the date, April 28, 1993, in the Baton Rouge, LA area. For purposes of this discussion, if agrochemicals were scheduled for application in the field on or about April 29, the applications could be delayed until May 2 based upon this 7-day forecast, because the probability of rainfall increases from April 29 through May 1, but decreases from May 2 through May 5. Since this 7-day forecast is available from the NWS each day, the application could be moved up (e.g., to May 1) if the predicted rainfall probability significantly decreases for May 1 in the Apr 30 NWS forecast.

The daily and 7-day weather NWS forecasts will also be used in making daily decisions regarding subirrigation of the water table managed plots in the project following the procedures outlined by Fouss and Cooper (1988). That is, if significant rainfall is predicted within two to three days from the date when subirrigation is to be started, then pumping may be delayed depending upon the actual soil-water status. Other operational procedures for the project are outlined in detail by Willis et al. (1991).

Table 1. Example 7-day Forecast from National Weather Service

```
==============================================================================
FOXS61 KWBC 280000
MRF-BASED OBJECTIVE GUIDANCE    4/28/93   0000 UTC

BTR   APR 28| APR 29| APR 30| MAY 01| MAY 02| MAY 03| MAY 04| MAY 05| CLIMO
MN/MX     81| 57  81| 61  79| 59  79| 61  82| 63  83| 63  84| 63  84| 61  82
POP12      6| 29  32| 35  42| 51  39| 33  23| 19  22| 17  19| 17  18| 18  18
CLDS      48| 51  77| 69  94| 78  81| 62  67| 55  65| 50  59| 50  55| 45  53
POP24       |    48 |    60 |    66 |    45 |    33 |    28 |    28 |    28
-----------
Where:  BTR   = Baton Rouge, LA
        MN/MX = Min./Max Air Temperature (deg. F.)
        POP12 = 12 Hr. Percent Probability of Precipitation (a.m. and p.m.)
        CLDS  = Cloud Cover (% Opacity)
        POP24 = 24 Hr. Percent Probability of Precipitation
        CLIMO = Climatological Mean (centered on day 4)
------------------------------------------------------------------------------
```

Data Acquisition and Drainage Sampling Systems

Four electronic, microprocessor-based, data-logger/controller systems (one per replication) is used on the project to continuously measure/record experimental variables and parameters with various sensors (e.g., water table depth, soil temperature, rainfall, etc.), and to automatically operate all drainage pumps and irrigation valves to control the water table depth (for details see Willis et al., 1991, 1992).

Runoff from each plot is routed through an H-flume where it is automatically measured and sampled by a microproccess-controlled, refrigerated system. The runoff samples collected are proportional to runoff rate/volume, and are analyzed for nutrient, pesticide, and sediment content. Subsurface drain effluent samples for the similar analyses are collected by an orifice-type device as water is pumped from the outlet water level control riser pipes for each center drainline; the sample volume is approximately 0.5 % of the total effluent. A composite effluent sample is collected for each storm event; a minimum 6-hour period with no more than 2.5 mm of rainfall defines the start of a new storm event.

Porous-cup soil-water samplers are installed midway between drains in each plot, at depths of 0.3, 0.6, 0.9, and 1.2 m. These cups are used to sample water for both unsaturated and saturated conditions. In addition 2.5 cm diameter piezometer wells are installed to depths of 1.5, 1.8, and 2.1 m for collecting soil-water samples under saturated conditions. These samples are analyzed for nutrient and pesticide content. Willis et al., 1991, gives complete details on the analysis procedures used.

Model Development/Validation Phases

Developing effective and efficient water management systems and determining how to operate them in a commensurate or integrated manner with fertilizer, pesticide, and tillage practices to reduce or eliminate environmental pollution potential, requires knowledge of soils, agrochemical transport, climate, crops, the response of the systems to control or management, and how the various modules or components relate. Thus, measured values of soil physical properties, soil-water dynamics, runoff and subsurface drainage, agrochemical utilization or movement, climate variables, crops response, and management systems responses are needed to develop dynamic simulation models that fully describe the interactions of the different variables and types of control involved, and to predict what will happen if different types and sequences of rainfall events occur when the soil is and is not drained or irrigated prior to the events.

The water management simulation models developed and validated will be merged, linked, or run in parallel (exchanging input/output predictions) with other models describing the movement of plant nutrients and pesticides within the soil profile, thus simulating the total water-fertilizer-pesticide management system. The development and validation of models that predict the transport of nutrients and pesticides in soil (and thereby the evaluation or their pollution potential) require an understanding of a number of processes, and the interactions of these processes. Also required is knowledge of how changes in various chemical and physical factors affect those processes. Since much of the required information is poorly characterized or understood, existing models currently depend on simplifying assumptions and lumped parameters that decrease prediction accuracy and flexibility. This on-going research will extend current knowledge concerning the processes and factors affecting nutrient and pesticide transport in soil, and will lead to the development of water management practices for reducing chemical leaching, and improve use efficiency of nutrients and pesticides in shallow water table soils.

COMMENTS AND DISCUSSION

The integrated water-fertilizer-pesticide management system has the potential for achieving environmentally sound and efficient crop production technology for shallow water table soils in humid climate regions. The study should lead to a strategy for managing water table depth to enhance plant fertilizer use efficiency, thereby reducing fertilizer needs and potential pollution by fertilizers. Further, there should be an opportunity to reduce pollution by pesticides through management of pesticide applications and water table depths as dictated by prevailing and predicted weather conditions.

The validated simulation models developed as a part of this research will serve as an 'expert knowledge base' for the creation of future 'decision support models' to aid the farmer or farm manager in day-to-day decisions concerning operation of the integrated water-fertilizer-pesticide management systems. These decision support models will almost certainly utilize the daily and/or 7-day weather forecasts as inputs to generate advisory information concerning operational changes in water management systems, and scheduling fertilizer and pesticide applications, to minimize the potential for agrochemical losses in runoff and subsurface drainage.

Knowledge obtained from this research will be applicable to chemical transport in general, and to the development of management techniques for wet and shallow water table soils in many regions. The project may serve as a prototype to the design and conduct of research on developing integrated water-fertilizer-pesticide management systems for the major agricultural crop production areas, thus providing the technology to reduce the potential of contaminating surface and groundwater resources with nutrients, pesticides, and other organic chemicals.

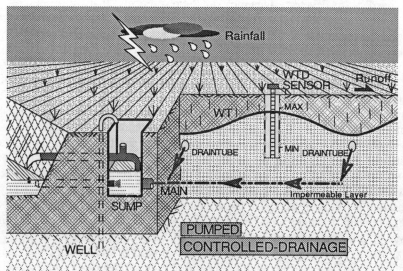

Figure 1. -- Schematic of pumped-sump water table management system in the controlled-drainage mode of operation.

REFERENCES

1. Acrement, G. J., L. J. Dantin, C. R. Garrison, and C. G. Stuart. 1989. Water resources data for Louisiana, water year 1988. U.S. Geological Survey. Baton Rouge, LA. Rept. No. USGS/WRD/HD-89/262.
2. Calhoun, H. F. 1988. Survey of Louisiana groundwater for pesticides. Louisiana Dept. Agric. Forestry. Baton Rouge, LA.
3. Carter, C. E. 1990. Personal communication on unpublished research results. USDA-ARS, Soil and Water Research Unit, Baton Rouge, LA.
4. Cavalier, T. C., and T. L. Lavy. 1987. Eastern Arkansas groundwater tested for pesticides. *Ark. Farm Res.* 36:11.
5. Dial, D. C., and C. Kilburn. 1980. Groundwater resources of the Gramercy area, Louisiana. Louisiana Dept. Transportation Development, Water Res. Tech. Rept. No. 24.
6. Fouss, J. L., R. W. Skaggs, J. E. Ayars, and H. W. Belcher. 1990. Water table control and shallow groundwater utilization. ASAE Monograph *Management of Farm Irrigation Systems*. pp. 783-824.
7. Fouss, J. L. and J. R. Cooper. 1988. Weather forecasts as a control input for water table management in coastal areas. TRANS. of the ASAE. 31(01):161-167.
8. Morgan, C. O. 1961. Groundwater conditions in the Baton Rouge area, 1954-1959. Louisiana Geol. Survey, and Louisiana Dept. Public Works, Baton Rouge, LA. Water Res. Bull. No. 2.
9. Poole, J. L. 1961. Groundwater resources of East Carroll and West Carroll parishes, Louisiana. Louisiana Dept. Public Works, Baton Rouge, LA.
10. Schwab, G.O., D. D. Fangmeier, W. J. Elliott, and R. K. Frevert. 1993. *Soil and Water Conservation Engineering*. 4th Ed. John Wiley and Sons, New York, NY.
11. Skaggs, R. W. 1980. Drainmod-reference report: Methods for design and evaluation of drainage-water management systems for soils with high water tables. USDA-SCS. South National Technical Center, Fort Worth, Texas. 330pp.
12. Southern Region Climate Center. 1994. Personal communication with Dr. K. D. Robbins, Assoc. Dir., Southern Region Climate Center, Louisiana State Univ., Baton Rouge, LA, re: 7-day forecast issued daily by the National Weather Service running the NGM Model.
13. U. S. Dept. Agric. 1987. *Farm drainage in the United States: history, status, and prospects.* G.A. Pavelis (Ed.). Misc. Pub. No. 1456.
14. Whitfield, Jr., M. S. 1975. Geohydrology and water quality of the Mississippi River alluvial aquifer, northeastern Louisiana. Louisiana Dept. Public Works, Baton Rouge, LA. Water Res. Tech. Rept. No. 10.
15. Williams, W. M., P. W. Holden, D. W. Parsons, and M. N. Lorger. 1988. Pesticides in groundwater data base: 1988 interim report. U.S. Environ. Protection Agency, Washington, DC.
16. Willis, G. H., Fouss, J. L., Rogers, J. S., Carter, C. E., and Southwick, L. M. 1991. System design for evaluation and control of agrochemical movement in soils above shallow water tables. ACS Symposium Series 465, *Groundwater Residue Sampling Design*, Chapter 11. pp. 195-211.
17. Willis, G. H., Fouss, J. L., Rogers, J. S., Carter, C. E., and Southwick, L. M. 1992. Status of the water table management-water quality research project in the lower Mississippi valley. Proceedings of the Sixth International Drainage Symposium, *Drainage and Water Table Control*, ASAE Publ. No. 13-92, pp. 210-218.

FEASIBILITY OF SITE-SPECIFIC NUTRIENT AND PESTICIDE APPLICATIONS[1]

J. C. Hayes, A. Overton, J. W. Price[2]

ABSTRACT

The need to reduce chemical usage in agricultural production is well established, but agricultural production must be maintained. Yields can be maintained with reduced chemical inputs if applications are matched with specific site conditions. Instrumentation is currently available that will permit precision application of nutrients and pesticides on a site-specific basis. This paper discusses the feasibility of using this high technology in field situations. Methods discussed include integration of geographical information systems, global positioning systems, and soil survey information with either manual or computer-assisted controls. The paper develops a rational approach to determining how much expense is allowable in a given farm situation. An example demonstrates the viability of using precision application in a specific situation.

INTRODUCTION

Often people view farming as a low technology occupation. A look at the overall efficiency and production of the American farmer serves to disprove this stereotype. Farmers have been quick to adopt practices that currently allow a single producer to feed large numbers of persons. Historically, this production has come at a great cost. From an environmental viewpoint, agricultural production traditionally has been the largest contributor to nonpoint source pollution. Thrusts by the USDA's Soil Conservation Service and others have led to large-scale adoption of various forms of conservation tillage.

Recently, agriculture has come under fire as a potential cause of nonpoint source pollution from chemicals. Concerns have been expressed that surface or ground water may be contaminated by nutrients such as nitrogen or pesticides. While evidence of significant problems is scant, the perception that chemicals are being misapplied may lead to additional regulations or restrictions on chemical usage. This in turn leads to additional costs for production and ultimately leads to higher food costs. Efforts such as the Chesapeake Bay Project have coordinated agency and agricultural personnel to reduce chemical usage to a specific water body. One suggestion is that nutrients be applied based on expected yield at a given location in the field. Equipment required to vary applications range from very simple, manually operated systems to extremely complex, computer controlled sensors and controls.

The farm community has control technology available that can vary application rates of fertilizers, pesticides, and seed on a site-specific basis. But how can a potential user determine the extent to which this relatively high technology should be adopted? Obviously in many cases, costs can outweigh benefits. This paper provides an example demonstrating the reasoning that may be used to decide how automated the system should be.

---

[1] This work was funded in part under Southern Regional Project S-249, entitled "Impact of Agricultural Systems on Surface and Groundwater Quality" and in part by the S. C. Agricultural Experiment Station. Mention of specific products is for information only and does not indicate the exclusion of others that may be suitable.

[2] J. C. Hayes, Professor; A. Overton, Research Assistant; and J. W. Price, Graduate Research Assistant; Department of Agricultural and Biological Engineering, Clemson University, Clemson, SC.

LITERATURE REVIEW

Site-specific application, or prescription farming, is using information about current field conditions to determine the amounts of fertilizer applied to a crop as a function of location. To take full advantage of field data, data (such as soil types and yield goals) must be stored along with location in the field. The farmer applying fertilizer must also know where he is in the field so the amounts can be changed when needed. A geographical information system (GIS) is an ideal method for storing this kind of information, and a global positioning system (GPS) is ideal for pinpointing locations on the field. The purpose of this paper is to describe how GIS and GPS combination would be useful to a farmer for aiding in distributing fertilizers on a field. It will give a brief description of what a GPS and GIS are and what they are capable of doing.

A GIS is a tool for spatially evaluating problems. Some examples of spatial problems would include showing buffer zones around highways or identifying areas within 20 meters of a polluted stream. A GIS can be thought of as a graphical database. They can be used to aid fire fighters in placement of fire perimeter information and aid in dispatching (Schulman,1991), evaluating evacuation plans in case of a nuclear power plant disaster, and finding ideal locations for building factories.

Typical GIS programs provide the capability of storing data as images by which each separate area (also known as a polygon) on the image represents a separate value. Some examples of what these values can represent is land use (forest,urban,crop, ...), local tax rates, soil types, soil qualities, and many more. Associated with each image or map, there are attributes which are usually contained in a table. The attributes relate to specific polygons on the image. Once the data is in a database, it can be manipulated to form new maps showing statistics or results of a mathematical models, such as runoff models. GIS programs can produce output in the form of maps that show results of the models. These maps could be used by a computer or a person to find specific areas meeting certain criteria. In the case of fertilizer application, the map could represent distribution amounts as a function of soil.

The purpose of a GIS in prescription farming is to store data spatially. Output helps farmers to locate areas in a field that require specific amounts of fertilizer depending on a set of criteria. Such input data could be soil types, nutrient deficiencies, and pest populations. From this data and other data, crop yields for each zone (soil type) can be estimated and used to optimize fertilizer and pesticide spray plans.

The question of cost of an appropriate GIS must be addressed. How much could the user spend on the GIS and produce a yield increase significant enough to pay for the system? GIS software comes in numerous packages that have the basic qualifications needed. An example of three such packages that are available at a variety of prices are IDRISI[3], SPANS[4] and ARC/INFO[5]. The software costs range from approximately $600 to $22,000. Each system has particular advantages and disadvantages for this type of application. IDRISI requires significant knowledge of operating systems and possibly a programming language. SPANS is slightly easier to setup and contains more automating capabilities. SPANS has a number of options which permit additional capabilities. A third software package that could be implemented is called ARC/INFO. ARC/INFO is available in versions for either workstations or microcomputers and can be made more user friendly and with added automation possibilities.

[3]IDRISI, Graduate School of Geography, Clark University, Worcester, MA.
[4]SPANS, TYDAC Technologies, Ontario, CANADA.
[5]ARC/INFO, Redlands, CA.

A Global Positioning System or GPS is a system that uses satellites to tell the operator where he is on earth. The system was set up by the U.S. military. Accuracies of a GPS varies depending on the unit used. Accuracy ranges from 180 m to 1cm. Typical GPS's give the longitude and latitude and elevation on a nearly continuous basis. A GPS can be placed on spraying equipment to tell computers on board the location of the receiver. Combining this information and recorded information in a GIS, the fertilizer amounts can be adjusted on the go. GPS is used for tracking spray trucks, tractors and combines. Receivers used for this type of operation cost from $5000 to $10,000 and had accuracies of nearly 15 m (Reichenberger, 1992a).

There are two modes of global positioning systems. One is normal mode which relies solely on information from satellites to provide location information. Normal mode has accuracies of about the width of an average street (Hurn, 1989). The second mode is called differential mode. This mode requires a base station that is located at a known point to correct for errors found in normal mode. With this mode of operation, accuracies of one centimeter are possible (Hurn, 1989) Base stations are expensive since they require a second GPS unit to be placed.

The technology is readily available to implement a remarkably accurate prescription farming system (PFS). The question that remains is whether or not the cost of a PFS is worthwhile. Different levels of accuracy and cost must be experimented with to find an optimum system. Cost could be minimized by letting the operator make changes in application rates manually according to a map based on a county soil survey or a farmer's knowledge of a field. A further improvement might be the use of a map output from a GIS program. Additional levels of automation might depend on a GPS unit that could help the equipment operator know when to change application amounts or even work with the GIS to completely automate these changes.

MATERIALS AND METHODS

A field located on the Pee Dee Research and Education Center near Florence, South Carolina was selected as an example location for possible adoption of site-specific applications. The field was selected because information about its soils, topography, and size was readily available. The field is in an area that has been used for row crops. Figure 1 shows the field shape and the soils included in the field. The field is typical of the region in that the fields are irregularly shaped and they contain a variety of soil types. Each of these conditions adds to the complexity of evaluating the field for adoption of site-specific methodologies.

The county soil survey delineated the field into representative soils. Areas of each soil were then estimated. Table 1 contains the soil types along with yields based on information from county soil surveys. The yields indicated are optimum yields assuming good management. These yields were used in the analyses in order to eliminate weather conditions from the analyses. Corn was selected as an appropriate crop for comparison purposes in this example since it is commonly grown in the region and because it has relatively high nutrient requirements. Nutrient requirements for corn were estimated using Clemson University Extension Service publications.

Additionally, the chemical handbook (Extension Service, 1994) provides recommendations on which pesticides used with corn require a rate that varies with soil type. Three herbicides were indicated for which application rates should be reduced based on soil conditions. It should be noted that the soils at the Pee Dee REC generally have very little organic matter so that little difference in herbicide as a result of organic matter is anticipated.

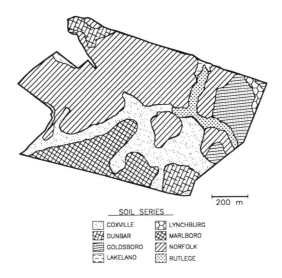

Figure 1. Field Shape and Soils for Example Situation at Pee Dee REC.

Table 1. Soils contained in example field.

Soil type	Area (ha)	Yield[a] (t/ha)
Coxville	16.1	7.41
Dunbar	4.4	7.74
Goldsboro	6.7	8.40
Lakeland	0.5	3.69
Lynchburg	1.0	7.74
Marlboro	11.1	6.72
Norfolk	40.2	7.41
Rutlege	3.3	NA[b]

[a] Obtained from soil survey.
[b] Not suited to growing corn according to soil survey.

The major expense in fertilizer for corn in this area is related to application of nitrogen fertilizer. In order to estimate the difference in costs, ammonium nitrate with an analysis of 33.5-0-0 and having a local price of $231 per metric ton was selected. For the purposes of comparison, corn has a value of $110 per metric ton. Based on this price and the yields shown in Table 1, there is a potential difference in gross income of up to $518 per ha for the different soils.

An estimate of nitrogen application was obtained using information provided in Parks (1989) so that

$$NREQ = (16.67 * YLD) + 22.45 \qquad (1)$$

where NREQ is nitrogen required in kg/ha and YLD is crop yield sought in t/ha.

Four possible scenarios were identified for comparison using the return from yield and nitrogen cost relationships.
1. Nitrogen applied on a site-specific basis using the yields shown in Table 1 to determine nitrogen required. Yield controlled by soil.
2. Nitrogen applied using the rate necessary for the highest yield (8.40 t/ha) for any soil shown in Table 1. Yield controlled by soil.
3. Nitrogen applied using a weighted average based on area of each soil. Yield limited by either the soil or nitrogen application.
4. Acreages of the Norfolk and Lakeland soils are reversed so that the large area had a low yield relative to the rest of the site. Nitrogen applied using a weighted average based on these areas. Yield limited by either soil or nitrogen application.

ANALYSIS OF RESULTS

In each scenario, returns from yield were deemed to be limited by soil or by nitrogen application. In scenarios 1 and 2, yield was considered to be limited only by the soil yield based on soil survey. In scenarios 3 and 4, yield was limited by the lesser value of the soil or yield produced by the given nitrogen application. In each case, the total return was compared with the costs of nitrogen inputs. Please note that other nutrients were not considered in this example for simplicity, but they could be accounted for in like manner if so desired. For this reason, a mixed fertilizer is not considered.

Scenario 1

Scenario 1 represents the situation in which prescription farming is practiced. Nitrogen is applied on a site-specific basis dependent upon the soil series. Total value of harvested corn would be expected to be $65, 087. Nitrogen cost is $8043. Subtracting these two values yields a difference of $57,044.

Scenario 2

Scenario 2 is a situation in which nitrogen is applied based on the maximum yield of any soil series in the field. Yield is still limited by soil so nitrogen is applied that is not expected to be utilized. Total value of harvested corn will be $65, 087 as in Scenario 1, but nitrogen cost will be $8980. Subtracting these values yields a difference of $56,107 or about $950 less than in Scenario 1.

Scenario 3

Scenario 3 uses a nitrogen application based on the weighted average yield by soil series for the entire area. In this case, yield is limited by either soil or nitrogen. Areas in which the yield expected from the soil is greater than the weighted average yield are assumed to be limited by nitrogen application. Total value of harvested corn in this situation is $64,057. This is approximately $1000 less than the income from Scenarios 1 or 2. Scenario 3 requires nitrogen costing $8051. which is less than Scenario 2, but greater than Scenario 1. The difference between income and nitrogen cost for Scenario 3 is $56,006, approximately $1000 less than Scenario 1 and about the same as Scenario 2.

Scenario 4

Scenario 4 is unique because it is a hypothetical situation in which it is assumed that the areas associated with the Lakeland and Norfolk soils were reversed so that the Norfolk is considered to have an area of 0.5 ha and Lakeland has an area of 40.2 ha. All other soils have the same acreage as previously shown in Table 1. Using this assumption and the weighting procedure as described in Scenario 3, the harvested corn would have a value of approximately $40,616, and the nitrogen cost would be $6317. The difference resulting is $34,299 or some $22,000 less than obtained in Scenario 3. Obviously, this is substantially less than any of the previous scenarios. This fact illustrates the need for a rational method to analyze local conditions.

CONCLUSIONS

Several conclusions are apparent from this analysis.
1. Obviously, it would be rash to purchase a system such as described herein without substantial analysis of local soil and crop conditions. Significant variation is found in the results depending on which type scenario is projected.
2. While the example field used in this example produces a difference of about $12.50 per hectare, this difference could justify a simple system or rethinking of a producer's nutrient scheme. Larger field sizes could justify more extensive control systems.
3. Large variations in yields from one portion of a field to another might also provide justification for more control. This was indicated by Scenario 4.

Reliable yield information is difficult to obtain for field situations such as that shown in Fig. 1. Other means of obtaining representative data should be sought. The use of soil survey yields as described herein is only one possible method. A better method might be to evaluate actual yield records for different parts of a field. With advanced planning, this can be done by monitoring the yields in a combine using a sensor such as described by Agricultural Research (1993). Computer spreadsheets can then be incorporated with actual costs and return information to better estimate scenarios that would be appropriate in an actual farm situation.

Finally, it is important to recognize that control systems may vary from quite simple, manual controls to completely automatic controls. Operator control of application rates may provide most of the benefits of the automated systems with little extra costs. Additionally, costs associated with microcomputer equipment and electronic sensors have dropped significantly in costs over the last several years. As this trend continues, use of such equipment in field applications may become routine.

REFERENCES

1. Agricultural Research. 1993. Fine-tuning agricultural inputs. Agricultural Research 41(1): 16-18.

2. Extension Service. 1994. Agricultural chemicals handbook. Cooperative Extension Service, Clemson University, Clemson, SC. 1083 p.

3. Hurn, J. 1989. GPS: A guide to the next utility. Trimble Navigation, Sunnyvale, CA. 76 p.

4. Parks, C. L. 1989. Corn: Soil fertility for corn. Cooperative Extension Service Info. Leaflet 34, Clemson University, Clemson, SC. 4p.

5. Reichenberger, L. 1992a. Accuracy from afar. Top Producer, A Farm Journal Publication, 9(2): 16.

6. Reichenberger, L. 1992b. Profiting from precision. Top Producer, A Farm Journal Publication, 9(2): 12-13.

7. Soil Conservation Service. 1986. Soil survey of Horry County, South Carolina. USDA-SCS. 137 p.

8. Schulman, R. D. 1991. Portable GIS: From the sands of desert storm to the forests of California., Geo Info Systems, 1(8): 25-33.

SUSTAINABLE, ENVIRONMENTALLY SOUND POTATO PRODUCTION IN NORTHERN MAINE

J. C. McBurnie[*]

ABSTRACT

Since 1990, a multidisciplinary group of researchers has been conducting an integrated study of the potato agro-ecosystem. The goal has been to identify management practices that can be incorporated into or modified for the current cropping system. The intent is to reduce the system's environmental impacts while maintaining productivity. Part of our work has been to show that environmental quality and agricultural productivity are not mutually exclusive and to demonstrate to area producers that sustainable agriculture is not necessarily synonymous with organic farming. In this paper I will present an overview of the research program, its evolution during the first four years, and some of its successes as well as its shortcomings.

KEYWORDS. Sustainable agriculture, Potato production, Environmental protection, Profitability.

INTRODUCTION

The Aroostook County Potato Production Tradition

Aroostook County is the northernmost county in the State of Maine. Although 85% of the region's land area is forested, it constitutes the largest potato production region in the State. The County has a long history of potato production, dating back to the 1780's, roughly 65 years after potatoes were introduced in this country (1719). When Maine was the nation's leading producer, more than 92 000 ha were cultivated annually; the current production area is approximately 32 000 ha.

The potato crop is chemically intensive in terms of both nutrient and pest management requirements. Potatoes will consume most of the plant available soil nutrients and return very little to the soil in the form of organic matter. A major contributor to the State's and the County's decline in production has been this mining of the soil nutrients. For the greater part of potato production history in Aroostook County, agriculture was a monoculture. Lacking an economically viable rotation crop and having a limited amount of production land, potatoes were grown year after year. As this monoculture continued, the need for supplemental nutrients and the dependence on commercial fertilizers grew. Meanwhile, the quality of the soil resource continued to decline.

---

[*]J. C. McBurnie, Assistant Professor, Department of Bio-Resource Engineering, University of Maine, Orono, ME.

Potato production also relies heavily on the use of pesticides. The potato plant is susceptible to a host of diseases and insects, which have the potential to quickly and completely decimate an entire crop. Chemical pest control is only marginally successful as many of the insects, and in particular the Colorado potato beetle, adapt easily and become tolerant of or resistant to certain pesticides. This requires annual changes the insecticides used. Furthermore, disease control often requires frequent application of fungicides. The amount and variety of chemicals applied represent a very real environmental risk with regard to air and water quality.

One final, although minor, problem for potato producers in Maine was that the region is rich in history and tradition related to the potato industry. As such, there was, and still is, resistance to change in some sectors. Farming is often practiced as it has been for generations, with the only visible changes being in the size and sophistication of farm tractors and implements. This attitude is not as prevalent as it once was, however it does persist in some sections of the region.

Facing Realities

Declining productivity and evidence of environmental damage attributable to agricultural activities have led to a realization that business as usual is not acceptable. It is now understood that environmental stewardship and industry viability are not and cannot be mutually exclusive. This realization is not a recent occurrence.

For the past four to five decades, agencies such as the Soil Conservation Service and the Cooperative Extension Service have promoted and disseminated information about soil and water conservation, best management practices, and integrated pest management. There was early resistance to these programs, but as time passed, the programs were gradually implemented. Positive effects were tangible and could be pointed out by county agents and area conservationists trying to promote land stewardship, and resistance waned.

Environmental and economic pressures are increasing at an ever quickening pace and producers are anxiously looking for comprehensive solutions. Still, resistance exists for any "new" system or practice, and significant research and demonstration is often required before these new methods will be considered for production agriculture. Producers recognize that there are problems and that changes must be made, but they are generally unwilling to "jump on the bandwagon" without a considerable amount of "proof".

Sustainable Agriculture vs. Organic Farming

Early in the development of Sustainable Agriculture (SA) programs across the country, there was a perception that these programs advocated the elimination of all commercial chemical inputs from crop production systems. Sustainable agriculture and organic farming were thought to be synonymous terms. This connotation led to some hesitancy and resistance in some tradition-oriented crop production regions such as Aroostook County. In some regions of the country there may be some validity to these statements, however, in the majority of SA programs nothing could be further from the truth.

Most Sustainable Agriculture programs are a sensible combination of "organic" and "conventional" agronomic practices. The ultimate intent of sustainable agriculture is to maintain economic viability while minimizing, but not excluding, the use of commercial fertilizers and pesticides. The goal of minimizing chemical use is two-fold: to reduce the potential for negative environmental impacts and to simultaneously improve the efficiency of chemical use, thus reducing production costs.

Project Concepts

From the early days of project development, it was understood that it would be beneficial for the sustainable agriculture project to serve as a showcase of systems research and interdisciplinary activity. The demonstration value and research value were recognized as being equally important. Furthermore, we realized that the proposed research did not lend itself to short-term analyses; it was initially planned to run for a minimum of ten years.

The project team also realized that it would be impossible to evaluate every possible factor that might influence the potato agro-ecosystem. After a system diagram was developed, we convened to identify areas in which we would concentrate. The criteria for selection included investigator expertise, concerns and interests of the potato industry, perceived importance to the potato agro-ecosystem, and potential level of impact on the environment. It was anticipated at that time that other issues of interest and importance would be incorporated as additional expertise and funding became available.

EXPERIMENTAL DESIGN AND EVOLUTION

Initial Development

The original plan was quite ambitious, calling for forty-five 0.4 hectare plots. Treatments included a complete random block design consisting of five replications of one conventional and two alternative crop management systems. Treatments involved three-year rotations, so each treatment had a total of three possible entry points. The three crop management systems were as follows:

> Conventional: Use synthetic pesticides and fertilizers as dictated by best management practices and integrated pest management recommendations.
>
> Biological: "Natural" and cultural pest control and fertilization; limited use of synthetic insecticides, herbicides and fertilizers. Because there was no plant pathologist associated with the research team, diseases were to be treated by conventional, synthetic fungicides.
>
> Pest Ecology: Pest controls similar to the biological system, but with higher insect and weed thresholds, although lower than crop failure thresholds.

This seemed to be workable since the Experiment Station had recently acquired an 32 hectare farm to add to its research holdings. For the 1990 growing season, the farm was planted in millet to establish a baseline field condition. Preliminary surveying was carried out to start the process of plot location.

Program Growth

The project was modified even before the beginning of the first field season. The first modification was in the overall treatments used. The original treatments were retained but they were split into sub-treatments based on rotation crops, soil amendments and varieties. Rotation was reduced to a two-year cycle with the following rotation crops/soil amendments:

> Clover underseeded barley; no amendment
>
> Clover underseeded barley; 14 t/ha waste potato compost, 33 t/ha cattle manure
>
> Clover underseeded winter wheat; 14 t/ha waste potato compost, 33 t/ha manure

The varieties evaluated were Atlantic, a disease and stress tolerant cultivar, and Superior, a disease and stress sensitive cultivar. These varieties were selected after consultation with area growers. Replications were reduced to four.

The experimental design was altered one final time prior to planting. The third rotation subtreatment was dropped. Furthermore, after closer site inspection, it was found that a considerable amount of the research farm was unsuitable for research application. After land with ledge outcroppings, trees, roads, excessive slopes, poor drainage, and excessive weeds were removed, approximately 9 ha remained for system and component studies. Plot size was reduced to 0.06 ha. Nine component studies were also planned.

The objectives for the system plot studies were:

1. To study the impact of nutrient management, soil amendments, soil properties, insect damage, weed competition, and cropping system management on crop response and productivity, and nutrient fate and transport,

2. To describe interactions among soil fertility, plant vigor and performance of Colorado potato beetles and their natural mortality factors on potatoes,

3. To determine effects of rotation, tillage, and weed management on abundance and species of weed seeds in the soil and of growing weeds, and

4. To conduct an economic evaluation of the consequences of the alternative cropping strategies.

Addressing Change

The experimental design is constantly being reevaluated to address flaws, shortcomings and new areas of interest. This is done during annual retreats for the principal investigators and research assistants. Preliminary results are presented and the following season's research plans are described. Through evaluation of system and component studies, new issues and problems are identified. Any problems with the system study are discussed and resolved by the research team. Issues of lesser importance can be dropped from the main study. In addition, the group responds to feedback and concerns from the industry. Most recently, a plant pathologist was added to the research team.

Because of difficulty in controlling weeds and insects biologically in the pest ecology management system, extensive crop damage and yield reduction were experienced in 1991. This led to selection of "new" main treatments. The conventional and biological treatments were kept as system "extremes" and the pest ecology treatment was replaced by a reduced input treatment in which conventional pesticides were used at higher application thresholds and synthetic fertilizer use was reduced by increased application of manure (45 t/ha); the biological plots received 66 t/ha of manure. Compost application to all amended plots was increased to 22 t/ha. Rotation crops were changed for the soil amendment subtreatments to increase organic nutrient availability.

The Current Program

At this time, no changes are anticipated for the system experimental design. The current system maintains the pest management strategies of conventional, reduced input, and biological control. The varieties studied are still Atlantic and Superior. The soil management systems consist of conventional (barley rotation crop, unamended, 1344 kg/ha commercial fertilizer) and amended (pea/oat/vetch/clover green manure mixture rotation crop, amended with 22 t/ha compost and 44 t/ha manure, 672 kg/ha commercial fertilizer). Both of these subtreatments receive 50 kg/ha sidedressed N fertilizer prior to tuber initiation.

STATUS

Preliminary Results

At this time promising responses are beginning to be fleshed out. Many of the physical changes will take considerably more time before conclusive results can be reported. There is an indication that the reduced input system could successfully replace the conventional system, as the yield differences between these treatments were not statistically significant. Visual analyses of crop vigor and yield data suggest that soil amendments can enhance the quality and quantity of potatoes, especially in growing seasons when available moisture is a limiting factor. Economic analyses are inconclusive, which is expected in light of the changing experimental design. Another field season should provide sufficient information to begin to develop more meaningful economic comparisons.

Weed biomass was significantly higher in the potato phase of the biological pest management system than in either the reduced input or conventional system. For the rotation phase, there were differences between the amended and unamended plots in the reduced input and conventional systems, but not the biological system; total weed biomass was low in the barley crop, higher in the green manure mixture.

For the original design year (1991), weed and insect populations and foliar damage were significantly higher for the biological and pest ecology management systems than for the conventional system (figures 1 - 3). This was an exceptionally dry growing season and it was hypothesized that nutrient and moisture stress may have made the plants more susceptible to attack. Differences were smaller for years with adequate moisture and in plots receiving soil amendment. Change in the experimental design (Management systems) also reduced differences.

Figure 1. Conventional Pest Management Plot (1991).

Figure 2. Biological Pest Management Plot (1991).

Figure 3. Pest Ecology Pest Management Plot (1991).

Regional Acceptance

The research program was initiated during the 1991 growing season. It met with only minimal skepticism and cynicism. This is largely due to the involvement of and consultation with industry leaders and representatives of agricultural support agencies (SCS, UMCE, ASCS). It is also due to the level of commitment being made to this region and this industry, which is unprecedented in recent history. It is also evident from the time and effort put into the project development that this commitment to systems and interdisciplinary research is more than just lip-service.

The project is now widely accepted and well-regarded by growers in the region. It is often a focal point of industry tours, Farm Days open to the general public, visits by State and Federal legislative leaders, and media coverage. This respect continues to improve as we respond to the needs of the industry. The influx of research dollars into the local economy in the forms of supplies and local labor has also brought positive exposure.

Future Direction and Planning

The project is now very stable, but by no means static. Its early success has facilitated the awarding of extramural funding to support investigations of water management, soil erosion, insect pest control, and disease identification and control for the potato cropping system. In addition, a wide variety of smaller component studies have been linked to the systems study. We are currently trying to add further expertise in plant pathology and soil science, especially in the area of nutrient speciation. As the experimental design remains more consistent through the years, the database will be able to be used with more confidence. Furthermore, the economic analyses will be more meaningful when applied over the course of the entire study period.

There is a consensus that we have identified the "proper" systems research program, so it is unlikely that there will be any wholesale changes in the future. It is anticipated that the study will run at least ten years, but in the fickle world of extramural, government funding it is impossible to state this with complete confidence.

SUMMARY AND CONCLUSIONS

A multidisciplinary, long-term potato agro-ecosystem project was initiated in 1990 with field experimentation starting in 1991. The overall project objective was to develop a sustainable potato cropping systems for northern Maine while minimizing environmental impacts through reduction of chemical and energy inputs. The project has been evolving since its inception and remains dynamic to this day. New research issues and technical expertise are added yearly. Despite its flexibility, the program remains focused on the ultimate objective, sustainable and environmentally compatible agriculture.

The research program has quickly gained the interest and support of local potato producers as well as receiving national and international attention. Most of the local support is derived from the inclusion of comments, criticisms and suggestions from industry representatives and growers in general in the development and refinement of the research "plan".

Through the system study and the associated component studies, progress is being made toward reducing the reliance on external chemical and energy inputs. We, by no means, have all the answers and will continue to explore alternative cropping practices that will protect the environment while maintaining the producers' "bottom line". Economic analyses are just now being undertaken, as the experimental design "stabilizes" and sufficient field data become available.

ACKNOWLEDGMENTS

This project is being supported by funds from the USDA Special Grants program and the Maine Agricultural and Forest Experiment Station. I would like to acknowledge the talents and efforts of this ambitious multidisciplinary team's members: Dr. Gregory Porter, Department of Plant, Soil and Environmental Sciences; Dr. Michele Marra, Department of Resource Economics and Policy, Dr. Matthew Liebman, Sustainable Agriculture Program; Drs. A. Randall Alford, Francis Drummond, and Eleanor Groden, Department of Entomology; and Dr. Bacillio Salas, Department of Plant Biology and Pathology. The contributions of the many graduate research assistants, research technicians and student workers must also be recognized. The diligent efforts of former project manager (now farm superintendent) Malcolm Brown and current project manager Leslie Cumming are also gratefully acknowledged. This paper is published as MAFES Miscellaneous Publication No. 1782.

THE AGRICULTURE/ENVIRONMENT INTERFACE:
LOCATING THE RELEVANT LITERATURE
Stephanie C. Haas

ABSTRACT

Key indexes and journals in the area of environmentally sound agriculture were identified, using citation counts and electronic ranking of journals in the Dialog system.

From a citation count of the references in the 1991 Proceedings for the Environmentally Sound Agriculture Conference, three journals were identified as publishing widely across the entire agriculture/environmental spectrum: *Agronomy Journal*, *Journal of Environmental Quality*, and *Journal of Soil and Water Conservation*.

For eight specific areas primary indexes and journals were identified using various search strategies in the online Dialog databases. These analyses confirmed the importance of the *Journal of Environmental Quality* and *Journal of Soil and Water Conservation* and provided guidelines for searching appropriate indexes/abstracts.

Based on total records retrieved for topics involving water quality issues, such as agrochemicals, BMPs, point and non-point source pollution, *Water Resources Abstracts* was the primary index. Wetlands topics should be searched in both *CAB* and *Water Resources Abstracts*. For air pollution, *Energy Science and Technology* appeared key, and for animal wastes, *CAB*. However, in each instance, multiple abstracts/indexes needed to be searched to provide assurance that some degree of comprehensiveness had been achieved.

INTRODUCTION

The scientific literature covering the agriculture/environment interface consists of a rich mixture drawn from agriculture, economics, engineering, chemistry, physics, meteorology, urban studies, and resource management. At the present time, there is no core literature for the discipline of sustainable agriculture. Therefore, it is of critical importance that researchers in this area be familiar with the multiplicity of indexing tools and journals from which they will be able to ferret out relevant literature. The purpose of the present study was to identify these key journals and indexes and to alert researchers to potential problems associated with research in this discipline.

METHODOLOGY

Two analyses were done. First, citation counts were completed on all papers published in the 1991 Proceedings for the Environmentally Sound Agriculture Conference, April 16-18, Orlando, Florida. For each topic, cited journals were analyzed. The topics included pesticides/herbicides, wildlife systems, nutrients, sustainable agriculture, animals wastes, wetlands, erosion control, environmentally sound agriculture, and the urban/agriculture interface. As with any type of citation analysis, the underlying caveat questions whether the references truly reflect the research universe.

Second, seven search strategies were were run in the agricultural, pollution and environmental sections of File 411, the master index to all Dialog databases. This defined the databases having the most records on a particular topic. The searches addressed four topics defined in the Proceedings: pesticides/herbicides, wildlife, wetlands, and animal wastes. Additional searches were done on BMPs and point and non-point source pollution. The searches were constructed so that key concepts had to appear in the same field of the record, i.e., in the title, in the descriptors(subject headings), or in the abstract. With that specificity, it was felt that the records retrieved would be relevant to the topic sought.
Then the same searches were run in the databases identified as having the most records, duplicates were eliminated, and the results ranked by journal. Data on agrochemicals and water pollution was available from a previous study by Haas and Clark (1992).

Caveats for rankings and other electronic manipulations always relate to the appropriateness of the search strategies constructed. The findings and recommendations of this study should be viewed much as the findings of any preliminary field study.

RESULTS AND DISCUSSION

Citation Count of Conference Proceedings

The analysis of the Proceedings from the Environmentally Sound Agriculture Conference are summarized in Table 1.

Table 1. Cited References in the 1991 Proceedings for Environmentally Sound Agriculture

Section Title	Articles	Total References	Journal References
Pesticides/Herbicides	9	67	40
Wildlife Systems	5	79	24
Nutrient Management	9	60	20
Wetlands	4	16	1
Sustainable Agriculture	6	82	30
Animal Wastes	19	145	41
Erosion Control	8	74	22
Environmentally Sound Ag.	17	155	50
Urban/Agriculture Interface	15	119	58
Total	96	813	287

Of significance to researchers is the finding that only 287, or 35%, of the total 813 references were to journals or to regularly issued proceedings or transactions. Sixty-five (65%) of the references were to other types of publications including books, dissertations, conference proceedings, and technical reports. Many of these publications are hard to identify and locate.

Of the 124 referenced serial titles, three journals were referenced extensively and in multiple topic areas: *Agronomy Journal* (13), *Journal of Environmental Quality* (28), and *Journal of Soil and Water Conservation* (18.) Two others were associated with a particular area: *Biocycle* was referenced eleven times in Urban/Agriculture Interface, and *Journal of Range Management* was referenced nine times in papers in the Wildlife Systems section.

Database Searching and Journal Ranking

The second series of analyses were conducted in the Dialog databases and one non-Dialog database Wildlife Review/Fisheries Review. Produced by the U.S. Fish and Wildlife Service, this database was expected to have significant coverage of wildlife/agriculture articles.

Table 2 indicates the databases searched and the total number of records retrieved in each. This total number includes not only records to articles in journals and conference proceedings, but technical reports, book chapters, etc. NTIS indexes technical reports, not journals, resulting from federally sponsored research. The absence of numbers under certain indexes indicates that the coverage was minimal and that the same entries would probably be found in the databases listed.

Table 3 indicates the key journals in which articles relevant to sustainable agriculture topics were published. While it is extremely common for several databases to include the same article, the numbers listed represent the composite of unique records from all Dialog databases searched. Since most of the databases started in the late 60s and were current as of October 1993 (give or take a month), one can make the rough interpretation that in the last 20 years, the *Journal of Soil & Water Conservation* has published 76 articles related to sustainable agriculture, 20 of which focused on wildlife, 22 on BMPs, etc. A complete list of the indexes searched and their coverage is given in Appendix I.

Table 2. TOTAL RECORDS IN EACH DATABASE BY TOPIC

Topic	AGRICOLA	AGRIS	ASFA	BIOSIS	CAB	CA (Chem Abst)	El Compendex	Energy Sci & Tech	Enviroline	GEOBASE	Life Sciences	NTIS	Poll Absts	Toxline	Water Res Abst	Wildlife Review/ Fish Review
Agrochemicals**	105				86	822*			130				31		700	
Air Pollution	89			246	202	195		390*				157		331		
Animal Wastes	77	103		89	242*		86	73							162	
BMPs	43		43		93		48		56			45	54	96	124*	
Non-Point Source	225	184		161	223		165		225				217	322	547*	
Point Source	24		41	49	58				41	50			44	62	107*	
Wetlands	260			195	319*			183	246	230		182			302	
Wildlife	305			260	558			231	353	181		920* (1)		253	395	238

a ** Agrochemical study covers the years 1985-90.
b * Indicates the highest number of records found.
c (1) This figure may be inflated by the occurrence of the agency name U.S. Fish and Wildlife Service.

Table 3. KEY JOURNALS IN SUSTAINABLE AGRICULTURE FIELDS

Total	Key Journals	Air Pollution	Animal Wastes	Wildlife	BMPs	Point Source	Non-Point Source	Wetlands
76	Journal of Soil & Water Conservation			20 (7)	22	5	4	24
66	Journal of Environmental Quality	15	35		8		8	8
28	Journal of the Water Poll. Control Fed.		13			7		
26	Biological Conservation			19 (6)				7
22	Environmental Conservation			22				
21	Ambio			12				9
20	Water Resources Bulletin			7	13			
18	Transactions ASAE		14					
18	Journal Air Pollution Control Assoc	18					4	
17	Trans N American Wildlife Nat Res Conf			(17)				
15	Water Research		8				7	
12	Journal of Wildlife Management							12
12	Ecos	12		12				
11	Atmospheric Environ. Part A.							
11	New Scientist			11				11
11	Environmental Management							10
10	Wetlands		10					
10	Meieriposten							
10	Defenders			10				
9	ES&T	9						
9	Agriculture, Ecosystems & Environ	9						
8	Journal of Environmental Mgt.							8
8	Archives of Environmental Contam.			8				
7	Journal of Dairy Science		7					
7	International Wildlife			7				
7	Environment			7				
7	Bulletin Int'l. Dairy Federation		7					
5	Journal of Great Lakes Research						5	
4	Ground Water Monitoring Review						4	

a ( ) In Wildlife column indicates records found in the Wildlife River/Fisheries Review database. These figures are not included in the total.

Table 4 contains the major journals covering the field of agrochemicals and water pollution, which were indexed by *Water Resources Abstracts* between 1985-90. This data is taken from the Haas and Clark study (1992).

Table 4. Major journals containing articles on agrochemicals and water pollution published between 1985-90 indexed by *Water Resources Abstracts*. (n=700)

References/%	Title
109(16%)	Bulletin of Enviornmental Contamination and Toxicology
50(7%)	Ecotoxicology and Environmental Safety
48(6.9%)	Environmental Pollution
45(6%)	Environmental Toxicology and Pollution
29(4%)	Aquatic Toxicology
23(3%)	Chemosphere
19(2.7%)	Archives of Environmental Contamination and Toxicology
18(2.6%)	Water Research
12(1.7%)	Journal of Environmental Science and Health B.Pesticides
12(1.7%)	Science of the Total Environment

A review of the index/abstract coverage of particular journals raised some concern over the comprehensiveness and consistancy of indexing practices. As Table 5 indicates, the same strategy on animal wastes pulled five unique records from AGRICOLA, eleven from BIOSIS, ten from CAB, etc.-- all from *Journal of Environmental Quality*. To assure retrieval of all thirty-five unique records from this journal, five databases were searched. Similar scattering was noticed in all of the searches done.

Table 5. Distribution of Unique Animal Wastes Citations Throughout Abstracts/Indexes

Journal	Total Cites	AGRICOLA	BIOSIS	CAB	Compendex	Enviroline	Water Res Abst.
J. Env. Qual.	35	5	11	10	3		5
Trans. ASAE	14		5		3		6
J.Wat.Poll.Cont.Fed.	13	2	2				9
Meieriposten	10	3		7			
Water Research	8					4	4
Bull. Intl. Dairy Fed.	7		1	4			2
Agric. Wastes	5			1	2	1	
Deut. Molkerei-Zeitung	5			5			

Because both the citation analysis of the Proceedings and the Dialog searches indicated that much of the relevant literature is from non-serial sources, it became important to more fully delineate the other types of information used.

In order to explore these other sources of information, the results of searching *Water Resources Abstracts* for the topic Best Management Practices (BMPs) were tabulated. This database contained the most records for BMPs (124). The breakdown by information type follows:

Publication Type	Citations	%
Journal articles	43	37%
Technical reports	27	23%
Conference proceedings	28	24%
Society publications	10	9%
Books	8	7%
Dissertations	1	.8%

In light of these findings, the indexes which focus on technical reports and conference proceedings need to be more fully explored. A cursory search on best management practices in three additional databases: Conference Papers Index, Federal Research in Progress, and the GPO Monthly Catalog, retrieved an additional thirty-five records which were unique to these databases.

CONCLUSION

In summary, the current study has identified key indexes and journals in the area of sustainable agriculture, but researchers must recognize that: 1) at the present time, there is no definable literature of sustainable agriculture and relevant literature is not only scattered, but often appears in non-journal sources, 2) searching multiple databases is critical to assure finding relevant articles, even those published in the same journal, and 3) access to technical report literature and conference proceedings appears essential. Additional research must further explore coverage of sustainable agriculture literature in technical report and conference proceedings databases.

REFERENCES

1. Haas, S.C. and M.Clark. 1992. Research Journals and Databases Covering the Field of Agrochemicals and Water Pollution. Science & Technology Libraries 13:57-63.

2. Proceedings of the Conference on Environmentally Sound Agriculture (What can be done now?), April 16-18, 1991, Orlando, Florida. A.B. "Del" Bottcher and others, editors. Gainesville, FL : Office of Conferences, University of Florida.

3. Smidt, Dale. 1993. Subject Guide to the Environment. Palo Alto, CA : DIALOG. (Part of the Tools & Applications packet prepared by DIALOG.)

APPENDIX I. Selectively searched DIALOG databases

Agricola (1970-Oct. 1993)
AGRIS International (1974-Sept. 1993)
Aquatic Science & Fisheries Abstracts (1979-Sept. 1993)
BIOSIS (Biological Abstracts, 1969-Dec. 1993)
CA Search (Chemical Abstracts, 1967-Nov. 1993)
CAB Abstracts (1972-Oct. 1993)
EI Compendex (1970-Dec. 1993)
Energy Science & Technology (1974-Oct. 1993)
Enviroline (1970-1993)
GEOBASE (1980-Nov. 1993)
Life Sciences Collection (1978-Sept. 1993)
NTIS (1964-Dec. 1993)
Pollution Abstracts (1970-Nov. 1993)
ToxLINE (1965-Oct.1993)
Water Resources Abstracts (1968-Oct.1993)

COOPERATIVE EXTENSION SERVICE
NATIONAL WATER QUALITY DATABASE

C. E. Burwell, E. Fredericks

ABSTRACT

The Cooperative Extension Service National Water Quality Database is the result of a cooperative agreement between ES-USDA and Purdue University. It's purpose is to bring together in one easily accessible location all water quality educational materials generated be State Cooperative Extension Services for search and retrieval be those professionals working in the water quality area.

The database currently includes over 2100 citations of water quality materials, and over 400 full-text documents delivered to the searcher via e-mail services.

Users may access the National Water Quality Database directly using Telnet connections, with the use of a dial-up modem, as well as directing batch searches and retrievals via e-mail capabilities and the use of Almanac.

DESCRIPTION OF WATER QUALITY DATABASE

The National Water Quality Database includes over 2100 water quality and waste management educational materials now in use throughout the country by State Cooperative Extension Services. Included are publications, fact sheets, computer software, and visuals such as slides and videos. Each state water quality coordinator determines which materials from his state will be included. Total updates are carried out annually, and inclusions are made continuously as new materials are received.

The database is divided into eight major categories: conservation; drinking water quality, nutrient management, pest management, waste management; wells; testing; and public policy. Searches are made on any word, or words included anywhere in the citation, or the full-text document (if available).

ACCESSING THE NATIONAL WATER QUALITY DATABASE

The database may be searched for key words of interest and the documents can be returned to the user via the e-mail system. It is available to anyone who has access to a computer, since it can be accessed directly from telnet, from a dial-up modem, or through the e-mail system by sending messages for searches and publications.

Located on the Purdue University Cooperative Extension Service Reference File (CERF), the database can also be located via the numerous Gopher systems now in use. CERF allows one to search for these water quality extension materials based on a number of keywords. It is capable of operating in interactive menu-driven mode, or in batch modes. In interactive mode, one has the ability to (a) view matched documents on screen, and (b) request and receive matched documents in a variety of formats through electronic mail. The batch mode provides almost identical facilities taking into account the nature of interaction.

TELNET CONNECTIONS

To use telnet connections to access the database the connection is: **telnet hermes.ecn.purdue.edu**
The login is **cerf**; and the password is **purdue**
You will be asked for your internet address, and to provide your terminal type. Once the main menu appears, you can search the National Water Quality Database by selecting option 1.

DIAL-UP MODEM CONNECTIONS

Generic Parameters. Set your system to operate at a speed of 1200, 2400, or 9600, baud (depending on your modem's capabilities), 7 data bits, 1 stop bit, and even parity.

The telephone number to reach dial-up connection to the National Water Quality Database is **317-494-8350**.
The computer will request that you Enter Service. Your response is **cerf** Your password here is **demo**
At this point enter **telnet hermes.ecn.purdue.edu**
The login is **cerf**; and the password is **purdue**
You will be asked for your internet address and terminal type. Once the main menu appears, you can search the National Water Quality Database by selecting option 1.

ELECTRONIC MAIL CONNECTIONS

To receive the Water Quality Guide, which describes the procedure for receiving water quality documents, send a message to **almanac@ecn.purdue.edu** In the body of the message include the command **send wq guide**

To receive the Water Quality Catalog, which lists all full-text documents presently available, send a message to **almanac#ecn.purdue.edu** In the body of the message include the command **send wq catalog**

To search for a Water Quality reference materials via e-mail, send a message to **almanac@ecn.purdue.edu** In the body of the message, enter the command **cerf all long keyword 1 keyword 2 keyword 3**
All can be replaced with ANY (matches all publications containing one of the keywords you include). All will match all publications matching all the keyword you enter). Long can be replaced with short, which will return the matches in a one-line format.

To order a document via e-mail send a message to **almanac@ecn.purdue.edu** In the body of the message include the command **send cerfcat index_number** The index number will have been returned to you in either the water quality catalog or the matches that you searched on.

Catherine E., Burwell is an Extension Specialist with the Purdue Cooperative Extension Service.
Olden Fredericks is the Team Leader for the Cooperative Extension Service National Water Quality Database Project.

USDA's WATER QUALITY PROGRAM - ENVIRONMENTALLY SOUND AGRICULTURE

Fred N. Swader, Larry D. Adams, and James W. Meek

ABSTRACT

The U.S. Department of Agriculture (USDA), and U.S. agriculture are moving rapidly to reduce agricultural nonpoint sources of water impairments. National data submitted to the Congress indicate that 90 percent of the assessed surface waters meet designated uses. USDA programs have aggressively encouraged farmers to develop and implement voluntary practices to reduce agricultural nonpoint source pollution. The effort has resulted in dramatic reductions in the use of agricultural chemicals in designated project areas, with little or no changes in yield levels, and in greatly reduced loadings to the environment.

KEYWORDS: runoff; impairment; farming; nitrates; sediment

INTRODUCTION

The U.S. has made tremendous progress in addressing the various aspects of water pollution. As a result of the Clean Water Act, industrial discharges have been controlled by permits; raw sewage discharges have been reduced by the construction of sewage treatment plants; and phosphorus discharges have been greatly reduced by a combination of technology, education and laws (e.g., the widespread banning of phosphorus-based detergents). This progress has resulted in the resurrection of Lake Erie (once proclaimed "dead"), the reclamation of the Cuyahoga River (which once caught fire), and the improvement of water quality in many major rivers.

Some very real secondary problems remain, such as the management of land applications of sewage sludges. Many application schemes have been approached from a disposal perspective; without consideration of the rates of application of nitrogen. At the same time agriculture is being "encouraged" to control the application of fertilizers and animal manures to reduce the environmental impacts of excess nitrogen.

Combined sewer overflows still occur in over 1000 communities, and at variable intervals, discharge untreated sewage directly into surface waters. The high costs of remediation preclude programs to deal with the problem. Fiscal stresses at all levels of government reduce the likelihood that the problem will be effectively addressed in the near future. This fiscal reality overshadows the situation, the clear identification of the problem, and its known, but expensive, solutions.

The U.S. Environmental Protection Agency (EPA) estimates that about one-third of our assessed surface waters do not meet designated uses, and that agriculture is the source of 60 percent of this impairment, largely from diffuse or nonpoint sources (NPS). EPA notes that the estimates are applicable only to the assessed one-third of our surface waters; that these data do not constitute a representative sample of the nation's waters; and that "States are generally constrained by diminishing resources and competing needs to monitor most often on those waters

---

Fred N. Swader is the Executive Secretary of the USDA Working Group on Water Quality; Larry D. Adams is a Special Assistant for External Liaison; James W. Meek is the EPA's resident representative in the USDA, in Washington, DC.

with known or suspected problems" (EPA, 1990). These data are often inappropriately extrapolated to National estimates.

Restoring the chemical and biological integrity of the nation's waters is a laudable goal. It is quite clear that the estimates of the sources and extent of the impairment have not been adequately documented, nor have achievable and realistic levels of remediation been defined in chemical, biological or ecological terms. EPA's national "Pesticides in Well Water" survey provided data, reliable on a national level, that indicate that there is no massive national problem; that pesticides are likely to be <u>detected</u> in only four percent of our water wells; that the likelihood is somewhat higher in rural wells than in urban ones; and that nitrate contamination is more widespread (EPA, 1992). The "detection" of chemicals in water does not imply anything about health effects or legal limits; detection levels are usually well below such effect levels, and the disparity among these levels is sure to increase as our detection capabilities develop. The survey also reported a number of "statistical associations" between water contaminants and human health problems, but pointed out that such associations do not establish any cause-and-effect relationships

The lack of monitoring data has been identified as a key barrier to state and local efforts to control nonpoint source pollution (General Accounting Office, 1990). There are few credible data on the scope and impacts of nonpoint source pollution, or on the effectiveness of potential solutions.

AGRICULTURE

There is little doubt that agriculture contributes to the degradation of water resources in some situations. But the real extent is unknown (and perhaps unknowable). It is quite possible to make generic statements about the problem, but USDA must be concerned with practical solutions to identified site-specific problems. Farmers, ranchers and other landowners may be convinced to alleviate identified NPS problems, but they can not afford generic approaches to generic problems.

Given this situation, some important questions are as yet unanswered:
- What are the agriculture-associated water quality problems?
- How do we separate agricultural NPS pollution from geologic sources?
- Which are the most important agricultural NPS water quality problems?
- How should we address them in a realistic manner?

We have one of the world's best, most aggressive, most progressive water pollution control programs; it is not unreasonable to ask ourselves what more we want, how much more we can afford, and how it should be spent. We have made significant progress; the remaining problems are the tough (i.e., expensive) ones.

A pragmatic approach must recognize both that there are problems, and that they are being addressed. Many States are conducting successful interagency programs to address agricultural NPS pollution. More could always be done, given the necessary resources and policy direction. The difficulty lies in the reality that resources are in short supply.

The USDA does not routinely collect data to assess water quality conditions; that role is assumed by the U.S. Environmental Protection Agency, the U.S. Geological Survey, some other federal agencies, and many state agencies. The USDA uses information from such assessments, and from its own and cooperative research programs to develop program priorities, and to shape program thrusts.

Programs

The USDA has developed and implemented a number of programs to address matters of agricultural contamination from both point and non-point sources. In the 1970's, considerable research emphasis was placed on the point-source aspects of the treatment and handling of animal wastes, and the influence of agriculture on water contamination. This resulted in many new handling and treatment processes. (Willrich and Hines, 1967; Cornell University, 1970; Porter, 1975; American Society of Agronomy, 1977; Loehr, 1977).

Formal programs to address NPS remediation began with the joint USDA-EPA Model Implementation Program, initiated in 1977. This program funded NPS projects in seven states to demonstrate interagency cooperation, accelerated farmer adoption of BMPs, and the effects of those BMPs on water quality. Interagency cooperation was achieved, and the farmer-adoption of BMPs was significantly accelerated over the life of the projects; water quality impacts were, and are, the most difficult to document.

In the early 1980's, the USDA's experimental Rural Clean Water Program established projects in 21 states, with funding of $70 million for the 10-year life of the program. Significant adoption of BMPs occurred in all of the areas; some were reflected in improved water quality. A description of the Program and the individual projects, which were selected in conjunction with officials at the State level, is available (Meek, et al, 1992). These projects were successful in stimulating the farmer-adoption of specified BMP's, and in demonstrating the effectiveness of deliberate programs to stimulate such adoption.

But the adoption of even the most pragmatic BMPs is not always reflected in short-term changes in water quality. For example, the impacts of 10 years of monitoring and extensive BMP adoption covering 74 percent of critical acres, resulted in no significant reduction in nutrient loads in either tributary streams or in St. Albans Bay, Lake Champlain, VT (Clausen, et al, 1992). Their model shows that reducing nutrient exports may take many years, depending on the amount of input reduction, and on initial field concentrations.

In the late 1980's, there was a considerable degree of concern in the Congress about the contamination of water by agricultural chemicals, and the need for efforts to enhance or protect the nation's water resources from such contamination. The resulting USDA Water Quality Program is based on the following guiding principles:

o The Nation's water resources must be protected from pollution by fertilizer and pesticides without jeopardizing the economic viability of U.S. agriculture;
o Pollution must be halted; and more environmentally benign farm production practices must be instituted;
o Farmers must be responsible for changing production practices to avoid polluting water resources.

The USDA program cooperates with the U.S. Geological Survey, the U.S. Environmental Protection Agency, and the National Oceanic and Atmospheric Administration. The challenge, is to conduct biological, physical and chemical research; to address the management of chemicals for crop production; to develop alternative cropping systems; to educate, demonstrate to, and assist farmers in making appropriate changes in production practices; and to monitor implementation of improved management practices and systems.

Since the program was launched in 1990, the USDA has received an earmarked appropriation increase of some $80 million per year to carry out this
program. These resources, along with redirected resources from participating USDA agencies, have funded the following efforts:

- Management Systems Evaluation Area (MSEA) projects in five Midwestern states, representing the Corn Belt, designed to evaluate current production systems, to develop new ones, and to transfer the information to farmers;
- Sixteen (16) Demonstration projects, to accelerate the transfer of research results to farmer adoption;
- Seventy-four (74) Hydrologic Unit Area projects, to accelerate the application of BMP's in identified critical hydrologic units;
- Seventy-one (71) Water Quality Special projects, designed to accelerate adoption of improved management practices, especially through cost-share programs;
- Fifty-nine (59) in-house research projects to assist the MSEA efforts, and to increase understanding of the movement and fate of agricultural chemicals in the environment;
- Ninety (90) cooperative research projects, in which University researchers are investigating various aspects of agricultural chemical management, movement and fate.

Many of these efforts are being conducted with active collaboration with the U.S. Geological Survey, the U.S. Environmental Protection Agency, State water quality agencies, and units of State and county government. Protocols have been established to involve a wide array of local, State, and Federal inputs to the selection of such projects, and to assure that they reflect State and local priorities for enhancing or protecting water quality (USDA, 1993).

PROGRESS

These cooperative projects are now operational and are producing significant results. Through FY91, they resulted in the adoption of water quality practices by 10,000 producers on 552,000 acres of farmland; and reduced applications of nitrogen by 2.7 million pounds, phosphorus by 1.7 million pounds; and pesticide applications by 239,000 pounds. In addition, over 9,000 people received water quality training in FY90, including more than 4,000 people from non-USDA agencies. These projects have since gathered momentum, and the accomplishment numbers are expected to increase substantially.

The USDA and its State cooperators are achieving results; but there is not yet any realistic water quality standard against which these results can be evaluated. The situation is still that most USDA program results must be reported in terms of "surrogate parameters," rather than terms of water quality improvements (Stewart et al, 1975). The USDA role is one of encouraging and assisting farmers to adopt management practices and systems that should enhance or protect water quality. Since the connections between NPS pollution and the quality of the receiving waters remain diffuse, the impacts of agricultural programs can be reasonably documented by recording the extent of farmer adoption of such practices.

The USDA's Water Quality Program has responded to the results of EPA's National Pesticide Survey, which indicated some potential problems of nitrate in groundwater. In 1992, USDA allocated some $700,000 for research on analyses for and management of nitrogen from soils, manures, and fertilizers. We have also encouraged our state staffs and counterparts to address the complex, field-and-crop-specific management practices for nitrogen management.

The "Pre-Sidedress Nitrogen Test" (PSNT) (Magdoff, 1991) for rapid, on-site assessments of the status of plant nitrogen in corn, and its use by farmers from Vermont to Iowa, is reducing the application of excess nitrogen fertilizers in corn production. While the procedure is being used on only a small proportion of the total corn acreage, and while PSNT is not universally applicable to corn growing areas, it is an indicator of the kinds of technology that will likely reduce total use of nitrogen fertilizers in US agriculture by an estimated 20% in the next decade, and greatly reduce nitrogen loadings to water resources.

A position paper on nitrate fertilizers (USDA, 1990) and a status report on the extent of the nitrate contamination of groundwater (Fedkiw, 1991) have been developed. These have been widely distributed, to help agency personnel and farmers to understand that there are identified areas where nitrate contamination of water is a problem. (And conversely, to help policy makers to understand that it is not an ubiquitous National problem.)

While American agriculture is addressing the matter of excess nitrogen fertilizer use and subsequent loadings to water resources, it must be recognized that much of the farmer adoption is in reaction to the economic benefits of careful management of nitrogen fertilizers, rather than to allegations of adverse impacts on human health.

The USDA has also conducted a separate "Conservation Reserve Program," (CRP) which has retired some 35 million acres of highly erodible lands from crop production. Grasses have been planted on 28 million of these acres, trees on 2 million acres, and the remaining 5 million acres have been converted to windbreaks, filter strips, wildlife and wetland areas. These conversions will reduce sediment loadings to surface waters by 210 million tons per year; phosphorus loadings by 66 percent, and nitrogen loadings by 75 percent (Iivari, 1992). The CRP may generate $3.5 billion to $4 billion in water quality benefits (Ribaudo, 1989).

Other USDA programs are aimed at reducing the use of agricultural chemicals, and the loadings of such chemicals to the environment. USDA's program to eradicate the boll weevil, for example, has essentially eliminated the pest from Virginia, North Carolina and South Carolina (USDA, 1992). Programs to foster the use of integrated pest management, integrated crop management, and sustainable agriculture also contribute to reducing the environmental loadings of agricultural chemicals.

U.S. agriculture is continuing to address the matters of pesticide nutrient and contamination of the nation's water resources. But the USDA, and the scientific, academic and policy communities must address these matters with some sense of reality; to be sure that we are reacting to real (rather than perceived) problems; and that we are addressing problems to which agriculture contributes, for which there are practical solutions, and that will result in enhancing or protecting water quality in some demonstrable ways.

The USDA represents the interests of U.S. Agriculture, of its productivity, and of its $18 billion annual positive balance in international trade, and provides leadership for an appropriate balance between efficient agricultural production and the legitimate demands for enhancing and protecting environmental quality. The agricultural community is moving toward an environmental ethic: "The time is right for everyone to work toward a new agricultural ethic that places equal emphasis on production and environmental protection"; but "Environmentalists need to recognize that there are limits on the speed and degree to which agricultural programs can be altered to serve environmental goals" (CAST, 1992).

Agriculture neither has nor expects license to pollute water resources. USDA has developed, and continues to implement programs that encourage farmers to voluntarily adopt realistic, environment-compatible practices. American agriculture is making progress in changing both attitudes and production practices.

American farmers are becoming more environmentally sensitive; and are adopting improved management practices, as they become convinced of the desirability to do so, whether the incentive is a reduction of input costs, or compliance with environmental ethics or regulations. But they can hardly be faulted for a limited response to generic allegations and unsupportable estimates of the problems. If a cause-and-effect relationship between their agricultural operations and adverse impacts on water quality can be presented, operators usually can be convinced to make significant changes in their practices. If such changes are not forthcoming, State or local regulations often soon follow. But the crucial element is a convincing connection between agricultural practices and water quality. In the case of agricultural NPS contamination, such convincing connections are -- by definition -- hard to establish.

At the extreme, abandoned farm lands may result in reduced loadings of agricultural NPS pollutants to our nation's waters. Their likely replacements -- housing developments, shopping malls, roads and parking lots -- may not enhance environmental quality (Horton and Eichbaum, 1992).

Pollution prevention is now a topic of conversation in both the Administration and the Congress. It is generally acknowledged to be more efficient than treating pollution. There are two associated difficulties; one is that we have no definite end points against which to gauge progress; the other is that it is impossible to prove that one prevented that which never happened. The latter is especially critical in times of tight budgets.

It is our collective responsibility to assist ourselves ("..a government of the people," etc.) in setting realistic expectations for environmental programs. We must remember that there are no "no-cost solutions"; that the nation's financial resources are limited; that someone must always pay the bill; that NPS pollution is diffuse and hard to identify as to source; and that site-specific solutions cannot be applied to generic problems.

Summary

American agriculture has made tremendous progress in enhancing our water quality; farmers continue to respond as a result of incentives, input cost reductions, and environmental concern. The USDA continues to conduct programs to reduce agricultural NPS pollution; USDA agencies continue to provide research, education, and technical and financial assistance.
In many cases, we are now on the relatively flat part of the "response curve;" additional increments of water quality will be increasingly difficult and expensive to attain.
As the U.S. continues to develop reasonable priorities and goals, agriculture will continue to respond with strategies and programs to achieve realistic levels of water quality.

REFERENCES

1. American Society of Agronomy, 1977. <u>Soils for Management of Organic Wastes and Waste Waters</u>. Madison, WI, 650p.

2. CAST, 1992. <u>Water Quality - Agriculture's Role</u>. Council on Agricultural Science and Technology, Ames, IA, 103p.

3. Clausen, J.D., D.W. Meals and E.A. Cassell, 1992. Estimation of lag time for water quality response to BMP's, in <u>The National Rural Clean Water Program Symposium</u>. U.S. Environmental Protection Agency, Washington, DC, EPA/625/R-92/006, 173.

4. Cornell University, 1970. <u>Relationship of Agriculture to Soil and Water Pollution</u>. Cornell University, Ithaca, NY, F. Swader, editor, 270p.

5. EPA, 1990. National Water Quality Inventory: 1990 Report to Congress. Washington, DC, 214p.

6. EPA, 1992. National Pesticide Survey: Update and Summary of Phase II Results. EPA 570/9-90-021, 6p.

7. Fedkiw, J., 1991, <u>Nitrate Occurrence in US Waters (and Related Questions)</u>, U.S. Dept. of Agriculture, Washington, DC.

8. General Accounting Office, 1990. <u>Water Pollution: Greater EPA Leadership Needed to Reduce Nonpoint Source Pollution</u>. GAO/RCED 91-10. Washington, DC, 56p.

9. Horton, T., and W. Eichbaum, 1992. <u>Turning the tide: saving the Chesapeake Bay</u>. Island Press, Washington, DC, 324p.

10. Iivari, T.A., 1992. Conservation reserve program, conservation practices and water quality. in <u>Beltsville symposium XVII: Agricultural Water Quality Priorities, A Team Approach to conserving Natural Resources</u>. USDA-ARS, Beltsville, MD.

11. Loehr, R.C., editor, 1977. <u>Food, fertilizer and Agricultural Residues</u>. Ann Arbor Science Publishers, Inc., Ann arbor, MI, 727p.

12. Magdoff, F., 1991, Understanding the Magdoff Pre-Sidedress Nitrate Test for Corn. <u>Journal of Production Agriculture</u>, <u>4</u>, 297.

13. Meek, J., C. Myers, G. Nebeker, W. Rittall and F. Swader, 1992. RCWP - the federal perspective. in <u>The National Rural Clean Water Program Symposium</u>. U.S. Environmental Protection Agency, Washington, DC, EPA/625/R-92/006, 289.

14. Porter, K.S., editor, 1975. <u>Nitrogen and Phosphorus: Food Production, Waste, The environment</u>. An Arbor Science Publishers, Inc., Ann Arbor, MI, 372p.

15. Ribaudo, M.O., 1989. <u>Water Quality Benefits from the Conservation Reserve Program</u>. USDA-Economic Research Service. Agricultural Economic Report No. 606.

16. U.S. Department of Agriculture, 1992. News release 1181-92, USDA, Washington, DC. Stewart, B.A., D.A. Woolhiser, W.H. Wischmeier, J.H. Caro, M.H. Frere, 1975. <u>Control of Water Pollution from Cropland</u>. Vol II, EPA-600/2-75-026b. U.S. Environmental Protection Agency, Washington, DC.

17. USDA, Working Group on Water Quality, 1990, <u>Water Quality and Nitrate: Agricultural Sources of Nitrate and Approaches</u>, U.S. Department of Agriculture, Washington, DC.

18. USDA, 1993. <u>Working Group on Water Quality: Progress Report</u>. USDA, Washington, DC, 16p.

19. Willrich, T.L., and N.W. Hines, editor, 1967. <u>Water Pollution Control and Abatement</u>. Iowa State University Press. Ames, IA, 194p.

ANALYSIS OF ON-FARM BEST MANAGEMENT PRACTICES IN THE EVERGLADES AGRICULTURAL AREA

L.M. Willis, S.B. Forrest, J.A. Nissen, J.G. Hiscock, P.V. Kirby[*]

ABSTRACT

An integral component of the Everglades restoration plan includes modifications in farming and pumping practices in the Everglades Agricultural Area (EAA) in order to improve the present water quality and quantity of one of Florida's principal freshwater resources. On-farm best management practices (BMPs) are cost-effective methods by which agricultural practices can be modified to reduce nutrient loadings from EAA runoff into the Everglades. The methods and results of the analysis used to evaluate the cost-effectiveness of BMP applications in the EAA are presented.

KEYWORDS. Best management practices, Everglades Agricultural Area

BACKGROUND

The Everglades is a major freshwater marsh ecosystem which originally extended from Lake Okeechobee to the Florida Bay. One of the byproducts of formation of the Everglades is rich organic soils. The economic potential of these soils for agricultural purposes was realized in the 1920s when canals were constructed to drain approximately 700,000 acres of marsh land for cultivation. The EAA, located immediately south of Lake Okeechobee, is responsible for a significant percentage of the sugarcane and winter vegetables grown in this country, and contributes over $1 billion dollars to the local economy. However, nearly 90 years of systematic alterations in the quality, quantity, and timing of the flow through the Everglades has led to measurable and deleterious changes in the ambient flora and fauna. Productive farming in the EAA has contributed to changes in the water quality flowing from the EAA, southward into the Everglades, and into the Florida Bay. Notable are the nutrient changes observed in the water quality. Phosphorus (P), for example, has been determined to be a limiting nutrient encouraging a dramatic invasion of cattails into the Everglades region. In addition to increased nutrient loadings, flow of water through the Everglades has been altered due to increased development in the area. As a consequence, there have been changes in the Everglades ecosystem.

In an attempt to restore water flow and nutrient loadings to levels more closely resembling natural conditions, the South Florida Water Management District (District) adopted the Everglades Surface Water Improvement and Management (SWIM) Plan (1992) in March 1992. The SWIM Plan called for (1) the construction of wetlands treatment systems to remove nutrients from the EAA drainage water prior to discharge into the Everglades, and (2) P load reductions of at least 25 percent by initiation of BMPs on farms in the EAA. Subsequently, the District hired a multidisciplinary consulting team led by Brown and Caldwell (BC) to evaluate alternative treatment technologies and the potential cost/benefit analysis of implementing BMPs on farms within the EAA.

---

[*] L.M. Willis, Project Engineer, Brown and Caldwell; S.B. Forrest, Project Engineer, Brown and Caldwell; J.A. Nissen, P.E., Senior Project Manager, Brown and Caldwell; J.G. Hiscock, P.E., Water Resources Engineer, Mock, Roos & Associates; P.V. Kirby, P.E., Agricultural Engineer, Mock, Roos & Associates.

OBJECTIVE

Improvements in water quality are increasingly viewed as critical to sustained, natural assimilation of anthropogenic nutrient contributions. On-farm BMPs in the EAA are considered a viable means of reducing nutrient loadings. This paper presents an evaluation of BMP effectiveness in reducing P discharges in the EAA, as well as least-cost combinations of BMPs to achieve varying levels of nutrient load reduction. Our evaluation of these practices included:

- Establishing baseline conditions;
- Evaluating BMP effectiveness in reducing nutrient loads;
- Estimating BMP implementation costs; and
- Formulating least-cost combinations of BMPs to achieve 25, 35, and 45 percent P reductions on sugarcane, vegetable, and sod farms in the EAA.

INDIVIDUAL BMP EVALUATION

Research conducted by the University of Florida Institute For Agricultural Studies (IFAS), under the direction of Drs. A.B. (Del) Bottcher and Forrest T. Izuno, to study the impact of BMPs on nutrient discharges in the EAA was a primary source of information for this investigation. The BMPs proposed by IFAS—water management, fertility, and aquatic cover crop practices—as well as the sediment control BMPs proposed by the U.S. Sugar Corporation, form the basis of the BMPs considered in this analysis.

BMP Description

A brief characterization of the BMPs considered in this evaluation is included in this paper. More detailed descriptions of the individual BMPs, including applicability in the EAA and compatibility with other BMPs, are defined by IFAS (1992) and Brown and Caldwell (1993).

Water Management BMPs. The water management BMPs are practices which reduce the net water discharged off-farm, resulting in reduced P loads in the EAA drainage water. There are two general types of water management practices: (1) water table management, and (2) retention of drainage on-farm.

Water table management practices involve the development of drainage and irrigation schedules to maintain a water table which provides optimal crop production and simultaneously minimizes water quality impacts. Upward water table fluctuations can adversely affect crop growth by saturating a portion of the root zone, while downward fluctuations can result in increased soil mineralization and nutrient releases. Storage of drainage water on-farm is another means of reducing the P load in the EAA drainage water by reducing the volume of drainage water being pumped into the District canals. The water table management BMPs evaluated in this study are presented in Table 1.

Table 1 Water Management BMPs

Water Management BMP	BMP features	BMP results
Pump schedule	• Pumping schedule modifications as recommended by FSCL	• Reduced pumping volumes • Standardized pumping criteria
IFAS temporal control	• Flow balance/water table optimization over time	• Field-calibrated water table response customized per farm
IFAS spatial control	• Farm-wide water table uniformity at any given time	• Improved farm hydraulics
IFAS temporal/spacial control	• Combination of above two BMPs	• Water table control farm-wide and over time
Storage in fallow fields	• Extension of fallow flooding practice	• Retained excess drainage water during/after storm events
Storage of vegetable drainage in sugarcane fields	• Co-managing vegetable and sugarcane runoff	• Reduced vegetable drainage load • Sugarcane lands receive vegetable runoff nutrients
Storage in on-farm reservoirs	• Retention of excess drainage water	• Drainage/irrigation water storage • Decreased production land

Fertility BMPs. Practices that reduce the quantity of fertilizer applied can result in reduced P loss during and after storm events. Table 2 presents fertility BMPs evaluated.

Table 2 Fertility BMPs

Fertility BMP	BMP features	BMP results
Calibrated soil testing	• Establish and optimize relationship between soil P levels, P in fertilizer, and crop yields	• Optimize fertilizer use and effects
Banding	• Location of fertilizer near crop roots	• Decrease in total fertilizer application
Prevention of misplaced fertilizer	• Training and equipment maintenance	• Efficient fertilizer use • Decrease in total fertilizer application
Split applications of fertilizer	• Two or more fertilizer applications during growing season	• Efficient fertilizer use • Lower net application rates

Aquatic Cover Crop BMPs. Aquatic cover crops provide increased storage capacity for receiving drainage water and excess rainfall. Additionally, nutrient mineralization and soil subsidence are reduced by aquatic crop production. As of this time, rice is the only aquatic crop commercially produced in the EAA, and thus, was the only aquatic cover crop BMP considered in this evaluation.

Sediment Control BMPs. Improved on-farm sediment control practices should reduce the amount of particulate being discharged into the canals and, consequently, should reduce nutrient loading in the EAA drainage water. The U.S. Sugar Corporation (1992) identified 16 sediment control practices as possible BMPs. Because there is little information regarding their cost and effectiveness, we combined the individual sediment control practices into three groups. Table 3 presents aquatic cover crop and sediment control BMPs.

Table 3 Aquatic Cover Crop and Sediment Control BMPs

BMPs	BMP features	BMP results
Aquatic Cover Crop		
Rice Production	• Rice production on fallow sugarcane and vegetable land	• Increased drainage water storage capacity • Reduced soil subsidence, mineralization; disease and weed control
Sediment Control		
Field erosion control	• Field tilling practices, grassing banks and ditch entrances, cover crops	• Reduced soil losses
Ditch sediment control	• Practices to slow ditch drainage	• Reduced soil losses
Canal sediment control	• Sediment traps, canal cleaning	• Reduced soil losses • Improved canal hydraulics

Land Uses Considered

Three different land uses were considered for the evaluation of BMPs in the EAA. Sugarcane, the most predominant land use, represents approximately 460,000 acres, or 89 percent of the total land area in the EAA. Vegetables and sod account for about 25,000 and 22,000, respectively, or about 5 and 4 percent of the total land area. The remaining 2 percent of the EAA land is devoted to pasture, citrus, and other agricultural and nonagricultural practices (Burns & McDonnell, 1993).

Applicability of the individual BMPs to sugarcane, vegetable, and sod lands was an important consideration because not all BMPs could be implemented for each of the land uses evaluated in this study. For example, banding of fertilizer is not recommended for crops with a continuous root system, such as sod, and crops whose roots extend laterally between rows, such as ratoon cane and celery. Therefore, for this analysis, the banding BMP was applicable for only 25 percent of the sugarcane land per year (only the plant cane), 90 percent of the vegetable land (excluding celery), and was not applicable for sod land. For each BMP, applicability to the three land uses was determined and then factored into the evaluation of BMP effectiveness in reducing P loads.

Baseline Conditions

Identification of historic farm practices was necessary to determine the P reduction potential from implementation of BMPs. Therefore, the potential impact of BMPs was evaluated against conditions that existed in the EAA during the period 1979 to 1988. This was accomplished by assessing the percentage of farm acreage in the EAA, by land use category, that had achieved full or partial implementation of the various BMPs during the historical period of record. On land where full implementation of a BMP had occurred previously, no additional benefit of P reduction from that BMP can be gained. Similarly, on land where a BMP had previously been partially implemented, only a portion of the full benefit of that BMP can be realized.

BMP Effectiveness

Estimates of BMP effectiveness in reducing P loads were developed based on (1) modeling of pre- and post-BMP conditions using hydraulic and water budget models, (2) results of research performed by IFAS (Bottcher and Izuno) on farms in the EAA, (3) results of field experiments and modeling reported by the FSCL, and (4) discussions with staffs of the Agricultural Extension Service, the Soil Conservation Service, IFAS, and the District. The length of time that a crop is in production, as well as BMP applicability to a crop type, were factored into the annual effectiveness estimates.

Effectiveness of Water Management BMPs. Two different models were used to analyze the effectiveness of the pump schedule BMP in reducing P loads. The Advanced Inter-Connected Pond Routing (AdICPR) model, developed by Streamline Technology, was used to predict pre- and post-BMP drainage from a 1,920-acre model farm resulting from a 5-year, 24-hour design storm event. The results of the modeling indicated that for such storm events, drainage volume from sugarcane farms could be reduced about 20 percent and drainage volume from vegetable farms could be reduced by about 7 percent using the pump schedule BMP. It was assumed that reductions in the volume of water pumped off sod farms would fall between those of sugarcane and vegetable farms.

A water budget model, developed by Melaika and Bottcher (1988) for the S-5A basin of the EAA, was modified by Brown and Caldwell's subconsultant Mock, Roos & Associates to evaluate the impact of the pump schedule BMP on a broader scale. Using daily rainfall data and estimates of daily evapotranspiration, the model predicted daily irrigation demand and drainage volumes in four of the EAA basins between 1980 and 1988. These results indicated an overall average annual reduction in the water volume pumped off-farm of approximately 24 percent.

Effectiveness of the IFAS temporal and spatial control BMPs was based on the results of the pump schedule BMP modeling and the results of IFAS field experiments conducted in the EAA. It was estimated that IFAS temporal control procedures, which rely on water table response curves for individual farms, can achieve additional benefits above the pump schedule BMP. Capital improvements for IFAS spatial control can achieve further P reductions, and the combination IFAS temporal/spatial control BMP can achieve even higher P reduction potential. Estimates of the effectiveness of BMPs involving storage of drainage on-farm were derived from information presented by IFAS (1992).

Effectiveness of Fertility BMPs. Estimates of the effectiveness of fertility BMPs were based on (1) the anticipated reduction in the quantity of fertilizer applied, and (2) the reduction on P load anticipated from reduced fertilizer application. Reductions in fertilizer quantities were estimated from information presented by IFAS (1992) as a result of research experiments in the EAA, and discussions with representatives of the Agricultural Extension Service. Reductions in P load were assumed to be approximately 40 percent of reductions in fertilizer applied, again based on the results of field experiments by IFAS on farms in the EAA.

Effectiveness of Aquatic Cover Crop BMPs. The effectiveness of rice as an aquatic cover crop to remove P from drainage water is well recognized, but not well documented. The research conducted by IFAS indicates that rice production on fallow vegetable land can achieve an additional benefit of approximately 5 to 20 percent P reduction when compared to simply storing water in fallow fields. Considered as an individual BMP, including the benefits of water retention on-farm, rice production on vegetable land can reduce P loadings from drainage water by as much as 30 percent during the 4 or 5 months of the year that it occurs. This translates to an annual load reduction of approximately 15 percent, since about half of the annual drainage occurs during the fallow period. Reductions from sugarcane land are expected to be less because of the lower concentration of P in the drainage water and the shorter fallow period.

Effectiveness of Sediment Control BMPs. Limited data exists to document the effectiveness of sediment control practices in reducing P loads from farms in the EAA. In theory, retention of soil and sediments on-farm should reduce sediment buildup in the District's canals and, eventually, should result in reduced discharge of particulate P from the EAA. P reduction percentages for field, ditch, and canal sediment control BMPs were estimated taking into account benefits associated with water table management as well as solids capture and containment on-farm.

Estimates of BMP Cost

BMP cost estimates were calculated based on definition of the capital improvements and annual operation and maintenance (O&M) activities necessary for implementation of the individual BMPs. Estimates of capital costs were prepared for each BMP reflecting the research and development, equipment purchase, and/or design and construction of improvements needed to implement the practice. All capital costs were amortized at 8 percent interest over a 20-year period to identify an equivalent annual cost per acre. Estimates of annual O&M costs for each of the BMPs were also prepared, taking into account the labor, fuel, equipment maintenance, and sampling and monitoring costs associated with their implementation. Where appropriate, cost savings for reduced fuel consumption and fertilizer application were included. Care was taken to reflect only those O&M costs which would be incremental increases in cost over and above pre-BMP farm practice.

BMP Cost-Effectiveness

The cost and effectiveness of the individual BMPs were estimated taking into account the previous degree of implementation in the EAA during the period from 1979 to 1988. This was accomplished by allocating no additional benefit or cost where the BMP had been fully implemented in the past, and only allocating a portion of the benefit or cost where the BMP had been partially implemented in the past. Using these estimates, the cost-effectiveness of the individual BMPs by land use was calculated. For all land uses, the fertility BMPs and the rice production BMP were the most cost-effective, actually saving money or having minimal net cost. The water table management BMPs were generally the next most cost-effective category of BMPs. The on-farm retention and sediment control BMPs appeared to be the least cost-effective categories of BMPs.

BMP ALTERNATIVES

Least-cost combinations of BMPs were created to achieve P load reductions of 25, 35, and 45 percent on each of the three land uses using the estimated cost-effectiveness of the individual BMPs. BMPs were selected starting with the BMP with the lowest cost per acre per year, and continuing to add the BMP with the next lowest cost until the net effectiveness of the combined BMPs satisfied the P reduction goal. For example, the least-cost combination of BMPs to achieve 25 percent P reduction on sugarcane lands included calibrated soil test, rice production, banding fertilizer, prevention of misplaced fertilizer, and pump schedule.

Table 4 summarizes the expected cost of the BMP alternatives for the three primary land uses in the EAA and the EAA as a whole. Costs are presented in terms of dollars per acre per year and in terms of dollars per pound of P removed. For sugarcane land, the unit costs of BMP implementation to achieve 25 percent and 35 percent reductions in P loading are estimated to be about $3 and $5 per pound of P removed, respectively. However, the unit cost is estimated to increase significantly to about $47 for a P reduction of 45 percent. For vegetable land, significant cost savings can be realized at the 25 and 35 percent reduction levels. A unit cost of approximately $59 per pound of P removed is estimated to achieve the 45 percent reduction level for vegetable land. BMP costs for sod land are appreciably higher, ranging from about $56 to $219 per pound of P removed. Weighting these BMP costs to the current land use distribution allows an average cost across the entire EAA to be estimated. To achieve a 25 percent reduction in P load, the result is an overall BMP cost of approximately $0.58 per pound of P removed. For the 35 and 45 percent reduction levels, this average cost increases to about $9 and $57, respectively.

Table 4 Summary of Costs for BMP Alternatives

P load reduction, percent	Cost of BMP implementation							
	Dollars per acre per year				Dollars per pound P removed			
	Sugar-cane	Vegetables	Sod	EAA	Sugar-cane[a]	Vegetables[b]	Sod[c]	EAA
25	0.51	(16.62)	10.43	0.08	3.52	(31.51)	56.38	0.58
35	1.07	(8.08)	36.74	2.07	5.27	(10.94)	141.85	9.35
45	12.23	57.23	72.80	16.59	46.86	58.96	218.62	57.29

[a] Costs based on assumed annual P loading of 0.58 pounds P/acre/year.
[b] Costs based on assumed annual P loading of 2.11 pounds P/acre/year.
[c] Costs based on assumed annual P loading of 0.74 pounds P/acre/year.

CONCLUSIONS

The evaluation of BMPs for achieving varying levels of P load reduction in the EAA indicates that 25 and 35 percent P load reductions are possible at low cost by implementing BMPs. Higher load reductions are possible from BMP implementation, but at greater cost.

REFERENCES

1. Brown and Caldwell. April 30, 1993. Evaluation of On-Farm Best Management Practices. Final Draft Report. Prepared for the South Florida Water Management District.

2. Burns & McDonnell. January 1993. "Adjustments to EAA Discharges Due to Implementation of Best Management Practices." Draft Technical Memorandum to the South Florida Water Management District.

3. Hutcheon Engineers. April 1992. Phosphorus Reduction Strategies—Evaluation of Proposed Modified Pumping Practices. Prepared for the Florida Sugarcane League.

4. IFAS, University of Florida, Florida Cooperative Extension Service. July 1992. Procedural Guide for the Development of Farm Level Best Management Practice Plans for Phosphorus Control in the Everglades Agricultural Area.

5. Melaika, N.F. and Bottcher, A.B. 1988. "Irrigation and Drainage Management Model for Florida's Everglades Agricultural Area." Transactions of the American Society of Agricultural Engineers. Vol. 31, No. 4.

6. South Florida Water Management District. March 1992. Surface Water Improvement and Management Plan for the Everglades—Planning Document.

7. Streamline Technology. Advanced Inter-Connected Pond Routing Model.

8. United States Sugar Corporation. August 25, 1992. "Best Management Practices for On-Farm Phosphorus Reductions Through Sediment Control." Presented at a meeting of the Scientific Advisory Group for the Everglades.

Save Energy, Resources, and Money with IFAS Bahiagrass Pasture Fertilization Recommendations

Sid Sumner, Wayne Wade, Jim Selph, Pat Hogue, Ed Jennings, Pat Miller, Travis Seawright, Mark Kistler, Greg Weaver* Gerald Kidder, Findlay Pate**

Abstract

Bahiagrass is the most commonly used forage on Florida ranches. With 2.5 million acres in production, it serves as the main forage for Florida's beef industry, currently ranked tenth in the nation. Fertilizer represents the largest expenditure of dollars and energy for maintaining bahiagrass pastures.

From 1986 to 1990, a field study was conducted on nine ranches in a nine county area in Central Florida. The study was conducted by a multidisciplinary group of University of Florida faculty from the Institute of Food and Agricultural Sciences (IFAS). The goal was to evaluate the efficiency of various fertilization programs recommended and used at the time. Data on forage quantity and quality were collected.

A major finding of the study was the lack of any increased yield or quality from phosphorus or potassium fertilization when the nitrogen fertilization rate was 50 lbs./acre/year. The average N fertilization used by ranchers had been determined by surveys to be 54 lbs. N/acre/year.

Using data from the study in conjunction with data from the literature, University of Florida/IFAS recommendations for bahiagrass pasture were revised to reflect cost and energy inputs. Since surveys had shown that few ranchers felt they could afford to fertilize for maximum forage production, the new phosphorus and potassium recommendations were keyed to the level of nitrogen fertilization. Producers select one of three management options based on the approximate amount they plan to spend on fertilizer. Each of the three options gives different recommendations based first on N input level then on P and K soil test level.

In 1991, with assistance from an Energy Extension Service Grant, a twenty minute video was developed to teach producers, fertilizer salesmen, and others how to use the revised recommendations. Producers using these recommendations have reported cutting fertilizer costs by up to 50% without lowering production. In addition, finite natural resources and energy are saved and chances of negative environmental impact from over-fertilization are reduced.

Keywords: Bahiagrass, Forage, Fertilizer, Pasture, Energy.

---

*The authors are county extension faculty at Polk, Hillsborough, Desoto, Highlands, Pasco, Okeechobee, Manatee, Sarasota, and Hardee Counties, respectively, and participants in the South Florida Beef-Forage Program.
** Kidder and Pate are University of Florida Extension Soils Specialist and Ona Research Center director respectively.

Introduction

There are about 2.5 million acres of bahiagrass pasture used for beef production in Florida. A major expense of maintaining this resource is its annual fertilization. Recognizing this expense, and the importance of good fertilization practices, the Florida Cattleman's Association recommended in 1985 that the University of Florida, IFAS, reevaluate the fertilization needs of pasture grasses.

A three-year research study conducted at the Ona Agricultural Research and Education Center in the early 1960's (McCaleb et al., 1966) showed that bahiagrass yield was not increased by phosphate (P_2O_5) fertilization, and a response to potash (K_2O) fertilization was not obtained at rates higher than 24 pounds per acre (lb/A) annually, even with 120 lbs nitrogen (N)/A applied as a split application. Based on soil test values reported in the study, IFAS fertilizer recommendations called for annual applications of 48 lb of P_2O_5 and 96 lb of K_2O/A (Jones et al., 1974). Later modifications of IFAS recommendations indicated that 40 lb P_2O_5 and 80 lb K_2O/A should have been applied annually (Whitty et al., 1977).

Research at the Beef Research Unit near Gainesville (Blue, 1970) showed that around 70% of the P applied to a limed Leon fine sand pasture over an 18-year period had remained in the surface soil. Further study (Rodulfo and Blue, 1970) showed that bahiagrass responded to added P_2O_5 when grown in the surface horizon of a virgin soil, but did not respond to P_2O_5 when grown in the surface horizon of soil from previously-fertilized pasture.

Considering evidence that the P_2O_5 and K_2O requirements of bahiagrass need to be evaluated under conditions present on commercial ranches that have been in production for many years, a field study was conducted with the following objectives: 1) to determine if bahiagrass pasture responds to P_2O_5 and K_2O fertilization when N fertilization is 60 lb/A/yr, a rate commonly used by ranchers (IFAS, 1986); and 2) to compare the response of bahiagrass pasture when fertilized according to IFAS standard recommendations based on soil tests with the response of bahiagrass fertilized at lower rates of N, P_2O_5 and K_2O.

Methods

In 1986, one site in each of nine south Florida counties was selected. Each site was a bahiagrass pasture on which a cow/calf management system had been in effect for more than 10 years.

At each site, five 50 x 100 ft areas were selected and assigned one of five fertilization treatments. These were: 1) no fertilizer; 2) 60 lb N/A applied in March; 3) 60 lb N, 45 lb P_2O_5 and 45 lb K_2O/A applied in March; 4) 60 lb N/A applied in March and 60 lb N/A applied again in September; and 5) 60 lb N, 90 lb P_2O_5 and 45 lb K_2O/A applied in March and 60 lb N and 45 lb K_2O_5 applied in September. Nitrogen, phosphate, and potash were applied as ammonium nitrate, superphosphate, and potassium chloride, respectively. Treatment 5 represented University of Florida, IFAS standard recommendations (Whitty et al, 1977) for fertilizing bahiagrass pasture based on test of soil samples from each site when the demonstration was initiated.

Soil samples were obtained from each treatment area immediately prior to fertilization in March and September each year of the demonstration. Each soil sample consisted of a composite of five 6-inch deep cores from each treatment area. Soil samples were analyzed for pH and for Mehlich-I extractable P, K, calcium (Ca), zinc (Zn), copper (Cu), magnesium (Mg), and manganese (Mn).

Two 4 x 8 ft wire cattle-exclusion cages were placed on each 50 x 100 ft treatment area in March. Cages were positioned on an area where the bahiagrass had been previously staged to a 2-inch stubble height, if needed, with a plot harvester. Forage from a 20-sq-ft area inside and outside each cage was harvested to a 2-inch stubble every 30 to 60 days from April or May through December. On each harvest date, each cage was moved to a pasture area harvested outside that cage, thus cages were moved around the 50 x 100 ft treatment areas throughout the year.

Total fresh forage harvested inside and outside each cage was weighed and sampled for analysis. Dry matter content was determined on samples dried in a forced-air dryer at 60°C. Dry matter yield was calculated from fresh weight data and dry matter content. Crude protein content and total digestible nutrients (TDN) were determined with a near-infrared analyzer. Forage samples were ashed at 600°C and acid digested to determine, P, K, Ca, Mg, Zn, Mn, Cu and Fe.

The field study was initiated in March 1987 and completed in December 1989.

Results

Two important terms are used in this paper to describe the type of forage harvested during the study. Regrowth forage is bahiagrass harvested inside an animal exclusion cage which had grown from a 2-inch stubble since the last harvest. Available forage is bahiagrass harvested outside the cage and is forage actually available to the grazing animal. Yield data were obtained from regrowth harvests.

There was a consistent increase in forage yield to 60 lb of N/A applied in March over the no fertilizer treatment. Over three years the treatment receiving 60 lb N/A averaged 1,760 lb more dry matter per acre annually than the treatment receiving no fertilizer. It presently costs about $20/A to apply 60 lb of N, including $4 per acre spreading costs. This expense appears justifiable, costing about $23 for each ton of additional dry forage produced.

In comparison to 60 lb of N/A only, a positive response in dry matter yield was obtained when 45 lb of P_2O_5 and 45 lb of K_2O/A were applied in March along with 60 lb of N/A. However, the increased production was only 400 lb of dry forage per acre annually. It costs about $14/A for the P_2O_5 and K_2O and approximately $72 for each additional ton of dry forage produced.

Applying 60 lb of N/A in March and then again in September produced an average of 480 lb more dry matter per acre than one 60 lb N application in March. It would cost about $20/A for the second N application, and the cost for each additional ton of dry forage would be about $84. Several research studies have shown a linear response in dry matter yield of bahiagrass to increasing rates of N fertilization, even when N was applied as split applications (Blue, 1966; Blue and Graetz, 1977). However, these studies did not evaluate a situation in which one half of the N was applied as a second application as late as September, a practice used on some ranches in Florida because of heavy summer rains.

In comparison to two applications of 60 lb N/A, a positive response was obtained in dry matter yield with the addition of 90 lb of P_2O_5 in March, and 90 lb/A of K_2O equally split between March and September, along with 120 lb of N. The increased yield averaged 700 lb more dry forage per acre annually than the 120 lb of N/A alone. It presently costs about $29/A for the P_2O_5 and K_2O applied, thus costing approximately $82 for each additional ton of dry forage produced.

Increased yield due to the application of N in March was immediate, and continued throughout the summer period. Early spring growth of pasture forage is important because of low forage availability after the winter months, and demands by cows which are usually nursing calves and being rebred. Typical low spring rainfall was experienced in all three years of this field study, and yet substantial responses in forage growth and forage quality to N fertilization were obtained both in April and in May. This points out the importance of applying N fertilizer to bahiagrass pasture as early as February or March.

The response to N application in September was also immediate but limited only to the September or October harvests. Forage growth in general was reduced after October, because of shorter days, so a response to N fertilization might not be expected. The results of this field study document the poor response of bahiagrass to N applied in September, and suggest that N should be applied to bahiagrass as a single application in the spring, but if split, the second application should be well before September. Research data developed previously at Gainesville (Blue, 1966; Blue and Graetz, 1977) support this conclusion.

When averaged across all harvests in a season, crude protein content of bahiagrass regrowth forage increased with increasing rates of N fertilization but increases were relatively small. Crude protein increases were most pronounced immediately following N application in March and September and rapidly diminished within 4 to 8 weeks. Short-term increases in crude protein content of the magnitude observed would be important in spring grass when cows grazing this forage are usually nursing young calves and being rebred.

Nitrogen fertilization also increased TDN of bahiagrass, but increases, when averaged over the entire year, were relatively small. Increases in TDN were most evident immediately following N application. Fertilization with P and K had little effect on crude protein content and TDN of bahiagrass.

Forage quality values for available forage responded to fertilization in a manner similar to that for regrowth forage. However, available forage was lower in crude protein content and digestibility than regrowth forage, and from July through the fall this difference became progressively larger. The crude protein and TDN requirements for a brood cow nursing a calf and having average milking ability are about 10% and 58% of the dry matter, respectively (National Research Council, 1984). During the spring, summer, and early fall, cattle would selectively graze bahiagrass having quality similar to regrowth forage which would come close to meeting the requirements of lactating brood cows for crude protein and TDN. However, in late fall and winter when bahiagrass stops growing and forage availability becomes limited, the quality of forage eaten by cattle would be similar to that shown for available forage harvested in October and December. This forage would only meet the needs of dry, pregnant cows, which are about 8% and 54% of the dry matter, respectively (National Research Council, 1984).

Fertilization with P_2O_5 and K_2O increased P and K content of bahiagrass regrowth forage, and the degree of increase was related to the amount of P_2O_5 and K_2O applied.

Dietary P levels recommended by the National Research Council (1984) for the types of beef cattle grazing in Florida range from 0.18% of the dry matter for dry cows to 0.23% of the dry matter for lactating cows of average milking ability (most Florida brood cows), and to 0.29% of the dry matter for lactating cows with superior milking ability. Phosphorus levels in regrowth forage were highest in 1987. Only one site had average P levels below that recommended for most beef cattle and that was in treatments not fertilized with P_2O_5. Levels of P in bahiagrass were lowest in 1988, and average P levels of treatments not receiving P_2O_5 at two sites were slightly below that required by most lactating cows.

The P content in available forage was lower than the P content in regrowth forage. The P level was particularly low in available forage in the fall and winter. These levels would cause P deficiency in lactating brood cows not supplemented with P. A deficiency in P could have a negative effect in rebreeding.

Although a mineral supplement containing

P is recommended for all grazing cattle in Florida, mineral supplementation would be more critical if pastures are not fertilized with P_2O_5. A mineral supplement similar to one commonly recommended for Florida (Cunha et al., 1964) would satisfy the P needs of cattle, and would be more economical than fertilizing bahiagrass to provide P nutrition for cattle.

The National Research Council (1984) recommends a dietary K level for beef cattle of 0.5 to 0.7% of the dry matter. Bahiagrass K levels were below this range at several sites in 1988. Other minerals were present in bahiagrass forage in adequate amounts as recommended by the National Research Council, with the exception of Cu. The National Research Council recommends that cattle diets contain 4 to 10 ppm of Cu. Forage copper levels were at or below the lower end of this range in many cases. Possibilities of a Cu deficiency for cattle grazing Florida pastures has long been recognized, so the addition of this element to the mineral supplement is recommended routinely (Cunha et al., 1964).

Soil P values were very low (< 10 ppm) to low (10 to 115 ppm) at eight sites and medium (16 to 30 ppm) at only one site. Soil K values were very low (<20 ppm) to low (20 to 35 ppm) at six sites, medium (36 to 60 ppm) at two sites and high 61 to 125 ppm) at only one site. All soil parameters were variable among sites and with there being no obvious relationships between any parameter and bahiagrass yield. Fertilization treatment also had no effect on any soil parameters. These data indicate that soil testing as now commonly used to manage the fertilization of Florida bahiagrass pastures is of limited value. This could be because soil test data and plant response relationships were developed with annual crops and bahiagrass is a deep-rooted perennial plant.

Conclusion

From data developed in this field study in conjunction with other data from the literature, the following recommendations are presented for fertilizing established bahiagrass pasture in Florida. These recommendations support revised University of Florida IFAS recommendations (Kidder et al., 1990).

1. With the annual application of 60 lb or less N/A bahiagrass pasture, do not apply any P_2O_5 and K_2O for at least 3 years. The field study is continuing and future recommendations of 60 lb/A of N only may be extended to periods longer than 3 years.

2. For the most efficient use of the fertilizer budget, only after 60 lb of N have been applied to every bahiagrass acre to be used for grazing should consideration be given to applying P_2O_5 and K_2O. At N rates of 100 to 120 lb/A, apply 25 lb/A of P_2O_5 and 50 lb/A of K_2O if these plant nutrients test low for the soil. Do not apply P_2O_5 and K_2O if these nutrients test medium or higher for the soil.

3. When applying up to 120 lb of N/A, it appears to be most efficient to apply all of the N as a single application in the spring. If a split application is used, the second should be applied before the first of July.

4. Apply N fertilizer to bahiagrass pasture in February or March. Bahiagrass produces growth in the early spring, so a response in both forage growth and forage quality to N fertilizer will be obtained. Bahiagrass should continue to benefit into the growing season from an early N application.

5. A mineral supplement containing P and trace elements should be available to all cattle grazing bahiagrass pastures, especially those grazing pastures not fertilized with P_2O_5.

Movement of Fall-Applied Nitrogen During Winter in Western Minnesota

J.A. Staricka, G.R. Benoit, A.E. Olness, J.A. Daniel, and D.R. Huggins[*]

ABSTRACT

Fall application of nitrogen (N) has advantages (e.g. cheaper prices, fewer time conflicts) but also allows more time (6 to 9 months) for loss. In addition, during winter in temperate climates, soil freezing influences the movement of water and solutes. This 3-yr study examined the effect of soil freezing and topography on N movement during winter and early spring. The study site was located on side slopes (<2% slope) of a 2-ha depression typical of Great Plains prairie pothole topography. Two tillage systems (chisel or moldboard plow) and two N rates (130 or 260 kg ha^{-1} applied as urea before tillage) were investigated. Vertical movement was assessed using soil samples down to 1.2 m taken during December, January, February, and after thaw (April or May). Lateral (down-slope) movement was assessed using additional soil samples taken after thaw at 3 and 9 m downslope of the application area. Results from three winters indicate little movement of N out of the tilled zone. More than 45% of the N was in the upper 0.3 m of soil and more than 65% was in the upper 0.6 m. Net movement of the center of N mass from December to thaw was 82 mm upward in 1990-91, not significant in 1991-92, and 63 mm downward in 1992-93. The change in 1992-93 was likely due to loss of N from the surface soil by denitrification or immobilization rather than downward movement of N. During the first winter, lateral (downslope) N movement was greater than 3 m but less than 9 m. No down-slope movement of N occurred during the second and third winter.

Keywords. Nitrogen fertilizer, Soil freezing, Tillage.

INTRODUCTION

Fall application of N for corn is common in the western Corn Belt. Nearly 21% of the N applied to corn is fall-applied, and nearly 24% of corn receives at least some of its N during the previous fall (Taylor and Vroomen, 1989). Reasons for fall application include cheaper prices and fewer time conflicts, however, fall application allows more time (6 to 9 months) for N loss.

In temperate climates, fall-applied N is exposed to soil freezing, which influences physical, chemical, and biological processes occurring in the soil (Miller, 1980; Edwards and Cresser, 1992). Water vapor and liquid may migrate upward to the freezing front and into the frozen soil (Gray and Granger, 1986; Benoit, et al., 1988). Liquid water flow may result in the migration of soluble salts as well (Gray and Granger, 1986). During spring, the soil surface thaws before the subsoil, thus a layer restrictive to infiltration exists below the permeable surface. Precipitation during this time may infiltrate the surface soil and then flow down-slope laterally through the surface soil, enhancing the down-slope movement of solutes.

---

[*]J.A. Staricka (Post-Doctoral Associate, University of Minnesota), G.R. Benoit [retired] and A.E. Olness (Soil Scientists, USDA-Agricultural Research Service): North Central Soil Conservation Research Laboratory, Morris, MN. J.A. Daniel (Geologist, USDA-Agricultural Research Service): National Agricultural Water Quality Laboratory, Durant, OK. D.R. Huggins (Assistant Professor, University of Minnesota): Southwest Experiment Station, Lamberton, MN.

The objectives of this study were to monitor fall-applied N for vertical movement during winter and lateral movement during spring as related to soil freezing.

METHODOLOGY

This study was conducted for three winters in west-central Minnesota (45° 41' N; 95° 48' W). At the location, soil freezing commonly begins in November. Maximum frost penetration occurs near the end of February and generally averages 1.2 m in depth. The frozen soil begins to thaw from both the top and bottom a few weeks later and generally is completely thawed by mid-April.

The experimental plots were located on side slopes (<2% slope) of a 2-ha depression. Maximum relief is 1.2 m. Hamerly clay loam (fine-loamy, mixed, frigid, Aeric Calciaquoll) occurs on the side slopes and Parnell silty clay loam (fine, mixed, noncalcareous, frigid Typic Argiaquoll) is in the depression bottom. Before the experiment, the area had been treated uniformly for more than 10 years. A split-split plot design with four replications was used. Main plot (18 × 61 m) treatment was tillage system (chisel or moldboard plow during fall). Tillage was performed perpendicular to the contour for the entire length of the slope. Subplot (6 × 18 m) treatment was N application rate (130 or 260 kg ha^{-1} applied as urea immediately before tillage). N treatment applications were restricted to subplots located midslope in a line parallel to the contour. Treatments were applied three consecutive falls. Wheat (*Triticum aestivum* L.) was grown the summer before the experiment, and corn (*Zea mays* L.) was grown during the two intervening summers.

Soil cores (1.2 m deep in 0.1 m increments) were taken during December, January, February, and after thaw (late April or early May). During April, samples were also taken 3 and 9 m downslope of the application area. The results were analyzed as two independent studies. Results from samples obtained on the four dates from within the N application area will be referred to as the 'Date Study' and samples obtained during April from the three slope positions after thaw will be referred to as the 'Position Study'. To examine temporal changes in the Date Study, sampling date was treated as a sub-subplot (Gomez and Gomez, 1984). Likewise, to examine spatial changes in the Position Study, sampling position was treated as a sub-subplot. Sample analysis included bulk density, water content, and mineral N (ammonium-N plus nitrate-N) concentration.

Analysis of variance was performed on N concentration and depth to center of N mass. Depth to center of N mass, Z_N, is a characterization of the depth distribution of N and was calculated as

$$Z_N = \sum_{i=1}^{12}(N_i Z_i) \bigg/ \sum_{i=1}^{12} N_i$$

where N_i = mass of N in depth increment i and Z_i = the midpoint depth of increment i. Analysis of variance was performed using the SAS GLM procedure (SAS Institute, 1985).

After chemical analysis of samples taken during the first winter indicated large variances in the tilled zone, three additional cores of the upper 400 mm were taken in each plot during the second and third winters. Data were averaged across quadruplicate cores prior to analysis of variance.

Depth of snow cover and soil freezing were measured with CRREL frost tubes, and soil water content was measured with neutron probe. These measurements were taken weekly during November to March and three times a week during March to May.

RESULTS AND DISCUSSION

The first two winters had average air temperatures (November to February, inclusive) warmer than the 100-yr mean of -8.6 °C, but the third winter was colder than the 100-yr mean (Table 1). Despite colder temperatures during the third winter, the deeper snow cover resulted in similar frost penetration depths during the second and third winters. At the onset of freezing, the soil was wetter the second season than the first or third season, but all three years, the soil water content was near midway between the permanent wilting point and field capacity.

Date Study

N concentration did not vary (*prob.* < 0.05) between N rates or tillage treatments, or among samples dates in 1990-91, but varied between tillage systems and between N rates in 1991-92 and among sample dates and between N rates in 1992-93 (Table 2a). This indicates there was no change in N concentration between the December and Thaw sampling dates in 1990-91 and 1991-92. During 1992-93, N concentration was less at the Thaw sampling date than at the other sampling dates (discussed later).

At all dates in all three winters, more N was recovered from the high N rate treatments than from the low N rate treatments. The additional N mass recovered from the high N plots accounted for 36% of the difference in application rates in 1990-91, 50% of the cumulative difference in 1991-92, and 52% of the cumulative difference in 1992-93. The differences in N concentration between N rate treatments were 2.78 mg N / kg soil in 1990-91, 8.02 in 1991-92, and 11.73 in 1992-93.

The increase in the difference between N rate treatments, as well as the overall increase in N concentrations, was probably from residual N remaining from the previous year. Based on University of Minnesota guidelines, the low N rate should have provided sufficient N for the corn crop. In 1991, grain yields were 9.17 Mg/ha for chisel tillage and 9.82 Mg / ha for moldboard tillage. In 1992 the yields were 3.91 Mg / ha for chisel tillage in 1992 and 6.03 Mg / ha for moldboard tillage. In 1993, no grain yields were obtained. There was no difference in grain yield between the two N rates in 1991 or 1992, further suggesting the occurrence of excess soil N in plots receiving the high N rate. The increasing amounts of soil N each winter also suggest little loss of N occurred during the summer.

In 1991-92, N concentration was greater for the moldboard treated plots than the chisel treated plots (Table 2b).

Center of N mass depth averaged 410 mm in 1990-91, 462 mm in 1991-92, and 408 mm in 1992-93, indicating most of the N remained near the surface. In 1990-91, 52% of the N was in the upper 0.3 m of the soil and 67% was in the upper 0.6 m. In 1991-92, the amounts were 43% for the upper 0.3 m and 62% for the upper 0.6 m. In 1992-93, the amounts were 48% for the upper 0.3 m and 72% for the upper 0.6 m.

Analysis of variance indicated a date effect in 1990-91 and 1992-93 and significant tillage × date and tillage × date × N rate interactions in 1991-92. All three winters, the center of N mass initially moved upward toward the surface and then moved downward

(Table 3a). The switch in direction occurred in February during the first winter and in January during the second and third winters. The net overall movement (i.e. December to Thaw) was downward in 1990-91, not significant in 1991-92, and upward in 1992-93.

The smaller temporal change in 1991-92 compared to 1990-91 may have been due to the shallower frost penetration in 1991-92 compared to 1990-91 (Table 1). Also, soil water redistribution patterns differed between the first two winters (Staricka et al., 1993). Movement of water toward the freezing front during mid-winter was more evident in 1990-91 than in 1991-92 and downward movement of water after soil thawing was more evident in 1991-92 than in 1990-91 . This may have contributed to the difference in N movement between these two winters. The relationship between water and N movement during the 1992-93 winter is being investigated.

In 1992-93, the decrease in N concentration occurred in the upper 300 mm of the soil; there was no change in concentration at the lower depths. This suggests that the downward movement of the center of N mass was due to loss of N from the surface (by denitrification or immobilization) rather than the downward movement of N.

In 1991-92, the difference between tillage systems was greater than the difference among dates (Table 3b). Temporal changes in depth to the center of N mass were inconsistent among tillage × N rate combinations.

Position Study

Soil N concentration decreased as distance from the application area increased (Table 4). In 1991-92 and 1992-93, N concentration was also less in the low N rate plots than in the high N rate plots and the position × N rate interaction was also significant. During the first winter, similar N concentrations at the application area and 3-m downslope positions suggest slight lateral (downslope) movement of N from the application area. This did not happen in the other two winters.

The significant effect of N rate during the second and third winters is probably related to the increased levels of residual N discussed above. Differences in soil N concentration between N rate treatments were greatest in the application area and decreased as distance from the application area increased. Although not statistically compared, N concentrations decreased each year at positions down-slope of the application area. This was probably due to crop uptake and insufficient replenishment by fertilizer.

Frozen soil increases the flow of surface water downslope resulting in a sudden rise in the water table in depression bottoms (Daniel et al., 1994). Ponding of water occurred in the bottom in the depression while the soil was still frozen. No corresponding increase in N concentration downslope of the application area was observed. Since the ponding of water occurred in early spring, while soil was still frozen, the water flow likely occurred as runoff rather than lateral flow through the soil. Runoff across a frozen soil surface would be unlikely to move N downslope.

Depth to center of N mass differed among sampling positions in 1990-91 and 1992-93 but did not differ among any treatment in 1991-92 (Table 5). In 1990-91 and 1992-93, depth to center of mass increased as distance from the application area increased. This is likely due to a greater N concentration near the soil surface where N was applied. Why the trend did not hold for the 1991-92 winter is unexplained.

CONCLUSIONS

Limited movement of soil N occurred during the winter when the soil was frozen, indicating soil freezing had only a minor effect on either vertical or lateral redistribution of N. The magnitude and direction of the vertical movement of N seemed related to water movement patterns during the first two winters (the 1992-93 data are still being analyzed). Thus a better understanding of water movement during winter may improve our understanding of N movement and thus our management of N. The majority of down-slope movement of water in the spring occurred while the soil was still frozen and thus did not move the soil N downslope.

ACKNOWLEDGEMENTS

The authors thank "Frost Crew" members Julie Cady, Lynette Howe, Scott Larson, Eric Poissant, and Danny Struxness for their work in the field under less than ideal conditions. The authors also thank Jerry Somsen for performing the statistical analysis.

REFERENCES

Benoit, G.R., R.A. Young, and M.J. Lindstrom. 1988. Freezing induced field soil water changes during five winters in West Central Minnesota. Trans ASAE 31:1108-1114.

Daniel, J.A., G.R. Benoit, and J.A. Staricka. 1994. Depressional focused recharge due to soil frost and landscape position. *Submitted to* Journal of Ground Water.

Edwards, A.C., and M.S. Cresser. 1992. Freezing and its effect on chemical and biological properties of soil. Adv. in Soil Sci. 18:59-79.

Gomez, K.A., and A.A. Gomez. 1984. Statistical procedures for agricultural research. 2nd ed. John Wiley & Sons. New York.

Gray, D.M., and R.J. Granger. 1986. *In situ* measurements of moisture and salt migration in freezing soils. Can. J. Earth Sci. 23: 696-704.

Miller, R.D. 1980. Freezing phenomena in soils. pp. 254-299. *In* Hillel, D. Applications of Soil Physics. Academic Press. New York.

SAS Institute, Inc. 1988. SAS/STAT™ User's Guide, Release 6.03 edition. SAS Institute, Inc. Cary, NC. 1028 pp.

Staricka, J.A., G.R. Benoit, J.A. Daniel, D.J. Fuchs, and D.R. Huggins. 1993. Movement of soil water during winter. Vol. 1, pp. 464-467. *In* Agricultural Research to Protect Water Quality: Conference Proceedings. 21-24 February 1992. Minneapolis, Minnesota. Soil and Water Conservation Society. Ankeny, IA.

Taylor, H. and H. Vroomen. 1989. Timing of fertilizer applications. pp 40-45. *In* Agricultural Resources Inputs: Situation and Outlook. US Dept. of Agric. Economic Research Service. AR-15. August 1989.

Table 1. Meteorological Data.

Winter	Avg air temp[a]	Avg snow cover[a]	Fall soil water content	Frost Depth			
				Max	Dec	Jan	Feb
	°C	mm	m³ / m³	------------------ m ------------------			
1990/91	-7.3	14	0.213	1.07	0.24	0.98	1.04
1991/92	-6.1	24	0.340	0.73	0.35	0.43	0.73
1992/93	-9.5	190	0.229	0.78	0.20	0.67	0.76

[a]November to February, inclusively.

Table 2a. Date Study: N concentration to 1.2 m depth averaged by sampling date and N application rate.

	1990-91		1991-92		1992-93	
Date	Low N	High N	Low N	High N	Low N	High N
	---------------------------- mg N / kg soil ----------------------------					
December	11.0	16.8	13.9	21.7	19.2	28.1
January	14.7	16.5	17.3	24.6	23.2	33.7
February	12.1	13.9	17.0	25.8	23.7	39.3
Thaw	11.5	13.6	16.3	24.4	15.1	27.0
$LSD_{0.05}$	NS	NS	NS	NS	4.3	4.3

Table 2b. Date Study: N concentration to 1.2 m depth averaged by tillage system and N application rate for the 1991-92 winter.

Tillage	Low N	High N
	-- mg N / kg soil --	
Chisel	14.1	20.7
Moldboard	18.2	27.7

Table 3a. Date Study: N center of mass depth in 1.2 m cores averaged by sampling date.

Date	1990-91	1991-92	1992-93
		mm	
December	462	461	401
January	452	440	368
February	351	473	398
Thaw	380	475	464
$LSD_{0.05}$	63	NS	41

Table 3b. Date Study: N center of mass depth in 1.2 m cores averaged by sampling date, tillage system, and N application rate for the 1991-92 winter.

	Chisel		Moldboard	
Date	Low N	High N	Low N	High N
		mm		
December	449	390	449	556
January	338	424	528	470
February	526	437	492	437
Thaw	422	470	507	501
$LSD_{0.05}$	95	95	95	95

Table 4. Position Study: N concentration to 1.2 m depth averaged by sampling position and N application rate.

Position	1990-91		1991-92		1992-93	
	Low N	High N	Low N	High N	Low N	High N
	mg N / kg soil					
Application area	11.5	13.6	16.3	24.4	15.1	27.0
3-m down slope	12.4	12.5	9.7	12.0	8.4	10.7
9-m down slope	7.7	9.1	7.7	8.7	6.7	6.8
$LSD_{0.05}$	3.2	3.2	2.7	2.7	3.4	3.4

Table 5. Position Study: N center of mass depth in 1.2 m cores averaged by sampling position.

Position	1990-91	1991-92	1992-93
	mm		
Application area	380	475	464
3-m down slope	455	442	512
9-m down slope	506	450	523
$LSD_{0.05}$	49	NS	42

NITROGEN MANAGEMENT, IRRIGATION METHOD, AND NITRATE LEACHING IN THE ARID WEST: AN ECONOMIC ANALYSIS USING SIMULATION

Gilbert D. Miller, Jay C. Andersen[1]

ABSTRACT

This study[2] concerns the economic impact of reducing the amount of nitrate leached from the root zone on irrigated cropland in the arid West. Irrigation management, nitrogen source, and application methods were evaluated on the basis of maximization of economic return and the minimization of nitrate leached out of the root zone. A simulation model was used to determine crop response and nitrate leaching for the various management practices. The results of the analysis indicate the following: (1) soil characteristics affect the amount of nitrate that leaches and are important in determining the proper irrigation management techniques; (2) proper irrigation management can reduce the amount of nitrate that leaches from the root zone; (3) the profit-maximizing levels of irrigation are close to the estimated ET; (4) applying nitrogen in two closely timed applications did not significantly reduce the amount of nitrate leached; (5) changing from ammonium nitrate to urea or anhydrous ammonia may increase returns to management and reduce the amount of nitrate leached; (6) fertigation may increase returns to management, reduce nitrate leaching, and/or reduce the total amount of nitrogen applied for farmers who use sprinklers; and (7) helping irrigators understand the relationships between nitrogen and water management may improve their economic return and reduce their impact on the environment.

KEYWORDS. Nitrogen management, Nitrate leaching, Simulation model, Economic analysis.

INTRODUCTION

Groundwater quality and fertilizer costs are important considerations in irrigated agriculture. Nitrogen fertilizers can be associated with the level of nitrates in groundwater. The source of nitrogen, quantity and timing of application, and the quantity and timing of irrigations might affect how much nitrate reaches the groundwater (Saliba, 1985; Newcomer, 1986). This study concerns the economic incentives associated with nitrogen source, application method, and irrigation management, and how these choices affect nitrate movement in the soil profile. We sought to identify costs and returns for corn silage production associated with a reduction in the amount of nitrate leached out of the root zone. We considered different nitrogen and irrigation management options on three soil types under two weather scenarios.

METHODOLOGY AND SIMULATION PROCEDURES

We used NTRM, A Soil-Crop Simulation Model for Nitrogen, Tillage, and Crop-Residue Management (Shaffer and Larson, 1987) to model corn growth and nitrate leaching. NTRM uses daily weather data, tillage events, and location and source of nitrogen in the

---

[1] Miller and Andersen are members of the Department of Economics, Utah State University, Logan, Utah 84322-3530.

[2] This research was supported by the Utah Agricultural Experiment Station--Project 411, and by a grant entitled Water Quality Initiative--Phase II from Federal Extension Service.

soil profile in simulating the soil-crop interactions. NTRM simulates crop responses to various soil types when soil characteristics are changed. We selected three soils (fine sandy loam, silt loam, and silty clay) based on water-holding capacity and other characteristics. The profile depth was 1.68 m for each soil simulated. Soil characteristics were obtained from soil survey data for the eastern part of Box Elder County, Utah (Chadwick et al., 1975).

The two simulated weather patterns were based on weather data for Corinne, Box Elder County, Utah. Average precipitation data for 30 years (1951 to 1980) were used with 1985-86 (near average) temperature data, from April 1 to March 31 (USC). The simulated high precipitation (1.6 times the 30-year average) conditions were based on actual temperature and precipitation data for April 1, 1982 to March 31, 1983. All precipitation was treated as rain. Although Crowder et al. (1985) discuss procedures to model snowmelt, it is difficult to model snowmelt because many factors (sudden temperature changes, frozen or thawed soil, snow movement by wind, etc.) affect it. Thus, effective precipitation during the winter period was probably less accurate than during the growing season.

Corn was selected because its high nitrogen requirement increases the risk of nitrate leaching. Planting density was 88,888 plants per hectare. Management practices (except irrigation, nitrogen source, and application method) were assumed to be similar for each soil type and weather condition. Management practices typify those now used in the area. The simulations assume corn was planted on May 15 and harvested on October 1. Two major tillage events were simulated with NTRM. The first, on May 14, was seedbed preparation and for some simulations to incorporate the fertilizer into the top 52 mm of the soil. The second tillage event on October 3 was plowing after harvest. NTRM does not simulate other cultural and management practices used in the study.

Nitrogen was applied at the rate of 224 kg/ha of elemental N. Ammonium nitrate (NH_4NO_3), anhydrous ammonia (NH_3), and urea ($CO(NH_2)_2$) were selected as nitrogen sources because they are the most widely used nitrogen fertilizers in the study area. Single applications were applied on May 14, the day before planting. When split applications were used, 50% was applied on May 14 and 50% was applied on June 29. When fertigation (adding fertilizer to irrigation water) was simulated, the amount of nitrogen applied was divided equally among irrigations. The initial level of residual nitrogen was assumed to be 46 kg/ha, of which 6 kg/ha was nitrate. We assumed residual nitrogen was evenly distributed in the top 0.3 m and that there was no nitrogen below 0.3 m. Expected yield with this available nitrogen (270 kg/ha) was 85 metric tons of silage per hectare (James and Topper, 1989), a yield substantially higher than the county average of 50.4 metric tons of silage per hectare (UASS, 1989), but one that allowed us to evaluate the effects of high target yields on nitrate leaching.

Irrigation schedules were based on water rights of the Bear River Canal Company, the major supplier of irrigation water for eastern Box Elder County. Farmers have a weekly water turn of about 51 mm per week per hectare (BRCC, 1991). Depending on the cropping pattern, an individual farmer can apply 152, 102, or 76 mm every two weeks or apply 76 or 51 mm every week, based on the water-holding and infiltration capacities of each soil type. We run simulations using these irrigation schedules. Irrigation schedules were developed using estimated evapotransporation (ET) requirements for irrigation levels that resulted in nitrate leaching under weekly or biweekly irrigations.

We used furrow irrigation when 152 or 102 mm irrigations were simulated to calculate irrigation costs. We used sprinklers (center pivot) to apply 76 and 51 mm on fine sandy loam and silt loam and to apply 51 mm on silty clay in calculating irrigation costs. We did not simulate the application of 76 mm on fine sandy loam or silt loam. Applying 76 mm weekly was overirrigating, adding unnecessary expense, and applying 76 mm biweekly was underirrigating, causing substantial yield reductions, thus reducing returns. We assumed that water was uniformly distributed over the field, which does not occur under field conditions. Irrigations started June 22 each year and ended by September 8 each year. We did not consider salt balance in this study because irrigation water quality is high in the Bear River Canal system (James and Jurinak, 1986), and little drainage is required to maintain the salt balance.

Costs of tillage and other cultural and management practices were calculated using the crop budget generator, Cost and Returns Estimator (CARE), developed by the USDA Soil Conservation Service (SCS, 1988) and corroborated by a producer panel. Interest on operating capital was 12% annually, starting from the day of the field operation until October 31. A land charge equal to the annual cash rental value for each soil type was included in the budgets.

RESULTS AND DISCUSSION

Soil attributes affect the amount of water and nitrate that leach through the soil profile (Deer et al.). Soil texture (water-holding capacity) is a major factor in the leaching of water and nitrate from the root zone (Tindall et al.). Figure 1 shows the effect of soil type on returns to management and the amount of nitrate that leached out of the root zone for fine sandy loam, silt loam, and silty clay when 152 mm irrigations were applied biweekly. The spread of points for a soil type are the result of different nitrogen sources and application methods.

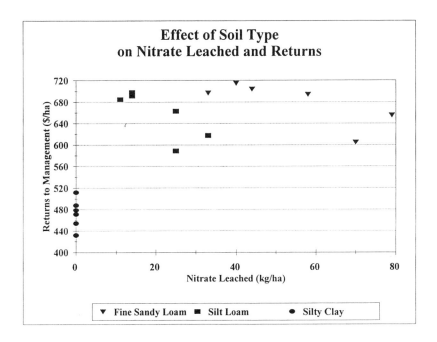

Figure 1. The Effects of Soil Type on Returns to Management and Nitrate Leached.

It is commonly believed that practices that maximize profits are more detrimental to the environment than practices that are thought to be less intensive. To examine the validity of this belief, we compare economic returns associated with specific practices with the amount of nitrate leaching out of the root zone. A change in technology or management practice that increases returns to management was considered as an incentive to change. While a change in technology or management practice that reduces returns to management was considered as a disincentive to change. An improvement in environmental quality was defined as a reduction in the amount of nitrate leached per acre. Because of the ambiguous relationship between the amount of nitrate leached and the concentration level of nitrates in the leachate (water leached through and out of the root zone), nitrate concentration was not used as a measure of environmental quality.

Our discussion of the results focuses on fine sandy loam, because it is most susceptible to leaching as demonstrated in fig 1. A single application of ammonium nitrate and applying 152 mm of water biweekly with 30-year average precipitation served as the control or reference treatment. These are typical management practices for the study area.

Figure 2 shows the effects of nitrogen source and application method on returns to management and the amount of nitrate that leaches out of the root zone. The 30-year average precipitation and 152 mm biweekly irrigations are used in fig 2. A farmer could improve both his/her return to management and reduce the amount of nitrate leached by changing sources of nitrogen from ammonium nitrate to either anhydrous ammonia or urea. Anhydrous ammonia is more hazardous to handle, so we recommend using urea. Under field conditions, we feel that it is unlikely that one could perceive a difference in yield between using anhydrous ammonia and urea when both are applied properly. Split applications of the fertilizers reduced the amount of nitrate leached but lower net returns when ammonium nitrate and urea were used. Split applications of urea resulted in the lowest amount of nitrate leaching in both weather scenarios simulated, while returns to management were close to the maximum. These changes could increase environmental quality with a small reduction in returns to management.

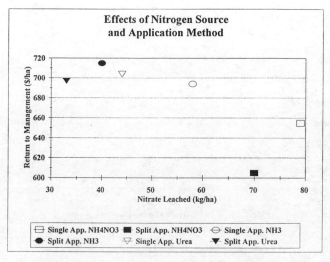

Figure 2. The Effects of Nitrogen Source and Application Method on Returns to Management and Nitrate Leached.

The amount of nitrate leached under the various nitrogen sources depends on temperature and soil moisture (Tisdale et al., 1985). Low soil moisture in the upper 51 mm of soil slowed the transformation of urea to ammonium and ammonium to nitrate in the simulations. Higher soil moisture might have accelerated the transformation to nitrate, increasing the likelihood that more nitrate would have leached below the root zone, thereby decreasing the environmental benefits associated with a change in nitrogen sources. Some farmers in the study area have begun to use split applications of liquefied urea on row-crops. They seem satisfied with the economic results. The environmental results of the change have not been measured.

Returns associated with split applications of anhydrous ammonia were at or near the maximum for most irrigation and weather conditions simulated, which largely reflected a low cost of application. The cost of the applicator was included in the price of the anhydrous ammonia. The only additional costs associated with split applications were the tractor and the operator's labor. The farmer incurred full ownership and maintenance costs of the equipment used to apply ammonium nitrate or urea.

Only with overirrigation was there much reduction in nitrate leaching associated with a change from single to split applications of nitrogen fertilizers. Figure 3 illustrates the effect of irrigation management on nitrate leaching and returns to management. When 152 mm irrigations were timed by estimated ET, the amount of nitrate leached out of the root zone was greatly reduced from the levels resulting from biweekly irrigations.

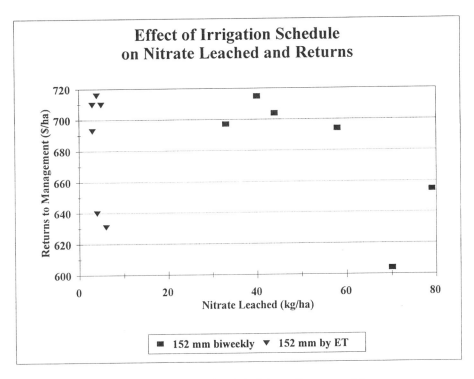

Figure 3. The Effect of Irrigation Schedule on Returns to Management and Nitrate Leached.

Nitrogen source and application method had a greater effect on returns to management than it had on the amount of nitrate leaching when irrigations were near ET. Changing the institutional framework of the water right to allow for on-demand water use coupled with an educational effort on water management might improve both environmental quality and returns.

Fertigation using furrow irrigation results in nonuniform application of nitrogen, tail-water runoff, and other problems. Nonuniform water application was considered the major source of variability in nitrogen application. Soil variability also contributes to variability in irrigation and fertilization. In addition, nitrogen in runoff may enter rivers, streams, or other bodies of water. Our simulations assumed that fertigation uniformly applied both nitrogen and water. Even with these idealized conditions, fertigation with 152 and 102 mm irrigations was only economically competitive with the other application methods when irrigations substantially exceeded ET requirements.

The simulation indicate that farmers who use sprinklers might improve profitability with fertigation. The highest returns to management using center pivot (51 mm water applications) in both weather regimes involved fertigation. Fertigation can provide nitrogen as the plants need it. Thus, reducing the amount of nitrate that is susceptible to leaching (Newcomer, 1986). The simulations indicated that fertigation could reduce the amount of nitrogen necessary to achieve desired yields by reducing the amount of nitrate that leaches out of the root zone.

CONCLUSIONS

The results of the analysis indicate the following: (1) soil characteristics affect the amount of nitrate that leaches and are important in determining the proper irrigation management techniques; (2) the profit-maximizing irrigation schedule differed for each soil; (3) at the profit-maximizing irrigation level, some nitrate was leached from the root zone on the fine sandy loam; (4) proper irrigation management can reduce the amount of nitrate that leaches from the root zone; (5) the profit-maximizing levels of irrigation are close to the estimated ET. However, the amount of water applied per application and the timing of application varied by soil type; (6) weather had little effect on the profit-maximizing level of irrigation per application for each soil type; (7) on coarse soils, the profit-maximizing level of irrigation was likely to increase leaching due to either precipitation during the nongrowing season or excess irrigation early during the next growing season when plant roots cannot reach and utilize the nitrate; (8) applying nitrogen in two closely timed applications did not significantly reduce the amount of nitrate leached; (9) changing from ammonium nitrate to urea or anhydrous ammonia may increase returns to management and reduce the amount of nitrate leached; (10) fertigation may increase returns to management for farmers who use sprinklers, reduce nitrate leaching, and/or reduce the total amount of nitrogen applied; (11) protecting groundwater from excessive nitrate leaching may be compatible with increasing returns, only when all leaching was prohibited was there a conflict; and (12) helping irrigators understand the relationships between nitrogen and water management may improve their economic return and reduce their impact on the environment.

REFERENCES

Bear River Canal Company (BRCC). 1991. Personal interview with manager, Tremonton, Utah. January 22.

Chadwick, R.S., M.L. Barney, D. Beckstrand, L. Campbell, J.A. Carley, E.H. Jensen, C.R. McKinlay, S.S. Stock and H.A. Stokes. 1975. "Soil survey of Box Elder County, Utah eastern part." U.S. Dept. of Ag., Soil Cons. Serv., and U.S. Dept. of the Interior, Fish and Wildlife Serv., BLM in cooperation with Utah Ag. Exper. Station. Washington D.C. June.

Crowder, B.M., H.B. Pionke, D.J. Epp and C.E. Young. 1985. "Using CREAMS and economic modeling to evaluate conservation practices: An application." *J. of Envir. Qual.* 14(3).

Deer, H.M., R.W. Hill, N.B. Jones and T.A. Tindall. "Utah water quality, Utah groundwater." Series EC425.1. Cooperative Extension Service, Utah State Univ., Logan, Utah.

James, D.W. and J.J. Jurinak. 1986. *Irrigation resources in Utah: Water quality versus soil salinity, soil fertility, and the environment.* Research Bulletin 514. Utah Ag. Exper. Sta., Utah State Univ., Logan, Utah. December.

James, D.W. and K.F. Topper (eds.). 1989. *Utah fertilizer guild.* Cooper. Exten. Serv., and Soils, Plant and Water Anal. Lab., Utah State Univ., Logan, Utah. November.

Newcomer, J.L. 1986. "Learning to cope." *Solutions* (July/Aug.):52-55. Solutions Magazine, Inc. 339 Consort Drive, St. Louis, MO.

Saliba, B.C. 1985. "Irrigated agriculture and groundwater quality--a framework for policy development." *Amer. J. Ag. Econ.* (December):1231-37.

Shaffer, M.J. and W.E. Larson. 1987. "NTRM, a soil-crop simulation model for nitrogen, tillage, and crop-residue management." Conservation Research Report 34-1, U.S. Dept. of Ag.

Soil Conservation Service (SCS), U.S. Dept. of Ag. 1988. "Users manual cost and return estimator (CARE)." Contract No. 54-6526-7-268, prepared by Midwest Ag. Res. Assoc., Inc., Lino Lakes, Minnesota. June 6.

Tindall, T.A., R.W. Hill, H.M. Deer and R. Peralta. "Utah water quality, fertilizer impact on groundwater in Utah." EC425.4, Cooper. Ext. Serv., Utah State Univ., Logan, Utah.

Tisdale, S.L., W.L. Nelson and J.D. Beaton. 1985. *Soil fertility and fertilizers*, (4th ed.). New York: MacMillian Pub. Co.

Utah Agricultural Statistics Service (UASS). 1989. "1989 Utah agricultural statistics, Utah Department of Agriculture annual report enterprise budgets." U.S. Dept. of Ag., Utah Ag. Stat. Serv., Salt Lake City, Utah.

Utah State Climatologist (USC). Selected years. "Weather data from Corinne, Utah. Dept. of Plants, Soils, and Biomet., Utah State Univ., Logan, Utah.

PREDICTION OF NITROGEN LOSSES
VIA DRAINAGE WATER WITH *DRAINMOD-N*

M.A. Brevé, R.W. Skaggs, J.E. Parsons, and J.W. Gilliam[*]

ABSTRACT

The environmental impacts of agricultural drainage have become a critical issue in many areas. There is a need to design and manage drainage and related water table control systems to satisfy both production and water quality objectives. The model *DRAINMOD-N* was used to study long-term effects of drainage system design and management on two poorly drained eastern North Carolina soils. Hydrologic results presented herein indicate that increasing the drain spacing, decreasing the drain depth, and/or improving the surface drainage reduces drainage while it increases runoff. Results also show that the use of controlled drainage (CD) reduces subsurface drainage and increases runoff. Water quality results indicate that increasing the drain spacing, decreasing the drain depth, and/or improving the surface drainage reduces predicted nitrate-nitrogen (NO_3-N) drainage losses and net mineralization, while increasing denitrification and runoff losses. Controlled drainage causes a predicted reduction in mineralization and drainage losses and an increase in denitrification and runoff losses.

The ideal drainage design and management combination is one that will optimize profits and minimize environmental impacts. Simulated results indicate that the most economical drainage system may not fully minimize nitrate-nitrogen losses. If an environmental objective is integrated, minimizing NO_3-N losses to receiving streams for instance, simulated results show that there is a strong incentive to decrease drainage intensity by increasing the optimum drain spacing and/or decreasing the optimum drain depth. However, a cost-benefit analysis indicates such design modifications are associated with a decrease in profit. Similarly, the use of controlled drainage could also reduce the nitrate-nitrogen losses, but with another sacrifice in profit. The overall simulation results have shown the practical applicability of *DRAINMOD-N* to identify drainage design and management combinations that could potentially satisfy both production and environmental objectives for a specific crop, soil and climatic condition.

INTRODUCTION

The environmental impacts of agricultural drainage have become a critical issue in many areas. Research has clearly shown that improved subsurface drainage increases losses of some pollutants, such as nitrogen, but decreases losses of others, such as phosphorus (Bottcher et al., 1981; Gilliam, 1987; Skaggs et al., 1994). In addition, the design and management of drainage and associated water table control systems has a substantial effect on drainage water quality (Skaggs and Gilliam, 1981; Skaggs et al., 1994). The challenge in current drainage research is to develop methods to design and manage drainage and related water table control systems to satisfy both production and water quality objectives.

Computer simulation models are useful to evaluate the effects of certain agricultural practices on the fate and movement of agricultural chemicals. Simulation models have been developed to describe the performance of drainage systems, including predicting the effects of system design on crop yields and hydrology (Feddes, 1987; Skaggs, 1987, 1991). Many models

---

[*]Professor, Escuela de Agricultura de la Región Tropical Húmeda (EARTH), Costa Rica, William Neal Reynolds Professor and Distinguished University Professor, Assistant Professor, Biological and Agricultural Engineering Department, and Professor, Soil Science Department, North Carolina State University, Raleigh, NC.

have also been proposed to predict the movement and fate of nutrients and pesticides. However, only a few of these models can be applied to quantify long-term effects of drainage system design and management on losses of agricultural chemicals in shallow water table soils.

DRAINMOD-N was developed to simulate to movement and fate of nitrate-nitrogen (NO_3-N) in artificially drained soils (Brevé et al., 1993). DRAINMOD-N can be applied over long-term simulation periods and is capable of predicting the effects of different factors that influence NO_3-N losses. The objective of this paper is to illustrate the use of DRAINMOD-N to evaluate long-term effects of drainage system design and management on the fate and movement of nitrate-nitrogen in artificially drained soils.

MODEL DESCRIPTION

As the name implies, DRAINMOD-N is based on the water balance calculations of DRAINMOD (Skaggs, 1978). It uses modifications described by Skaggs et al. (1991) to determine average daily soil-water fluxes and water contents by breaking the profile into increments and conducting a water balance for each increment. In the saturated zone, vertical fluxes are linearly decreased from Hoodghout's drainage flux at the depth of the water table to zero at the impermeable layer depth. In addition, a water content profile is generated using soil-water characteristic data, based on the assumption of hydrostatic conditions above the water table at the end of the day. This approach for computing fluxes and water contents proved to be reliable for shallow water table soils as indicated by comparisons with numerical solutions to the Richards equation for saturated and unsaturated flow (Skaggs et al., 1991; Kandil et al., 1992; Karvonen and Skaggs, 1993).

DRAINMOD-N is based on a simplified version of the nitrogen cycle. It considers only the nitrate pool. The processes considered are rainfall deposition, fertilizer dissolution, net mineralization, denitrification, plant uptake, and runoff and drainage losses. These processes can be represented by the advective-dispersive-reactive (ADR) equation:

$$\frac{\partial(\theta C)}{\partial t} = \frac{\partial}{\partial z}(\theta D \frac{\partial C}{\partial z}) - \frac{\partial(qC)}{\partial z} + \Gamma \tag{1}$$

where C is the NO_3-N concentration [M L^{-3}], θ is the volumetric soil-water content [L^3 L^{-3}], q is the vertical soil-water flux [L T^{-1}], D is the coefficient of hydrodynamic dispersion [L^2 T^{-1}], Γ is a source/sink term used to represent additional processes (plant uptake, transformations, etc.), z is the coordinate direction along the flow path [L], and t is the time [T].

Assuming z is positive in the downward direction and water flows downward in the soil profile, Equation 1 is approximated as follows:

$$C_i^{l+1} = \frac{C_i^l \theta_i^l}{\theta_i^{l+1}} + \frac{\theta_{i+1}^l D_{i+1}^{l+1}(\frac{C_{i+1}^l - C_i^l}{\Delta z}) - \theta_i^l D_i^{l+1}(\frac{C_i^l - C_{i-1}^l}{\Delta z})}{\Delta z} \frac{\Delta t}{\theta_i^{l+1}} + \frac{(q_i^{l+1} C_{i-1}^l - q_{i+1}^{l+1} C_i^l - q_{s_i}^{l+1} C_i^l) \Delta t}{\theta_i^{l+1} \Delta z} + \frac{\Gamma \Delta t}{\theta_i^{l+1}} \tag{2}$$

where l and $l+1$ indicate the previous and new time steps, respectively, i corresponds to the layer where the concentration is being estimated, Δz and Δt are the space and time discretizations, respectively, and q_s, the difference between the vertical fluxes entering and leaving the corresponding layer, is the lateral flux going to the drain. Total nitrate-nitrogen losses in subsurface drainage are estimated by adding up the lateral transport from each layer in the saturated zone. An average concentration in drainage water is approximated by dividing the total lateral mass transport by the estimated drainage rate.

For upward flow, the solution is similar to Equation 4, except that the product of $q_i^{l+1} C_{i-1}^l$ becomes $q_i^{l+1} C_i^l$ and $q_{i+1}^{l+1} C_i^l$ becomes $q_{i+1}^{l+1} C_{i+1}^l$. In addition, the q_s term vanishes, unless water is flowing from the drains into the soil profile as it sometimes happens for controlled drainage (CD) or subirrigation (SI). In that case, the model assumes the water flowing into the domain has a prescribed NO_3-N concentration.

DRAINMOD-N uses functional relationships to quantify processes other than NO_3-N transport, as follows:

$$\Gamma = \Gamma_{dep} + \Gamma_{fer} + \Gamma_{mnl} - \Gamma_{rnf} - \Gamma_{upt} - \Gamma_{den}$$

(3)

where Γ_{dep} stands for rainfall deposition, Γ_{fer} for fertilizer dissolution, Γ_{mnl} for net mineralization, Γ_{rnf} for loss in surface runoff, Γ_{upt} for plant uptake, and Γ_{den} for denitrification. A detailed description of each functional relationship is given by Brevé et al. (1993).

MODEL APPLICATION

Procedure

DRAINMOD-N was applied to evaluate long-term effects of several drainage and related water table control systems on nitrate-nitrogen losses from two eastern North Carolina (NC) soils. Simulations were conducted for a 20-year period (1971-1990) of corn production at Plymouth, NC. The soils used in the simulations are Portsmouth sandy loam (fine-loamy, Mixed, Thermic *Typic Umbraquults*) and Tomotley sandy loam (fine-loamy, Mixed, Thermic *Typic Ochaquults*). The physical and chemical properties of each soil are listed in Table 1.

The field was assumed to have a drainage system consisting of parallel, 10-cm diameter, corrugated plastic drains. Several design and management treatments were simulated for each soil. The drainage designs evaluated (Table 1) consisted of three drain depths (0.75, 1.0, and 1.25 m), eight drain spacings (10, 15, 20, 25, 30, 40, 50, and 100 m), and two surface conditions (0.5 and 2.5 cm depressional storage). The management treatments included conventional drainage, controlled drainage at 50 cm during the summer season (May 15 to August 15), and controlled drainage both in the summer (at 50-cm) and in the winter (at 40-cm between November 1 and March 15). Combining all treatments yielded 288 20-year simulations.

Detailed inputs for the corn production practices and NO_3-N transport and transformation variables are listed in Table 1. The corn production practices used in the simulations are characteristic of eastern North Carolina. The reaction rate coefficients were obtained from ranges published in the literature (Schepers and Mosier, 1991; Pierce et al., 1991).

A cost-benefit analysis was performed to determine the optimum drain spacing-drain depth-surface condition-water management combination for each soil. Annual costs included production, drainage system and maintenance costs. Annual corn production costs ($557/ha) were based on extension material prepared by North Carolina State University. Drainage system costs were based on estimated prices of drain tubing installation ($2.6/m), surface drainage grading ($247/ha) and control structure costs ($55/ha) amortized over 30 years at a 10% interest rate.

Table 1. Summary of inputs for *DRAINMOD-N*.

		Tomotley	Portsmouth
1.	**Soil Properties:**		
	θ_{sat} (cm³ cm⁻³)	0.46	0.37
	θ_{wp} (cm³ cm⁻³)	0.21	0.12
	Bulk Density (g cm⁻³)	1.4	1.6
	Organic-N in top soil (μg g⁻¹)	2000	2000
	K_{mnl} (d⁻¹)	3.0E-05	3.5E-05
	K_{den} (d⁻¹)	0.30	0.35
	Lateral Sat. Hyd. Cond. (m d⁻¹)	0.96 (0-30 cm)	3.60 (0-30 cm)
		0.24 (30-110 cm)	0.48 (30-100 cm)
		0.72 (110-170 cm)	1.92 (100-215 cm)
2.	**Drainage System Parameters:**		
	Drain Depth (m)	0.75, 1.0, 1.25	
	Drain Spacing (m)	10, 15, 20, 25, 30, 40, 50, 100	
	Surface Storage (cm)	0.5, 2.5	
	Effective Drain Radius (cm)	1.5	
3.	**Controlled Drainage Paremeters:**	Weir Depth (cm)	
	Summer (Su) only (May 15-Aug 15)	50	
	Summer and Winter (Wi, Nov 1-Mar 15)	50 (Su), 40 (Wi)	
4.	**Corn Production Parameters:**		
	Desired Planting Date	April 15	
	Length of Growing Season (d)	130	
	Potential Yield (kg ha⁻¹)	11000	
	NO₃-N Content of Plant (%)	1.55	
	Max. Effective Root Depth (cm)	30	
	N-Fertilizer Input (kg ha⁻¹)	50 (April 15) + 100 (May 22)	
	Depth Fertilizer Incorporated (cm)	10	
5.	**Other Parameters:**		
	Dispersivity (cm)	20	
	NO₃-N Concentration of Rain (mg L⁻¹)	0.8	

Annual maintenance costs consisted of a fraction (2%) of the annual amortized drainage system cost. An additional $20/ha was included in the annual maintenance costs for systems with good surface drainage. Annual income figures were based on a maximum potential yield of 11000 kg/ha and a corn market price of $0.10/kg.

Results and Discussion

Detailed results for the hydrologic components are tabulated for Portsmouth in Table 2. Trends for Tomotley were similar to Portsmouth. Simulation results show that, for any equivalent design and management scenario, subsurface drainage is greater and surface runoff lower for the Portsmouth soil than for the Tomotley soil. This can be explained by the greater lateral hydraulic conductivity of the Portsmouth soil. Results also indicate that increasing the drain spacing reduces subsurface drainage while it increases surface runoff and ET (Table 2). Increasing the drain depth increases drainage and decreases ET and runoff (Table 2). Furthermore, improving surface drainage by filling potholes and grading the surface to reduce depressional storage causes an increase in surface runoff, a decrease in subsurface drainage, and a slight increase in ET (Table 2). Controlled drainage reduces subsurface drainage and increases surface runoff, as compared to conventional drainage (Table 2). The magnitude of these changes increases with the intensity of CD, as a more pronounced effect is observed when controlled drainage is used during both the summer and winter seasons (Table 2).

Because the economic analysis indicates that net profit is greater for systems with poor surface drainage conditions, only those results are presented in Figure 1. Cost-benefit results indicate that the optimum drainage system design and management scenario for corn production in the two eastern North Carolina soils studied is a conventionally drained system with a 1.25-m

Table 2. Average annual values of hydrologic components predicted by DRAINMOD for a Portsmouth sandy loam at Plymouth, NC. Values are averages predicted for a 20-yr period (1971-1990) in which the average annual rainfall = 132.1 cm.

Drain Spacing (m)	Drain Depth (m)	Good Surface Drainage, S=0.5 cm				Poor Surface Drainage, S=2.5 cm			
		R.Yld (%)	ET (cm)	Drn (cm)	Rnf (cm)	R.Yld (%)	ET (cm)	Drn (cm)	Rnf (cm)
				Conventional Drainage					
10	0.75	85.8	78.0	50.2	4.2	84.7	78.1	52.6	1.7
15	0.75	85.7	78.7	49.1	4.6	84.3	78.8	51.7	1.9
20	0.75	85.1	79.4	47.7	5.3	83.7	79.5	50.8	2.0
25	0.75	83.9	80.1	46.2	6.1	82.2	80.3	49.9	2.3
30	0.75	82.0	80.9	44.6	6.9	79.1	81.0	48.8	2.5
40	0.75	77.4	82.2	41.5	8.7	71.3	82.6	46.7	3.2
50	0.75	74.1	83.4	38.7	10.3	63.9	83.9	44.8	3.7
100	0.75	59.2	86.2	27.6	18.5	35.9	87.8	36.8	7.7
10	1.00	84.3	72.6	57.4	2.5	83.9	72.7	58.9	0.9
15	1.00	84.5	73.0	56.7	2.7	84.0	73.1	58.3	1.0
20	1.00	84.6	73.6	55.8	3.1	84.0	73.7	57.6	1.2
25	1.00	84.7	74.3	54.6	3.5	83.9	74.4	56.7	1.3
30	1.00	84.6	75.1	53.4	4.0	83.5	75.2	55.7	1.6
40	1.00	83.3	76.9	50.2	5.3	80.9	77.1	53.3	2.1
50	1.00	79.6	78.6	47.0	6.8	74.9	79.0	50.9	2.6
100	1.00	65.1	84.4	33.4	14.6	44.4	85.5	41.4	5.5
10	1.25	83.1	70.2	61.0	1.4	83.1	70.2	61.8	0.6
15	1.25	83.1	70.3	60.8	1.6	83.1	70.3	61.7	0.6
20	1.25	83.2	70.5	60.4	1.7	83.1	70.5	61.5	0.6
25	1.25	83.3	70.8	59.8	2.0	83.2	70.9	61.0	0.7
30	1.25	83.4	71.3	58.9	2.4	83.2	71.3	60.4	0.8
40	1.25	83.5	72.5	56.8	3.2	82.7	72.6	58.8	1.2
50	1.25	82.9	74.2	53.9	4.4	81.1	74.5	56.4	1.6
100	1.25	69.8	82.1	39.0	11.4	52.4	83.2	44.8	4.5
				Controlled Drainage in Summer					
10	0.75	86.1	78.9	48.9	4.6	84.7	79.0	51.5	1.9
15	0.75	85.2	79.4	47.9	5.1	83.7	79.5	50.8	2.1
20	0.75	84.0	80.1	46.5	5.8	82.2	80.2	49.9	2.3
25	0.75	82.3	80.8	45.0	6.6	79.2	80.9	48.9	2.5
30	0.75	79.9	81.5	43.4	7.5	74.7	81.7	47.9	2.8
40	0.75	74.7	82.8	40.3	9.3	65.4	83.1	45.9	3.4
50	0.75	70.7	83.8	37.5	11.1	57.4	84.4	44.1	3.9
100	0.75	56.9	86.4	26.8	19.2	33.9	88.0	36.3	8.1
10	1.00	86.4	74.0	55.4	3.0	85.6	74.1	57.2	1.2
15	1.00	85.9	74.4	54.8	3.3	84.9	74.4	56.7	1.3
20	1.00	85.3	74.9	53.8	3.7	84.1	75.0	56.0	1.4
25	1.00	84.3	75.6	52.7	4.2	82.2	75.6	55.2	1.6
30	1.00	83.2	76.3	51.4	4.8	79.7	76.4	54.1	1.9
40	1.00	80.5	78.0	48.2	6.2	73.9	78.2	51.8	2.4
50	1.00	76.1	79.6	44.9	8.0	66.2	80.0	49.4	3.1
100	1.00	60.7	84.8	31.7	15.9	38.3	86.1	40.0	6.3
10	1.25	85.5	71.6	59.1	1.9	85.1	71.6	60.3	0.7
15	1.25	85.2	71.6	58.8	2.1	84.7	71.7	60.2	0.7
20	1.25	84.7	71.9	58.4	2.4	84.1	71.9	59.9	0.8
25	1.25	84.1	72.2	57.7	2.7	83.0	72.2	59.4	0.9
30	1.25	83.5	72.6	56.8	3.1	81.5	72.7	58.8	1.1
40	1.25	81.9	73.9	54.5	4.2	77.6	74.0	56.9	1.6
50	1.25	80.2	75.6	51.3	5.6	72.8	75.9	54.4	2.2
100	1.25	63.8	83.0	36.4	13.1	44.0	84.2	42.6	5.6
				Controlled Drainage in Summer and Winter					
10	0.75	86.1	81.1	45.3	5.9	84.7	81.2	49.2	2.0
15	0.75	85.2	81.4	44.2	6.7	83.7	81.5	48.6	2.2
20	0.75	84.0	82.0	42.5	7.8	82.2	82.1	47.8	2.4
25	0.75	82.1	82.5	40.9	9.0	78.9	82.6	46.9	2.8
30	0.75	78.8	83.0	39.3	10.0	73.7	83.3	45.9	3.2
40	0.75	74.6	84.0	36.3	12.1	65.1	84.4	44.2	3.7
50	0.75	70.5	84.8	33.0	14.5	57.2	85.5	42.2	4.6
100	0.75	56.3	86.9	21.4	24.0	33.7	88.6	33.6	10.1
10	1.00	86.4	77.5	50.7	4.2	85.6	77.5	53.6	1.2
15	1.00	85.9	77.7	49.8	4.9	84.9	77.8	53.2	1.3
20	1.00	85.3	78.3	48.2	5.8	84.1	78.4	52.4	1.6
25	1.00	84.3	78.7	46.8	6.8	82.2	78.8	51.7	1.9
30	1.00	83.1	79.4	45.1	7.8	79.7	79.6	50.5	2.2
40	1.00	80.3	80.8	41.4	10.2	73.7	81.0	48.3	3.0
50	1.00	75.2	82.0	37.5	12.8	64.9	82.5	45.8	4.0
100	1.00	60.7	85.7	23.9	22.6	37.8	87.3	35.2	9.8
10	1.25	85.5	75.3	54.3	2.9	85.1	75.3	56.4	0.7
15	1.25	85.2	75.3	53.7	3.4	84.7	75.4	56.3	0.8
20	1.25	84.7	75.8	52.5	4.1	84.1	75.8	55.7	0.9
25	1.25	84.1	75.9	51.5	5.0	83.0	76.0	55.2	1.2
30	1.25	83.5	76.5	49.9	6.0	81.5	75.7	55.3	1.4
40	1.25	81.9	77.7	46.1	8.5	77.6	77.9	52.2	2.2
50	1.25	80.0	79.1	42.1	11.1	72.6	79.5	49.5	3.3
100	1.25	63.8	84.5	26.3	21.5	43.1	85.9	37.2	9.2

124

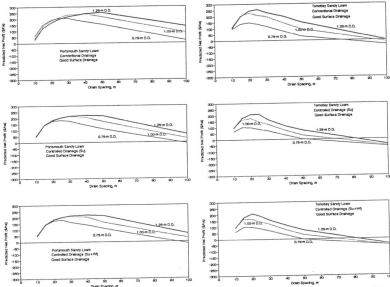

Figure 1. Predicted average annual net profit ($/ha), as affected by surface condition, drain depth, drain spacing and system management, for corn production in Portsmouth and Tomotley soils in eastern North Carolina.

drain depth and poor surface drainage conditions (Fig. 1). The optimum spacings are 40 and 25 m for the Portsmouth and Tomotley soils, respectively. These results are consistent with those reported by Skaggs and Nassehzadeh-Tabrizi, 1986 for both soils. The predicted net profits associated with these two optimum systems are $282/ha for the Portsmouth soil and $280/ha for the Tomotley soil (Fig 1). The economic analysis further shows that the use of controlled drainage does not result in the optimum system for either soil. This is attributed to a slight decrease in simulated relative yields resulting from CD and to an increase in the annual drainage system costs related to the cost of the control structures. These results are in contrast to experiences in the field which have shown that CD, if properly managed, can result in an increase in corn yields (Evans, R.O., pers. comm., 1993). The open ditch outlets typically used in eastern NC have more storage than was considered in these simulations. This enables CD to have a greater effect in conserving water and relieving drought stresses than considered here. Despite these limitations, the maximum predicted profits associated with controlled drainage are $239/ha for the Portsmouth soil ($43/ha reduction in profit as compared to the optimum system) and $211/ha for the Tomotley soil ($69/ha reduction in profit). Simulated results also show that the intensity of controlled drainage does not have a substantial impact on simulated corn yields, and, thus, on net profits (Fig. 1).

Effects of drainage system management and drain spacing and depth on the nitrogen budget components are shown in Tables 3-4 for both the Portsmouth and Tomotley soils. Because of space limitations, only data for poor surface conditions are presented. Simulation results show that, for any equivalent design and management scenario, denitrification, net mineralization, rainfall deposition into the soil profile, plant uptake, and drainage losses are greater for the Portsmouth soil than for the Tomotley soil. Likewise, greater runoff losses are associated in general with the Tomotley soil. Simulated results also indicate that increasing the drain spacing reduces NO_3-N drainage losses and net mineralization, but it increases NO_3-N runoff losses and denitrification for both soils. On the other hand, increasing the drain depth increases drainage losses and net mineralization, and it decreases runoff losses and denitrification.

Table 3. Predicted annual nitrogen budget for corn production on a Portsmouth sandy loam with poor surface drainage (S=2.5 cm) at Plymouth, NC. Values are averages for a 20-yr period (1971-90).

Drain Spacing (m)	Drain Depth (m)	Net Mineralization (kg/ha)	Plant Uptake (kg/ha)	Denitrification (kg/ha)	Drainage Loss (kg/ha)	Runoff Loss (kg/ha)
Conventional Drainage						
10	0.75	72.0	112.5	83.0	20.6	0.1
15	0.75	71.9	112.6	82.5	20.5	0.2
20	0.75	71.7	111.8	82.9	20.4	0.2
25	0.75	71.6	110.7	84.2	20.3	0.2
30	0.75	71.3	108.0	87.6	20.0	0.2
40	0.75	71.1	98.0	98.7	19.2	0.3
50	0.75	70.8	87.6	110.6	18.7	0.3
100	0.75	70.3	48.2	156.1	16.3	0.7
10	1.00	74.5	107.6	77.6	33.5	0.1
15	1.00	74.1	108.6	77.8	32.6	0.1
20	1.00	73.8	109.6	78.5	31.3	0.1
25	1.00	73.5	110.4	80.0	29.6	0.1
30	1.00	73.2	110.9	81.8	27.4	0.1
40	1.00	72.5	109.8	87.0	23.8	0.2
50	1.00	72.1	103.8	96.1	21.5	0.2
100	1.00	70.6	60.5	145.0	17.0	0.5
10	1.25	78.0	106.8	61.5	55.6	0.0
15	1.25	77.6	107.1	63.2	54.0	0.0
20	1.25	77.1	107.1	65.6	51.5	0.0
25	1.25	76.8	107.4	68.6	48.2	0.1
30	1.25	76.2	107.8	72.1	44.1	0.1
40	1.25	75.2	108.7	79.6	35.6	0.1
50	1.25	74.1	109.6	85.7	28.1	0.1
100	1.25	71.4	72.7	134.4	17.4	0.4
Controlled Drainage in Summer						
10	0.75	72.8	115.3	82.4	19.9	0.2
15	0.75	72.5	113.4	83.1	19.8	0.2
20	0.75	72.3	111.3	84.6	19.7	0.2
25	0.75	72.1	107.9	87.7	19.6	0.2
30	0.75	71.8	102.8	93.1	19.4	0.2
40	0.75	71.5	89.7	107.1	18.9	0.3
50	0.75	71.1	78.2	120.0	18.4	0.3
100	0.75	70.2	45.6	159.1	15.8	0.7
10	1.00	73.3	116.3	80.8	26.9	0.1
15	1.00	73.0	115.8	81.3	26.7	0.1
20	1.00	72.7	114.9	82.7	25.9	0.2
25	1.00	72.5	112.7	84.4	24.7	0.1
30	1.00	72.3	105.6	91.0	22.7	0.2
40	1.00	71.7	93.9	103.8	20.4	0.3
50	1.00	71.5	82.2	116.2	18.9	0.3
100	1.00	70.3	46.3	158.2	16.0	0.7
10	1.25	74.6	117.1	77.6	43.3	0.1
15	1.25	74.3	116.4	78.7	42.1	0.1
20	1.25	74.0	117.4	80.8	40.1	0.1
25	1.25	73.6	113.6	82.3	36.5	0.1
30	1.25	73.2	108.5	89.4	31.4	0.2
40	1.25	72.4	97.3	101.0	23.9	0.2
50	1.25	71.8	84.7	113.9	19.5	0.3
100	1.25	70.3	49.0	155.4	16.6	0.8
Controlled Drainage in Summer and Winter						
10	0.75	71.6	115.0	86.5	18.8	0.2
15	0.75	71.4	113.1	87.1	18.4	0.2
20	0.75	71.3	110.8	88.3	18.1	0.2
25	0.75	71.1	107.5	91.0	17.7	0.2
30	0.75	71.0	101.3	96.8	17.4	0.3
40	0.75	70.7	89.4	110.0	16.9	0.3
50	0.75	70.4	77.4	122.8	16.4	0.4
100	0.75	70.2	45.4	159.6	15.0	0.9
10	1.00	73.3	116.2	86.6	21.9	0.1
15	1.00	73.0	115.6	87.1	20.9	0.1
20	1.00	72.8	114.6	88.0	19.5	0.1
25	1.00	72.6	112.4	90.5	19.0	0.2
30	1.00	72.3	109.6	93.8	18.2	0.2
40	1.00	71.8	102.2	101.7	17.3	0.2
50	1.00	71.4	89.9	114.6	16.8	0.3
100	1.00	70.2	50.1	157.1	15.4	0.8
10	1.25	75.8	117.0	86.7	24.3	0.1
15	1.25	75.5	116.2	87.8	23.2	0.1
20	1.25	75.2	114.9	88.9	22.2	0.1
25	1.25	74.9	113.3	90.8	21.1	0.1
30	1.25	74.4	111.3	93.2	20.5	0.1
40	1.25	73.7	107.4	98.4	18.6	0.2
50	1.25	72.9	102.0	104.6	17.9	0.3
100	1.25	70.9	59.5	148.7	16.6	0.8

Table 4. Predicted annual nitrogen budget for corn production on a Tomotley sandy loam with poor surface drainage (S=2.5 cm) at Plymouth, NC. Values are averages for a 20-yr period (1971-90).

Drain Spacing (m)	Drain Depth (m)	Net Mineralization (kg/ha)	Plant Uptake (kg/ha)	Denitrification (kg/ha)	Drainage Loss (kg/ha)	Runoff Loss (kg/ha)
Conventional Drainage						
10	0.75	64.4	112.8	77.4	16.1	0.2
15	0.75	63.6	103.7	85.8	15.9	0.2
20	0.75	62.9	92.8	97.7	15.6	0.3
25	0.75	62.2	80.2	110.3	15.4	0.3
30	0.75	61.8	69.3	122.4	15.2	0.4
40	0.75	61.3	56.8	136.1	14.3	0.7
50	0.75	60.9	46.5	147.0	13.5	0.9
100	0.75	60.4	31.0	163.7	10.8	1.6
10	1.00	69.1	114.1	75.3	20.9	0.1
15	1.00	67.7	115.6	77.4	18.7	0.1
20	1.00	66.4	112.8	82.6	17.2	0.1
25	1.00	65.3	101.8	94.1	16.5	0.2
30	1.00	64.2	92.1	104.6	15.7	0.3
40	1.00	62.8	71.1	127.7	14.6	0.5
50	1.00	61.9	53.7	145.0	14.0	0.8
100	1.00	60.6	33.7	163.2	11.6	1.5
10	1.25	72.1	111.3	65.3	38.4	0.0
15	1.25	71.1	113.5	72.1	31.1	0.1
20	1.25	69.7	114.2	77.8	24.1	0.1
25	1.25	68.2	113.3	82.8	19.4	0.1
30	1.25	66.9	104.2	93.7	17.7	0.2
40	1.25	64.7	84.4	115.3	16.0	0.3
50	1.25	63.3	66.3	135.2	14.3	0.5
100	1.25	60.7	37.0	161.0	12.6	1.3
Controlled Drainage in Summer						
10	0.75	64.4	109.5	81.7	15.6	0.2
15	0.75	63.6	96.2	93.5	15.4	0.2
20	0.75	62.8	82.3	107.7	15.2	0.3
25	0.75	62.1	70.8	119.6	15.0	0.5
30	0.75	61.6	62.2	129.4	14.6	0.5
40	0.75	61.2	51.1	142.4	13.6	0.8
50	0.75	60.8	43.1	150.7	12.8	1.0
100	0.75	60.3	29.3	165.4	10.5	1.7
10	1.00	68.7	115.2	76.9	18.3	0.1
15	1.00	67.4	110.5	83.3	17.0	0.1
20	1.00	66.1	101.5	92.5	16.3	0.2
25	1.00	64.9	86.7	106.5	15.8	0.3
30	1.00	63.9	77.2	117.2	15.0	0.4
40	1.00	62.3	60.3	137.2	14.0	0.8
50	1.00	61.5	45.2	150.0	13.3	1.0
100	1.00	60.4	30.9	165.1	10.9	1.6
10	1.25	72.0	118.0	68.4	31.0	0.0
15	1.25	70.8	114.6	76.9	25.7	0.1
20	1.25	69.4	109.2	85.8	20.9	0.1
25	1.25	67.8	102.0	94.5	18.1	0.2
30	1.25	66.3	88.2	108.8	16.8	0.3
40	1.25	64.0	69.4	129.2	15.0	0.5
50	1.25	62.5	55.0	144.9	13.6	1.0
100	1.25	60.4	32.5	164.6	11.6	1.5
Controlled Drainage in Summer and Winter						
10	0.75	62.9	107.5	85.5	13.4	0.2
15	0.75	62.4	94.6	96.9	13.3	0.3
20	0.75	61.8	81.4	110.2	13.2	0.4
25	0.75	61.4	70.1	121.2	13.1	0.5
30	0.75	61.1	60.5	131.6	13.0	0.7
40	0.75	60.8	50.0	143.9	12.4	1.0
50	0.75	60.5	42.9	151.2	12.0	1.2
100	0.75	60.2	29.2	166.4	10.1	1.9
10	1.00	65.9	114.0	82.3	14.3	0.1
15	1.00	64.9	109.2	87.3	14.1	0.2
20	1.00	64.0	97.1	97.7	13.8	0.3
25	1.00	63.2	85.1	110.1	13.5	0.4
30	1.00	62.7	76.0	119.5	13.3	0.5
40	1.00	61.5	58.3	139.5	12.9	0.9
50	1.00	61.0	44.5	150.8	12.2	1.2
100	1.00	60.3	29.9	165.9	10.5	2.1
10	1.25	68.4	117.2	82.2	15.7	0.1
15	1.25	67.4	113.4	86.4	15.0	0.1
20	1.25	66.3	107.7	91.5	14.8	0.2
25	1.25	65.2	95.7	102.4	14.3	0.3
30	1.25	64.3	85.7	113.4	13.8	0.5
40	1.25	62.7	68.4	131.1	13.4	0.9
50	1.25	61.7	52.8	147.2	12.7	1.3
100	1.25	60.4	31.3	165.5	11.0	1.9

Furthermore, improving surface drainage decreases drainage losses and denitrification but increases runoff losses. The effect on net mineralization is not evident. Net mineralization hardly increases when surface drainage is improved. The use of controlled drainage practices is associated with a reduction in drainage losses and an increase in denitrification and runoff losses, as compared to the use of conventional drainage practices (Tables 3-4). The effect of controlled drainage on net mineralization is not clear, but a modest decrease in mineralization can be observed for the 1.0 and 1.25 m drain depths. The reduction in total (subsurface drainage plus runoff) NO_3-N losses by having CD in these two soils is more marked during the winter season. This can be explained by the climatic conditions of eastern North Carolina where substantial drainage occurs in the winter due to low ET.

Clearly, the ideal drainage design and management combination is one that will optimize profits and minimize environmental impacts. The results from a cost-benefit analysis indicate that the optimum (most economical) system for the Tomotley soil is a conventional drainage system with a 25-m drain spacing, 1.25-m drain depth, and poor surface drainage conditions (Fig. 1). In the case of the Portsmouth soil, a conventionally drained system with poor surface drainage, 40-m spacing and 1.25-m drain depth is the optimum choice (Fig. 1). However, these optimum systems would not necessarily fully minimize detrimental impacts on water quality. The total nitrate-nitrogen losses associated with these two optimum systems are 19.5 and 35.7 kg/ha/yr for the Tomotley and Portsmouth soils, respectively (Tables 3-4).

Figure 2 shows the effect of drain spacing and system management on net profit and total loss of NO_3-N for both soils. These results illustrate the benefit of simulation modeling and the complexity of designing drainage systems to meet both environmental and production objectives simultaneously. Although it was found that the optimum spacings were about 25 and 40 m for the Tomotley and Portsmouth soils, respectively, a smaller spacing is more typical in real situations because drain spacing recommendations are usually based on conservative designs. These conservative systems are more expensive than necessary, but they satisfy the production objectives as indicated by their profitability. It is obvious that by adopting an optimum spacing, profits can be maximized, and, also importantly, total NO_3-N losses can be reduced substantially.

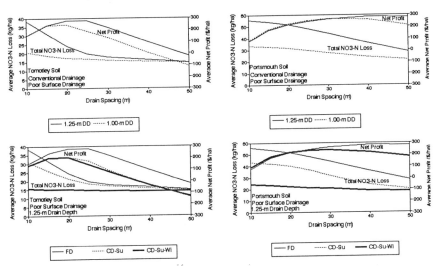

Figure 2. Predicted annual profit and total NO_3-N losses for a Portsmouth and Tomotley soils as affected by system management, and drain depth and spacing. Notation: CD=controlled drainage, FD=free drainage, Su=summer, and Wi=winter.

Furthermore, if an environmental objective is integrated, minimizing nitrate-nitrogen losses to receiving streams for instance, there is a strong incentive to decrease drainage intensity by increasing the drain spacing or decreasing the drain depth. For a Portsmouth soil with conventional drainage, poor surface conditions and a drain depth of 1.25 m, increasing the drain spacing from 40 to 50 m would reduce total NO_3-N losses by 21%, from 35.7 to 28.2 kg/ha (Table 3, Fig. 2). In the case of the Tomotley soil with conventional drainage, poor surface conditions, and a drain depth of 1.25 m, increasing the drain spacing from 25 to 30 m would decrease total NO_3-N losses by 8%, from 19.5 to 17.9 kg/ha (Table 4, Fig. 2). However, these changes are associated with a decrease in predicted profit. The estimated decrease in profit would be of $3.5/ha (from $282.1/ha to $278.6/ha, Fig. 2) for the Portsmouth soil and of $59.2/ha (from $280.2/ha to $221.0/ha, Fig. 2) for the Tomotley soil.

The nitrogen data presented in Tables 3-4 and Fig. 2 show that a substantial reduction in NO_3-N losses could be achieved by decreasing the drain depth of 1.25 m. If a 1.0 m drain depth is used, the projected reduction in total NO_3-N losses would be 33% and 14% for the Portsmouth and Tomotley soils, respectively. Nevertheless, this would result in a predicted loss in profit of about $19.8/ha for the Portsmouth soil (Fig. 2) and $90.2/ha for the Tomotley soil (Fig. 1). One other way of decreasing total NO_3-N losses is by improving the surface conditions, but such decrease is not substantial compared to the associated decrease in profit.

Simulated results also indicate that using controlled drainage could reduce total nitrate-nitrogen losses, with another sacrifice in profits, however. If controlled drainage is used in the summer only, total NO_3-N losses can be decreased by 32% for the Portsmouth soil (to 24.1 kg/ha) and by 6% for the Tomotley soil (to 18.3 kg/ha), as compared to the original conventionally drained system (Tables 3-4, Fig. 1). The decrease in profit associated with this water management modification is substantial for both soils ($62.1/ha and $107.9/ha for the Portsmouth and Tomotley soils, respectively). If controlled drainage is implemented during both the winter and summer seasons, total NO_3-N losses can decrease by 47% (to 18.8 kg/ha) for the Portsmouth soil and by 25% (to 14.6 kg/ha) for the Tomotley soil. Because the use of CD in the winter, if properly managed, may not affect the yields of the summer crop, the decrease in profit associated with this water management scheme (CD in summer and winter) is the same as that of the CD-in-summer scenario. This suggests that by having controlled drainage in the winter alone may not affect profits greatly, may decrease NO_3-N losses substantially, and, thus, may meet both the production and environmental objectives.

The overall simulation results have shown the practical applicability of *DRAINMOD-N* to identify drainage design and management combinations that could potentially satisfy both production and environmental objectives for a specific crop, soil and climatic condition. The implication of the results is that, from a societal point of view, it may become less expensive to pay greater prices for grain compared to treating the water to remove excessive NO_3-N and, therefore, mitigate detrimental environmental impacts.

SUMMARY AND CONCLUSIONS

The environmental impacts of agricultural drainage have become a critical issue in many areas. There is a need to design and manage drainage and related water table control systems to satisfy both production and water quality objectives. Simulation models are a powerful tool to evaluate the effects of different drainage design and management combinations on drainage water quality. One such a model, *DRAINMOD-N*, was used to study long-term effects of drainage system design and management on two poorly drained eastern North Carolina soils.

Hydrologic results indicate that, for any equivalent design-management combination, greater subsurface drainage and lower surface runoff are associated with the Portsmouth soil due to its greater lateral conductivity, as compared to the Tomotley soil. Increasing the drain spacing and/or decreasing the drain depth reduces drainage while it increases runoff. Improving surface drainage causes an increase in runoff and a decrease in subsurface drainage. The use of controlled drainage (CD) is associated with a reduction in subsurface drainage and an increase in surface runoff. Water quality results show that, in general, net mineralization, rainfall deposition, plant uptake, and drainage losses of nitrate-nitrogen are greater for the Portsmouth soil, while denitrification and runoff losses are greater for the Tomotley soil. Results also indicate that increasing the drain spacing and/or decreasing the drain depth reduces NO_3-N drainage losses and net mineralization, but increases denitrification and runoff losses. Improving surface drainage decreases drainage losses and increases denitrification and runoff losses. Controlled drainage is associated with a reduction in mineralization and drainage losses and an increase in denitrification and runoff losses.

The ideal drainage design and management combination is one that will optimize profits and minimize environmental impacts. Results from an economic analysis indicate that the optimum system for the Portsmouth soil, based on the design and management variables studied, is a conventional drainage system with a 25-m drain spacing, 1.25-m drain depth, and poor surface drainage conditions. In the case of the Portsmouth soil, a conventionally drained system with poor surface drainage, 40-m spacing, and 1.25-m drain depth is the optimum choice.
If an environmental objective is integrated, minimizing nitrate-nitrogen losses to receiving streams for instance, simulated results show that there is a strong incentive to decrease drainage intensity by increasing the drain spacing or decreasing the drain depth. However, a cost-benefit analysis indicates such design modifications are associated with a decrease in profit. Similarly, the use of controlled drainage could also reduce the nitrate-nitrogen losses, but with another sacrifice in profit. The implication of the results is that, from a societal point of view, it may become less expensive to pay greater prices for grain compared to treating the water to remove excessive NO_3-N and, therefore, mitigate detrimental environmental impacts. The overall simulation results have shown the practical applicability of *DRAINMOD-N* to identify drainage design and management combinations that could potentially satisfy both production and environmental objectives for a specific crop, soil and climatic condition.

REFERENCES

Bottcher, A.B., E.J. Monke, and L.F. Huggins. 1981. Nutrient and sediment loadings from a subsurface drainage system. *Transactions of the ASAE*, 24:1221-1226.

Brevé, M.A., R.W. Skaggs, J.E. Parsons, and J.W. Gilliam. 1993. *DRAINMOD-N*, a nitrogen model for artificially drained soils: Development. To be submitted to *Transactions of the ASAE*.

Feddes, R.A. 1987. Simulating water management and crop production with the SWACRO model. *Proc. Third International Workshop on Land Drainage*, A27-A40, The Ohio State Univ., Columbus, OH.

Gilliam, J.W. 1987. Drainage water quality and the environment. Keynote address. In *Proc. Fifth National Drainage Symposium*, ASAE Publ. No. 7-87, 19-28, St. Joseph, MI: ASAE.

Kandil, H., C.T. Miller and R.W. Skaggs. 1992. Modeling long-term solute transport in drained, unsaturated zones. *Water Resources Research* 28:2797-2809.

Karvonen, T. and R.W. Skaggs. 1993. Comparison of different methods for computing drainage water quantity and quality. In *Proc. Workshop on Subsurface Drainage Simulation Models*, The Hague: ICID (In press).

Pierce, F.J., M.J. Shaffer, and A.D. Halvorson. 1991. Screening procedure for estimating potentially leachable nitrate-nitrogen below the root zone. In *Managing Nitrogen for Groundwater Quality and Farm Profitability*, ed. R.F. Follet, D.R. Keeney and R.M. Cruse, 259-283. SSSA, Inc., Madison, WI.

Schepers, J.S. and A.R. Mosier. 1991. Accounting for nitrogen nonequilibrium soil-crop systems. In *Managing Nitrogen for Groundwater Quality and Farm Profitability*, ed. R.F. Follet, D.R. Keeney and R.M. Cruse, 125-138. SSSA, Inc., Madison, WI.

Skaggs, R.W. 1978. A water management model for shallow water table soils. Rept. 134, North Carolina Water Resour. Res. Inst., North Carolina State Univ., Raleigh, NC.

Skaggs, R.W. 1987. Design and management of drainage systems. Keynote address. In *Proc. Fifth National Drainage Symposium*, ASAE Publ. No. 7-87, 1-12, St. Joseph, MI: ASAE.

Skaggs, R.W. 1991. Drainage. In *Modeling Plant and Soil Systems*, 205-243, Agronomy Monograph No. 31, ASA-CSSA-SSSA, Madison, WI.

Skaggs, R.W. and J.W. Gilliam. 1981. Effect of drainage system design and operation on nitrate transport. *Transactions of the ASAE* 24:929-934.

Skaggs, R.W. and A. Nassehzadeh-Tabrizi. 1986. Design drainage rates for estimating drain spacings in North Carolina. *Transactions of the ASAE* 29:1631-1640.

Skaggs, R.W., M.A. Brevé and J.W. Gilliam. 1994. Hydrologic and water quality impacts of agricultural drainage. *Critical Reviews in Environmental Science and Technology* (In press).

Skaggs, R.W., T. Karvonen, and H. Kandil. 1991. Predicting soil water fluxes in drained lands. ASAE Paper No. 91-2090, St. Joseph, MI: ASAE.

MODELING EROSION FROM FURROW IRRIGATION IN THE WEPP MODEL

E. R. Kottwitz, J. E. Gilley*

ABSTRACT

The Water Erosion Prediction Project (WEPP) model employs hydrology, hydraulics and erosion components to estimate erosion resulting from furrow irrigation. The hydrology component uses a two-dimensional Kostiakov-Lewis infiltration equation. To reduce model execution time, parameters for the Kostiakov-Lewis equation are obtained from a least squares regression analysis of time-cumulative infiltration pairs developed from a physically based infiltration model. Hydraulics of all phases of a furrow irrigation event are modeled using kinematic wave theory. The erosion component estimates detachment, transport, and deposition along the entire length of the furrow. Net detachment occurs when critical shear stress of the soil is exceeded and sediment load is less than transport capacity. The model predicts net deposition when sediment load is greater than transport capacity. Inflow management options considered by the model include continuous, cut-back, and surge flow. The WEPP model can be used to identify irrigation, cropping, and management alternatives which best meet established erosion guidelines while providing reasonable irrigation system performance. **KEYWORDS.** Erosion models, Erosion control, Sediment transport, Sediment discharge, Furrows, Irrigation, Irrigation management.

INTRODUCTION

The use of irrigation has greatly increased agricultural productivity in the United States. Long term soil productivity, however, may be influenced by irrigation induced erosion. It has been estimated that approximately 3 million hectares of irrigated land in the United States are affected by soil erosion (Koluvek et al., 1993).

Erosion on furrow irrigated areas has long been recognized as a serious problem. Irrigation induced erosion may result in loss of soil productivity and decreased potential profits for farmers who irrigate (Carter, 1990). Erosion may also affect water quality (Deason, 1989). At present, reliable procedures for estimating irrigation induced erosion are not available.

The WEPP model was developed to provide erosion prediction technology for use in soil and water conservation and environmental planning and assessment. Recent advances in irrigation, hydrologic, hydraulic, and erosion sciences are incorporated into the model. The effects of climate, soil and topography on erosion, deposition and sediment transport are identified.

The WEPP model can be used to evaluate a broad range of irrigation, cropping, and management conditions which may not be practical or economical to examine in the field. This technology can be utilized to identify procedures for controlling furrow irrigation induced erosion which will result in improved water quality, sustained soil productivity, and increased farmer profits. The objective of this paper is to describe how furrow irrigation induced erosion is estimated in the WEPP model.

GOVERNING EQUATIONS AND METHODOLOGY

Hydrology

Infiltration into an irrigation furrow is a three-dimensional process. This process can be simplified to a two-dimensional form if infiltration opportunity time at each location within the furrow is known. Numerous models are available for estimating infiltration. The Green and Ampt (1911) equation is frequently used to predict one-dimensional flow. Considerable work has been performed to improve the original Green and Ampt equation and to predict appropriate parameters from physical characteristics of the soil (Rawls et al., 1983).

---

* E. R. Kottwitz, Research Engineer, Dept. of Biological Systems Engineering; and J. E. Gilley, Agricultural Engineer, USDA-ARS, University of Nebraska, Lincoln.

Fok and Chiang (1984) presented a two-dimensional infiltration function useful for furrow irrigated conditions (Fig. 1). To use their function, Green and Ampt infiltration parameters must be known. The function assumes two regions of one-dimensional horizontal infiltration, one region of one-dimensional vertical infiltration, and two regions of two-dimensional infiltration. The two-dimensional regions are assumed to be quarter ellipses. Using these assumptions, cumulative infiltrated volume per unit furrow length, Z, is given by

$$Z = \left[2 d I_x + b I_y + \left(\frac{\pi}{2}\right) I_x I_y \right] \Delta\theta \qquad (1)$$

where d = depth of flow; I_x = horizontal advance distance of the wetting front; b = bottom width of the furrow; I_y = vertical advance distance of the wetting front; and $\Delta\theta$ = net change in soil water content.

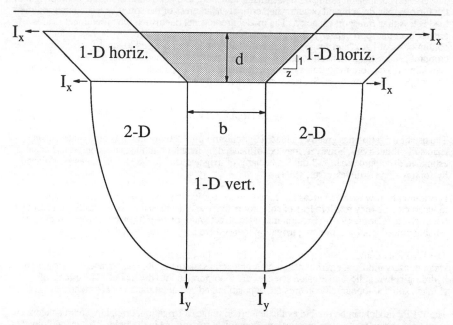

Figure 1. Two-Dimensional Infiltration Pattern.

The equation of Green and Ampt (1911), the basis for estimating vertical advance distance, is given as:

$$\frac{K_e \tau}{\Delta\theta h_y} = \frac{I_y}{h_y} - \ln\left(1 + \frac{I_y}{h_y}\right) \qquad (2)$$

where K_e = effective hydraulic conductivity; τ = infiltration opportunity time; and h_y = total head loss in the vertical direction. The equation for horizontal advance distance is given by Fok and Chiang (1984) as

$$I_x = \left(\frac{2 K_e h_x}{\Delta\theta}\right)^{1/2} \tau^{1/2} \qquad (3)$$

where h_x = total head loss in the horizontal direction. The value of h_x is assumed to be equal to h_y.

Because the hydraulic model component described below must be solved numerically, it is not computationally efficient to use the Green and Ampt infiltration function which also requires a numerical solution. To improve efficiency, parameters for the Kostiakov-Lewis infiltration function are calibrated to a set of time-cumulative infiltration pairs. The Kostiakov-Lewis infiltration function, which is often used in computer models, is given as

$$Z = k\,\tau^a + f_o\,\tau \qquad (4)$$

where k, a, and f_o are empirical parameters. The value of f_o is assumed to be equal to K_c times maximum wetted perimeter.

The largest infiltration opportunity time used for the time-cumulative infiltration pairs is the summation of all inflow durations. The numerical solution of the Green and Ampt infiltration function is used to identify vertical advance distance for this opportunity time. This vertical advance distance is then divided by the number of time-infiltration pairs. The resulting value is used to incrementally increase vertical infiltration distance. The time required for the wetting front to advance a given distance can be determined directly using a form of Eq. (2). From the calculated time, it is possible to identify the horizontal advance distance of the wetting front, then cumulative infiltration, using Eqs. (3) and (1), respectively. In this manner, all time-cumulative infiltration pairs can be determined from only one numerical solution of the Green and Ampt function.

The use of horizontal advance distance in equations derived by Fok and Chiang (1984) requires appropriate adjustments if horizontal wetting fronts meet prior to the completion of irrigation. Horizontal wetting fronts are assumed to meet when I_x equals a maximum horizontal distance, $(I_x)_{max}$, given by

$$(I_x)_{max} = \frac{W - b}{2} \qquad (5)$$

where W = distance between the center lines of irrigated furrows. A consequence of basing the definition of $(I_x)_{max}$ on b is that, for a trapezoidal furrow, some overlapping of the horizontal wetting fronts occurs. The water contained in the overlapping regions is assumed to move upward into ridges located between furrows. The volume of water in the overlapping regions is small compared to the total infiltrated volume. Thus, the effect of this assumption on cumulative infiltration is small. If horizontal wetting fronts are found to meet, the incremental vertical advance distance is assumed to occur over a representative width equal to W.

The Kostiakov-Lewis infiltration parameters are estimated using a least squares regression analysis of time-cumulative infiltration pairs. The technique is similar to that used by James et al. (1985) for exponential functions. This procedure weights errors obtained during the least squares analysis.

Hydraulics

In the WEPP model, the hydraulics of a complete irrigation event are simulated using principles of conservation of mass (continuity) and kinematic wave theory. The kinematic wave approach can be executed rapidly and has been shown to provide reasonable results for slopes greater than 0.1% (Walker and Humpherys, 1983). Rayej and Wallender (1985) decreased computer processing time and the amount of required memory by using variable time steps. A specified space interval can also be used to reduce processing time and memory requirements.

The continuity equation is given as

$$\frac{\partial Q}{\partial x} + \frac{\partial A}{\partial t} + \frac{\partial Z}{\partial t} = 0 \qquad (6)$$

where Q = flow rate; x = downslope distance; A = cross-sectional flow area; and t = time. A kinematic wave approximation of the momentum equation is used, resulting in the assumption that

$$S_f = S_o \tag{7}$$

where S_f = friction slope and S_o = furrow slope.

A relationship between Q and A is given by the Chezy equation which is written as

$$Q = c\, A\, R^{1/2}\, S_o^{1/2} \tag{8}$$

where c = Chezy roughness coefficient and R = hydraulic radius. A power function relationship between Q and A used in previous studies (Elliott et al., 1982; Walker and Humpherys, 1983; and Rayej and Wallender, 1987) is given as

$$Q = \alpha\, A^m \tag{9}$$

where α and m are empirical coefficients. The values for α and m depend on c, S_o, and the shape of the furrow (Walker, 1989). The model allows the user to specify values for α and m directly or to provide b, furrow side slope, z, and c, from which the model determines α and m. Procedures for determining α and m are similar to those used to identify the Kostiakov-Lewis infiltration parameters k and a.

Bassett et al. (1983) identified four phases of a typical irrigation event: advance, continuing, depletion, and recession (Fig. 2). Advance is that portion of the irrigation event during which water is supplied to the furrow but has not yet progressed to the end of the field. The continuing phase begins when water reaches the end of the furrow and concludes when inflow ceases. The period between cessation of inflow and the time at which flow area at the upper end of the furrow decreases to zero is the depletion phase. The recession phase is defined by the end of the depletion phase and the disappearance of water from the soil surface. If water does not reach the end of the furrow, the continuing phase does not occur and water may continue to move downslope during the depletion and recession phases.

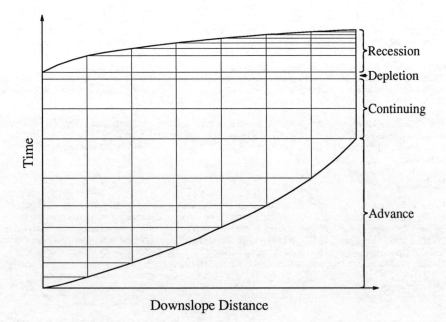

Figure 2. Furrow Irrigation Phases and Simplified Computational Grid.

Erosion

Methods for predicting erosion on croplands using process-based models have progressed gradually since mathematical descriptions of erosion mechanisms were introduced by Meyer and Wischmeier (1969). Foster and Meyer (1972) presented a relationship between detachment rate and the ratio of sediment load to transport capacity. Meyer et al. (1975) evaluated detachment and transport on rill and interrill areas, a critical step in the prediction of erosion due to furrow irrigation.

In the WEPP model, erosion is estimated using an approach identified by Nearing et al. (1989). This procedure uses a steady-state sediment continuity equation to describe the movement of suspended sediment. Two hydrologic variables, peak runoff rate, P_r, and effective runoff duration, are required input parameters. Effective runoff duration is the ratio of runoff volume to peak runoff rate.

The steady-state sediment continuity equation for suspended sediment is given by

$$\frac{dG}{dx} = D_f + D_i \tag{10}$$

where G = sediment load; D_f = furrow erosion rate; and D_i = interrill erosion rate. For irrigated furrows, $D_i = 0$. The furrow erosion rate is greater than zero if detachment is taking place and less than zero if deposition is occurring. The calculations for G and D_f are expressed on a per unit area basis.

Soil detachment occurs when shear stress exceeds the critical value for the soil and sediment transport capacity is greater than sediment load. For the detachment process

$$D_f = D_c \left(1 - \frac{G}{T_c}\right) \tag{11}$$

where T_c = sediment transport capacity. Detachment capacity, D_c, is given by

$$D_c = K_r (\tau_f - \tau_c) \tag{12}$$

where K_r = furrow erodibility; τ_f = shear stress acting on soil particles; and τ_c = critical hydraulic shear stress for the soil. Sediment transport capacity, calculated using the Yalin equation, is given by

$$T_c = k_t \, \tau_f^{3/2} \tag{13}$$

where k_t = transport coefficient.

Whenever possible, field measurements should be used to obtain estimates of K_r and τ_c. If in-situ measurements of K_r and τ_c are not available, regression equations which use soil physical properties can be employed to approximate these parameters.

Net deposition occurs when the sediment load exceeds transport capacity. For the deposition process

$$D_f = \frac{V_f}{q}(T_c - G) \tag{14}$$

where V_f = effective fall velocity for the sediment and q is flow rate per unit width. For a rectangular furrow

$$q = \frac{Q}{b} \tag{15}$$

The flow shear stress function, derived assuming that discharge varies linearly with downslope distance, is given by

$$\tau_f = \gamma \left(\frac{P_r \times S_o}{c} \right)^{2/3} \tag{16}$$

where γ = specific weight of water. The intermediate steps in calculating sediment load as a function of downslope distance are made using nondimensional equations. The final solution is then redimensionalized. Nearing et al. (1989) provide additional information related to the nondimensional erosion equations and their solution.

RELATED TOPICS

Inflow Management

Several approaches to furrow inflow management exist. Each inflow management technique has a unique set of advantages and disadvantages that must be considered in the selection or evaluation process. Inflow management approaches the furrow erosion model is capable of simulating include constant inflow, cut-back inflow, and surge inflow.

Constant inflow, the simplest of the three methods, is continuous application at a single flow rate. A cut-back inflow management option may also be employed. This irrigation option uses a continuous supply of water to the furrow but the flow rate is reduced when the water reaches the end of the field or after a selected delay period.

Surge irrigation is the most complicated of the three inflow management practices. Surge irrigation consists of a series of intermittent inflows to a furrow (Stringham and Keller, 1979). The surges may be of constant or variable duration (Israeli, 1988). During the advance surges, the final advance distances seldom correspond with the uniform space intervals initially established (Fig. 2). This difficulty is resolved by shifting the nearest grid boundary to the location corresponding to the final advance distance. Overlapping advance and recession waves of successive surges are not considered by the model.

Data Requirements

Variations in soil properties commonly occur within a single irrigation furrow (Baustista and Wallender, 1985). Thus, multiple overland flow elements (OFE's) are used in the model. An OFE is a section of the furrow which has homogeneous soil, cropping, and management practices.

The WEPP irrigation model components are designed to allow the user to provide empirical information that may be available. Although the model can determine parameters for the Kostiakov-Lewis infiltration function, the user may enter these values directly. The user may also enter parameters for the power function relationship between Q and A shown in Eq. (9).

Output Information

The WEPP model predicts runoff and erosion from a field for a specific set of conditions. By changing input parameters and running the model again, the user can quickly obtain results for different management options. Using this information, the model user is able to evaluate the effects of selected management alternatives on erosion.

Model output information includes flow area versus time, cumulative infiltration versus time, runoff hydrograph, and total runoff volume. Varying set times, flow rates, or slope lengths will affect irrigation system performance and result in different runoff characteristics. Erosion output information includes estimates of net soil loss along the entire length of the furrow as well as sediment delivery from the end of the furrow. By comparing output from various model runs, those management alternatives that meet existing erosion guidelines can be determined. The model user will then be able to identify management alternatives which provide reasonable irrigation system performance and meet established erosion guidelines.

SUMMARY

The furrow erosion model contains three principal components: hydrology, hydraulics, and erosion. The hydrology component uses a two-dimensional Kostiakov-Lewis infiltration function. The procedure of Fok and Chiang (1984) is used to obtain a two-dimensional infiltration pattern from effective hydraulic conductivity, total head loss, and change in soil water content. Adjustments are made for overlapping of the horizontal wetting fronts. Parameters for the two-dimensional Kostiakov-Lewis infiltration function are calibrated to closely match results obtained using the procedure of Fok and Chiang (1984). This reduces the need for computationally inefficient numerical procedures.

The hydraulics component is based on a kinematic wave approximation of the complete hydrodynamic method and uses specified space intervals. A power function relationship between flow rate and flow area is assumed. All phases of the irrigation event (advance, continuing, depletion, and recession) are simulated. For conditions where a continuing phase does not exist, the model estimates lower end advance or recession during the depletion and recession phases.

The erosion component is based on procedures reported by Nearing et al. (1989). This approach uses a steady-state assumption for the sediment continuity equation. Required hydrologic variables are peak runoff rate and effective runoff duration. Soil detachment, transport and deposition processes are considered.

The model can be used with constant inflow, cut-back inflow, and surge inflow management practices. Including the surge inflow option requires that the model be able to shift the boundaries of the computational grid to match the progressive distances of the advance surges. Overlapping advance and recession waves of successive surges are not considered by the model.

Data requirements are limited to commonly available physical parameters. If the Kostiakov-Lewis infiltration function or the power function relationship between flow rate and flow area are known, fewer physical parameters are required. The irrigation furrow is divided into overland flow elements which have homogeneous soil, cropping, and management characteristics.

Runoff characteristics predicted by the model include flow area versus time, cumulative infiltration versus time, runoff hydrograph, and total runoff volume. Estimates of net soil loss along the entire length of the furrow and sediment delivery from the end of the furrow are also provided as output. Through repeated use of the model, the user is able to determine which management alternatives provide reasonable irrigation system performance and meet established erosion guidelines.

REFERENCES

1. Bassett, D. L., D. D. Fangmeier and T. Strelkoff. 1983. Hydraulics of surface irrigation. In M. E. Jensen (ed.) Design and operation of farm irrigation systems, revised printing. ASAE, St. Joseph, Michigan.

2. Baustista, E. and W. W. Wallender. 1985. Spatial variability of infiltration in furrows. Transactions of the ASAE, 28(6): 1846-1851, 1855.

3. Carter, D. L. 1990. Soil erosion on irrigated lands. In B. A. Stewart and D. R. Nielsen (eds.) Irrigation of Agricultural Crops. Agronomy Monograph 30, American Society of Agronomy, Madison, Wisconsin.

4. Deason, J. P. 1989. Irrigation induced contamination: how real a problem? Journal of Irrigation and Drainage Engineering, ASCE, 115(1): 9-20.

5. Elliott, R. L., W. R. Walker and G.V. Skogerboe. 1982. Zero-inertia modeling of furrow irrigation advance. Journal of Irrigation and Drainage Engineering, ASCE, 108(3): 179-195.

6. Fok, Y. S. and S. H. Chiang. 1984. 2 D infiltration equations for furrow irrigation. Journal of Irrigation and Drainage Engineering, ASCE, 110(2): 208-217.

7. Foster, G. R. and L. D. Meyer. 1972. A closed form soil erosion equation for upland areas. In H. W. Shen (ed.) Sedimentation. Ft. Collins, Colorado.

8. Green, W. H. and G. A. Ampt. 1911. Studies in soil physics. I. The flow of air and water through soils. Journal of Agricultural Science, 4: 1-24.

9. Israeli, I. 1988. Surge irrigation guide. Colorado State University Cooperative Extension Bulletin 543A.

10. James, M. L., G. M. Smith and J. C. Wolford. 1985. Applied Numerical Methods for Digital Computation. Harper and Row, Publishers, New York.

11. Koluvek, P. K., K. K. Tanji and T. J. Trout. 1993. Journal of Irrigation and Drainage Engineering, ASCE, 119(6): 929-946.

12. Meyer, L. D., G. R. Foster and M. J. M. Romkens. 1975. Source of soil eroded by water from upland slopes. In Present and Prospective Technology for Predicting Sediment Yields and Sources. U.S. Department of Agriculture, Agricultural Research Service, ARS-S-40.

13. Meyer, L. D. and W. H. Wischmeier. 1969. Mathematical simulation of the process of soil erosion by water. Transactions of the ASAE, 12(6): 754-758.

14. Nearing, M. A., G. R. Foster, L. J. Lane and S. C. Finkner. 1989. A process based soil erosion model for the USDA water erosion prediction project technology. Transactions of the ASAE, 32(5): 1587-1593.

15. Rawls, W. J., D. L. Brakensiek and B. Soni. 1983. Agricultural management effects on soil water processes. Part I. Soil water retention and Green and Ampt infiltration parameters. Transactions of the ASAE, 26(6): 1747-1752.

16. Rayej, M. and W.W. Wallender. 1987. Furrow model with specified space intervals. Journal of Irrigation and Drainage Engineering, ASCE, 113(4): 536-548.

17. Stringham, G. E. and J. Keller. 1979. Surge flow for automatic irrigation. Proc. 1977 Irrigation and Drainage Specialty Conference, ASCE, Albuquerque, New Mexico.

18. Walker, W.R. and A.S. Humpherys. 1983. Kinematic-wave furrow irrigation model. Journal of Irrigation and Drainage Engineering, ASCE, 109(4): 377-392.

19. Walker, W. R. 1989. Guidelines for designing and evaluating surface irrigation systems. Food and Agriculture Organization of the United Nations Irrigation and Drainage Paper 45.

IS CONSERVATION TILLAGE A SUSTAINABLE AGRICULTURAL PRACTICE?

P.G. Lakshminarayan, Aziz Bouzaher, and S.R. Johnson [*]

ABSTRACT

Conservation tillage is known to be an environmentally sound agricultural practice for controlling soil erosion. However, use of conservation tillage has been associated with increased use of chemicals, posing potential water quality problems. The challenge is to identify conservation tillage practices that could achieve soil and water quality simultaneously. An approach to addressing this issue is to conduct field studies and monitoring for a variety of physical and management conditions. This approach, is in general, not practical. In this paper we propose a novel approach based on building metamodels (response functions) for outputs (soil and water quality indicators) generated through simulation of process models, using experimental design techniques. These metamodels are then used to predict soil and water quality and for infinite number of site-specific conditions and statistically aggregate them to regional levels.

We applied this methodology to predict soil erosion and nitrate-N runoff/leaching from conventional and conservation tillage practices for two major cropping systems (continuous corn and corn-soybean) in a selected watershed within the U.S. Corn Belt. The results indicate that conservation tillage, which reduces erosion and nitrate-N runoff, results in increased leaching of nitrate-N, especially in a corn-soybean rotation. An important conclusion is that conservation tillage, combined with proper nutrient management including soil testing and taking adequate credit for N fixed by soybeans, could become a key sustainable agricultural practice.
KEYWORDS. Erosion, Nitrates, Metamodels, Simulation, Tillage.

INTRODUCTION

Nonpoint Source (NPS) pollution of ground and surface water resources is a national concern. Agriculture is the major source of nonpoint contamination of rivers and lakes (USEPA 1992). Because of continuing NPS pollution, and despite more than a decade of policy, research, and intervention, the relationship between agricultural production and environmental quality is still the subject of ongoing debate. The conflicts among environmental objectives, such as soil erosion and water quality, make the management of agricultural resources more difficult. For instance, conservation tillage that substitutes chemical weed control for tillage may lead to increased water quality problems (Hinkle 1983). Alternative best management practices are being developed to mitigate the NPS pollution threat. Proper management of any system requires an assessment of environmental impacts of alternatives being considered.

Mounting environmental concerns and problems of sustainable agricultural productivity increases, linked to improvements in environmental quality, underlie the need for added research and methods for supporting innovative integrated economic and environmental management. With the growing focus on ecosystems and policy trade-offs the tools for analysis must have the capacity to address regional as well as local and farm-level

---

[*] P.G. Lakshminarayan, Assistant Scientist, Aziz Bouzaher, Resources and Environmental Policy Division Head, and S.R. Johnson, Director and Distinguished Professor, Center for Agricultural and Rural Development, Iowa State University, Ames, IA 50011.

management. A simple and scientifically valid statistical approach for extrapolation and aggregation of these local and site-specific attributes to regional levels is proposed. The approach, *metamodeling*, uses statistically validated response functions fitted to the outputs of process models (Bouzaher et al. 1993).

This paper is organized into three sections. Section 2 briefly describes the simulation method, sampling design, and metamodeling. The results are summarized in Section 3.

EMPIRICAL MODELS AND METHODS

Monitoring and simulation modeling are two approaches to assessing water quality. Monitoring is the first-best option, but the usefulness of monitoring data depends on the design and implementation of the monitoring effort. For regional soil and water quality assessments, a large network of monitoring stations is required, which makes monitoring impractical because of time and resource limitations. An alternative approach is to use process models. These are mathematical models that describe and simulate the management, physical, chemical, and biological processes impacting the real-life system being modeled. Recently, process models have been used widely to estimate ex ante nonpoint pollution impacts of alternative production and management practices. These models consider site-specific attributes including land use patterns and management practices. The availability of superior computing capabilities has enabled these models to simulate the real-life processes in significant detail. Combining the process models, statistical design, and metamodeling, the spatial distribution of environmental impacts can be approximated adequately.

Simulation Plan

The process model used in this analysis is EPIC (Erosion Productivity Impact Calculator) developed by USDA (Jones et al. 1991). EPIC was designed to analyze alternative cropping systems and to project their socioeconomic and environmental impacts with specific reference to soil erosion, water quality, and crop productivity. Soil erosion and nitrate-nitrogen runoff/leaching are obtained by simulating EPIC according to an experiment designed to capture the range of management, soil, and weather parameters in the study region.

On a watershed scale, even the simulation approach is unmanageable and impracticable because of extensive simulation runs required to cover different combinations of soil, climate, hydrology, and management alternatives. This extensive coverage is required to capture the heterogeneity of the physical environment as well as the agricultural practices so that a meaningful site-specific assessment is possible. Alternatively, a spatial sampling design, which will reduce the simulation runs considerably and retain the statistical validity of aggregation, is adopted.

The SOILS-5 (Soil Interpretation Records [SIRS]) database from which the soil sample was drawn includes layered (soil profile) information. A typical soil in this database is characterized by physical and hydrologic factors, such as % clay, % sand, permeability, organic matter, pH, bulk density, % slope, k-factor, available water, and hydrological groups A to D (classified on the basis of rate of infiltration, with group A having the maximum infiltration and group D having the minimum infiltration). A stratified random sampling with a complete factorial design was used. Five soil factors —% clay, bulk density, permeability, pH, and k-factor —were used for stratification. The selected factors were stratified into three levels, as high, medium, and low, and 4 soils were sampled from each of the 15 strata (3 levels and 5 factors) *without replacement*.

In summary, the resulting sample was *balanced* (soils in all levels of each property are represented at similar proportions), *self-weighting* (each cropped acre in the watershed had equal probability of selection), and *representative* of the population of soils in the watershed. EPIC also requires daily weather data such as precipitation, maximum and minimum temperatures, solar radiation, and relative humidity. The simulations were conducted over 15 years using historical weather data from EPIC weather stations. Preliminary calibration runs suggested that the environmental indicators reached steady state after 8 to 10 years; therefore, 15 years is a reasonable time for predicting long-term consequences.

To evaluate the environmental impacts of alternative BMPs, the simulation plan includes alternative management and cultural practices. Continuous corn and corn-soybean rotation are the two major cropping systems simulated. Four tillage practices based on residue cover were simulated in EPIC —conventional tillage with fall plow, conventional tillage with spring plow, reduced tillage, and no-till. The reduced and no-till practices represent "conservation tillage." Optimum nitrogen (N) application and the crop residue cover rates were obtained from *Tillage Update* published by the Resource and Technology Division, USDA. The schedule of operations — management, tillage, and harvest — were taken from FEDS (Firm Enterprise Data System) budget.

The simulation methodology was applied in a major watershed in the U.S. Corn Belt. Specifically, the hydrologic area representing the water resources aggregate subarea 703 (WRC 1970) is the area of study. This watershed comprises most of central and eastern Iowa. It has nearly 25 million acres of cropland (6% of the national cropped area). In 1990 it produced $4.3 billion worth of crops, representing about 5% of total U.S. crop receipts. Corn and soybeans are the two major crops and account for 21 million acres.

Metamodeling

Metamodeling is a statistical method used to approximate outcomes of a process model. While the complex simulation model is a tool to approximate the underlying real-life system, the analytic metamodel attempts to approximate and aid in the interpretation of the simulation model and ultimately the real-life system (Kleijnen 1987). A metamodel is a regression model explaining the input-output relationship of a process model. Ideally, it would be preferable to experiment with the real-life system rather than a simulation model of the system. In that case we would have a statistical model of the system rather than a metamodel. This approach is not adopted because it would mean incurring the cost and delay of waiting, in this case for 15 years of weather to present itself to the real-life system.

RESULTS AND DISCUSSION

In this section, EPIC-simulated soil erosion and water quality indicators are summarized followed by the metamodels and predictions for site-specific physical and management conditions. The results are validated with actual measurements wherever possible.

Summary of Simulation Output

The summary statistics —mean, standard deviation, minimum, and maximum —of EPIC-simulated soil loss, nitrate-N runoff/leaching for alternative management systems are shown in Table 1. Soil erosion, as measured by the Modified Universal Soil Loss Equation (MUSLE), is reported. The amount of soil erosion is smaller for conservation tillage practices compared with conventional tillage. The mean soil loss for no-till continuous corn was 90 percent lower and for the corn-soybean rotation it was 70 percent lower.

Table 1. EPIC Simulated Long Term Average Soil Erosion and Nitrate-N Runoff and Leaching by Cropping Systems for Alternative Tillage Practices

Indicator / Cropping System	Tillage Practice	Mean	Standard Deviation	Minimum	Maximum
Soil Erosion		(tons/ha)			
Continuous Corn	Fall Plow	69.28	93.27	0.58	670.26
	Spring Plow	64.55	85.30	0.47	609.39
	Reduced Till	48.27	101.78	0.20	761.92
	No-Till	6.49	22.51	0	156.34
Corn-Soybeans	Fall Plow	90.05	103.83	0.99	721.27
	Spring Plow	80.13	89.65	0.88	614.37
	Reduced Till	70.80	113.02	0.61	842.89
	No-Till	26.71	51.22	0.16	382.50
Nitrate-N in Runoff		(mg/L)			
Continuous Corn	Fall Plow	7.37	3.44	2.00	14.00
	Spring Plow	7.14	3.39	2.00	14.00
	Reduced Till	6.05	3.39	2.00	13.00
	No-Till	1.66	1.65	0.00	10.00
Corn-Soybeans	Fall Plow	5.86	2.42	2.00	11.00
	Spring Plow	5.74	2.36	2.00	11.00
	Reduced Till	5.77	2.56	2.00	12.00
	No-Till	4.79	2.71	1.00	11.00
Nitrate-N in Percolate		(mg/L)			
Continuous Corn	Fall Plow	2.12	3.99	0.22	23.00
	Spring Plow	2.05	3.68	0.20	20.00
	Reduced Till	2.87	3.70	0.28	19.00
	No-Till	2.36	2.67	0.46	16.00
Corn-Soybeans	Fall Plow	1.73	2.28	0.00	13.00
	Spring Plow	1.84	2.13	0.34	11.00
	Reduced Till	2.19	2.37	0.42	15.00
	No-Till	9.43	10.72	0.60	42.00

Simulated soil erosion indicates that corn-soybean rotation is the most erosive cropping system. Erosion from a corn-soybean rotation is often greater than from continuous corn because of loss of residue cover after soybean harvest, exposing the topsoil to the impact of raindrops and the deterioration of the soil aggregate stability associated with soybean cropping (Corak and Kaspar 1990).

Mean concentration of nitrate-N in runoff is lower under conservation tillage practices compared with conventional tillage, while the concentrations in percolate are higher for conservation tillage systems. This result is supported by the actual measurements at the experimental sites in Nashua watershed in northeast Iowa. Surface runoff measurements at this site reveal that the average concentration of nitrate-N in runoff was greatest from the moldboard-plowed plots than from the conservation tillage plots. Average annual measurements of nitrate-N in runoff in the Roberts Creek watershed in northeast Iowa, with 49 percent row crop (mostly corn under fall plow), was 8 parts per million (mg/L) (Seigley et al. 1993).

A long-term weighted average concentration of nitrate-N in surface water was estimated as 6.1 mg/L (assuming that the concentration in runoff is an approximation for surface water concentration). The weights are the historical proportions of alternative tillage practices. Keeney and DeLuca (1993) reported an average flow of nitrate-N of 5.6 mg/L in the Des Moines River during the period 1980-91. The average leaching of nitrate-N in a continuous corn system was as high as 2.87 mg/L under reduced tillage, and in corn-soybean rotation it was as high as 9.43 mg/L under no-till. A USGS (1993) study of near-surface aquifers in Iowa reported concentrations in the range of 0.05 to 12 mg/L.

Estimated Regression Metamodels

Metamodels were fitted to EPIC-simulated soil erosion, nitrate-N runoff/leaching. Regression model development requires close examination of the data so that the prior information contained in the data is fully utilized. Data diagnosis is necessary to avoid model mis-specification and bias. Therefore, it is a good practice to examine the distributions and residual plots from the ordinary least squares model to decide whether the data require any transformation. Experience in building metamodels for herbicide leaching strongly suggests the need for data transformation (Bouzaher et al. 1993).

An initial linear model for soil erosion indicated that the error terms are nonrandom, suggesting heteroskedasticity (nonconstant variance). This problem was resolved by taking a cubic-root transformation of the data, resulting in the following model:

$$(\widehat{\text{soil loss}}_{CC})^{1/3} = -5.79 + 0.20(\%\text{slope}) + 3.86(K_f\text{factor}) + 0.07(\text{OM}) - 0.25(\text{pH})$$
$$+ 0.01(\text{rainfall}) + 0.06(\text{RCN}) - 0.25(\text{c-prac}) - 0.12(\text{residue})$$
$$\text{Adjusted } R^2 = 0.82 \text{ and RMSE} = 0.68.$$

$$(\widehat{\text{soil loss}}_{CS})^{1/3} = -5.87 + 0.23(\%\text{slope}) + 5.23(K_f\text{factor}) + 0.07(\text{OM}) - 0.20(\text{pH})$$
$$+ 0.01(\text{rainfall}) + 0.05(\text{RCN}) - 0.53(\text{c-prac}) - 0.06(\text{residue})$$
$$\text{Adjusted } R^2 = 0.91 \text{ and RMSE} = 0.43.$$

The coefficients in these models were significant at 5 percent. The subscripts on the dependent variable represent the cropping system (CC is Continuous Corn and CS is Corn-Soybeans). K_f-factor is the soil erodibility factor, OM is the percentage of organic matter, RCN is the runoff curve number that captures the effect of hydrology and soil cover complexes in controlling runoff, and c-prac is a binary regressor capturing the difference between straight row and contour systems. The residue cover is included to capture the tillage effects.

Soil erosion increases with slope, erodibility, organic matter, rainfall, and RCN. Higher organic matter content of the soil implies greater microbial activity, reducing soil compaction and thereby increasing erosion. The RCN increases for soils with lesser infiltration capacity, which explains the positive sign on this coefficient. Soil pH has a negative sign, implying reduced erosion of alkaline soils because of increased compaction of soils with higher pH. The residue cover, as expected, reduces soil erosion.

Nitrate-N runoff/leaching is influenced by several factors including soil, weather, hydrology, and agronomic factors. Identifying a simple relationship explaining nitrate-N runoff/leaching is very useful for modeling purposes. The linear model fitted to the untransformed nitrate-N runoff data is:

$$(\widehat{\text{nitrate-N runoff}}_{CC}) = -1.21 + 0.07(\%\text{slope}) + 0.08(\%\text{clay}) + 0.29(\text{OM}) + 0.18(\text{perm})$$
$$- 0.01(\text{rainfall}) + 0.22(\text{RCN}) + 1.06(\text{c-prac}) - 0.25(\text{residue})$$
$$\text{Adjusted } R^2 = 0.73 \text{ and RMSE} = 1.88.$$

$$(\text{nitrate-}\hat{\text{N}}\text{ runoff}_{cs}) = -3.93 + 0.07(\%\text{slope}) + 0.05(\%\text{clay}) + 0.31(\text{OM}) + 0.12(\text{perm})$$
$$- 0.01(\text{rainfall}) + 0.22(\text{RCN}) + 0.78(\text{c-prac}) - 0.03(\text{residue})$$
$$\text{Adjusted } R^2 = 0.84 \text{ and RMSE} = 1.04.$$

The parameters were significant at 5 percent. Factors that influenced runoff such as slope, organic matter, and RCN have positive signs, implying increased nitrate-N runoff.

A linear model fitted to the untransformed data on nitrate-N leaching did not produce a good fit. The distribution of the error term was skewed, suggesting heteroskedasticity. Data transformation did not solve the problem. A close examination of the data showed that a large number of observations were at or near zero, suggesting an exponential distribution. It is quite common to have such a distribution for chemical and nutrient leaching (Bouzaher et al. 1993). Therefore, an exponential model $Y(x_i; \beta) = \exp(X\beta)$ was fitted as follows:

$$(\text{nitrate-}\hat{\text{N}}\text{ leaching}_{cc}) = -18.11 + 0.24(\%\text{slope}) + 0.21(\%\text{clay}) + 0.51(\text{OM})$$
$$+ 0.24(\text{perm}) + 4.19(\text{bulk dens.}) + 0.01(\text{rainfall})$$
$$- 0.03(\text{RCN}) - 0.29(\text{c-prac}) - 0.12(\text{residue})$$
$$\text{Adjusted } R^2 = 0.72 \text{ and RMSE} = 1.84.$$

$$(\text{nitrate-}\hat{\text{N}}\text{ leaching}_{cs}) = -7.79 - 0.00(\%\text{slope}) + 0.03(\%\text{clay}) + 0.21(\text{OM})$$
$$+ 0.07(\text{perm}) + 1.66(\text{bulk dens.}) - 0.01(\text{rainfall})$$
$$- 0.08(\text{RCN}) - 0.90(\text{c-prac}) + 0.18(\text{residue})$$
$$\text{Adjusted } R^2 = 0.64 \text{ and RMSE} = 4.04.$$

The coefficients of these models were all significant at 5 percent, except for slope. This nonlinear model produced a good fit for the data. The negative sign on the RCN coefficient suggests that higher RCN (meaning lower infiltration) results in lower leaching. On the contrary, residue cover increases leaching.

Prediction and Spatial Distribution

Using the estimated metamodels, soil erosion and nitrate-N runoff and leaching were predicted for every unique soil and weather observation in the watershed described previously. Thus, the metamodel-predicted estimates account for site-specific variations in soil, weather, and hydrologic parameters. These estimates are summarized in the form of cumulative frequency distributions, which give the probability that a given soil under a given technology will exceed the appropriate benchmark for an indicator. For example, 1T or 2T soil loss tolerance could be the erosion benchmark and the drinking water standard of 10 mg/L of nitrate-N could be the water quality benchmark. This measure, "probability that a soil is at-risk," is more intuitively interpreted as a measure of the spatial distribution of risk, and is useful for targeting.

Figure 1 shows the cumulative frequency distributions of soil erosion and nitrate-N runoff/leaching under alternative tillage practices for continuous corn and corn-soybean rotations. Using an erosion benchmark of 2T (10 tons/ha), from the distributions in section 1.1 of Fig. 1, we see that nearly 80 percent of the soils are at risk under continuous corn with conventional tillage and only 40 and 15 percent of the soils are at risk under reduced and no-till practices. Corn-soybean rotation is more erosive than the continuous corn reiterating the field study results.

The proportion of at-risk soils for nitrates in runoff decreases as we move from conventional till fall-plow to no-till. The tillage differences are more pronounced in

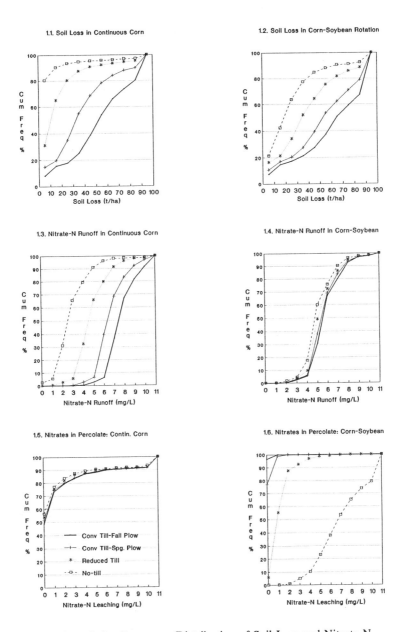

Figure 1. Cumulative Frequency Distribution of Soil Loss and Nitrate-N Runoff/Leaching Under Continuous Corn and Corn-Soybean Rotation.

continuous corn than in a corn-soybean rotation. This could be explained by the smaller difference in tillage impacts on soil erosion under corn-soybean rotation. On the other hand, the tillage impacts are more pronounced in the case of nitrate-N leaching from corn-soybean rotation than from continuous corn. The higher leaching of nitrates under corn-soybean rotation could be explained partly by the nitrogen-fixing capacity of soybeans. Therefore, by developing a BMP that takes adequate credit for the N fixed by soybeans and reduces the amount of fertilizer application, the potential leaching problem of a corn-soybean rotation could be controlled. This is an interesting finding because it asserts that conservation tillage is an environmentally sustainable agricultural practice.

REFERENCES

1. Bouzaher, A., P.G. Lakshminarayan, R. Cabe, A. Carriquiry, P. Gassman, and J. Shogren. 1993. Metamodels and Nonpoint Pollution Policy in Agriculture. Water Resources Research, 29(6):1579-1587.

2. Corak, S.J. and T.C. Kaspar. 1990. Fall-Planted Spring Oats: A Low-Risk Cover Crop to Reduce Erosion Following Soybeans. Research Report. Leopold Center for Sustainable Agriculture, Iowa State University, Ames, Iowa.

3. Hinkle, M.K. 1983. Problems with Conservation Tillage. Journal of Soil and Water Conservation, 38(3):201-206.

4. Jones, C., P. Dyke, J.R. Williams, J.R. Kiniry, V. Benson, and R. Griggs. 1991. EPIC: An Operational Model for Evaluation of Agricultural Sustainability. Agricultural Systems, 37:127-137.

5. Keeney, D.R. and T.H. DeLuca. 1993. Agricultural Contribution of Nitrate to the Des Moines River: A Preliminary Report. Proceedings: Agricultural Research to Protect Water Quality, Soil and Water Conservation Society, Ankeny, Iowa.

6. Kleijnen, J.P.C. 1987. Statistical Tools for Simulation Practitioners. Marcell Dekker, Inc., New York.

7. Seigley, L.S., G.R. Hallberg, R.D. Rowden, R.D. Libra, J.D. Giglierano, D.J. Quade, and K. Mann. 1993. Agricultural Land Use and Nitrate Cycling in Surface Water in Northeast Iowa. Proceedings: Agricultural Research to Protect Water Quality, Soil and Water Conservation Society, Ankeny, Iowa.

8. U.S. Environmental Protection Agency. 1992. Managing Nonpoint Source Pollution: Final Report to Congress on Section 319 of the Clean Water Act, 1989. EPA-506/9-90, Washington, D.C.

9. U.S. Geological Survey. 1993. Hydrologic, Water Quality, and Land Use Data for the Reconnaissance of Herbicides and Nitrates in Near Surface Aquifers of the Midcontinental United States, 1991. Open-File Report 93-114, Iowa City.

Nutrient and Sediment Removal by Grass and Riparian Buffers

J. E. Parsons, J. W. Gilliam, R. Muñoz-Carpena, R. B. Daniels, T. A. Dillaha[1]

ABSTRACT

This study compares the effectiveness of grass filters and riparian buffers for trapping sediment and nutrients from 1991 - 1992 on two sites in North Carolina - one in the Piedmont region and the other in the Coastal Plain region. Each site contains 4 m and 8 m lengths of grass and riparian buffers. Runoff from the field edge and each buffer plot is continuously monitored for quantity and discrete automatic water samplers are used to obtain water quality samples.

Runoff hydrographs and sediment concentrations have been collected on 48 storms at the Coastal Plain site and 50 storms at the Piedmont site. Nutrient concentrations and loads have been analyzed on 20 storms from the two sites. In nearly all storms, grass buffers filtered in excess of 50% of the sediment from the agricultural source area. In many cases, a 4 m grass buffer was sufficient to reduce field runoff sediment by greater than 50%. Riparian buffers reduced sediment yield but generally the reduction was less consistent. Chemical filtration of the agricultural runoff by the grass and riparian buffers also occurred. Soluble nutrients such as ortho-phosphorus had smaller reductions than sediment bound constituents. **KEYWORDS.** Vegetative filter strips, Erosion, Runoff.

INTRODUCTION

Concerns about pollution from agricultural sources has generated much interest in the use of vegetative filter strips (VFS). The U. S. Department of Agriculture - Soil Conservation Service (USDA-SCS) has guidelines for installation of VFSs as a best management practice (BMP) for non-point pollution control. Many states are considering or adopting regulations requiring VFSs to protect surface water sources from non-point pollution sources. Many of the recommendations on the design of the VFSs are poor at best. Dillaha et al. (1987) surveyed a number of existing filter strips in eastern Virginia and found many of these were ineffective a few years after their installation.

For our discussions, a VFS is defined as any vegetated area which separates a non-point pollution source area from a body of water such as a stream or lake. Non-point pollution sources can be as varied as urban housing developments, industrial parks, and agricultural fields. VFSs can refer to either constructed or naturally vegetated or riparian areas.

Research in Georgia, Maryland and North Carolina have reported very similar conclusions on the effectiveness of natural and riparian buffers for removing pollutants from agricultural source areas. This research has examined the removal of sediment, nitrogen (N) and phosphorus (P) by buffer zones. In the North Carolina Coastal Plain, Jacobs and Gilliam (1985) observed that nitrate-N decreased from levels greater than 10 mg-N/L to less than 1 mg-N/L while passing through a 50 m riparian zone. Only 5 kg/ha/year of the estimated 35 kg/ha/year of N entering the riparian zone left the watershed in stream flow. Lowrance et al. (1984) found similar results in

---

1. J. E. Parsons, Assistant Professor, Department of Biological and Agricultural Engineering, J. W. Gilliam, Professor, Department of Soil Science, R. B. Daniels, Visiting Professor, North Carolina State University, Raleigh, NC, R. Muñoz-Carpena, Research Scientist, Centro de Investigacio'n y Technologia Agrarias, Tenerife, Spain; T. A. Dillaha, Professor, Agricultural Engineering Department, VPI&SU, Blacksburg, VA.

the Georgia Coastal Plain; of the 52 kg/ha/year of N entering the riparian ecosystem, only 13 kg/ha/year exited the system in stream flow. In Maryland, Peterjohn and Correll (1984) estimated that 45 kg/ha/year of nitrate-N was removed by the riparian zone. The N reductions in all three studies were observed in the shallow subsurface flow.

Sediment leaving agricultural fields is removed from the surface flow by vegetative and riparian buffers. Cooper et al. (1987) estimated that approximately 90% of the sediment leaving agricultural fields was deposited in the riparian zones of watersheds in North Carolina. Most of the sediment was deposited within 100 m of the field edge. In Georgia, Lowrance et al. (1986) also measured sediment accumulation in riparian zones and concluded that these are important sinks for sediment. This work indicates that relatively narrow buffers adjacent to streams may be effective for sediment removal.

Phosphorus is removed primarily via surface flow through the vegetative and riparian buffers. Much of the research indicates that removal is less effective than either N or sediment. In North Carolina, Cooper and Gilliam (1987) estimated that about 50% of the P entering riparian areas was trapped. Lowrance et al. (1984) also measured lower retention of P in Georgia riparian zone studies. New Zealand watershed - riparian studies by Cooke (1988) also measured less P removal than N removal. Although P trapping efficiency of riparian areas is less than that for N or sediment, this removal is still very important since P is often the limiting nutrient in many freshwater bodies.

Research on grass VFSs have reported high sediment trapping efficiencies as long as the surface water flow is shallow and the VFSs are not innudated with sediment. If either of these fail, then trapping efficiency decreases dramatically. Barfield et al. (1979) found this to be especially the case at higher runoff rates. A number of short-term experimental studies have quantified the effectiveness of grass VFSs in reducing sediment and nutrients in runoff (Dillaha et. al. 1987, 1988; Magette et al., 1989). In these studies, VFSs were effective for removing sediment and sediment-bound nutrients with trapping efficiencies often exceeding 50% if the flow was shallow. Dissolved nutrients were not removed as effectively and researchers often report higher concentrations in the VFS effluents than in the influent runoff (Dillaha et al., 1989; Magette et al., 1989). This was attributed to the limited assimilative capacity of the VFSs, thus, releasing nutrients that were deposited there in previous storm events.

The objectives of this paper are (1) describe the field project and data acquisition system, (2) present analysis of the effectiveness of field edge grass buffers and riparian zones for trapping sediment and surface water transported nutrients and (3) examine the relationship between rainfall events and buffer zone sediment trapping efficiency.

METHODS

<u>Field Sites</u>

The vegetated filter research sites are located in the Piedmont and Coastal Plain of North Carolina. These sites are not only representative of the two major agricultural areas of North Carolina and many Atlantic Coastal states but also areas where most urban nonpoint source contaminates originate. The Piedmont site is in Wake County on NCSU Research Farm Unit 9 which has topography and Appling-Cecil soils typical of the lower Piedmont. The Coastal Plain site is on the Cunningham Research Farm near Kinston, North Carolina with topography and Norfolk-Rains Soil Association typical of the middle and lower Atlantic Coastal Plain. The layout of the Piedmont site is given in Fig. 1.

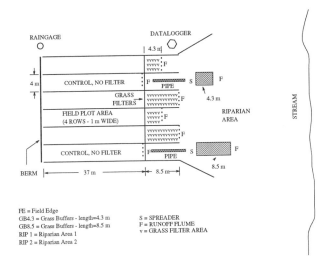

FIGURE 1. Layout of the Piedmont experimental site.

At both study sites the cultivated area is 27.4 m by 36.6 m. This width allows four cultivated rows for each of six runoff plots. The slope of the runoff plots are approximately 3.6% and 1.9% for the Piedmont and Coastal Plain sites, respectively. Two buffer rows are on the outside of the plots. The crop rows are up and down hill and bedded to prevent runoff from crossing from one cultivated plot to the next. Plastic edging prevents runoff cross over in the lower 3 meters of the cultivated part and throughout the grass and riparian filters. The slope of the grass filters at the Piedmont site range between 4% and 6%. For the Coastal Plain site, the slope of the grass filters are approximately 1%. The riparian buffers at the Piedmont are located on the steep hillslope just above the stream floodplain have approximately a 15% slope, while the slopes of the Coastal Plain riparian buffers are about 1%.

Gutters collect runoff from two plots at the field edge, and at the end of two 4.3 m and 8.5 m grass filters. Pipes carry the runoff to HS flumes (0.15 m) for volume measurement and sampling. Spreaders deliver the runoff to the upslope portion of the two riparian plots (Fig. 2). Gutters at the downslope end of the riparian plots collect the runoff. Similar to the grass buffers, HS flumes (0.15 m) measure the runoff in the riparian buffers.

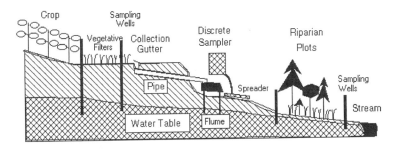

FIGURE 2. Cross-sectional view of the measuring system.

Instrumentation

A Campbell CR10[2] datalogger at each site monitors rainfall and surface runoff and activates the water samplers. A tipping bucket rain gage records rainfall totals and 5-minute intensities. The datalogger monitors the HS flume stages at 30 second intervals using a potentiometer - float assembly to measure water levels changes. Recorded changes in flume stage allow measurement of flow volume by comparison of flow volume with stage-discharge relationships for each flume. During a runoff event, these changes trigger the discrete samplers, American Sigma Designs[2], to pump a 500 ml sample from the flume outflow for water quality analysis.

Sample Analyses

A suite of samples is taken from each plot during major events for chemical analysis. The events chosen for chemical analysis represent critical periods such as major storm or runoff after planting and fertilization. If possible, sediment chemistry is determined monthly during the growing season. About 1/3rd of the samples from selected runoff events are analyzed for total silt + clay, Kjeldahl nitrogen (TKN), NH_4, NO_3, total phosphorus (TP), PO_4-P (ortho-phosphorus (OP)) in solution and Cl. The samples selected are on the rising limb of the hydrograph, at the hydrograph peak, and those on the falling limb. If the hydrograph has more than one peak, all peaks are sampled using the same criteria. By interpolating the concentrations, sediment and chemical mass flows are computed.

RESULTS AND DISCUSSION

Runoff, sediment and chemical analyses have been completed on a number of storms events at each site. Earlier reports on the project have described some of the 1990 and 1991 data collection and results (Parsons et al. 1990, 1991). This early data was collected when the field edge filters were a combination of fescue, weeds and crab grass. During the fall of 1990, the grass filters were reseeded with fescue at both sites. This paper will concentrate on events after the establishment of the fescue during mid-1991 at the Coastal Plain site. The results for the Piedmont site are similar and only the riparian buffers are considered here.

Table 1 presents a storm summary of some typical rainfall events which produced runoff during 1991 and 1992. The storm summary includes the storm erosivity index calculated using the 5-minute rainfall intensities (Storm Erosivity Index) and the maximum 30-minute rainfall intensity (Daily Erosivity Index) (Cooley 1980).

For the Coastal Plain site, a comparison of the amount of sediment yield from the grass buffers versus the agricultural source area is presented in Table 2. In all cases sediment yield of the grass buffers were reduced substantially as compared to the yields from the field edge plots. For the larger storm events, the runoff volume from the field edge was plots was generally much larger than the volume from either of the grass buffers. In general, for smaller runoff events, little or no sediment was transported through either of the grass buffers. However, occasionally there are smaller runoff volumes from the field plots with larger runoff volumes through the grass buffers. For example, on day 218, 1992, the runoff volume from the field edge collector was 1.5 m^3 and the volume from the 4.3-m grass buffer was 1.7 m^3. The sediment yields from the grass buffer was still 50-60% less than the field runoff sediment loads. Similar reductions were seen for storms at the Piedmont site.

---

2. Mention of tradenames does not constitute endorsement of a particular product by North Carolina State University or its cooperators. Other products may be suitable.

Table 1: Summary of Some Selected Storms at the Coastal Plain Site.

Year	Day	Rainfall mm	Maximum Rainfall Intensity mm/h	Storm Erosivity Index	Daily Erosivity Index	Maximum Storm Erosivity Index
				MJ-mm/ha-h		
1991	219	22	48.8	138.8	106.7	27.50
	263	81	48.8	448.2	775.7	27.50
1992	161	54	70.1	309.6	419.6	41.57
	176	98	100.6	678.6	1021.5	62.54
	208	25	97.6	183.3	133.9	60.44
	216	94	414.0	691.3	966.1	304.51
	218	113	67.1	684.9	1272.0	39.52
	226	34	48.8	186.6	208.4	27.50
	290	39	100.6	281.3	257.1	62.54

Table 2 presents chemistry comparisons for selected storms at the Coastal Plain site. Total Kejdahl nitrogen loadings from the grass buffers were reduced by greater than 50% for all storms. In most storms, the 8.5-m grass buffers yield less TKN than the 4.3-m buffers. The only exception was the storm on day 216, 1992. The TKN loading from the 8.5-m grass buffer was larger than the 4.3-m buffer; but this value was still substantially lower than the TKN load from the field edge plot.

Ortho-phosphorus show similar trends as the TKN results (Table 2) for the shorter grass buffer (4.3-m). The longer grass buffer (8.5-m) yielded larger amounts of OP than either the field edge plots or the 4.3-m buffer. The only exception to this case was the storm event on day 208, 1992. On this date, both grass buffers yielded more OP than the field edge plots with the 4.3-m buffer the largest. This is probably due to the low assimilative and storage capacity of the grass buffers.

Total phosphorus filtration by the grass buffers was better that for the OP (Table 2). Reductions in TP loads by the grass buffers were similar to those for sediment load. However, on day 216, 1992, the TP load from the 8.5-m buffer was larger than that for the 4.3-m buffer. Both buffers yielded less than the loads from the agricultural source area.

A summary of the effect of the 4.3 m riparian filter in the Coastal Plain on reducing sediment is presented in Table 3. The longer riparian filter data is not presented because installation of the flow measurement and sampling equipment did so much damage to the native vegetation that it did not recover. The 4.3 m filter contained a very thick growth of dog fennel (a large weed which is common in this area at field - forest interface). Although the growth appeared to be very vigorous, a close inspection showed stems were much more widely separated than in grass filters and offered little resistance to flow. This weed is an annual and the cover essentially disappeared during the winter. The difference in cover between the grass and dog fennel is reflected in large differences in sediment removal (Compare data in Tables 2 and 3). However, even the though the riparian filter contained relatively poor vegetative cover, more than 50% of the sediment was removed from the runoff during most storms. The same situation exists with regard to total P and TKN removal. It is a bit surprising to us that the TKN and total P removals are almost as high as sediment removal since it is the smaller sediment particles which are not trapped in the filter.

However these data indicate that even a poorly vegetated filter which is relatively short (4.3 m) can be valuable for water quality protection.

Table 2: Sediment Yield and Chemistry Comparisons for Selected Storms for the Coastal Plain Site.

Year	Day	Set[a]	Field Edge	Grass Buffer 4.3 m	Grass Buffer 8.5 m	Field Edge	Grass Buffer 4.3 m	Grass Buffer 8.5 m	Field Edge	Grass Buffer 4.3 m	Grass Buffer 8.5 m
			Runoff Volume (m^3)			Sediment (kg)			TKN (g)		
1991	219	2	1.3	0.4	0.5	5.7	0.7	1.0	5.5	1.1	1.8
	263	1	0.5	-	0.1	2.5	-	0.4	3.4	-	0.1
1992	161	1	0.6	0.0	0.3	5.3	0.0	0.0	3120	106	0.0
	176	1	2.2	3.0	2.3	12.9	8.3	3.4			
		2	2.8	2.1	3.0	13.8	3.1	4.3			
	208	1	1.3	1.0	0.8	10.2	3.6	2.4	9.1	4.6	2.5
	216	1	1.2	0.7	0.5	7.5	2.0	1.0	8.8	1.0	2.3
	218	2	1.5	1.7	1.4	9.4	3.0	4.7			
	226	1	1.3	1.1	1.0	6.4	2.5	1.7			
		2	1.0	0.7	1.5	5.3	0.5	0.4			
	290	1	0.9	0.3	0.0	18.4	0.7	0.0			
			OP (g)			TP (g)					
1991	219	2	0.3	0.4	0.7	2.2	0.7	0.9			
	263	1	0.2	-	0.0	1.3	-	0.1			
1992	161	1	0.2	0.0	0.0	1.2	0.0	0.0			
	208	1	0.8	1.2	1.0	4.1	3.1	1.9			
	216	2	0.9	0.3	1.0	4.1	0.6	1.6			

a. The set is either 1 or 2 which is the group of grass buffers (See Fig. 1)

Table 3: Sediment Yield and Selected Chemistry Comparisons for the Coastal Plain Riparian Buffer.

Year	Day	Source Field Edge	Riparian Zone 4.3 m	Source Field Edge	Riparian Zone 4.3 m	Source Field Edge	Riparian Zone 4.3 m	Source Field Edge	Riparian Zone 4.3 m
		Runoff Volume (m^3)		Sediment (kg)		TP (g)		TKN (g)	
1991	263	0.5	0.4	2.5	1.0	1.3	0.7	3.4	1.5
	267	0.4	0.3	2.4	2.3	0.9	1.1	2.5	2.4
1992	161	1.3	0.6	5.3	0.8	1.2	0.5	3.1	1.2
	172	0.6	0.3	4.7	1.1	1.6	0.6	4.0	1.6
	208	2.3	2.2	10.2	8.3	4.1	5.7	9.1	10.6
	290	0.9	0.9	18.4	1.2				
	331	1.2	0.7	13.4	1.6				
	345	0.6	0.4	4.9	0.6				

The effectiveness of the forested riparian filters at the Piedmont site is shown in Table 4. Because these filters are located on a relatively steep slope, there was a much greater tendency for the flow to channelize as it moved down the slope. However, note that infiltration was large which lead to relatively high, although very variable, removals of all constituents from the surface runoff water. Thus, even though the small stream which flows below this site drains several agricultural fields, the water is generally clear with a low concentration of both N and P because the stream is protected by a riparian buffer throughout its entire reach.

Table 4: Sediment and Chemistry for Selected 1991 Storms for the Piedmont Riparian Buffers.

Day	Source Field Edge	Riparian Zone 4.3 m	Source Field Edge	Riparian Zone 8.5 m	Source Field Edge	Riparian Zone 4.3 m	Source Field Edge	Riparian Zone 8.5 m
	Runoff Volume (m^3)		Runoff Volume (m^3)		Sediment (kg)		Sediment (kg)	
170	1.8	1.5	2.2	0.5	6.5	0.3		
183	0.8	0.7	1.3	0.5	14.9	0.8	17.3	0.7
226	2.3	1.5	2.6	1.7			3.8	0.9
262	2.1	0.6	1.6	0.5	3.2	0.4		
	NO$_3$ (g)		NO$_3$ (g)		TKN (g)		TKN (g)	
170	7.5	3.0			12.5	1.9		
183	7.2	3.5	10.6	3.3				
226	0.5	1.6	2.5	1.5	0.5	2.3	4.4	5.0
262	2.9	2.5			3.8	1.3		
	OP (g)		OP (g)		TP (g)		TP (g)	
170	0.6	0.8			5.6	1.5		
183								
226	0.3	1.0	0.6	0.7	0.3	1.7	2.3	1.9
262	2.3	1.2			3.1	1.3		

The data presented in this report seem to indicate that the planted grass filters are somewhat more effective as filters than natural vegetation in riparian buffers. This may be true for surface runoff but it should be remembered that many researchers working in the field believe that forested areas are more effective in removing nitrate from subsurface flows.

ACKNOWLEDGEMENT. This study is supported in part by USDA-SCS, US-EPA, NC-WRRI, and Southern Region Project S249. Thanks also to Mr. Charles A. Williams, Mr. William Thompson and Ms. Bertha Crabtree for their help.

REFERENCES

1. Cooke, J.G. 1988. Sources and sinks of nutrients in a New Zealand Hill Pasture Catchment. II. Phosphorus. Hydrol. Processes 2:123-133.

2. Cooley, K. R. 1980. Chapter 2. Erosivity "R" for individual design storms. *In*: Knisel, W. G. (ed.). CREAMS: A field-scale model for Chemicals, Runoff, and Erosion from Agricultural Management Systems. U. S. Department of Agriculture, Science and Education Administration, Conservation Research Report No. 26. pp. 386-397.

3. Cooper, J.R. and J.W. Gilliam. 1987. Phosphorus redistribution from cultivated fields into riparian areas. Soil Sci. Soc. Am. J. 51:1600-1604.

4. Cooper, J. R., J.W. Gilliam, R.B. Daniels and W.P. Robarge. 1987. Riparian areas as filters for agricultural sediment. Soil Soc. Am. Proc. 51:416-420.

5. Dillaha, T.A., R.B. Reneau, S. Mostaghimi, V.O. Shanholtz, and W.L. Magette. 1987. Evaluating nutrient and sediment losses from agricultural lands: Vegetative filter strips. Annapolis, MD: U. S. Environmental Protection Agency, Chesapeake Bay Liaison Office; 93 p.; CBP/TRS 4/87.

6. Dillaha, T.A., J.H. Sherrard, D. Lee, S. Mostaghimi, and V.O. Shanholtz. 1988a. Evaluation of vegetative filter strips as a best management practice for feedlots. J. of the Water Poll. Cont. Fed. 60:1231-1238.

7. Dillaha, T.A., R.B. Reneau, S. Mostaghimi, and D. Lee. 1989. Vegetative filter strips for agricultural nonpoint source pollution control. Trans. of the ASAE 32(2):491-496.

8. Jacobs, T. J. and J. W. Gilliam. 1985. Riparian losses of nitrate from agricultural drainage waters. J. Environ. Qual. 14:472-478.

9. Lowrance, R.R., R.L. Todd and L.E. Asmussen. 1984. Nutrient cycling in an agricultural watershed: Streamflow and artificial drainage. J. Environ. Qual. 13:27-32.

10. Lowrance, R.R., J.K. Sharpe and J.M. Sheridan. 1986. Long-term sediment deposition in the riparian zone of a Coastal Plain Watershed. J. Soil and Water Cons. 41:266-271.

11. Magette, W.L., R.B. Brinsfield, R.E. Palmer   ood. 1989. Nutrient and sediment removal by vegetated filter strips. Trans. of the ASAE 32(2):663-667.

12. Parsons, J. E., R. B. Daniels, J. W. Gilliam and T. A. Dillaha. 1990. Water Quality Impacts of Vegetative Filter Strips and Riparian Areas. Am. Soc. of Agric. Eng. Paper #90-2501.

13. Parsons, J. E., R. B. Daniels, J. W. Gilliam and T. A. Dillaha. 1991. The Effect of Vegetation Filter Strips on Sediment and Nutrient Removal from Agricultural Runoff. Proc. for the Environmentally Sound Agriculture Conference. Volume 1, pp. 324-332. April 16-18, 1991. Delta Orlando Resort, Orlando, FL.

14. Peterjohn, W.T. and D.T. Correll. 1984. Nutrient dynamics in and agricultural watershed: Observations on the role of a riparian forest. Ecology. 65:1466-1475.

TERRACES TECHNOLOGICALLY OBSOLETE OR WILL TERRACES STILL HAVE A FUTURE?

D. R. Speidel*

ABSTRACT

Terraces and contour farming have been used by agriculturists for centuries. Terraces have been the mainstay of Soil Conservation Service practices since the "dirty thirties." Terraces control gully and ephemeral erosion, but are not cost effective in many resource regions. Only in fields that have severe ephemeral and gully erosion problems, not solved with rotations and residue management, can these high cost systems be justified. Environmental planners must use terrace systems judiciously to solve the most serious erosion problems with limited funds. Policy makers must be willing to set priorities to solve unique water quality problems which justify the public support necessary to fund application of this conservation practice.
KEYWORD. Terraces.

OVERVIEW

Conservation of the soil resource is accomplished by application of practices to achieve maximum land treatment at minimum cost. These resource systems, when employed with consideration of the producers ability to manage and the effects on the environment, create holistic systems that assure protection of all resources used by people.

Technological advances challenge conservation planners to be informed of current changes in order to make the most suitable recommendations in resource management. The objective of this paper is to determine if current terrace systems can be justified as a component of holistic land treatment systems. This will be discussed in terms of the individual practice costs compared to the benefits achieved by installing terrace systems when compared to alternative practices. A cost and benefits analysis of two case studies using terraces as the primary erosion control practice will be developed to determine the current cost of terrace systems. The benefits of terraces will be demonstrated through the review of relevant research.

Parallel terraces, when first introduced 30 years ago, met with great success. Studies by the USDA, Agricultural Research Service (ARS) demonstrated that soil erosion was reduced by a factor of 25. The cost of installation was comparable to alternative erosion control methods and the practices were acceptable to the producer (Spomer et al., 1973). In the past decade, technology has made alternative methods of erosion control possible, such as residue management. At the same time the cost of terrace installation has increased by a factor of five while the value of the commodities produced increased only three-fold (USDA - Nebraska Agricultural Statistics, 1989).

Soil erosion is a real cost to society. The ARS has developed the Erosion Productivity Impact Calculator (EPIC) to predict future soil productivity (Benson, 1989). Using EPIC, ARS has determined that many of our soils will be depleted 20 to 35 percent in 100 years. The value of lost annual income as soils become severely eroded ranges from $33.65 ha^{-1} ($13.63/ac) to $44.69 ha^{-1} ($18.10/ac) on loess soils in western Iowa (USDA - Western Iowa Rivers Study, 1990) and similar depletion on Missouri glacial till soils (USDA - Impact of Soil Erosion, 1987). Eventually, a soil will become so depleted it is unprofitable to farm. At this point the field must revert to a less valuable use. While water quality may also be damaged from non-point pollution it is beyond the scope of this paper to investigate the benefits of controlling non-point pollution.

---

* D. R. Speidel, District Conservationist, USDA-SCS, Stanton, NE 68779 (402-439-2213).

CASE STUDIES

In 1964 level terraces were constructed on Ida Silt Loam in Treynor Watershed No. 4 located in Pottawattamie County, Iowa. The cost of installation was $96.17 ha^{-1} ($38.95/ac). When amortized at 8 percent over 20 years an annual cost of $8.77 ha^{-1} ($3.56/ac) is determined. The cost of terraces in the Treynor Watershed is the basis for comparing the cost of current systems (Spomer et al., 1973).

In 1978 a level terrace system was built for modern eight-row farm equipment on Galva (Typic Hapludolls) silty clay loam, 4 to 8 percent in Sioux County, Iowa. Cost of the system was $212 ha^{-1} ($86/ac), 2708 m (8880 ft) at $2.56 m^{-1} ($.78/ft) protecting 32.4 ha (80 ac) in a 58 ha (144 ac) field. The annual cost of $20 ha^{-1} ($8/ac) is amortized for the affected acres. If the same system was built today the estimated annual cost would be $44 ha^{-1} ($18/ac). Maintenance costs have not been incurred since installation.

Erosion was controlled to tolerance levels for the soil on the entire field (based on calculations using the Universal Soil Loss Equation). Terraces and contour farming treated the 6 to 8 percent slopes. Crop residue levels were managed at 30 percent, and a continuous corn rotation treated the remaining 4 to 6 percent slopes. A grassed waterway was maintained to control the remaining ephemeral erosion below the terraces.

The benefits of controlling erosion have been determined by recent research (USDA - Western Iowa Rivers Study, 1990). A value of $21.42 ha^{-1} ($8.68/ac) was determined for economic damage of similar soils due to sheet and rill erosion and soil depletion. The same study investigated the damage due to ephemeral erosion and determined the cost to be $21.78 ha^{-1} ($8.82/ac). However, field observation of the topography of the first case study indicates $21.78 ha^{-1} was too high. Considering the footage of grassed waterways replaced by the terrace system and the lack of need to maintain the remaining grassed waterway, a cost can be derived for ephemeral erosion damages. The assumption used is that 610 m (2000 ft) of waterway would need replacement every five years. The current cost of $1.44 m^{-1} ($.44/ft) is used. The result amortized would be $7.41 ha^{-1} ($3.00/ac) shown in Table 1.

Table 1
Erosion Control Costs
Sioux County, IA

Erosion Damage and Depletion Costs ($)	14.07 ha^{-1}	(5.70/ac)[a]
Ephemeral Costs ($)	7.41 ha^{-1}	(3.00/ac)
Total Costs ($)	21.48 ha^{-1}	(8.70/ac)

[a] (USDA Western Iowa Study, 1990)
corn price $0.087kg^{-1}

Level terraces also have the unique characteristic of conserving water by allowing the runoff to percolate through the soil profile. Research on the benefits of moisture saved by terraces is not clear for the Midwest. The Treynor Watershed study did demonstrate that 80 percent of the runoff left the terraced watershed as base flow. The Treynor study assumed that only the terrace channel received runoff. The conclusion was that moisture conservation was not a significant benefit (Spomer et al., 1973).

However, earlier research on contour farming does show a moisture conservation benefit of 10 percent increased corn yield. These observations were from research stations and over 300 Midwest farms during 1939-44. Contour farming was credited with saving one inch of moisture and level terraces in the Great Plains with saving two inches of moisture (USDA - Inservice Manual, 1949).

Due to this information discrepancy, drawing a conclusion on the benefit of water conservation from level terraces is difficult. The Treynor Watershed is unique due to the steep slopes and very porous soils. With the use of conventional tillage, the effectiveness of contour farming may have been negated. Also, improved yields due to better management may have masked benefits of moisture conservation.

Personal observation in Sioux County indicates a different case in northwest Iowa. Each spring water has been observed standing in the rows as well as the terrace channels. It is possible the combination of mulch tillage with contour farming on fields with moderate slopes and a fine texture soil allows better moisture distribution. The combination of terraces, residue management and contour farming could be complementary on this soil type. Terraces control ephemeral erosion, residue management reduces erosion within the terrace interval and the level contour rows capture the runoff.

If there is a water conservation benefit, the general consensus is that 15.4 mm (one inch) of additional moisture results in a 752 kg ha^{-1} (12 bu/ac) increase in corn yield (Spomer et al., 1973). Review of 10 years of soil moisture and weather data from the Doon and Sutherland Extension station farms, located 12 and 20 km respectively from the study site, indicated a moisture deficient half the time each spring. Using this as a valid assumption, an annual yield benefit of 376 kg^{-1} (6 bu/ac) corn is determined. If only 10 percent of the terraced area received an increase in moisture a significant yield increase should be realized.

If no-till had been used as an alternative management practice, only 75 percent of the field would had been treated to sustainable levels. On the remaining 25 percent, soil loss would still exceed the tolerance level for the soil. After 100 years of erosion, only 90 percent of the potential productivity would be maintained (Benson et al., 1989). This, however, still meets the goal established by the Food Security Act of 1985.

The costs associated with no-till as an alternative management practice are considered minimal today (Jolly et al., 1983). Machinery normally used by today's producers can also be used for no-till. Increased cost from using different chemicals is generally offset by reduced labor and machinery expense. The real issue is whether the producer can manage no-till systems.

The second case study was made in Stanton County, Nebraska, where a 46.6 ha (115 ac) field was treated with terraces and no-till. Dominate soils were Crofton (Typic Ustorthents) silty clay loam, 11 to 16 percent slopes, and Moody (Udic Haplustolls) silty clay loam, 2 to 6 percent slopes. Extensive ephemeral erosion prompted the producer to install terraces. Due to the extreme topographic undulations and soil texture, tile outlet terraces were installed. Grassed waterways and no-till were used for the lower slopes not terraced.

In 1992 and 1993, 2132 m (6990 ft) of terraces and 672 m (2205 ft) of a system of underground PVC pipe with risers in the terrace channels were installed at a cost of $17,686. Additional work is planned with an estimated cost of $5000. The total project will treat 19 ha (47 ac) of cropland.

If tile outlet terraces were first installed in the Treynor Watershed, the cost would have been an additional $2000 (Spomer et al., 1973). Amortized at 8 percent, the complete system's cost would be $12.10 ha^{-1} ($4.90/ac).

The amortized annual cost for the system in Stanton County today at 8 percent over 20 years is $185 ha^{-1} ($75/ac). Maintenance costs, estimated at 5 percent, are $74 ha^{-1} ($30/ac).

Table 2
Erosion Control Costs
Stanton County, NE

Erosion Damage and Depletion Costs ($)	32.54 ha^{-1}	(13.18/ac)[a]
Ephemeral Costs ($)	21.78 ha^{-1}	(8.82/ac)[b]
Total Costs ($)	54.32 ha^{-1}	(22.00/ac)

(a) (USDA Impact of Soil Erosion, 1987)
(b) (USDA Western Iowa Rivers Study, 1990)

The tile outlet terrace system with no-till reduces soil loss to a rate capable of retaining 90 percent of the soil's productivity potential in 100 years. A change in the rotation to include close-sown small grains or legumes would be required to bring the soil loss to an acceptable level for the specific acres. Ephemeral erosion and off-site sediment damage is controlled by the terrace system. If no-till was used alone, the USLE predicts the soil loss rate would double. The result: in 100 years only 80 percent of the soil productivity potential would remain. Furthermore, the ephemeral erosion damages would continue each year.

These calculations are comparable to the results of a study in the Maple Creek Watershed by the ARS (USDA - Maple Creek MIP, 1982) and the University of Nebraska (El'Osta, 1990). This economic study concluded providing federal financial assistance to individual producers to install and maintain practices was justified based on the gross public benefit of controlling soil erosion. The study demonstrated that costs of practices exceed the benefits to the individual compared to only the cost of filling in the gullies and allowing 20 percent depletion of the soil.

The pubic benefit of controlling the cost of sediment damage to navigable streams and public reservoirs plus the benefit of retaining at least 10 percent greater yield exceed the cost of installation and operation of sediment basins in 66 percent of the hypothetical cases analyzed (El'Osta, 1990).

The economic benefit of improved water quality of Maple Creek was not analyzed by the university study. However, sediment is a significant problem in the local stream (USDA - Maple Creek MIP, 1982). Sediment basins controlled 97 percent of the sediment. Only 1 to 3 g L^{-1} were measured at the tile outlet compared to 123 g L^{-1} in the local stream during runoff events. An interesting point observed was that sediment measured 30 to 40 g L^{-1} until the settling pool formed around the terrace riser. The basin did not affect the content of soluble nutrients in the runoff. This points out the need for mutual supporting practices to control the various erosion and agricultural chemical problems on steep sloping cropland.

DISCUSSION

No-till is considered less costly when compared to terraces. On steep slopes, to be effective, terraces must be complemented by no-till or a higher level cropping sequence which includes legumes or close-sown crops. The most significant benefit of terraces compared to no-till is the control of ephemeral erosion. However, the benefit of controlling ephemeral erosion does not necessarily offset the cost of installing and maintaining terrace systems. This is especially true of high cost tile outlet terrace systems. Terrace systems with costs exceeding benefits to the field treated must be judiciously applied.

To justify terraces today off-site benefits must be realized. Terraces do an effective job of controlling sedimentation (USDA - Treynor Watershed, 1986). Agricultural chemicals transported by sediment are controlled by terraces (Hanway and Laflen, 1974). Since terraces can normally be justified only by off-site benefits, society should not expect the individual to carry the total burden of installation of terrace systems which benefit society as much or more than the individual. Furthermore, society should expect the planner who recommends terraces to be able to support the cost of such systems to society if terrace systems are supported by public funds.

More research is needed to determine the water quality benefits of tile outlet and level terraces. Research is also needed to develop systems of practices that protect the entire resource base from depletion and not just treat 80 or 90 percent. The planner must take into consideration the cost of the system and on-site and off-site benefits of terrace systems. The conservation practices recommended to the producer must be based on the cost and benefits and the producer's ability to manage the practice. The proposal should be a complete holistic system taking into account the producer's needs and abilities and enhances the environment off-site as well as protection of the resource base. Terraces should be considered in cases where ephemeral erosion and high value off site benefits can be clearly realized. Use of no-till and, if necessary, legume-based rotations which fully protect the resource base need to be part of this system to ensure effective terrace maintenance programs are in place on steep sloping fields.

Where the proposed terrace system is needed to meet these requirements, the cost of a modern system can be justified.

BIBLIOGRAPHY

Benson, V.W., O.W., Rice, P.T. Dyke, J.R. Williams and C.A. Jones. 1989. *Conservation impacts on crop productivity for the life of a soil*. Journal of Soil and Water Conservation, Ankeny, IA. 44:600-604.

El'Osta, H. 1990. Unpublished thesis. *Terraces installation and maintenance costs, Stanton County, Nebraska*. University of Nebraska, Lincoln, NE.

Hanway, J.J. and J.M. Laflen. 1974. *Plant nutrient losses from tile-outlet terraces*. Journal of Environmental Quality, Madison, WI. 3:351-356.

Jolly, R.W., W.M. Edwards and D.C. Erbach. 1983. *Economics of conservation tillage in Iowa*. Journal of Soil and Water Conservation, Ankeny, IA. 38:291-294.

Spomer, R.G., W.D. Shrader, P.E. Rosenberry and E.L. Miller. 1973. *Level terraces with stabilized backslopes on loessial cropland in the Missouri Valley: a cost effectiveness study*. Journal Paper No. J-7221, Iowa Agriculture and Home Economics Experiment Station, Ames, IA.

U.S. Department of Agriculture, Soil Conservation Service. 1949. *Inservice Manual USDA extracts from experiment station annual reports and technical bulletins*. Coshocton, OH.

U.S. Department of Agriculture, Soil Conservation Service. 1982. *Maple Creek Model Implementation Project - Final Report*, Lincoln, NE.

U.S. Department of Agriculture, Agriculture Research Service. 1986. Data Summary *Treynor Watersheds*. North Central Watershed Research Center, Columbia, MO.

U.S. Department of Agriculture, Soil Conservation Service. 1987. *Impact of soil erosion on crop yields in Missouri (1982-1985)*. Columbia, MO.

U.S. Department of Agriculture, Statistical Reporting Service. 1989. *Nebraska Agricultural Statistics*, Nebraska Department of Agriculture, Lincoln, NE.

U.S. Department of Agriculture, Soil Conservation Service. 1990. Special Report *Western Iowa Rivers Basin Study - Erosion of Ida and Monoa Soils*. Des Moines, IA

AN ECONOMIC AND ENVIRONMENTAL EVALUATION OF CONSERVATION COMPLIANCE ON A WEST TENNESSEE WATERSHED REPRESENTATIVE FARM

R. G. Bowling, B. C. English*

ABSTRACT

Soil and water conservation continue to be important environmental policy concerns. The Conservation Compliance provision of the Food Security Act of 1985 was enacted, in part, to reduce sediment from erosion and to improve water quality. This provision discourages production on highly erodible cropland; producers who use highly erodible land cannot participate in commodity programs unless they adopt a conservation plan. Conservation Compliance could affect the cost of program participation. A representative farm was modeled under alternative management practices using linear programming. The management practices included current production practices and SCS recommended practices. For each of the management practices, soil erosion was estimated using the Erosion-Productivity Impact Calculator. Profitability was compared across the options with and without constraints on soil erosion.

The implementation of Alternative Conservation Systems would reduce erosion on the representative farm by approximately 9,151 tons and net returns by $916 from the current farm practices. If soil erosion were constrained to less than 5-tons per acre, the erosion level would decrease by approximately 12,264 ton, and net returns by $25,247 from the current practices. Alternative Conservation Systems allow farm operators to reduce erosion substantially, but at the same time be cost effective for a given situation.

KEYWORDS: Soil Conservation, Conservation Compliance, Linear Programming.

INTRODUCTION

Despite 60 years of federal efforts and billions of dollars spent for soil conservation, erosion persists as one of this country's major conservation problems. In addition to productivity losses, sediment from soil erosion has been considered the greatest single pollutant in U. S. surface waters, with cropland being the major contributor (NCAES, 1982). As soil erodes, the runoff carries with it such pollutants as fertilizer residues, insecticides, herbicides, fungicides, and animal waste, which, in excess, degrade water quality (Batie, 1983). These pollutants impose costs on society in terms of the costs of removing or correcting damage caused by increased health risks and the aesthetic cost of a degraded natural environment. Erosion-related pollutants are estimated to have imposed a cost of $6.1 billion per year, with cropland's share being $2.2 billion (Clark, et al., 1985).

The Food Security Act of 1985 represented some major changes in farm legislation with soil conservation and USDA program benefits becoming linked for the first time in history. It demonstrated a movement toward the consideration of broad social goals as well as those arising from the farm sector. Conservation Compliance is the most sweeping of the conservation provisions in the Conservation Title of the 1985 Food Security Act and the 1990 Food, Agriculture, Conservation, and Trade Act, not only in scope, but in its potential to encourage the nation's farmers to control soil erosion on highly erodible cropland. About one acre in three of all U. S. cropland--an estimated 140 million acres--is subject to the policy (Soil and Water Conservation Society, 1992).

The Conservation Compliance Provision was enacted, in part, to reduce sediment from erosion and to improve water quality. This provision discourages production of crops on highly erodible cropland. Crop production on highly erodible land without a locally approved soil conservation plan may disqualify farmers for certain USDA program benefits. Compliance is intended to reduce soil erosion by placing restrictions on the methods used to produce crops. Thus, Conservation Compliance adds additional complications to farmers' production decisions.

---

*R. G. Bowling, Research Assistant; B. C. English, Professor in Department of Agricultural Economics, The University of Tennessee, Knoxville.

The impact of Conservation Compliance is dependent on participation in commodity programs, since only participating farms are affected. Conservation Compliance, by forcing reductions in soil erosion, will add to the costs of program participation. The objective of this study is to analyze how Conservation Compliance will affect farmers' net return and soil erosion levels. To accomplish this objective, a linear programming model of a representative Beaver Creek Watershed farm was developed. The effects of Conservation Compliance were tested for the representative farm.

REVIEW OF LITERATURE

The methodology of this study was built on several earlier studies reviewed. Gillespie, et al., (1990), Brooks and Michalson (1983), Domanico, et al., (1986), and Hunter (1981) analyzed soil management practices using a budgeting approach, as in this study. Budgets were developed to represent the costs and returns for management practices. Domanico, et al., (1986) used linear programming to evaluate alternative management practices. The model was run to determine the trade-offs between income and rates of soil loss. This study uses a similar approach to evaluate the costs of reduction in soil loss under Conservation Compliance at the farm-level. The linear programming model in this study is based on models used by Domanico, et al., (1986), Huang and Uri, (1991), and Mims, et al., (1989) consisting of four parts: 1) crop production activities varying by crop grown, tillage practices, and rotations, 2) farm sales and purchases, 3) government programs, and 4) environmental impacts. Crowder, et al., (1985) used CREAMS to estimate losses of soil and nutrients, and then incorporated the coefficients into a representative farm linear programming model to analyze the trade-off between income and reductions in soil and chemical losses. A similar approach was used in this study using the Erosion Productivity Impact Calculator (EPIC) to obtain soil loss coefficients to be used in a linear programming model.

DATA

The Beaver Creek Watershed, located in Fayette, Haywood, Shelby, and Tipton counties in Southwestern Tennessee, was selected for this study because it has a high potential for nonpoint source pollution. A serious erosion problem exists in this area due to the erosive nature of the soils and the extensive use of the soils for clean cultivated row crop production. Farm operators in the watershed were surveyed to determine crop mix, tillage practices, participation with government programs, as well as other farm and farmer characteristics. This survey information, along with other data, were used to represent current practices.

The programming model was constructed for a representative 1,200 acre farm located in the Beaver Creek Watershed. For this analysis, only information from full-time farm operations was used. The crop mix consists of cotton, soybeans, soybeans/wheat double cropped, and grain sorghum. Based on survey findings, the farm had a 550 acre cotton base, 150 acre grain sorghum base, and 100 acre wheat base. Approximately 75% of the cropland is rented. All farm operators reported participation in government programs. The representative farm is made up of five soil types. These soils include Grenada-Loring-Memphis, and Falaya-Collins (USDA, 1987). Soils are excellent for agricultural production but in general are highly erodible (USDA, 1991).

For the conservation requirement of the 1985 Farm Bill to met, farmers must reduce erosion to a level at or below that which would occur under the Alternative Conservation System defined the by Soil Conservation Service. Alternative Conservation Systems (ACS) must achieve a substantial reduction in existing soil loss levels or potential soil losses, but at the same time be cost effective for the given situation (USDA, 1988). ACS, in some cases, incorporate physical structures such as terraces. These structures are site-specific to the geographical location of the farm and will not be included in the analysis. This study will evaluate the implementation of cropping sequence, conservation tillage, contour farming, crop residue, and cover crops.

For both the current and ACS, soil erosion was estimated using the Erosion-Productivity Impact Calculator (EPIC). Soil losses obtained from EPIC were incorporated into the farm model. Enterprise budgets were also developed for the crop practices and incorporated in the farm model. Data for the budgets were obtained from the *Tennessee Farm Planning Manual* (Johnson, 1991), the survey, *Conventional and Sustainable Agriculture Enterprise and System*

Budgets (Johnson, 1992), and personal conversations with SCS, ASCS, and Extension specialists. Deficiency payments were the only government benefits considered in this study.

METHODS

The linear programming model used in this study maximizes net returns over cash operating costs, subject to resource limiting, input purchasing, and output selling constraints. The objective function is specified such that the cost of production excluding land ownership costs, family labor, insurance, tax rates, and farm debt, is subtracted from the income of selling crop products and government payments. The model is constrained by the land, labor, and crop acreage base available. Commodity programs complicate production decisions through price and income supports, cropping restrictions, and temporary set-aside requirements (Glaser, 1986). To participate in government programs a farm operator must comply with the acreage reduction program and triple base.

There are two constraints that restrict the amount of land available to the producer. The first constraint reflects the amount of land used in a cropping practice, with the addition of acreage reduction, normal flex, and optional flex acreage, minus the land rented is less than or equal to the amount of land owned. The second constraint restricts the operator's ability to rent land. The operator can rent five land types. The amount of each land type rented is constrained to the amount of the land available of each type.

The labor constraint reflects the labor requirements for each crop practice, normal flex acreage, and optional flex acreage. Labor requirements were specified in four periods per year. Labor can be supplied by the family or purchased at $5.25 an hour. There was no limit placed on purchased labor.

The crop acreage base constraint restricts the amount of base available for each commodity crop to the amount of base owned. The constraint reflects the amount of base used in a crop practice, with the addition of acreage reduction, normal flex, and optional flex acreage, minus the base rented is equal to the amount of base owned.

To be eligible for payments on commodity crops, the producers must comply with the Acreage Reduction Program. Each commodity crop is subject to a certain percentage of its base to be put into the acreage reduction program. The acreage reduction constraint reflects the amount of base required for the acreage reduction program minus the percent of acreage reduction required on base rented is equal to the amount of acreage reduction acreage required on the amount of base owned.

For 1991-1995 program crops, deficiency payments will be based on 85 percent of the crop acreage base. The 15-percent difference is referred to as normal flex acreage. The normal flex acreage can be planted into other crops without losing their base. The normal flex acreage constraint reflects the amount of base required for normal flex minus the percent of normal flex required on base rented is equal to the amount of normal flex required on the amount of base owned. Producers also have the option of flexing an additional 10 percent of the farm's base called optional flex acreage. The optional flex acreage constraint reflects the amount of base for optional flex minus the percent of optional flex on base rented is less than the amount of optional flex on the amount of base owned.

The crop production constraints require all production to be sold. The crop production is transferred to the sale activities.

The base acreage minus acreage reduction and normal flex is eligible for deficiency payments. If optional flex acreage is placed into another crop, the base acreage minus acreage reduction, normal flex, and optional flex is eligible for deficiency payments. The deficiency payment constraint reflects the remaining base acreage times the program yield is transferred to the deficiency payment activity.

Each cropping practice results in different amounts of soil loss. The soil loss constraint is a transfer row. The soil loss level of each crop practice, normal flex practice, and optional flex practice is transferred to the soil loss activity. The soil loss activity reflects the total soil loss of the farm. A second soil loss constraint was evaluated. The amount of soil loss was restricted to less than 5 tons per acre. Under the constraint crop practices with an erosion level greater or equal to 5 tons per acre were not active.

RESULTS

The model was first simulated using the Beaver Creek Watershed survey results. A base farm income and soil erosion level was obtained from the linear programming model. Second, the farm model was simulated with the Alternative Conservation Systems, to find the crop mix with the highest net return. The soil loss under the ACS was also determined. An additional constraint was used in this study to limit the amount of soil erosion per acre to below 5 tons. With this constraint, the number of production alternatives available to the producer were reduced. In each case, it was assumed the producer remained in the commodity programs and therefore, maintaining the farm base acres was required.

Current Farm Practices

The farm size was developed from the Beaver Creek Watershed survey, a 1,200 acre row crop farm. The choice of crop enterprises (cotton, soybeans, grain sorghum, and soybeans/wheat) was held constant, as was the amount of family labor available. The model was designed to maximize net returns.

Under current practices, net returns for the representative farm was $171,196. The optimal crop mix for the farm is shown in Table 4.1. The base farm consisted of 412 acres of cotton, 103 acres of grain sorghum, 80 acres of soybeans/wheat double cropped, and 403 acres of soybeans. An additional 135 acres of cotton were planted as a result of the commodity crop flex acres. All crops were grown continuously. Sixty-eight acres of land were put into the Acreage Reduction Program. All available land was rented. The farm used 109 hours of hired labor in period two (April, May, and June). Total soil loss on the farm as a result of current farm practices was 16,184 tons, approximately 14 tons per acre. 1,061 acres (94%) of the land planted eroded at rates higher than 5 tons per acre, and 826 acres (73%) eroded at rates greater than 10 tons per acre.

Alternative Conservation Systems

With the implementation of Alternative Conservation Systems, the net return was $170,280, $915 less than base net return. The crop acreage mix remained the same, although the practices differed. Cotton was grown on the contour with residue on 547 acres including 135 acres of flex. Grain sorghum was grown on the contour with a cover crop on 103 acres. No-till soybeans were grown on 403 acres. Soybeans/wheat double cropped were grown on 80 acres. All available land was rented. The farm used 20 hours of hired labor during period four (October, November, and December). Total soil loss as a result of ACS was 7,030 tons or approximately 6 tons per acre. This is a decrease of 9,154 tons from the current practices. 747 acres (66%) of the land planted eroded at rates higher than 5 tons per acre, and 292 acres (26%) eroded at rates greater than 10 tons per acre.

Soil Loss Constraint

With a soil loss constraint of less than 5 tons per acre, the net return would be $145,949, a decrease of $24,331 from the ACSs and a decrease of $25,247 from the current practices. The crop mix remains the same, yet more soil conservation measures were added. No-till cotton with a cover crop was grown on 412 acres. Grain sorghum was grown on the contour with a cover crop on 103 acres. No-till soybeans were grown on 403 acres. Soybeans/wheat was double cropped on 80 acres. The flex acreage (135 acres) was put into soybeans/wheat double cropped. Also, the most erosive land (type 5), was put into the ARP. All land was rented. 290 hours of hired labor were used during period two (April, May, and June), and 105 hours were used during period four (October, November, and December). This was a result of additional soil conservation measures. The total soil loss was 3,919 tons, a decrease of 3,110 tons from the ACSs and a decrease of 12,264 tons from the current. All land was under 5 tons of erosion per acre.

Non-Compliance

If producers decided not to comply with Conservation Compliance, they would no longer be eligible for government benefits. However, the model indicated that net returns were the highest among all alternatives. Without participation, net returns would be $173,817, an increase of $2,621 from the current. Cotton was grown on all the land. According to the survey, this would not be realistic because a farm operator would diversify to reduce risk. For example, if the yield for cotton was decreased by a third, the resulting net return would be $47,085. With non-compliance the resulting soil loss would be 20,000 tons, approximately 17 tons per acre. All land exceeded 5 tons per acre, and 984 acres (82%) exceeded 10 tons per acre.

Comparison of the Alternatives

The implementation of Alternative Conservation Systems would have little effect on net return, but would significantly decrease soil erosion. However, movement to a 5 ton per acre erosion limit would result in a large decrease ($24,000) from that solution using the ACS. The trade-off curve between net returns and soil erosion is shown in Fig. 1. The farm operator could comply with a small change in net return. If compliance regulations were stricter and allowed less soil erosion, cost for the farm operator to comply would increase. At high erosion rates, the incremental cost of reducing erosion to current compliance levels is rather small, while the cost of reduction to a lower level of erosion will be more expensive. Cost of production increases at the more restrictive level and gross sales decline. If these changes are viewed by the producer as prohibitive, the farm operator would likely withdraw from federal programs altogether rather than cease production. Presumably the farmer would continue to create erosion at the non-compliance rate. Using current ACS, however, the farmer may be allowed to reduce the erosion rate significantly without economic hardship. The primary difference in the base, compliance, and non-compliance scenarios is tillage practice. Land tenure remains constant under the scenarios. The land use in the base farm and under compliance are the same. All base acreage would be used to maintain benefits. The only benefit from non-participation would be the opportunity to change planting patterns and increase the acres planted of a profitable crop for increased net returns. Under current and ACS, the crop flex acres would be put into cotton. If soil erosion constraints were set at less than 5 tons per acre, the flex acres would be put into soybeans/wheat double cropped. Also, no-till would be practiced on all land. Under ACS, the amount of labor used is less than current practices. This is due to the increased use of no-till. No-till requires less labor due to the decrease in number of field operations (Jolly, et al., 1983). Increased labor would be used with stricter erosion control due to the added soil conservation measures such as cover crops. If the farm operator decided not to comply, cotton would be produced conventionally on all land. Increased cotton production would also increase the labor use and erosion from the current level.

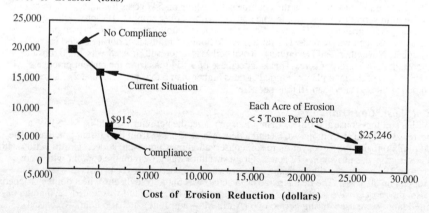

Figure 1. Trade-off Curve Between Net Returns and Soil Erosion For Representative Farm.

CONCLUSIONS

Under the assumptions of this study, a farm operator could comply with little change in net return. Under compliance, soil erosion would decrease by more than 9,000 tons. ACSs allow the farm operator to reduce erosion substantially while being cost effective. Currently, the farm is projected to plant and harvest in ways that result in 826 acres exceeding 10 ton per acre of erosion. Under compliance, this is reduced to 292 acres. This reduction in erosion is achieved by changing the cost of production by less than $1,000.

Richardson, et al. (1989) also found that implementation of the Conservation Compliance Provision using ACSs resulted in either small or no change in net farm income. Yet, if farm operators were forced to reduce erosion to less than 5 tons per acre, they would experience large losses in net farm income. Grumbach(1983) also found that if higher, less stringent soil loss limits could be established, there would be greater participation and at least some improvement in conservation.

In areas of high erosion, Alternative rather than Basic Conservation Systems appear to result in reduced erosion levels. While ACSs allow more erosion per acre, they are less expensive to adopt. Using the current strategy of voluntary participation, the farm operator might employ the ACSs. If Conservation Compliance standards are more restrictive, the cost of compliance for farm operators will increase. If the compliance costs exceed USDA program benefits, farmers likely will avoid compliance and forego the program benefits, resulting in higher erosion levels.

REFERENCES

1. Batie, Sandra S. *Soil Erosion--Crisis In America's Croplands*. Washington, D.C.: The Conservation Foundation. 1983.

2. Brooks, Robert O., and Edgar L. Michalson. *An Evaluation of Best Management Practices In the Cow Creek Watershed, Latah County, Idaho*. University of Idaho Agricultural Experiment Station. Research Bulletin No. 127. 1983.

3. Clark, Edwin H., Jennifer A. Haverkamp, and William Chapman. *Eroding Soils: The Off-Farm Impacts*. Washington, D.C.: The Conservation Foundation. 1985.

4. Crowder, Bradley, M., Harry B. Pionke, Donald J. Epp, and C. Edwin Young. "Using CREAMS and Economic Modeling to Evaluate Conservation Practices: An Application." *J. Environ. Qual*. Vol. 14, No. 3, 1985: 428-433.

5. Domanico, J. L., P. Madden, and E. J. Partenheimer. "Income Effects of Limiting Soil Erosion Under Organic, Conventional, and No-Till Systems in Eastern Pennsylvania." *Am. J. of Alternative Ag*. 1(1986):75-82.

6. Gillespie, Jeffrey M., L. Upton Hatch, and Patricia A. Duffy. "Effect of the 1985 Farm Bill Provisions on Farmers' Soil Conservation Decisions." *Southern J. Agr. Econ*. December, 1990: 179-189.

7. Glaser, L. *Provisions of the 1985 Food Security Act*. United States Department of Agriculture, Economic Research Service. Agr. Info. Bull. #498. Washington, D. C. 1986.

8. Grumbach, Alyson R. "Cross Compliance as a Soil Conservation Strategy: A Case Study of the North Fork of the Forked Deer River Basin in Western Tennessee." M.S. thesis. Virginia Polytechnic Institute and State University, May 1983.

9. Huang, Wen-Yuan, and Noel D. Uri. "An Analysis of the Marginal Value of Cropland and Nitrogen Fertilizer Use Under Alternative Farm Programs." *Agricultural Systems*. 35 (1991) 433-453.

10. Hunter, David Lee. "Economic Evaluation of Alternative Crop and Soil Management Systems For Reducing Soil Erosion Losses on West Tennessee Farms." Ph.D. dissertation. The University of Tennessee, Knoxville. December, 1981.

11. Johnson, Larry A. *Conventional and Sustainable Agriculture Enterprise and Systems Budgets For Southern Limited Resource Farms*. Project Report for Strategic Planning and Policy Analysis Staff, USDA Soil Conservation Service. December, 1992.

12. Johnson, Larry A. *Guide to Farm Planning*. The University of Tennessee Agricultural Extension Service. September, 1991.

13. Jolly, R. W., W. M. Edwards, and D. C. Erbach. "Economics of Conservation Tillage in Iowa." *J. Soil and Water Conserv*. 38(1983): 291-294.

14. Mims, Anne M., Patricia A. Duffy, and George Young. "Effects of Alternative Acreage Restriction Provisions on Alabama Cotton Farms." *Southern J. Agr. Econ*. December, 1989.

15. North Carolina Agricultural Extension Service. *Best Management Practices For Agricultural Nonpoint Source Control. III. Sediment*. In Cooperation with: U. S. Environmental Protection Agency and U. S. Department of Agriculture. August, 1982.

16. Richardson, James W., Delton C. Gerloff, B. L. Harris, and L. L. Dollar. "Economic Impacts of Conservation Compliance on a Representative Dawson County, Texas, Farm." *J. Soil and Water Conserv*. September/October, 1989: 527-531.

17. Soil and Water Conservation Society. *Implementing the Conservation Title of the Food Security Act: A Field-Oriented Assessment by the Soil and Water Conservation Society, Final Report 1992.* Ankeny, Iowa. 1992.

18. United States Department of Agriculture, Soil Conservation Service. *Beaver Creek USDA Water Quality Hydrologic Unit Area Work Plan.* January, 1991.

19. United States Department of Agriculture, Soil Conservation Service. 1988 *Section III. Technical Guide-Resource Management Systems, Basic Conservation Systems, Alternative Conservation Systems: Tennessee.* 1988.

20. United States Department of Agriculture, Soil Conservation Service. *The 1987 National Resources Inventory.* Computer Tape. 1987.

CONJUNCTIVE USE OF SURFACE AND GROUND WATER FOR CROP IRRIGATION IN THE WESTERN UNITED STATES

B. M. Crowder[*]

ABSTRACT

This paper discusses five artificial recharge and recovery projects in the western United States. Declining groundwater reserves and the shifting of water towards higher-value urban uses, environmental and wildlife purposes, and so forth have led agricultural producers to examine groundwater recharge and conjunctive use to manage crop irrigation water supplies. Conjunctive use of surface and ground water involves artificial recharge of aquifers and withdrawal during the irrigation season. Water is recharged when agricultural, urban, environmental, wildlife, and other demands are least on surface waters. Benefits of artificial recharge and groundwater storage include: (1) lower evaporation losses compared to surface storage, (2) avoidance of many costs and environmental damages associated with surface storage, and (3) slowing or reversing groundwater depletion and reduced pump lifts. Other benefits may include enhancing or restoring fish and wildlife habitats, flood control, and reuse of drainage water. Groundwater degradation generally is not a problem for artificial recharge projects, but where problems are encountered data have proven beneficial for planning future projects. **KEYWORDS.** Artificial groundwater recharge, Conjunctive use.

SUMMARY

Two primary purposes for artificially recharging agricultural aquifers are to: (1) meet peak demands, and (2) replenish an aquifer to meet future demands should surface and ground water supplies be depleted because of drought, over-pumping of groundwater, increased crop irrigation, and other supply and demand factors. Other significant objectives are flood control, wildlife enhancement, recreation, and reuse of irrigation drainage return flows.

Artificial groundwater recharge and storage, like surface storage, does not make new water available. However it may increase seasonal water supplies and improve the efficiency of irrigation water use. Irrigation districts typically recharge water from autumn through spring when demands are sharply curtailed and runoff is highest in most of the West. Most irrigation districts use existing water delivery systems for recharging water through earthen basins with permeable soils. Recovery involves pumping water from existing or project-dedicated production wells. Water is pumped from nearby wells or delivered to district irrigators through canals and laterals. Low-cost water is needed to efficiently recharge water for crop irrigation, a relatively low-value use. To minimize costs, most project sponsors use flood flows or seasonal excess stream flows for artificial recharge. There are now hundreds of projects throughout the West.

Widespread opportunities exist for recharge and conjunctive use, but each situation is unique. A project must be technically, economically, and financially feasible. The availability, cost, and reliability of a water supply for recharge are crucial factors. Delivery systems for both recharge and recovery must be planned. Finally, project sponsors must evaluate political and institutional acceptability by populations within and outside their water service area.

---

[*] B.M. Crowder, Economist, Bureau of Reclamation, P.O. Box 25007, D-5110, Denver, CO 80225. The views expressed are those of the author and do not necessarily represent the views of the Bureau of Reclamation.

Institutional constraints--permits, documentation, legal action, political action, and so forth--frequently stymie efforts to establish recharge projects. Endangered species and other environmental concerns are also mentioned as notable hurdles. While not insurmountable, these constraints add substantially to project development costs. Sometimes, restoration or enhancement of wildlife habitat provides significant benefits where spreading basins and in-channel recharge are used, and can increase project benefits and local willingness to pay.

Artificial recharge projects are perceived by crop producers as having significant net economic benefits, based on their willingness to pay for artificial recharge. Projects have been developed because of the cost effectiveness of artificial recharge relative to alternatives. Various mechanisms for cost recovery exist, but costs generally are recouped from general district revenues or by special assessments on those receiving benefits from a project.

INTRODUCTION

Five artificial recharge and recovery projects are discussed. Projects near Hollis, Oklahoma and York, Nebraska are in the High Plains States Ground Water Demonstration Program (High Plains Program). This program is a special study by the U.S. Department of the Interior, with technical support from the U.S. Environmental Protection Agency, to evaluate and demonstrate the potential for groundwater recharge in the High Plains and other western states. The other projects are in Bakersfield, California; Arvin-Edison Water Storage District, California (a Bureau of Reclamation water service contractor); and Julesburg, Colorado. Figure 1 shows project locations and Table 1 lists key project characteristics.

DISCUSSION OF HIGH PLAINS PROGRAM ARTIFICIAL RECHARGE PROJECTS

Blaine-Gypsum, Oklahoma[1]

This project is located outside Hollis, Oklahoma. The economy is based almost entirely on agriculture, and sustainable development requires stable and reliable sources of crop irrigation and drinking water. Sponsorship from Oklahoma's Water Resources Board (OWRB) to evaluate the water quality effects from injecting surface runoff into the aquifer has allowed the Southwest Water and Soil Conservation Association of Oklahoma (Southwest Association), the local sponsor, to participate in the High Plains Program.

Declining groundwater levels caused irrigation wells to run dry more than 40 years ago because of overdraft. Crop failures led Southwest Association farmers to experiment with artificial recharge to insure against crop loss, using gravity injection wells, as early as the 1950's. Their efforts demonstrated rapid response of wells to artificial recharge, with groundwater levels increasing throughout the area even as irrigation demands increased.

The karst aquifer, with layers of shale, gypsum, and dolomite, cannot be used for domestic purposes because of total dissolved solids (TDS), which average 2,800 to 3,000 mg/l, mostly magnesium sulfate and potassium sulfate. Recharge water is injected into confined aquifers and is comprised entirely of surface runoff and irrigation tailwater from cropland and adjacent lands. An injection well typically receives runoff from 160 to 240 ha of land. Costs for water quality monitoring were prohibitively expensive for the potential local benefits. The High Plains Program drilled and developed five injection wells, 16 monitoring

---

[1] Information based on interviews with R. Fabian, Oklahoma Water Resources Board, and P. Horton, Southwest Water and Soil Conservation Association, June 9, 1992.

Figure 1. Artificial Groundwater Recharge
and Recovery Projects

wells, and eight water level observation wells. A small earthen dam impounds water for one injection well and provides flood protection (primarily basement flooding) for homes nearby.

Federal cost sharing provides 80% of the funding. The local cost share includes land for the dam site, legal and related fees, and management and labor. The OWRB pays the balance of project costs from its general revenues. Total construction costs were over $320 thousand. Operation and maintenance (O&M) for the project is estimated to be less than $1,000 per month. The bulk of the costs are for water quality monitoring and analysis. A primary objective for the OWRB and the Federal agencies is to evaluate how much TDS in the aquifer are reduced by recharging agricultural runoff that average 200 mg/l TDS. Costs for

Table 1. Artificial Groundwater Recharge Projects for Crop Irrigation.

Project	Sponsor	State	Technology	Purpose	Water Source	Existing or Planned Annual Recharge[a] (10^9 m^3)
York	Upper Big Blue Natural Resources District	NE	Reservoir, spreading basins, injection	Crop irrigation & M&I	Runoff & industrial	0.6
Blaine-Gypsum	Oklahoma Water Resources Board	OK	Passive injection	Crop irrigation	Runoff	6.2
Bakersfield	Bakersfield Water Department	CA	Spreading basins	M&I and crop irrigation	Kern River, Central Valley Proj., & CA State Water Project	246.7
Arvin-Edison	Arvin-Edison Water Storage District	CA	Spreading basins	Crop irrigation	Central Valley Project	143.0
Julesburg	Lower South Platte Water Conservancy District	CO	Spreading basins	Crop irrigation	Unappropriated S. Platte River flows	3.4

[a] Annual recharge represents estimates of anticipated or targeted annual average recharge with the exception of the Bakersfield project, which is designed to recharge a maximum of 200 million m^3 of water annually.
M&I represents municipal and industrial use.

extensive monitoring would be prohibitive without Federal and State cost sharing; the four-year (1992-1995) monitoring program is estimated to cost $531 thousand.

A rural natural gas distribution system provides service for operating irrigation wells. Dues in the Southwest Association are based on a member's number of wells, and fees that are based on natural gas bills are used to support the Southwest Association's recharge efforts. Revenues depend on rainfall and other factors that affect the need to run irrigation pumps, and are used to further develop recharge capabilities in the service area. Strong local support, both for those activities being conducted independently by the Southwest Association and for the recharge and monitoring project under the High Plains Program, indicates a willingness to pay for the local costs of artificial recharge.

York, Nebraska[2]

The York project is in southeastern Nebraska, with a local economy based on agriculture. The Upper Big Blue Natural Resources District (District), the project sponsor, is the most irrigated district in Nebraska. Concerns about aquifer depletion and elevated groundwater levels of nitrates, atrazine, and other pesticides prompted development of a High Plains artificial recharge project. Project features include a dam and reservoir, spreading basins, water treatment facility, injection well, and monitoring well field. Three methods of recharge are being evaluated for effectiveness and practicality: (1) reservoir seepage, (2) spreading basins, and (3) injection after treating water to drinking water quality standards.

About 3.0×10^3 ha of land, nearly all irrigated cropland, drains into the recharge lake. The reservoir has 22 ha of surface area and normal storage of 380×10^3 m^3. Because runoff does not provide stable lake levels, the District acquires recharge water from a refrigerated storage facility. The cold storage water is native groundwater that is unchanged chemically before it reaches the lake via pipeline. Recharge is mainly "coincidental" through the lakebed. Water inflow and recharge are shown for December 1990 through April 1992:

```
Inflow:  Cold Storage pipeline = .975 x 10⁶ m³
                Watershed runoff = .296 x 10⁶ m³
                           Total = 1.271 x 10⁶ m³
                     Evaporation = .102 x 10⁶ m³
                         Storage = .037 x 10⁶ m³
                    Net Recharge = 1.132 x 10⁶ m³

Recharge Sources:      Reservoir = 1.087 x 10⁶ m³  (.385 x 10⁶ m³ annually)
                          Basins = .048 x 10⁶ m³   (.03 x 10⁶ m³ annually)
                   Injection well = .004 x 10⁶ m³  (.192 x 10⁶ m³ annually)
```

The Federal government provides about 80% of the project's costs, with the District covering the remainder. Development and construction were nearly $1.6 million, and O&M over a four-year period of demonstration was expected to be $200 thousand; the project has been extended for additional water quality monitoring. Significant wildlife habitat and recreation components of the project include a small visitor center, picnic facilities, fishing access, a boat dock, and wildlife and other plantings. Eight paired monitoring sites are used to evaluate the effects on water quality. Upfront costs for monitoring wells and associated data collection facilities were over $128 thousand, and annual data collection and chemical analysis costs for three years amounted to $38 thousand per year.

---

[2] Information based on interview with J. Bitner and J. Turnbull, Upper Big Blue Natural Resources District, York, Nebraska, May 14, 1992.

Concentrations of nitrate-nitrogen have been stable or declining in monitoring wells. However, cropland runoff during the spring of 1992 caused lake concentrations of atrazine to reach 167 ug/l. Little atrazine was detected in the aquifer, but injection was ceased because of concerns about atrazine. Because of those concerns, the York project added additional monitoring wells around the lake and the project was extended for two years beyond the original termination date to better understand the fate and transport of atrazine in the aquifer.

A lesson from York is that "coincidental" recharge is not covered by existing regulations to protect groundwater quality. Reservoirs and other impoundments are found throughout the Midwest and West, where substantial groundwater recharge occurs. An implication is that, if the lake is a significant source of atrazine in groundwater, other reservoirs and impoundments may be leaching pesticides and other chemicals to groundwater.

Another lesson is the value of the York project's recreation and wildlife features. Additional costs to provide these benefits are modest compared to total project costs, and they greatly enhanced users' willingness to pay for artificial recharge. Local (agricultural) users of groundwater are the beneficiaries of the project, and should experience reduced pumping lifts, but the project provides research, recreation, and other benefits to the District's members at large. Therefore, repayment comes from the District's property tax revenues.

DISCUSSION OF NON-FEDERAL ARTIFICIAL RECHARGE PROJECTS

Non-Federal projects (Fig. 1, Table 1) are integrated with existing local and Federal water supply projects. Each project is a pragmatic effort to cost-effectively recharge water. Therefore, compared to High Plains projects, monitoring and administrative costs are lower because they lack the research and demonstration objectives of the Federal projects.

Bakersfield, California[3]

Bakersfield lies at the southern end of California's San Joaquin Valley (Fig. 1). The city is experiencing rapid growth, and agriculture is the major water user throughout the San Joaquin Valley. Their artificial recharge project is an integrated urban-agricultural effort. Bakersfield contracts with users to recharge water in the recharge project. Facilities include an on-river settling and spreading basin and eight off-channel spreading basins. About 467 ha of recharge basins have been developed, with anticipation to develop another 79 ha.

Bakersfield has priority to spread water in its recharge facility. Spreading is permitted through contractual agreements with other districts and agencies. From 1978 to August, 1986, 9.39×10^6 m^3 of water were recharged and about 8.3×10^6 m^3 of water were banked. Many of the recharge basins were not completed by 1986, and the project is capable of recharging 2.5×10^6 m^3 of water per year now, when that much water is available. The Kern River (70% of the water from 1977 through 1986) is the primary source of recharge water. California State Water Project supplies were used for about 20% of the project's water, and the Bureau of Reclamation's Central Valley Project (CVP) provided the rest. Estimates of overdraft in the County during the 1987-93 drought are 24 to 30 m, or 6.0 or 7.0×10^9 m^3 (sustainable annual yield is about at 3.0×10^9 m^3). Primary benefits are increased reliability and stability of the water supply and savings in pumping lifts. Pumping lift benefits from the Bakersfield recharge project were estimated as $10-$12 per 10^3 m^3 of water, with lifts in the recharge area 27 to 37 m compared to 55 to 60 m in other areas.

---

[3] Information based on interviews with G. Bogart, Bakersfield Water Department; K.F. Totzke, Kern County Water Agency; April 5, 1993.

Vegetation is allowed to grow in the recharge basins. Grasses and other vegetation keep soil pore spaces open and reduce O&M. Primary O&M involves sterilizing levees and disking the river channel, with water delivered by gravity flow. The recharge facilities restore wildlife habitat and provide a valuable resting and foraging site for waterfowl and other migratory birds, and habitat for many resident species. Substantial recharge is accomplished while both providing wildlife benefits and minimizing O&M costs. Water use by plants is offset by the values of native habitat restoration, valuable in the arid San Joaquin Valley.

Construction expenditures spanned the period from 1977 to the present, and total about two million dollars. Annual operation and maintenance expenditures are estimated to be about $1 to 1.5 million, based on typical water deliveries for artificial recharge and the project's fees per volume of water. Contracts for recharge fees pay the full costs of Bakersfield's project. If a water entity pays its costs upfront, the cost to recharge water in Bakersfield's water bank is about $6.75 per 10^3 m^3. An O&M fee was established at $.405/m^3 for recharge in 1984, and this has been indexed upwards with inflation. A facilities improvement fee was set at $2.30/m^3 in the same year. The spreading fee currently is $4.05/m^3. All costs for the project are covered by user fees. To these fees, the costs for water would have to be added to determine the total cost for a user to obtain and recharge water in the water bank. The Bakersfield recharge project successfully accomplishes groundwater storage and recovery among area water users. Project O&M fees are maintained at levels which encourage users to store water rather than use their water unwisely or develop other uses for their water. Financing of economically viable projects is viewed as secondary relative to the need for State and Federal permits, National Environment Policy Act and California Environmental Quality Act compliance, public involvement, and the political process.

Arvin-Edison Water Storage District, California[4]

The Arvin-Edison Water Storage District (Fig. 1), east of Bakersfield, provides its agricultural contractors with stable, reliable water supplies for sustainable production. Though Arvin-Edison has an entitlement of 49 x 10^6 m^3 of "firm" water from the CVP, they received as little as 12.3 x 10^6 m^3 in the 1977 drought. Up to 386 x 10^6 m^3 of "non-firm" water is delivered during wet years. In addition, Arvin-Edison has an exchange agreement to reduce fluctuations in its water supply, assuring a minimum of 158.3 x 10^6 m^3 of water during most years. When water deliveries fail to meet irrigation demands, previously recharged water is pumped from groundwater wells to meet the deficiency.

Arvin-Edison has 53 x 10^3 ha of land. About 40 x 10^3 are developed for irrigated crops, with about 21 x 10^3 of those receiving surface irrigation. Water demand totals 185 to 222 x 10^6 m^3 for those 21 x 10^3 ha. Remaining acreage receives water from irrigation wells. Primary crops include vineyards, truck crops, potatoes, cotton, and citrus. In spite of the artificial recharge project, cropland has fallen steadily from over 45 x 10^3 ha since 1982.

CVP water service costs Arvin-Edison about $16.20 per 10^3 m^3. Existing delivery systems and other facilities deliver recharge water to their spreading basins. There are two spreading basins, comprising a total of 314 ha. Fifty production wells are operated within the spreading basins. Groundwater overdraft is reduced by providing water to land that previously was irrigated with groundwater. Artificial recharge, combined with reduced groundwater pumping, has stabilized declines in groundwater supplies. Groundwater overdraft exceeded 250 x 10^6 m^3 per year prior to the CVP. Overdraft now averages 12 x 10^6 m^3 per year, estimated with current land use to be 155 x 10^6 m^3 per year in the absence

---

[4] Information based on interview with C. Trotter, Arvin-Edison Water Storage District, April 5, 1993.

of artificial recharge. Average annual water table declines are 0.2 m, versus 2.6 m had artificial recharge not been developed. Eventually, there are plans to use 1.2×10^9 m^3 of underground storage capacity for regulating the District's water supply.

Arvin-Edison's recharge ponds have operated since 1966. Maximum recharge is about 5.66 m^3 per second. Prior to 1993, approximately 1.2×10^9 m^3 of water had been percolated through Arvin-Edison's spreading basins, while more than 3.7×10^9 m^3 of water were delivered to surface irrigators. A net increase of 538.2×10^6 m^3 of water was accumulated in underground storage as of February 1992. This accumulated storage was down from a maximum accumulation of 860×10^6 m^3 at the end of the 1986, prior to the drought.

All costs for artificial recharge in Arvin-Edison's conjunctive use program are paid by district irrigators. Composite water service charges to irrigators in the surface water service area total $64 per 10^3 m^3. Other service charges are levied on all lands in Arvin-Edison, and amount to $21.37 per ha in 1993. The project has demonstrated willingness to pay by farmers in the service area for conjunctive water management and groundwater storage.

Julesburg, Colorado[5]

The Lower South Platte Water Conservancy District (District) is located in the northeastern corner of Colorado (Fig. 1). Just west of Julesburg, a joint project between the District and the State of Colorado is recharging water through spreading basins to mitigate the effects of cropland irrigation pumping. Recharge takes place from October to April when flow requirements under a South Platte River agreement between Colorado and Nebraska are not restrictive. Water is recharged through spreading basins into the upper edge of the river alluvium. Because the aquifer is unconfined, recharged water moves down gradient towards farmlands which have reduced groundwater levels because of over-pumping.

The project site covers 17.4 ha, with about four ha in two recharge ponds. Recharge water originates in the South Platte River and is delivered to the recharge site through an existing ditch. The goal is to recharge 3.1 to 3.7×10^6 m^3 of water per year. Beneficiaries of the project are nearby farmers with about 60 wells who irrigate 400-600 ha of cropland. Besides raising groundwater levels and reducing irrigation pumping costs, the project's seasonal ponds provide waterfowl and wildlife habitat.

The total cost (not including administrative expenses) was approximately $80 thousand. About $60 thousand was for construction of recharge facilities and monitoring wells, paid by the State and completed in 1992. Monitoring wells are sampled quarterly for nitrates and a few other constituents. No significant water quality degradation concerns have arisen.

Irrigators pay O&M costs, including (1) $43.24 per ha of cropland to the District for its normal expenses, (2) $1.62 per 10^3 m^3 of water to the ditch company for recharge water ($5 to $6 thousand per year), and a special assessment of $3.71 per ha for groundwater pumping to cover O&M. An additional assessment for the project is an increase from $25 to $60 to permit new wells in the project area. Fees for recharge water and O&M are about $8 to $10 thousand per year. Should annual recharge average 3.1 to 3.7×10^6 m^3, the variable cost of recharge will be approximately $2.63 per 10^3 m^3 of water. The availability of inexpensive recharge water and the lack of Federal monitoring requirements allowed easy, inexpensive, and cost-effective artificial recharge for these farmers.

---

[5] Information based on interview with M. Law, Lower South Platte Water Conservancy District, October 20, 1992.

CONCLUSIONS

Opportunities for cost-effective artificial recharge and recovery depend on site-specific alternatives for acquiring and managing water. Calculating the cost per volume of water recharged and recovered is not meaningful for determining the efficiency of the practice. Project sponsors are concerned with meeting peak irrigation demands, political and institutional costs, financial costs, and system management. They must satisfy their water demands to continue cropping activities. A more useful measure of the efficacy of artificial recharge is a comparison of its costs relative to local supply alternatives. Depending on management problems and opportunities, artificial recharge may allow better management of available surface supplies without the attendant political, environmental, and financial problems typically associated with dam proposals.

To achieve economic efficiency and equity, costs for artificial recharge should be borne by beneficiaries so that they have appropriate incentives to efficiently use recharged and recovered water. The government's (Federal or state) role, at a minimum, should be protection of public lands and resources. Particularly, the government should assure compliance with environmental laws, regulations, and other institutional requirements. An additional role may be actions to enhance the efficiency and equity of water use among competing interests, such as encouraging districts to price water at its replacement cost.

Estimating benefits of artificial recharge and recovery often is difficult because of integration with water service and delivery systems. The marginal value of artificial recharge projects is at least as great as local costs, as expressed by users' willingness to pay for them. Hundreds of artificial recharge projects, with many more in development or construction phases, indicate a growing awareness by urban and agricultural users that the technology can contribute to their water management needs.

REFERENCES

1. The Arvin-Edison Water Storage District, *The Arvin-Edison Water Storage District Water Resources Management Program*. October, 1992.

2. Campbell, Jane and the City of Bakersfield, Department of Water Resources Staff, *City of Bakersfield 2800 Acre Groundwater Recharge Project*. City of Bakersfield, November, 1986.

3. Personal interview, R.J. Bitner and John Turnbull, Upper Big Blue Natural Resources District, York, Nebraska, May 14, 1992.

4. Personal interview, Gene Bogart, Bakersfield Water Department and Kane F. Totzke, Kern County Water Agency, Bakersfield, California, April 5, 1993.

5. Personal interview, Robert Fabian, Oklahoma Water Resources Board, Oklahoma City, Oklahoma, and Paul Horton, Southwest Water and Soil Conservation Association, Hollis, Oklahoma, June 9, 1992.

6. Personal interview, Marian Law, South Platte Water Conservancy District, Julesburg, Colorado, October 20, 1992.

7. Personal interview, Cliff Trotter, Arvin-Edison Water Storage District, Arvin, California, April 5, 1993.

Irrigation Storage Reservoirs as a Water Supply Solution for Upper Telogia Creek

William R. Reck[1]

ABSTRACT

The Upper Telogia Creek Watershed is located in Gadsden County, Florida and comprises an area of approximately 44,000 acres. Farmers within this area are facing severe water shortage problems. The local groundwater aquifers are not highly productive and most irrigation water is drawn from Upper Telogia Creek. During droughts, withdrawal from the creek has caused water shortages at several locations which has raised concern for the ecological health of Telogia Creek and for water availability for downstream farmers. Sediment, nutrients, and pesticide residue in runoff and irrigation tailwater were also considered as potential problems.

The Soil Conservation Service (SCS) initiated a study of the water supply and water quality problems of the Upper Telogia Creek Watershed. Alternative solutions to the problems were formulated, and the effects of these alternatives considered. It is anticipated that this study will be used to make decisions about the siting and design of facilities to store water, and to estimate water quality and water use efficiency improvements obtainable.

The study approach was the analysis of upland, on farm, irrigation storage reservoirs (ISR). The ISR would be used to store runoff water to be used for irrigation, thus reducing the draw upon the creek for irrigation purposes. Eighteen sample sites within the basin were evaluated using the computer program Simulator for Water Resources in Rural Basins (SWRRB). The analysis showed that the use of ISR's produced a net reduction in volume and frequency of pumping from the creek. Other benefits of ISR's considered include settling of sediment, and retention of nutrients and pesticides. BPM's for operation of ISR's were evaluated for flexibility to pump from the creek during high flow and reduce pumping during low flow. This type of BMP is expected to reduce the ecological stress to the stream system.
Installation of ISR's were calculated to reduce the total inflows to the creek by a small percentage during periods when creek levels are elevated by stormwater.

WATERSHED PROBLEMS

The 1992 Water Quality Assessment Section 305(b) Technical Appendix states: "Telogia Creek has severe nutrient and dissolved oxygen problems in the upper portions due to runoff from the Gretna Water Treatment plant spray fields. This facility is under a consent order to correct the problem. Nutrient and weed problems extend several miles downstream". Agricultural activities in the watershed can contribute sediments, nutrients, and pesticides to Telogia Creek by surface water runoff and irrigation tailwater which, with sufficient

---

[1] Agricultural Engineer, USDA Soil Conservation Service, Gainesville, Florida. "All SCS programs and services are offered on a nondiscriminatory basis without regard to race, color, national origin, sex, age, religion, maritial status or handicap."

quantity, will impair water quality. Problems can be greatly aggravated by low flows in the creek due to excessive pumping for irrigation.

The watershed encompasses almost 41,000 acres, of which about 3,330 acres (almost 10 %) are irrigated. The ground water aquifers in Gadsden County are not highly productive, and therefore almost all irrigation water is withdrawn from surface sources. During dry periods, when the creek is at its lowest, the pumping of irrigation water from the creek and its tributaries has been sufficient to essentially "dry up" the creek. This loss of volume in the creek causes severe stress on the ecosystem of the creek. During these times the diversity of macroinvertebrates, the basis of the food chain, is limited to species adapted to high water temperatures, and low dissolved oxygen. In addition, the productivity of the macroinvertebrates is reduced. A healthy ecosystem requires a broad based and productive food chain.

DESCRIPTION OF RECOMMENDED PLAN

The recommended plan is the installation of irrigation storage reservoirs (ISR) which will collect stormwater runoff and store it for use as supplemental irrigation. The reservoirs were evaluated for their effectiveness in reducing the volume of pumping from Telogia Creek and its tributaries.

Although the investigation did not find sediment, nutrients or pesticides to be a problem, the use of an ISR will reduce the amount of these pollutants which reach Telogia Creek. A significant amount of nutrients from agricultural surface runoff would also be collected in the reservoir and be reused through irrigation.

Since groundwater aquifers have limited productivity, most agricultural irrigation water in the basin is drawn from Telogia Creek and its tributaries. The instillation of an ISR would significantly reduce the dependence upon the creek for irrigation supply water and give the farmer the flexibility to store irrigation water for dry periods.

As an example, a 40 acre fall tomato field could reduce the volume of water pumped from Telogia Creek by 70% with the installation of a reservoir with 8 ac-ft of storage. This example was based upon average conditions for the watershed and effects of storage on individual fields will vary. By pumping into the ISR from Telogia Creek following a rainfall of greater than one inch (during periods when creek flows are elevated by stormwater), a farmer would reduce the volume pumped from the creek to irrigate the crop during drier periods. This reduction would reduce the stress on the creek. This type of strategy would not eliminate stress on the ecosystem of the creek, but would rather reduce the length of time this stress would occur.

INVESTIGATION AND ANALYSIS

Eighteen sample sites within the basin were evaluated using the computer model SWRRB for 5 reservoir sizes under 4 alternative BMPs to determine the reservoirs effect upon pumping from Telogia Creek. An example layout of an ISR can be seen in Figure 1. BMPs for operation of ISRs were evaluated for flexibility to pump from the creek during high flow and reduce pumping during low flow.

FIGURE 1

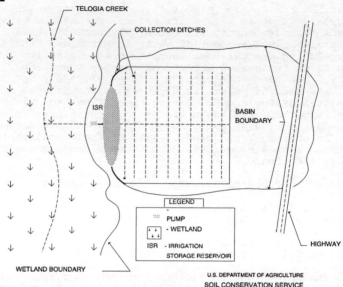

Physical Setting
Inventory information on field size, shape, location, elevations, and irrigation types were taken from Northwest Florida Water Management District permit applications and ASCS maps. The cropland in this watershed consist mainly of tomatoes grown with drip irrigation, sod and nursery grown with overhead irrigation, and 1600 acres of cotton, peanuts, beans, and corn grown under center pivot irrigation.

The investigation was based upon modeling of 16 sample sites within the basin. The drainage area of the sample sites ranged from 18 to 120 acres. The proposed ISRs were modeled for 6 fall, 3 spring, and 7 spring and fall tomato fields. The spring and fall tomato fields were fields with adjacent spring and fall fields or fields which were used for both seasons which could share one ISR. Tomatoes were chosen for the modeling process because of the large amount of permitted irrigation volume for tomatoes and the existence or plans for ISRs for the nursery operations within the watershed. Figure 2 shows the permitted irrigation volume by crop in the watershed.

Climate
The model SWRRB was run using historical climate data from Quincy, Florida. Rainfall and Temperature data were used for the years 1937 through 1989. Evaporation was calculated internal to the model using Ritchie's Equation. Evaporation was calibrated based upon available Class A pan data for the IFAS Quincy Experiment Station and Woodruff Dam. The monthly normal rainfall and potential evaporation for this basin is shown in Figure 3.

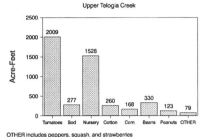

Figure 2: Permitted Annual Irrigation Volume
Upper Telogia Creek

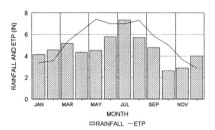

Figure 3: Monthly Rainfall And Potiential ET
Upper Telogia Creek

Water Budget

The SWRRB computer model used for this project is a continuous watershed scale model. This model was chosen for its ability to model adjacent fields with different land uses. Another major advantage of SWRRB was the ability to change the SCS curve number throughout the year which accounted for the change from tomatoes grown under plastic with a high curve number to a small grain winter cover crop with a lower curve number. The SCS curve number is a parameter dependent upon soil type, land use, and moisture condition which is used to calculate volume of surface runoff resulting from a storm event. SWRRB was used to calculate daily runoff, evaporation, nutrient and pesticide movement, sediment transport, crop growth and biomass accumulation, soil water movement, etc. A daily water budget was created using daily output from SWRRB. Reservoir storage was calculated daily based upon rainfall, runoff, calibrated evaporation, seepage, irrigation usage, overflow, as well as volume pumped from surface water. This daily water budget was used to sum annual water pumped from creek, number of time water was pumped from creek (Dry Pumping Days), and annual nutrient, pesticide, and sediment loads.

Irrigation was scheduled based upon IFAS recommendations (McClure). For fall tomatoes, irrigation was scheduled using 1/4 pan evaporation (PE) for the first three weeks after transplanting, 1/2 PE for second three weeks, 3/4 PE for third three weeks, and 1/2 PE for remainder of the season. For spring tomatoes, 1/2 PE was used to schedule irrigation throughout the growing season since IFAS does not have data to support a staggered approach similar to fall tomatoes for this growing season.

Reservoir Management Practices

Several reservoir management scenarios were modeled to determine there effects upon the creek.

RMP #1 was the management practice of filling the reservoir prior to the growing season. This could be accomplished very slowly with minimal effect upon Telogia Creek.

RMP #2 was the management practice of pumping into the reservoir from the creek during the growing season after a rainfall event which causes runoff. This practice would reduce stress on the stream since water would be pumped from the stream during periods when creek flows are elevated with storm water.

Original was the management practice of pumping from the creek only when irrigation water is needed. This option does not allow flexibility to pump prior to the season or when creek elevations are elevated. Combination of RMPs #1 and #2 was also evaluated to see the if these practices would have any additional effect together.

Results and Discussion
This section will show the analysis of the effectiveness of the system from several stand points with respect to reservoir size: volume pumped from the creek, overflow to creek, dry pumping from the creek, maximum dry pumping from the creek, effectiveness of BMPs, and sediment yield.

Volume Pumped From Creek
Volume pumped from the creek was calculated as the amount of water pumped from the creek for irrigation. This does not include water pumped from storage reservoir which has collected runoff water. Average volume of water pumped annually per acre of tomato cropland is presented in Figure 4. This figure shows average of all the sites modeled for the range of reservoir sizes modeled. There is a significant reduction in the volume of water pumped with a reservoir with as little as 3 acre feet of storage. Spring, spring/fall and fall tomato crops all showed similar percentages of reduction with increasing reservoir size.

In addition to volume pumped, the effect on pumping from the creek during drought was analyzed. It was found that the 1 year 7 day drought pumping was reduced 10 to 20% and similarly the 10 year drought pumping was reduced 5 to 15% depending upon reservoir size.

Annual Sediment Yield
Figure 5 shows the average annual sediment yield reduction for all sites modeled versus the reservoir size. The sediment yield reduction was different for each site modeled and was dependent upon soil type, slope, and channel configuration. There was a large reduction of annual sediment yield with the installation of a reservoir, but increasing reservoir size had only a small effect upon total sediment yield.

Figure 4: Average Annual Creek Pumpage
Upper Telogia Creek

Figure 5: Average Annual Sediment Yield
Upper Telogia Creek

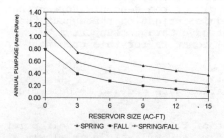

Annual Overflow
One concern for the installation of an ISR is that such a structure would not be effective since by capturing storm runoff water for use as irrigation, the reduction of stormwater runoff from the installation of an ISR would have to same effect as pumping from the creek to irrigate. In order to determine that installation of an irrigation storage reservoir does not cause this type of problem, overflow from the reservoir was considered in the analysis.

Upper Telogia Creek Watershed was divided into subbasins to determine the percentages of agriculture for different subbasins. The average subbasin which contains agriculture had an average percentage of cropland of 25%. There were a couple small subbasins which had up to 85% cropland. Annual overflow from cropland which has had a ISR installed will have overflow reduced by 50 to 70% depending upon the size of the reservoir. Since this water is captured during a runoff event, creek flows will be elevated. Even if the watershed is in drought conditions and the ISR captures all runoff for its collection area, the creek will still receive flow from 75% of the drainage area. Average annual overflow for all the modeled sites is shown in Figure 6 with respect to reservoir size.

Dry Pumping Days
Dry pumping days occur when water needed for irrigation is not available in the irrigation storage reservoir and must be pumped directly from the creek. Since these dry pumping days will likely occur during the driest periods, the design of a ISR should try to minimize the number of these days. Figure 7 shows average annual dry pumping days for spring tomatoes for each of the RMPs modeled which is discussed in more detail later.

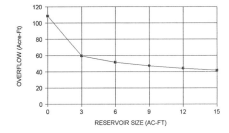

Figure 6: Average Annual Overflow
Upper Telogia Creek

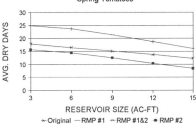

Figure 7: Average Dry Pumping Days
Upper Telogia Creek
Spring Tomatoes

Reservoir Management Practices
In order to evaluate the effectiveness of the reservoir management practices, the management scenarios were compared for their effectiveness in reducing the volume of water pumped from Telogia Creek and reducing the number of dry pumping days. Four scenarios were modeled: normal operation, RMP #1, RMP #2, and a combination of RMP #1 and RMP #2. The combination of RMPs had the highest volume pumped for all reservoir sizes and normal operation had the lowest volume pumped.

RMP #1 and original had roughly the same number of dry pumping days for each reservoir size. This shows that the benefit of pumping prior to the season is insignificant. RMP #2 and the combination of BMPs were also roughly the same for all reservoir sizes, but were significantly less than RMP #1 and original. This shows that RMP #2 has a significant impact upon the number of days that pumping occurs during droughty conditions.

Design
In order to aid in the design of these irrigation storage reservoirs, a relationship between average dry pumping days, reservoir size, and crop area was determined. A relationship was found between these three parameters based upon crop type. Dry pumping days was found to be linearly related to reservoir

storage volume (ac-ft) per acreage of cropland. Figure 8 and Figure 9 shows this relationship for fall tomatoes and spring tomatoes when RMP #2 is used.

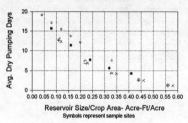

Figure 8: Reservoir Size Design
Upper Telogia Creek
Fall Tomatoes - RMP #2

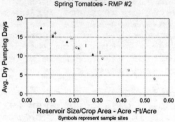

Figure 9: Reservoir Size Design
Upper Telogia Creek
Spring Tomatoes - RMP #2

By determining the desired average number of dry pumping days and reading from the appropriate graph the corresponding reservoir size per acre of crop land can be determined by multiplying reservoir size/crop area by the crop acreage, the storage volume of the reservoir can be determined. For example, if Farmer A has 40 acres of fall tomatoes and plans on using RMP #2, from Figure 8 we determine that for an average of 12 dry days per year he would need 0.19 acre-feet of storage volume for every acre of crop land. Using this information, Farmer A needs 7.6 acre-feet of storage. If Farmer B has the same plans for a spring tomato field, from Figure 9 he would need 0.28 acre-feet per acre for a total storage volume of 11.2 acre-feet. The value of RMP #2 can be seen by doing the same calculations for normal operation and comparing the answer. Farmer A (40 ac of fall tomatoes) would need 14 acre feet compared to 7.6 and Farmer B (40 ac of spring tomatoes) would need 21 acre feet of storage compared to 11.2 acre feet.

CONCLUSIONS

These results of this analysis indicate that the best method of reducing the stress of agricultural production upon Telogia Creek and its tributaries is the installation of ISRs. Since there is no data available on how much reduction in pumping volume is needed to secure the ecological health of the stream, the exact size of the needed ISR is not known. Once this information is estimated, Figure 10 should be used to determine the proper storage capacity based upon crop area irrigated. RMP #2, pumping into the reservoir when creek flows are elevated, is recommended despite the additional cost of pumping since this practice will add a measure of insurance for drought periods. It is not expected that installation of an ISR will eliminate the need for pumping from Telogia Creek, but rather reduce the dependence upon the creek for water and add flexibility to when pumping is necessary.

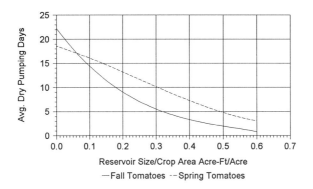

Figure 10: Design Curves for ISR
Upper Telogia Creek Watershed
With Use of RMP #2

Further information on the Upper Telogia Creek study is available from the SCS. This study information can be found in the *Upper Telogia Creek Watershed Cooperative River Basin Study, Gadsden County, Florida*.

REFERENCES

Florida Governor's Energy Office, 1993. Telogia Creek Irrigation Energy Conservation Demonstration on Mulched Staked Tomatoes.

Locascio, S.J., Olson, S.M., and Rhoads, F.M., 1989. Water Quantity and Time of N and P Application for Trickle-irrigated Tomatoes, Journal of the American Society for Horticultural Science, Vol. 114(2), March.

McClure, Scott, 1992. Knowing Tomato Water Needs Helps Growers Justify Usage, Florida Grove and Vegetable Management,

Northwest Florida Water Management District, Consumptive Use Permit Information for Telogia Creek Basin.

Rhoads, F.M., Irrigation Use by Mulched Staked Irrigated Tomatoes in North Florida, Institute of Food and Agricultural Sciences, University of Florida.

Rhoads, F.M. and Olson, S.M., 1990. Irrigation of Fall Tomatoes in North Florida, Institute of Food and Agricultural Sciences, University of Florida.

Smajstrla, A.G. and Locascio, S.J., 1990. Irrigation Scheduling of Drip Irrigated Tomato Using Tensiometers and Pan Evaporation, Proc. Fla. State. Hort. Soc. 103:88-91.

CHEMICAL MIXING CENTERS FOR LOADING AND MIXING OF CHEMICALS

Jesse T. Wilson, PE [1]

ABSTRACT

With the many chemicals used in Florida's agriculture and Florida's sandy soils, there is a potential for ground and surface water pollution resulting from the loading and mixing of chemicals. The Jackson Karst USDA Hydrologic Unit Area (HUA) located in Jackson County, Florida, was chosen to demonstrate best management practices (BMPs) to prevent the contamination of ground water from agricultural chemicals. The Jackson Karst HUA consists of over 45 325 ha (112,000 acres) of cropland characterized by karst topography with limestone at or near the surface. Due to the karst topography, intensive agriculture operations and high rainfall, contamination of the aquifer from agricultural chemicals was an identified problem. The use of chemical mixing centers (CMCs) was identified as an alternative best management practice (BMP) for landowners to use when loading and mixing chemicals. Standard and specifications were developed for the design and operation of CMCs in the Jackson Karst HUA. This paper describes the requirements for the planning and designing of CMCs.

PURPOSE

The use of CMCs provide an environmentally safe permanent structure for the loading and mixing of chemicals. CMCs are applicable to all areas of Florida where agricultural chemicals are loaded and mixed.

BASIC DESCRIPTION

A CMC provides an impermeable pad (concrete) at which pesticides (insecticides, herbicides, fungicides, etc.) are transferred from a transport or storage container to application equipment. The transfer may be as simple as pouring the pesticide into a spray tank or may involve the use of sophisticated pumping and metering equipment. In most cases the pesticide is mixed with a diluent, usually water, at the same site.

The typical sequence of events at a chemical mixing center is:

   1. Begin filling the pesticide application equipment tank (spray tank) with water.

   2. Open pesticide container(s) and pour into spray tank.

   3. Wash empty container and add container wash water to the spray tank.

   4. Continue filling the spray tank (usually 760 L to 1890 L (200 to 500 gal) in capacity) until full.

---

[1] Jesse T. Wilson is the State Conservation Engineer with the USDA Soil Conservation Service, Gainesville, Florida.

Differences from this typical sequence occur if the pesticide is mixed with oil rather than water (primarily for some applications of insecticides by aerial applicators), mixed with a fertilizer, or if the pesticide is transferred using a closed system involving pumping equipment and metering directly from the container to the application equipment.

SYSTEM COMPONENTS

A properly designed CMC include components necessary to properly manage chemical materials and prevent pollution of ground and surface water supplies. Components of a typical CMC include the following:

1. A sloped concrete pad of sufficient size to accommodate the spray equipment and contain the chemicals from the largest spray tank or rinsate tank in the event of a spill.

2. A collection sump, sump pump, and appropriate safety devices.

3. A permanent roofed structure over the concrete pad designed to keep out rainwater.

4. Adequate water supply for mixing chemicals, rinsing chemical containers and rinsate tanks, and for cleaning the concrete pad.

5. Water supply pump, pipeline, and back flow prevention devices.

6. Tanks for storage of rinse water.

7. Water hoses and accessories for rinsing and cleaning operations.

8. Emergency washing areas.

9. Were needed, a platform to facilitate the loading of chemicals in the spray tank.

DESIGN CONSIDERATIONS

Each chemical mixing center shall be designed to meet the needs of the user.

The planning, design, and construction shall insure that the CMC is sound and of durable materials commensurate with the anticipated service life, initial and replacement costs, and safety and environmental considerations.

Location. The CMC shall be located as follows:

1. Adjacent to or near the chemical storage building.

2. As far as practical from streams, ponds, lakes, wetlands, and wells with a minimum recommended distance of 30 m (100 ft).

3. As far as practical from known sinkhole and subsurface anomalies with a minimum recommended distance of 30 m (100 ft).

4. The CMC should be isolated from other buildings used to store feed, seed, petroleum products, livestock, and from residences. The CMC should be located downwind of these buildings.

5. It is recommended to locate the chemical mixing center above the 100 year flood plain elevation and as a minimum above the 25 year flood plain elevation. The CMC shall be elevated above the surrounding ground to prevent runoff water from entering the facility.

6. At sites that have not been used as stationary mixing/loading sites in the past.

Size and Capacity. The size of the concrete pad used for chemical mixing shall be as needed to accommodate the length, width, height and capacity of the largest sprayer used at the facility. The chemical mixing pad shall be sloped to allow for drainage of water and pesticide spills to a collection sump and curbed to prevent outside runoff water from entering the chemical mixing pad and for providing storage of chemical spills. The chemical mixing pad shall be sloped a minimum of 2% toward the sump. The chemical mixing pad including the sump shall have a minimum storage capacity of 950 L (250 gal) or equal to 1.25 times the largest storage or spray tank brought onto the pad.

The entrance to the chemical mixing pad shall be graveled, paved, or otherwise treated to provide a suitable entrance for the equipment and to prevent erosion and the tracking of sediment onto the chemical mixing pad.

Water Supply, Pump, and Pipe. A permanent water supply shall be provided for filling the sprayers, rinsing the chemical containers, spray tanks, and chemical mixing pad, and for emergency washing if an applicator is exposed to chemicals. A pipeline shall be installed for conveyance of water from the water supply to the chemical mixing center. Back flow preventers, antisiphon devices or a minimum 5 cm (2 in.) air gap shall be installed on all water supply lines from a well or other water source. If a pump and well are installed, it shall be located outside of the chemical mixing pad and meet the distance requirements listed under Design Considerations.

Enclosure. The chemical mixing center shall be roofed to prevent rainfall from entering the system. Enclosure supports shall not be installed within the concrete pad area. The enclosure shall have a minimum eave height of 3 m (10 ft) and shall have a minimum overhang of 30 degrees from the edge of the pad or 0.6 m (2 ft), whichever is greater to prevent rain from blowing in on the chemical mixing pad. Side walls may be constructed on one or more sides to reduce the amount of overhang required. Enclosed buildings shall be designed with proper ventilation.

Emergency Washing Area. CMCs shall include emergency washing facilities (eyewash/overhead shower) for use by the applicator in case of exposure to chemicals. The emergency washing area shall be conveniently located on the pad and easily assessable to the applicator.

Loading Platform. A loading platform shall be constructed where needed to facilitate the filling of the spray equipment. The platform shall be of sufficient height and size so as to provide a safe workable area.

Rinsate Storage Tanks. Rinsate storage tanks shall be provided to temporarily hold rinsates resulting from cleaning of the chemical mixing pad or sprayer if the operator intends to store or accumulate rinse water for use as a pesticide or as a diluent. The tanks shall be labeled with the type of chemicals and target crops. Tanks shall be fiberglass, polyethylene, or other durable material and have the capacity to meet the requirements of the operation plan. The rinsate tanks shall be located on the chemical mixing pad and near the sump.

Sump. The sump size shall be as small as practical but of sufficient size that it will easily accommodate the pump and provide easy access for the removal of sediment accumulated in the sump. The maximum size of the sump should be limited to a capacity of 190 L (50 gal) and covered with a metal grate for safety.

Sump Pump. The sump pump shall be a chemical resistant submersible pump and should create a minimum of turbulence within the sump. A filter shall be installed between the sump pump and sprayer or rinsate tanks. All electrical components shall be waterproof and explosive proof. The pump shall be manually operated.

Structural Design. The structural design shall consider all items that will influence performance, such as design analyses, methods, and assumptions, construction methods and quality control, and operational exposure, use, maintenance, and repair.

Minimum requirements for chemical mixing centers are specified as follows:

1. Steel. AISC Specifications for the Design, Fabrication, and Erection of Structural Steel for Buildings.

2. Timber. NFPA National Design Specifications for Wood Construction. The roofed structure shall be designed to withstand the applicable deadloads and wind loads.

3. Reinforcing steel. ASTM A-615, Grade 60

4. Concrete. A watertight concrete design shall be used to avoid leakage from the sump and chemical mixing pad. A minimum of 152 cm (6 in.) of well compacted granular subbase shall be placed prior to concrete placement. The minimum concrete thickness of slabs and sump shall be 127 mm and 203 mm (5 in. and 8 in.) respectively. The minimum reinforcement for slabs shall be equal to that of 10 by 10 gage, 152 mm by 152 mm (6 in. by 6 in.) welded wire fabric. Final pad and sump thickness and reinforcement shall be designed based on anticipated loading. The slab and sump shall be poured in one pour without expansion joints or openings. Portland cement Type I or II shall be used. The maximum size aggregate used shall be 25 mm (1 in.) The concrete slab shall have a minimum 28 day strength of 28 000 kPa (4000 psi). The slab shall be coated with a chemical resistant impervious sealant to protect the concrete from chemicals.

COSTS

Figure 1 and Figure 2 show a typical CMC installed in Florida agricultural operations. The construction costs for a typical 7.4 m by 8.0 m (24 ft by 26 ft) CMC is approximately $8500.00.

PLANS AND SPECIFICATIONS

Plans and Specifications for chemical mixing center should be prepared and describe all construction requirements. They shall be site specific and shall describe the requirements for installing a CMC to achieve its intended purpose. Plans and specifications should include construction plans or other similar documents. These documents are to specify the requirements for installing the practice, such as the kind, amount, or quality of materials to be used, or the timing or sequence of installation activities.

OPERATION AND MAINTENANCE

An Operation and Maintenance (O&M) Plan shall be prepared for all CMCs. The O&M Plan should specify all operation and maintenance needs necessary to properly operate and maintain the structure in accordance with the requirements of all local, state, and federal laws and regulations. A copy of the O&M plan shall be located at the chemical mixing center. The operation and maintenance (O&M) plan should include an inventory of chemicals used at the center and the methods proposed for handling of sediment, rinsate, and potential spills. An emergency response plan shall be part of the O&M plan. Warning signs, such as "WARNING, HAZARDOUS CHEMICALS, KEEP OUT", shall be posted at all entrances to the facility.

All pesticide spills shall be collected immediately and applied to the target crop at recommended pesticide application rates.

The sump should be thoroughly cleaned between the mixing and loading of different chemicals. The resulting rinsate can be applied as a dilute pesticide to a labeled site or used as make-up water for subsequent batches of pesticides that are labeled for the same crop. Sediment from the sump shall be removed with proper precautions taken to reduce exposure of the worker to any potential contaminants in the sediment. This sediment should be land applied to the target crop at a rate below the label recommendation. The sediment shall be removed prior to a switch from one crop to another crop. The sump shall be pumped dry at the end of each day of operation.

Normal winterization procedures to prevent damage to the facility and to chemical containers shall be performed when weather conditions dictate.

The center should be inspected periodically to insure back flow prevention devices are operating satisfactory. The concrete pad and sump shall be checked for leaks and cracks and repaired immediately. The concrete sealant shall be checked for wear and repaired as needed. The rinsate storage tanks shall be checked to insure proper labeling and methods for applying rinsate back to the land are being followed.

The United States Department of Agriculture (USDA) prohibits discrimination in its programs on the basis of race, color, national origin, sex, religion, age, disability, political beliefs and marital or familial status. (Not all prohibited bases apply to all programs). Persons with disabilities who require alternative means for communication of program information (Braille, large print, audio tape, etc.) should contact the USDA Office of Communications at (202) 720-5881 (voice) or (202) 720-7808 (TDD). To file a complaint, write the Secretary of Agriculture, U.S. Department of Agriculture, Washington, D.C. 20250, or call (202) 720-7327 (voice) or (202) 720-1127 (TDD). USDA is an equal employment opportunity employer.

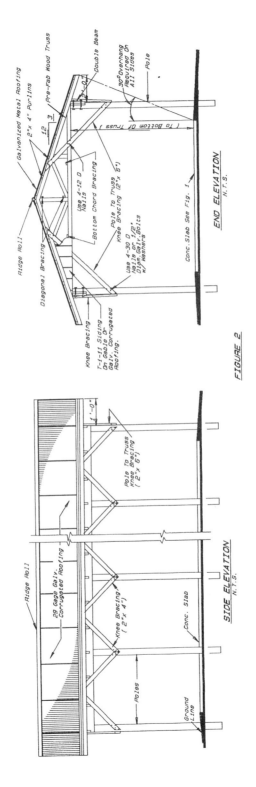

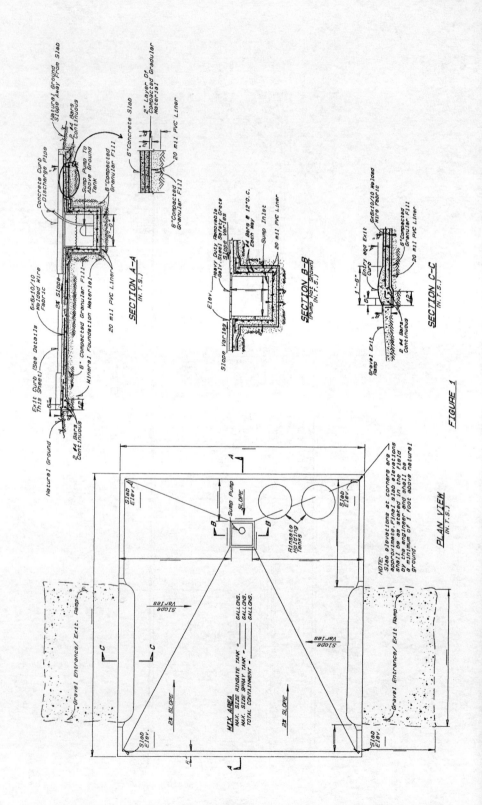

STATE OF THE ART AGRICHEMICAL HANDLING FACILITY: BUTLER'S ORCHARD

J.G. WARFIELD, C. M. GROSS. AND T.N. DAVENPORT[*]

ABSTRACT

Safe handling and use of agrichemicals is of paramount concern to farmers, ranchers, and the general public. To assist agriculturalists with this concern, The U.S. Department of Agriculture has introduced a special project to demonstrate how agrichemicals can be handled, stored, and used so that the environment and the agrichemical user are protected. Maryland is one of eleven states participating in this innovative USDA Agricultural Stabilization and Conservation Service special project to fund the construction of Agrichemical Handling Facilities (AHF). The AHF is a permanent structure with an impervious surface to increase the efficiency of agrichemical use and provide an environmentally safe area for the handling of on-farm agrichemicals, such as pesticides and fertilizers, that are used in spraying operations of orchards, vineyards and cropland. These state of the art facilities are specifically designed to USDA Soil Conservation Service conservation practice standards to accommodate the equipment and needs of an individual farm operation.

An AHF consists of five major components:

1. A specifically designed and treated concrete floor to contain accidental spills and/or rinsate water;
2. A sump to collect all liquid from the concrete floor;
3. A rinsate storage and transfer system;
4. Safety features such as water for clean up and an emergency eye wash/shower; and
5. A farm specific "Operation and Maintenance Plan", including agrichemical inventories, floor/site plans, emergency procedures and safety/handling protocols.

The first AHF built in Maryland, funded by the special project, is located at Butler's Orchard. Butler's Orchard is a 300 acre pick your own fruit, vegetable, cut flower and Christmas tree operation located in a rapidly urbanizing section of Montgomery County, Maryland. Aware of and very sensitive to environmental issues and public scrutiny, the Butler family opted to install this environmentally sound agricultural practice. The objective is to provide both economic and management savings in their pesticide spraying operation.

This paper details the planning, design, installation, and operation/maintenance aspects of the AHF, with emphasis on the environmental, economic, social, and management benefits related to the practice.

KEYWORDS. Agrichemical handling facility, agrichemicals, pesticides, pesticide containment, environment, environmental concern/safety.

---

[*]J.G. Warfield, District Conservationist, USDA- Soil Conservation Service, Derwood, MD; C.M. Gross, Water Quality Specialist, and T.N. Davenport, Civil Engineer, USDA Soil Conservation Service, Annapolis, MD.

INTRODUCTION

Awareness of and sensitivity to the nation's environmental problems are forcing society, including the agricultural sector, to re-evaluate current modes of operation. Agricultural producers as well as environmentalists and government agencies need to implement and promote management practices which enhance environmental protection while maintaining an abundant food supply and the economic vitality of agriculture. Safe agrichemical storage and handling is but one environmental issue facing agriculture today.

The reasons warranting the construction and use of an Agricultural Handling Facility, (AHF) are numerous. State, federal, and local regulations are in place or being developed, which are designed to better protect the environment and the consumer. These regulations include pesticide use restrictions, set backs/buffer strips, storage and transport regulations, ground and surface water protection plans, detailed record keeping requirements, as well as numerous safety concerns.

Agrichemical storage and handling practices have been targeted as a "point source" for potential groundwater contamination by federal and state legislation across the United States (Kammel and Walsh, 1989). In fact, more than 125 legislative bills are introduced by state and federal governments each year to regulate pesticides across the country. Nearly 40 percent of these are motivated by public concerns that agricultural chemicals contribute to the contamination of groundwater (Reichenberger, 1990). AHF's are a form of insurance, providing environmental safety by preventing spills from entering the soil, surface water, and ground water.

The USDA- Soil Conservation Service (SCS) has recognized the need to develop solutions to solve many of the environmental problems facing agriculture today. The Agrichemical Handling Facility (AHF) located at Butlers Orchard is one example of the state of the art technology being employed by agricultural producers for sustaining an environmentally sound and productive agricultural industry in an urbanizing United States.

By definition, the Agrichemical Handling Facility is a "permanent structure with an impervious surface to provide an environmentally safe area for the handling of on-farm chemicals, such as pesticides and fertilizers, that are used in spraying operations of orchards, vineyards, and cropland." An AHF shall be located a minimum of fifteen meters (50 feet) from a well and as near as practicable to the agricultural chemical storage area. The complete AHF system shall include all those components necessary to properly handle the mixing, loading and rinsate of any agrichemical mixture to prevent direct pollution to the environment. Proper use, management, and maintenance of the complete AHF system is key in operating this practice successfully. Each AHF has a farm operation specific "Operation and Maintenance Plan" developed prior to initial operation. This plan details general operational notes, operator/operation specific data, as well as emergency/safety and handling items.

Butler's Orchard: A Case Study

The first AHF built in Maryland is located at Butler's Orchard. This is a 300 acre family owned and operated "pick your own" fruit, vegetable, cut flower and Christmas tree operation located in a rapidly urbanizing area of Montgomery County, Maryland. The farm is 3 miles from the I-270 Technology Corridor at Germantown, 15 miles from the Washington Beltway, and 25 miles from the White House. Well over a million people live within that 25-mile radius. In addition to crop production, Butler's Orchard offers weekend harvest festivals, hayrides, and a farm market. They are used to having the public on their farm from Memorial Day until Christmas.

The Butler family now has two generations actively involved in management and production. They raise approximately 25 different crops from apples to zucchinis on highly erodible soils. The Butlers are long term cooperators with the Montgomery Soil Conservation District and are aware of and sensitive to environmental issues and concerns. Most of their cropland is under irrigation, and they are very concerned with water quality and quantity. For the last several years the Butler family has been studying methods to meet the environmental challenge and regulatory responsibility of using agricultural chemicals.

The goal of the Butler's Orchard agrichemical handling facility was to improve the safety and efficiency of storing, handling, mixing, loading, and application of agricultural pesticides. Good management is one of the key ingredients to evaluate before constructing an agrichemical handling facility. The current and future regulation of pesticides demand a commitment to detail. As in any successful agricultural operation, Butler's Orchard is well managed. This management is evident by their adept handling of hundreds of people and many crops every day.

The Butlers do all their own pesticide application. Due to numerous crops, small acreages, and critical timing, they have chosen not to utilize custom applicators. Integrated Pest Management (IPM) has reduced the amount of pesticides used, but has made timely applications and appropriate rates even more critical. Many of the chemical applications are made very early morning, late afternoon or evening, or at night.

Planning of the Facility

In the initial stages of planning, the Butlers had many decisions. The first concern was financial: even though this project was partially funded by Agricultural Stabilization and Conservation Service (ASCS) special project funds, the Butlers still had a significant financial commitment.

The AHF needed to be constructed in an area of the farm away from the public areas, yet convenient to the crops. The Butlers decided to build a new building recessed into a wooded area adjacent to the shop/storage/greenhouse complex. This would keep the new building out of public view and yet still be reasonably close to water and electricity, both of which were needed.

Locating the building on that site would also take pesticide storage and loading further away from a stream that flows through the property. Trees on three sides of the building shield it from sight and also shade the facility during the summer in order to maintain a uniform temperature.

The Butlers also decided to make the facility meet their needs: they incorporated a chemical storage room, a record keeping-safety equipment room, and a mixing room with the pesticide containment area where applicator loading, unloading, cleaning, and equipment repairs take place. They also needed a well lighted facility for their mixing and loading during darkness.

With the goal of safety and efficiency in mind, the Soil Conservation Service (SCS) team looked at SCS conservation practice standards and specifications for this practice (SCS, 1992). Typically, an AHF consists of five parts:

1. A specifically designed and treated concrete floor to contain spills and/or rinsate water;

2. A sump to collect all liquid from the concrete floor;

3. A rinsate storage and transfer system;

4. Safety features such as water for clean up and an emergency eye wash/shower; and

5. A farm specific "Operation and Maintenance Plan."

Design of the Facility (Figure 1)

The concrete pad was sized to accommodate the largest sprayer to be used on the farm. It was finished smooth enough so water would flow easily, yet rough enough to prevent the operator from slipping. It has a minimum two percent slope towards the sump. The pad also has a curb that will contain any potential spills on the pad. The curb around the exterior was designed to the convenience of the operator. All concrete was treated with a chemically resistant, non-vapor barrier forming coating to prevent contamination of the concrete.

The sump contains a chemically resistant sump pump to transfer any spillage/rinsate/wash water through a cartridge type filter (chosen for ease of cleaning and/or replacement) into a bank of valves that determines which rinsate storage tank receives the liquid. For safety, the sump is covered with a flush metal grate.

Four rinsate tanks were used to provide segregated storage of different products or to allow compatible products to be comingled. The four rinsate tanks were mounted in a heavy duty steel tank rack. All piping was PVC Schedule 40. The tanks were plumbed to a waterproof transfer pump located on the pad. This pump can transfer stored rinsate into the sprayer for use as make-up water for the next appropriate application. Any left over, unused pre-mixed product can also be stored separately to be used for later application. A 300-gallon fresh water tank mounted above the rinsate tanks provides clean water to fill the sprayer tank by gravity. This reduces the time spent filling the sprayer, therefore improving efficiency. A roof was built over the floor that will keep out rain, therefore eliminating the need to store potentially contaminated rainwater.

A mixing room, record storage/safety equipment room, and a secured pesticide storage facility complete the AHF system. Besides a sink and counter space, the mixing room contains a pre-mixing barrel. This barrel has a sump pump to circulate and mix wetable powder-type pesticide products and then can pump the solution into the sprayer. After that the barrel can be rinsed to avoid any contamination. The floor of the mixing room slopes to a low corner with an outlet to the sump on the loading pad. There is a 10 cm (four inch) concrete curb around this, and every room, to contain any spillage.

Other features in this room include a scale to weigh pesticides, a ventilation fan and a window. This window between the mixing room and the loading/rinsate pad allows the operator to keep close control over the loading of any premixed products into the sprayer tank, and allow someone on the outside to check on the operator.

The record keeping room has files and storage for protective gear. In keeping with restricted use pesticide regulations, the records of all pesticide applications are maintained in this room. The pesticide storage area is heated and ventilated. It has shelving along all walls. There is also a small sump in the floor.

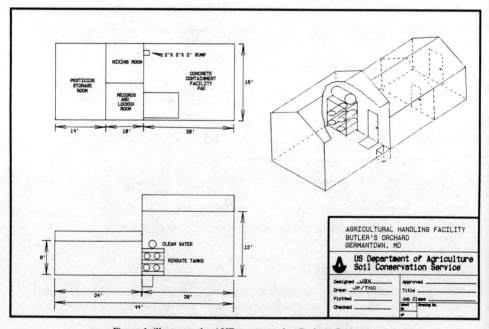

Figure 1, illustrates the AHF constructed at Butler's Orchard.

Safety Features of Agrichemical Handling Facilities

A complete AHF must be designed with adequate worker protection and environmental safety in mind. Many pesticides sold for agricultural use can cause severe illness, even death, if misused (Meyer and Daum, 1988). In order to reduce the risk of injury, human safety must be paramount in the design and operation of these facilities. The exact amount and type of safety equipment required will depend on the type of facility, number of employees, types of products handled, and regulations.

A permanent water supply must be provided to the AHF for emergency washing, filling the sprayers, and rinsing the chemical containers and the pad. Water lines connected to pesticide mixing and rinsate storage tank systems are susceptible to backflow of pesticides and fertilizers into the water system. Backflow preventers provide for a safe water supply by reducing the chance of back siphoning of pesticides or fertilizer into the well or other water source. It is recommended that this facility be located a minimum of 15 meters (50 feet) from a well. Antisiphoning devices, backflow preventers, and a method to allow winterizing of the pipelines should be installed on all water supply lines.

One of the biggest benefits of an AHF is that a shower and an emergency eyewash station should be installed for immediate and easy access. The emergency facilities should be located in an area with no hazards in the way of workers. Select emergency showers that provide a 114 liters/min. (30 gpm) capacity and eyewash equipment that provides 9.5 liters/min. (2.5 gpm) (MWPS, 1991).

A well designed facility will have adequate space for the storage of personal protection equipment. Personal protective equipment might include respirators, aprons, unlined protective gloves, unlined boots made of rubber, loose fitting coveralls with long sleeves, and goggles. Product labels specify the appropriate protective equipment and clothing to wear (MWPS, 1991). All personal protection equipment should be located in an accessible area, partitioned from the stored pesticides.

AHF's are a form of insurance, providing environmental safety by preventing spills from entering the soil, surface water, and ground water. The pad collects spills that occur during loading and mixing, contains any chemical leaks if a tank should rupture and retains rinse water from sprayer cleaning. Repairs and maintenance of application equipment should occur on the pad when possible. The SCS recommends that the containment volume for a pad covered by a roof be sized to contain 125 percent of the volume of the largest sprayer tank that will be located on the pad (SCS, 1991). The volume from a 2-year, 24 hour storm event over the entire pad should be added to the containment volume for structures not covered by a roof (SCS, 1991).

The complete AHF system can be used for storage of written documentation and should include Material Safety Data Sheets (MSDS) for all stored products, maintenance records, a record of chemical application inventory control, and the emergency action plan. Records should be kept for several reasons. As of May 10, 1993, certified private pesticide applicators must keep records of Federal restricted-use pesticide applications (USDA, 1993). Record keeping may also be required by insurance carriers to show how risks are minimized. An emergency response plan will assure that important information is immediately available if an emergency occurs. A copy of the emergency response plan should be provided to the local fire department. Each of these reasons will help minimize the risk to human health and environmental contamination.

OPERATION AND MAINTENANCE

The cost of constructing a well designed AHF system is an earnest financial investment that may exceed $10,000. Maintenance and management are essential and may be required by law on all systems to protect the investment and assure the facility functions as designed. The life of a containment facility can be substantially extended and its performance improved with regular maintenance (MWPS, 1991). Illustrated below is a sample Operation and Maintenance Plan.

OPERATION & MAINTENANCE PLAN
AGRICHEMICAL HANDLING FACILITY (AHF)
NAME_____
COUNTY_____

I. GENERAL OPERATIONAL NOTES

Test the AHF using clean water before the first use.

Keep the AHF clean at all items. The pad should be kept free of items not necessary for storing, mixing, loading, and clean-up operations.

Thoroughly inspect the AHF on a regular basis. The inspection should include, but is not limited to the pad, coatings on the interior surfaces of the pad and sidewalls, roofs, doors, access roads, ramps, hoses, pipes, backflow prevention devices, valves, connectors, filters, tanks, related plumbing material, safety equipment, electrical systems, access controls and runoff controls. Consult the Operation and Maintenance plan during inspections. Complete any needed repairs and replacements prior to using the AHF.

Immediately respond to any spills, leaks, accidents, or normal operational procedures from which pesticides or pesticide contaminated water come in contact with the pad.

Liquids shall not be allowed to remain in the sump.

Sump discharge water holding tank(s) should be labeled and then emptied as soon as possible. Liquid in the sump discharge water holding tank(s) should be land applied to the target crop at a rate below the label requirements or used as makeup water for the next chemical application if product is identical or compatible.

Do not drain rinse water or rinsate from the sprayer or pesticide containers onto the pad as a standard practice.

Sediment collected in the sump should be removed periodically to reduce buildup. This sediment should be land applied to the target crop at a rate below the label requirements. At a minimum, sediment should be removed prior to a switch from one crop to another.

Cross mixing of various pesticides or pesticide contaminated water shall be avoided except where allowed by the pesticide label.

All materials which come in contact with pesticides and pesticide contaminated material shall be handled as required by state regulations and pesticide labels.

Clean the pad by triple *rinsing* or power washing with a biodegradable detergent after use and after any spill requiring clean-up.

II. OPERATOR SPECIFIC DATA

The following information should be kept at the storage area/mixing facility, and your home. Use the emergency telephone numbers as appropriate for emergencies involving chemicals, whether they result from, fire, spillage, transportation accidents, natural disaster, or poisoning:

l) A list of emergency telephone numbers:

- police **911**

- fire **911**

- For poisonings: Maryland Poison Control Center **1-800-492-2414**

- For emergencies except poisonings: Maryland Emergency Response Division **410-631-3800**

- For pollution, toxic chemical & oil spills: National **1-800-424-8802**

- For general information: National Pesticide Telecommunications **1-800-858-7378**

- For significant spills: Chemtrec **1-800-424-9300**

2) An inventory showing the names and quantities of pesticides

3) Material Safety Data Sheets (MSDS) for each pesticide stored and/or used on site.

4) A floorplan of the AHF showing the location of pesticides stored.

5) A site plan indicating the AHF, other buildings, houses, sewers, wells, underground structures and pipes, livestock areas, direction of runoff and other sensitive areas or nearby environmental hazards. Update chemical inventories, floor plans and site plans as needed.

Give the local fire department a copy of items 2), 4), and 5) mentioned above. Provide the fire department copies of updated chemical inventories, floor plans and site plans.

III SAFETY AND HANDLING

Keep all chemicals out of reach of children, pets, livestock, and irresponsible people.

Post and maintain highly visible, weatherproof warning signs in accordance with state regulations and pesticide label requirements to indicate to anyone entering the AHF that chemicals are handled and stored there. Also post and maintain highly visible "NO SMOKING" signs.

Have clean-up materials and equipment (kitty litter, sawdust or absorbent material, plastic lined containers, small shovel, broom, dustpan, etc) readily available.

A fire extinguisher approved for chemical fires, and first-aid equipment, including the emergency eyewash and emergency shower, should be easily accessible.

Periodically inspect and maintain all safety equipment including fire extinguishers, backflow prevention devices, first aid equipment and other equipment (as required by pesticide label).

Empty pesticide containers shall be kept on the pad until the containers are returned to permanent/primary pesticide storage or are decontaminated in accordance with state regulations and pesticide labels.

Provide safety training for all personnel who will be using the AHF.

The inspection schedule for an AHF varies based on the type of facility and the geographic location. In all cases, the schedule must require thorough inspections on a seasonal or at least yearly basis. Outline a routine inspection procedure that includes a checklist to be followed. The inspection should include, but is not limited to the pad, coatings on the interior surfaces of the pad and sidewalks, roofs, doors, access roads, ramps, hoses, pipes, backflow prevention devices, valves, connectors, filters, tanks, related plumbing material, safety equipment, electrical systems, access controls and runoff controls (USDA, 1992).

Practice "good housekeeping" at the facility on a daily basis. The pad should be kept free of items not necessary for storing, mixing, loading, and cleanup operations (USDA, 1992). Immediately respond to any leakage or spill. The concrete pad can be cleaned by triple rinsing or power washing with a biodegradable detergent after use and after any spill requiring cleanup (USDA, 1992). Remove or repackage any small containers that develop leaks (MWPS, 1991). Liquids should not remain in the sump and must be used according to the operation and maintenance plan. Minimizing exposure of the concrete to pesticides and fertilizers increases the life of the protective coating and concrete.

Depending on the geographic location of the AHF, cold weather preparation may be one of the major components of the Operation and Maintenance Plan. Washdown of the pad for final season cleanup helps flush product from sumps and plumbing. The pumps and plumbing must then be protected from possible ice damage due to trapped water. It is also important that the facility manager devise a management plan that prevents excess storage of chemicals during the cold weather months. A well designed AHF system and proper planning will minimize the effects of weatherization.

Conclusion

The AHF system will provide for the containment and isolation of spillage from on farm agrichemical mixing, loading, unloading, and rinsing operations in order to minimize risk of contamination to humans and our environment.

REFERENCES

1. Midwest Plan Services. 1991. Designing Facilities for Pesticide and Fertilizer Containment. Iowa State University. Ames, Iowa.

2. USDA/SCS. 1992. Agrichemical Handling Facility (AHF), Maryland, (Code 596I).

3. USDA/Agricultural Marketing Service. 1993. Federal Pesticide Record Keeping Requirements for Certified Private Applicators of Federal Restricted-Use Pesticides. Pesticide Records Branch, Manassas, Virginia.

4. Daum, D.R. and D.J. Meyer. 1988. Pesticide Storage Building. Agricultural Engineering Fact Sheet. The Pennsylvania State University, University Park, Pennsylvania.

5. Reichemberger, L. 1990. Concrete Answers to Chemical Concerns. Farm Journal, Mid March 1990: 16-19.

6. Kammel, D.W. and P. Walsh. 1989. Technical Approaches for Reducing Legal Liability Associated with Agrichemical Storage. ASAE meeting Presentation Paper No. 892675.

The United States Department of Agriculture (USDA) prohibits discrimination in its programs on the basis of race, color, national origin, sex, religion, age, disability, political beliefs, and marital or familial status. (Not all prohibited bases apply to all programs). Persons with disabilities who require alternative means for communication of program information (braille, large print, audiotape, etc.) should contact the USDA Office of Communications at (202) 720-5881 (voice) or (202) 720-7808 (TDD).

To file a complaint, write the Secretary of Agriculture, U.S. Department of Agriculture, Washington, D.C., 20250, or call (202)-720-7327 (voice) or (202) 720-1127 (TDD). USDA is an equal employment opportunity employer.

PESTICIDE CONTAINMENT: AN ON-FARM SYSTEM

Robert V. Carter Jr.*

ABSTRACT

An on-farm pesticide loading facility to improve water quality has been designed and installed in Henderson County, North Carolina. This facility includes a roofed concrete slab sloping to a sump for collecting chemical spills. It is equipped with a storage building for storing farm chemicals. It also has elevated chemical rinsate tanks for storing spilled materials or wash water. A nurse tank is also elevated and is equipped with a device to prevent sprayer tanks from overflowing when being filled. The facility has a pressurized water system which supplies water for a nurse tank, container rinsing, and sprayer washdown and rinsing. Using this facility as a model, the Soil Conservation Service in North Carolina has developed an interim practice standard for agrichemical handling facilities. This new practice has also been approved for 75% cost sharing under the North Carolina Agriculture Cost Share Program, a state funded water quality program.

ON-FARM CHEMICAL HANDLING - IMPROVEMENT NEEDED

The manner in which agricultural chemicals are stored and handled varies greatly from farm to farm. All too often, however, chemicals are handled next to a water source where sprayers are loaded. Stories abound of fish kills, soil contamination, soil erosion at the loading site from heavy traffic, improper disposal of containers, improper storage of materials etc. Some of the problems associated with improper handling can be prevented by the installation of well designed agrichemical handling facilities. Such a facility has been designed in Henderson County, North Carolina. Henderson County, which is located in the mountain valleys of North Carolina, is a logical area for developing such a facility since it is one of the largest apple producing counties in the eastern US. Apple farmers in this area use large amounts of pesticides and tend to load these materials at the same location year after year, usually beside a pond or stream. With the increasing emphasis on both groundwater and surface water protection, many growers have begun installing pesticide storage buildings, mixing pads, sprayer wash sites etc. on their own. Many growers have also expressed interest in installing a complete well-designed facility for handling pesticides and other farm chemicals.

OBJECTIVES IN FACILITY DESIGN

In order to assist local growers in installing agrichemical handling facilities, the Soil Conservation Service in Henderson County began designing a system which would meet local grower's needs. The current practice evolved over a period of years during which extensive revisions in design took place. Originally a design was considered which would have provided an uncovered graveled pad for the grower to park his tractor and sprayer on while loading the sprayer. The pad would have been sloped away from the water source and would have directed runoff through a grassed waterway. The objective was to prevent spilled pesticides from flowing directly into the water source. This idea was soon abandoned due to groundwater considerations and the practice objectives eventually evolved to their current state.

---

*Robert Carter, District Conservationist, USDA, Soil Conservation Service, 140 Fourth Ave. West, Hendersonville, NC 28792

administered through the local soil and water conservation district. As a result of this first installation, approval has been authorized for cost sharing such practices statewide. The USDA, Soil Conservation Service has developed a state interim practice standard for Agrichemical Handling Facilities for North Carolina. This standard is based upon the design used for the McConnell facility.

COST

The cost of the first facility was approximately $20,000.00. About 75% of this amount was cost shared. The grower performed part of the work himself and contracted the rest.

FUTURE IMPROVEMENTS

Five other local growers have been approved to install facilities using cost share funds. Several modifications are being made to the designs for these facilities as a result of the experience gained from the first system. These modifications are listed below to provide insight into the learning curve experienced throughout this project:

    Block building substituted for frame chemical storage building for easier construction

    Electric valves and switches replaced with manually operated devices for simplicity

    Concrete curbs and humps or "speed bumps" placed around perimeter of the pad for better containment

    Deck support posts designed so they do not penetrate the concrete pad

    Floor drain in storage building eliminated for simplicity

    All electrical components to be inspected by local building inspector for safety

    Eyewash station and emergency shower added for safety reasons

    Appropriate hazardous materials warning signs on storage building for safety

As more facilities are built, further improvements in this design will surely take place.

SUMMARY

The facility installed in Henderson County, North Carolina has been well received by local growers. Several other growers across North Carolina have requested design and financial assistance for facilities. The concept of this unit is versatile and is being considered for several different types of farming operations including greenhouses, grain farms, nursery operations etc. Through the experience gained from this installation, other farmers will improve their handling of agrichemicals. As more of these facilities are installed and more farmers become aware of them, the overall level of care in handling agrichemicals will improve, and water quality will be protected.

Pesticide Contamination of Mixing / Loading Sites: Proposals for
Streamlined Assessment and Cleanup, and Pollution Prevention.

Michael V. Thomas*

ABSTRACT

An assessment of pesticide contamination at 49 mixing / loading sites at university research centers has been underway for several years. Seventy-eight percent of these sites have been found to have soil or ground water pesticide levels which exceed one or more recommended action levels. Many also exceed allowable levels under federal and/or state hazardous waste regulations. It is suspected that a large percentage of private farms have similar sites. Conventional methods of cleanup of on-farm sites are unlikely to result in environmental action because they are very expensive. The Florida Department of Environmental Protection is working with the Florida Department of Agriculture and Consumer Services (FDACS) and the U.S. EPA to streamline the assessment and cleanup of on-farm sites. Proposals include land application of non-hazardous materials, on-site bioremediation and other methods of on-site treatment.

INTRODUCTION

In the past decade, mixing and loading sites at several aerial applicators and some commercial pesticide applicators have come to the attention of the Florida Department of Environmental Protection (FDEP). Those sites which were found to be contaminated were addressed as conventional hazardous waste sites using the typical CAP/RAP (Contamination Assessment Plan/Remedial Action Plan) process. This procedure is extremely expensive and time-consuming, often taking five years and hundreds of thousands of dollars to clean up even a small site.

In 1986, the University of Florida Institute of Food and Agricultural Sciences (UF-IFAS) signed a Consent order with the FDEP regarding alleged improper disposal of pesticides at research and education stations around the state. Among the 130 potentially contaminated sites identified during the preliminary assessment were 49 mix / load sites at 26 research facilities. Table 1 shows the level and frequency of occurrence of the most commonly found pesticide residues in the soil at these sites. Table 2 presents similar information for ground water. Overall, 78 percent of the mix/load sites experienced contamination levels which exceed at least one recommended action level, which means that further investigation and possible cleanup are warranted. To date, the university has expended over three million dollars on Consent Order related assessments and emergency remediation, with only one site officially closed.

The results of the site investigations at these mix / load areas, which are seen as representing typical practices at small to medium sized farms, raised alarm flags at FDEP and FDACS. If the percentages hold for the nearly 40,000 private farms in Florida, the cost of investigation and cleanup at these sites would be staggering. As a result of these realizations, FDEP and FDACS have launched a concerted effort with the EPA to develop alternative methods for assessment and cleanup of mix / load sites at private farms. This is necessary because many of the sites contain materials which are now listed hazardous wastes under RCRA.

---

*Michael V. Thomas. Professional Engineer III, Agricultural Source and Water Well Management Section, Bureau of Drinking Water and Ground Water Resources, Florida Department of Environmental Protection. Tallahassee, FL 32399-2400.

Table 1.

Soil Contaminants Detected at > 10% of Mix / Load Sites					
Contaminant	% of Sites Where Detected	Average Level Detected	Maximum Level Detected	Typical Regulatory Level *	Units
DDE	59.2	1,136	19,000	3,500	µg/kg
DDT	42.9	4,408	128,000	3,500	µg/kg
Chlordane	40.8	129,054	12,000,000	900	µg/kg
Dieldrin	36.7	211	3,520	73	µg/kg
DDD	30.6	4,149	38,300	4,900	µg/kg
Endosulfan B **	26.5	4737	147,000	6.4	µg/kg
Endosulfan A **	26.5	10,276	329,000	6.5	µg/kg
Dichlorprop	20.4	142	1,170		µg/kg
Toxaphene	20.4	375,717	12,700,000	1,100	µg/kg
Dinoseb **	18.4	676	6,620	7.6	µg/kg
Acetone	18.4	3,414	42,000	365	µg/kg
Endosulfan Sulfate **	16.3	905	9,540	.043	µg/kg
Heptachlor Epoxide	14.3	41	105	260	µg/kg
Dicamba **	14.3	141	837	1,200	µg/kg
2.4-DB	14.3	229	713	11,000	µg/kg
Endrin	14.3	641	3,970	0.70	µg/kg
BHC,D	12.2	28	55		µg/kg
Dalapon **	12.2	63	299	120	µg/kg
Lindane	12.2	74	234	900	µg/kg
2,4-D **	12.2	415	3,150	1,200	µg/kg
Aldrin	10.2	91	576	69	µg/kg
PCB-1260	10.2	155	600	800	µg/kg
Copper	10.2	193	1,010	600	mg/kg
Heptachlor	10.2	4,289	38,300	260	µg/kg

* Typical cleanup goals that have been recommended by FDEP for unrestricted (residential) scenarios at hazardous waste sites. Due to the constant progression of knowledge in toxicology and risk assessment, these figures may change without notice. Contact the FDEP Bureau of Waste Cleanup for current formulas used in the calculation of cleanup levels.
** Leachability-based criteria, others are health-based.

Table 2.

Ground water Contaminants Detected at > 5% of Mix / Load Sites					
Contaminant	% of Sites Where Detected	Average Level Detected	Maximum Level Detected	Regulatory Limit	Units
Arsenic	16.3	37.72	160.	50	µg/L
Copper	16.3	55.23	435.	1000	µg/L
Lead	14.3	47.42	178.	15	µg/L
Methomyl *	12.2	6.51	15.40	200	µg/L
Alachlor	12.2	10.68	38.20	2	µg/L
Dieldrin *	10.2	.10	.40	.05	µg/L
Carbaryl *	10.2	1.38	2.27	200	µg/L
Dicamba *	10.2	7.31	27.00	200	µg/L
Endosulfan B *	8.2	1.59	4.82	.40	µg/L
Endosulfan A *	8.2	2.29	8.81	.40	µg/L
Chloroform **	8.2	3.22	4.18	100	µg/L
Hexazinone *	8.2	15.05	45.50	200	µg/L
1,2-Dichloropropane	8.2	118.78	480	5	µg/L
Barium	8.2	914.11	3960.	2000	µg/L
Carbofuran	6.1	.38	.66	40	µg/L
Mercury	6.1	1.50	3.40	2	µg/L
Atrazine	6.1	171.16	503.	3	µg/L

* FDEP policy guidance memo, not law, see caveat on Table 1.
** MCL for Total Trihalomethanes

PROPOSED ALTERNATIVES

The joint FDEP / FDACS proposal initially focuses on separating contaminated sites into two categories, A and B. Category A sites are where the soils contain only currently registered pesticides; or that contain pesticides that were registered when applied but are no longer registered and when analyzed show no detectable levels of listed hazardous wastes, or where TCLP testing indicates that the soil is not a hazardous waste due to toxicity characteristics. Under the proposal, the state will allow the landowner to land apply soils containing residues of pesticides meeting the category A criteria on a crop for which the pesticides were initially labeled. Under an interagency agreement, FDACS will be responsible for establishing and enforcing the procedures used in these remedial actions.

Category B sites are those where residues of pesticides which are no longer registered exist and where detectable levels of listed hazardous wastes are found, or for which a TCLP test indicates that the soil is a hazardous waste due to toxicity characteristics. Under current law, these materials cannot be land applied and require RCRA permits to treat as hazardous waste. The Department is continuing its efforts to work with the EPA to devise a farm exemption from RCRA permitting requirements for on-site treatment of category B pesticide containing soils using evaporation, biodegradation, or other on-site / in-situ methods as may be developed and approved by the state. Such an exemption would dramatically reduce the costs involved in remediation of these sites. This process should be implemented within the next year, following additional state / federal negotiations and a second interagency agreement.

In addition to streamlining remediation options, the Department is seeking ways to simplify the sampling and assessment phase of the process. There is a need to minimize the number of samples and individual analyses needed to determine into which category a site will fall and the extent of contamination. Mix / load sites are usually located away from residences, and ground water contamination is not always present, depending on the distance to water table, soil type, and the types and amounts of pesticides used. Because the focus of this effort is the removal of the source of contamination, it has been proposed that initial sampling be limited to soils. If analysis of the soils data indicates a significant probability of ground water contamination, then ground water monitoring wells would be required.

These steps should save considerable time and money in the cleanup of on-farm mix / load sites, while satisfactorily reducing any risk to the public health or to the environment. While the site must still be cleaned up to a risk level acceptable to the Department (currently 10^{-6}), these steps can help to minimize the cost.

PREVENTION INITIATIVES

The department is involved in promoting several initiatives to prevent pollution from pesticides. One example is the department's push for the use of impervious chemical mixing centers (CMC's) at pesticide mixing and loading operations. The use of CMC's is included in the best management practices (BMP's) for golf courses and citrus groves, and is being incorporated into newly issued BMP's. The design and use of CMC's has been discussed by several sources (Bucklin and Becker, 1991; Midwest Plan Service, 1991, 1992; Tennessee Valley Authority, 1992). The purpose of these centers is to retain all accidental spillage on an impervious surface from which it can be recovered and reused or properly disposed of.

The centers are designed to contain 125 percent of the largest storage or spray tank to be used. Storage should be provided for the retention of spillage and rinsate for use in subsequent batches of the same pesticide. Protection from washdown by and contamination of rainwater is also needed. Many CMC's also provide for a pesticide storage facility under the same roof.

The department is currently involved with the EPA and the United States Department of Agriculture-Soil Conservation Service (USDA-SCS) in a cost-sharing demonstration project for the construction and evaluation of ten CMC's at farms in Jackson County. Recently, the department obtained additional grant funding to cost-share in the building of twelve demonstration CMC's at farms and golf courses in the Indian River basin and the Tampa Bay area. Contract negotiations with FDACS are in progress to commence building the facilities in 1994.

The department's efforts at pesticide container recycling have moved out of the pilot stage and recycling centers for triple-rinsed pesticide containers are now available in many areas of the state. Many of these are operated through the county solid waste offices.

CONCLUSIONS

Historical practices at pesticide mixing and loading areas have resulted in the contamination of soils and ground water. Unpaved farm areas repeatedly used for the mixing and loading of pesticides may present a significant hazard to humans and to the environment, particularly to the surficial aquifer where one is present. Excessive levels of both current and banned pesticides and their degradation products have been found in the soil and ground water near mixing and loading areas. Neither the property owners nor the government can afford to clean up all of these sites using the current CAP/RAP process.

The state is working with the USEPA and is attempting to find more flexible and more economical ways to assess and remediate on-farm mixing and loading areas, and to prevent future contamination. Proposed methods discussed, including land application and on-site biodegradation with an exemption from RCRA permit regulations, represent a feasible alternative to current procedures while satisfactorily reducing any risk to the public health or to the environment.

REFERENCES

Bucklin, R.A. and W.J. Becker, 1991. Pesticide mixing-loading facility. Agricultural Engineering Department Special Series Report SS-AGE-20. University of Florida, Gainesville, FL. 10p.

Midwest Plan Service, 1991. Designing facilities for pesticide and fertilizer containment. MWPS-37. Midwest Plan Service, Ames, IA. 116p.

Midwest Plan Service, 1992. Proceedings of the National Symposium on Pesticide and Fertilizer Containment: Design and Management. MWPS-C1. Midwest Plan Service, Ames, IA. 160p.

Tennessee Valley Authority, 1992. Environmental handbook for fertilizer and agrichemical dealers. Tennessee Valley Authority, Muscle Shoals, AL

Agricultural Chemical Spill Cleanup in Minnesota:
A Site-Specific Approach

Sheila R. Grow, Camp Dresser & McKee Inc.

ABSTRACT

Accidental releases of agricultural chemicals pose serious threats to human health and the environment. The investigation and remediation of agricultural chemical spill sites requires a site-specific approach. In order to characterize contamination that may have resulted over a long period of time at an agrichemical facility, the investigation must identify the contaminants of concern, as well as determine where to investigate, and how to proceed with the investigation. An investigation of a contaminated site requires knowledge of past and present facility operations. State programs can help promote cleanups and reduce responsible party costs through: reimbursement programs, such as Minnesota's Agricultural Chemical Response and Reimbursement Account and Wisconsin's Agricultural Chemical Cleanup Program; establishing reasonable cleanup objectives; and developing programs to help remediation activities be more effective and efficient. Responsible parties can help reduce costs by selecting qualified cost-effective environmental consulting companies and by developing and implementing best management practices for the storage, handling, and transportation of agricultural chemicals to help prevent future incidents.

KEYWORDS: agricultural chemicals, incidents, remediation, reimbursement, cost-effective.

EXTENT OF ENVIRONMENTAL RISKS FROM AGRICULTURAL CHEMICAL INCIDENTS

Accidental releases of agricultural chemicals pose serious environmental threats to the soil, groundwater, and surface water in many states. Accidental releases may be sudden spills and/or more long-term releases due to past or ongoing agricultural chemical handling and management practices, particularly at agrichemical facilities. Long-term environmental impacts also may result from sudden spills of pesticides and fertilizers. In Minnesota, more than 200 sudden releases are reported each year and long-term releases have been identified at more than 90 agrichemical facilities (*1*).

Twenty-one of 26 states that responded to a survey conducted by the State FIFRA Issues Research Evaluation Group (SFIREG), "indicated an awareness of severe environmental impacts resulting from agrichemical handling sites" (*2*). Five states (Minnesota, California, Florida, Michigan, and Wisconsin) reported 82 percent of the known sites (*2*).

The Wisconsin Department of Agriculture, Trade and Consumer Protection and the Department of Natural Resources investigated potential agricultural chemical contamination at agrichemical facilities in Wisconsin. The October 1991 "Report on Wisconsin Pesticide Mixing & Loading Site Study" (*3*), reports on agricultural chemical contamination at 27 facilities randomly selected from approximately 530 licensed facilities operating at 600 to 700 locations. The report estimates that 86 to 99.9 percent of the Wisconsin sites may have soil contamination; 45 to 75 percent of the sites are likely to require soil remediation. According to the report, 56 to 86 percent of the sites may have groundwater contamination; 29 to 63 percent may exceed Wisconsin's Enforcement Standards (State groundwater quality standards) and may require remediation.

The Illinois Department of Agriculture and the Illinois State Geological Survey conducted a statewide study designed to help identify the extent of pesticide contamination in Illinois, which was not one of the five states with the highest number of known incident sites. The July 1993 "Agrichemical Facility Site Contamination Study" (*4*) reports on 49 facilities randomly selected from 1200 registered facilities in Illinois to investigate potential sources of contamination. Although, the report concludes "that the groundwater below many of the retail agrichemical facilities in operation today has not been extensively impacted by past facility operations", the cost for addressing soil contamination of affected sites is estimated at $42 to $96 million. This estimate does not include costs for remediation of groundwater.

CLEANUP OBJECTIVES

Cleanup objectives for agricultural chemical spill sites will depend on how contamination is defined and on the answer to the common question: how clean is clean? Because agricultural chemicals were designed to be applied to the soil, the documentation of the presence of an agricultural chemical in soil does not necessarily indicate a misuse of the chemical. The cost of incident remediation will depend on the cleanup objectives. State policies can help promote cleanup activities by establishing reasonable objectives. Cleanup objectives should be site-specific, flexible, and should consider the potential risks to human health and the environment for a particular site.

The Minnesota Department of Agriculture has adopted a reasonable approach. Although Minnesota has established drinking water standards, Health Risk Limits (HRL), for private drinking water supplies, the HRL are not necessarily used as cleanup objectives. The Minnesota Department of Agriculture determines cleanup objectives for agricultural chemical spill sites by identifying vulnerable or sensitive areas (such as wellhead protection areas) and evaluating the <u>use</u> of the groundwater, technological limitations, potential impacts, environmental benefit of the cleanup activities, and cost-effectiveness. The Minnesota Department of Agriculture uses label rates as guidelines for soil cleanup objectives for pesticides, even for off-label areas. Because fertilizer needs are site-specific and there is no one label rate, ambient or background concentrations of nitrogen may be used for perspective when developing reasonable objectives. Cleanup objectives may even change during the remedial activities based on technical achievability or cost.

CONDUCTING AGRICULTURAL CHEMICAL INCIDENT CLEANUPS: A MINNESOTA MODEL

SITE CHARACTERIZATION

The investigation and remediation of long-term agricultural chemical spill sites requires a site-specific approach. The purpose of an investigation of an agrichemical facility is to provide information regarding the agricultural chemicals that may be found at the site, determine the source and extent of any contamination, identify potential human and environmental effects of the contamination, and provide information to help determine the appropriate corrective actions.

Agrichemical facilities are often large and the characterization of potential contamination may be complicated. The Wisconsin study, for example, found no direct correlation between pesticide concentrations in the soil and pesticide concentrations in groundwater (*3*). The Illinois study found no correlation between the depth at which a pesticide was detected and the pesticide mobility and persistence characteristics (*4*).

In order to characterize contamination that may have resulted over a long period of time at an agrichemical facility, the investigation must identify the pesticides of concern, where to investigate, and how to proceed with the investigation. An investigation of a potentially contaminated site requires knowledge of past and present facility operations. Pesticide usage has changed over time. In addition, different regions of each state, as well as regions of the country, use different pesticides. Consequently, each site may have site-specific pesticide concerns. Various areas of a facility including mixing and loading areas, pesticide storage areas, fertilizer impregnation tower areas, interior surfaces of earthen dikes, surface water pathways, equipment parking and storage areas, and areas of suspected or known spills should be identified and sampled. The Illinois study suggests that samples collected from surface water drainage pathways may indicate which pesticides are likely to be present at the facility and therefore may help identify target compounds for that site (4).

Soil samples for analytical purposes should be collected during the investigation to help determine potential corrective actions. Soil samples may be either "composite" of "grab" samples; composite samples are composed of subsamples, while grab samples are from a unique area. For example, the Minnesota Department of Agriculture recommends collecting surface composite samples, with 3 to 6 subsamples, from within a 15-foot radius. Composite samples are collected to identify surface areas of concern that may require future excavation. Grab samples should be collected at depth in order to determine the depth of potential future excavation. In order to lower costs for collecting and analyzing soil samples, soil samples can be collected and frozen, and then analyzed in a phased approach; frozen soil samples can be selected sequentially for analysis depending on the results of the previously analyzed samples.

Groundwater monitoring wells should be installed during the investigation of most long-term incidents, especially if groundwater is anticipated to be near the surface or the site is in a sensitive area (e.g., fractured bedrock close to the surface). Surface water impacts also should be evaluated. Potential uses (e.g., human consumption, irrigation, and livestock needs) of the water resources should be identified.

ANALYTICAL CONSIDERATIONS

Analytical costs are a major expenditure during an investigation of an agrichemical facility, especially if each pesticide requires a unique analysis for each location. Pesticide analyses are complicated because, unlike volatile organic compounds (VOCs), there is not a single uniform Environmental Protection Agency (EPA) method that provides analysis for all pesticides of concern. EPA Method 8080, which is commonly requested during Superfund investigations, identifies cancelled products such as aldrin, chlordane, DDT, dieldrin, endrin, heptachlor, and polychlorinated biphenyls (PCBs), among other analytical parameters, but the cancelled pesticides may not be the most common pesticides or the pesticides of most concern at an agrichemical facility. However, the documented presence of cancelled products at an agrichemical facility may limit options for corrective actions, such as land application.

In response to cost concerns and a concern about reliable laboratory pesticide analysis, the Minnesota Department of Agriculture (MDA) worked with the MDA analytical laboratory to develop three lists of pesticides (5) that can be analyzed using three methodologies. The neutral pesticide list includes 20 pesticides, including those most commonly detected at Minnesota sites. There is also an acid list and a carbamate list for generally less common pesticides. Together, these lists have dramatically reduced the costs for investigations of agrichemical facilities. To further ensure the quality of the laboratory analysis during the investigation of agricultural chemical incidents, the Minnesota Department of Agriculture reviews the methodology of any laboratory that requests participation in the investigation.

INVESTIGATIVE TOOLS

Innovative technologies, or options other than laboratory analysis, also may be used to limit costs and provide timely information in the field. These technologies may include a field gas chromatograph and field test kits, such as immunoassay kits and a nitrogen meter. These technologies may be effective if the pesticide of concern (target compound) has been identified and there is a need for immediate results. Immunoassay tests may be used for screening contaminated areas during the investigation if a test for the target compound is available. When time is a consideration, it may be effective to have field gas chromatography available. A nitrogen meter may not provide accurate information, but the nitrogen meter can be a useful screening tool to identify relative levels of nitrogen.

REMEDIAL OPTIONS

Current treatment technologies for soil contaminated with agricultural chemicals include (*4,6,7,8*):
- Excavation and land application;
- Bioremediation, which may include land application, in situ bioremediation, slurry-phase biodegradation, soil mounds, and composting;
- Soil washing/solvent extraction;
- Solidification/stabilization;
- Incineration and thermal desorption; and
- Landfilling.

Generally, the most inexpensive treatment option is excavation and land application of soils contaminated with registered pesticides and fertilizer. Rock, gravel, or other debris should be separated before land-applying the excavated soil. Land application of soil containing fertilizer has not been a problem. However, high concentrations of pesticides, mixtures of incompatible products, availability of labeled areas, and the identification of banned or cancelled pesticides in the soil may make land application an undesirable treatment option. If acceptable to the state regulating authorities, it may be best to clarify which pesticides will be reported by the analytical laboratories and to avoid analyzing for pesticides that are not of regulatory concern.

Current treatment technologies for groundwater contaminated with agricultural chemicals include (*4,6,7,9*):
- Bioremediation;
- Granular activated carbon;
- Chemical oxidation and ultraviolet light oxidation; and
- Reverse osmosis.

Cleaning up groundwater contamination usually is expensive and difficult. Many pesticides do not volatilize and cannot be readily treated by methods commonly used for other contaminants, such as petroleum or solvents. The general effectiveness of "pump-and-treat" systems also is being questioned (*10*). If cleanup objectives allow, it may be more efficient to treat the water at the point of use or to provide an alternative drinking water supply, such as a new well or municipal water supply.

REIMBURSEMENT PROGRAMS FOR AGRICULTURAL CHEMICAL CLEANUPS

To help fund environmental remediation of sites affected by agricultural chemicals, Minnesota was the first state in the nation to develop a reimbursement fund, the Agricultural Chemical

(10) Mott, R., August 28, 1992, Aquifer Restoration Under CERCLA: New Realities and Old Myths: Environmental Reporter, Bureau of National Affairs, pp. 1301-1304.

(11) Minnesota Statutes Chapters 18D and 18E (1990), Agricultural Chemical Liability, Incidents, and Enforcement and Agricultural Chemical Incident Payment and Reimbursement.

(12) Minnesota Department of Agriculture, February 1993, Environmental Consulting Services Report (Draft), 25 pp.

(13) Agricultural Chemical Response Compensation Board and the Commissioner of Agriculture, November 1993, Report to the Minnesota Legislative Water Commission, p. 17.

(14) Wisconsin Statutes Section 2119b. 94.73 (1993), Agricultural Chemical Cleanup Program.

"PLANNED INTERVENTION": MERGING VOLUNTARY CAFO POLLUTION ABATEMENT WITH COMMAND-AND-CONTROL REGULATION

Larry C. Frarey and Ron Jones*

ABSTRACT

"Planned intervention" to abate pollution from concentrated animal feeding operations (CAFOs) describes an appropriate balance between voluntary best management practice (BMP) adoption and command-and-control environmental regulation. While voluntary pollution-abatement programs are favored by most agricultural interests, the presence of "bad actors" necessitates the threat of command-and-control intervention. Alternatives to "planned intervention" are "unplanned intervention" and "planned non-intervention." The former describes a wholly voluntary approach to CAFO pollution abatement; the latter is a reactive approach relying on *ad hoc* enforcement and large fines. "Planned intervention" requires close cooperation between conservation agencies, regulators and producers to insure efficient pollution abatement. The top ten milk-producing states in the United States exhibit varying degrees of "planned intervention" to address pollution from dairy CAFOs. Recently passed legislation in Texas expressly requires "planned intervention" by tying a voluntary pollution abatement program directed by the Texas State Soil and Water Conservation Board (TSSWCB) to command-and-control regulation for "bad actors." **Keywords**. Agricultural nonpoint source pollution, voluntary abatement, command-and-control regulation.

"PLANNED INTERVENTION," "UNPLANNED INTERVENTION" AND "PLANNED NON-INTERVENTION"

Although the United States Environmental Protection Agency (EPA) and its state counterparts regulate large CAFOs as point sources, pollution caused by CAFOs is both point source and nonpoint source (NPS) in nature. Discharge from animal confinement and process areas represents a point source of pollution, and is relatively amenable to traditional site inspection and control. However, under current production practices, manure solids and lagoon effluent applied to pasture or cropland may cause nonpoint pollution source in the presence of precipitation. While command-and-control regulation may be adequate for controlling point source pollution, it will likely prove insufficient for abating nonpoint source pollution. Rasmussen observes that

> [t]he widespread problem of nonpoint source pollution--runoff and deposition of air pollution to land and water--also underscores the limits of effective enforcement. Our society does not have the resources to police each citizen's behavior and lifestyle in order to prevent or punish our polluting habits. The lack of regulatory resources aside, an environmental police force is an affront to our concept of individual liberty.[1]

Many members of the agricultural community, including CAFO operators, argue that a voluntary NPS pollution abatement program will produce the same environmental benefits as command-and-control regulation at less cost to agricultural producers.[2] That program would consist of voluntary BMP adoption through conservation plans implemented under the auspices of USDA-associated conservation agencies. However, even the strongest supporters of voluntary agricultural pollution abatement concede that the most effective programs combine "voluntary" BMP adoption with the

---

* Respectively, Policy Analyst and Director, Texas Institute for Applied Environmental Research, Tarleton State University, Stephenville, Texas.

threat of more onerous regulation in the absence of sufficient water quality improvement: "Doing more of what has been done in the past to control agricultural nonpoint-source pollution is not going to suffice. Voluntary participation in soil conservation programs will not reduce agricultural pollution to socially acceptable levels. Mandatory controls of some sort are necessary and likely will be forthcoming."[3]

Once a potential CAFO pollution problem is recognized, a deliberate, well-planned combination of voluntary and regulatory efforts by policymakers to remedy the problem can be described as "planned intervention." The proper combination of voluntary and regulatory measures provides CAFO operators the freedom to implement necessary BMPs to achieve environmental goals within a reasonable time frame, while insuring that "bad actors" are not permitted to pollute indefinitely. Under "planned intervention," BMP adoption by CAFO operators is voluntary only to the extent that producers choose the pollution-abatement BMPs to implement and, within reason, the time frame for implementation. Voluntary BMP implementation under "planned intervention" does not mean that a CAFO operator has the freedom to reject environmental compliance outright. Thus, "planned intervention" simply inserts a voluntary, time-and-resources loop into an inflexible regulatory process (see figure below). That loop provides producers a period of time and, ideally, cost-share financial assistance with which to reach compliance with established environmental goals.

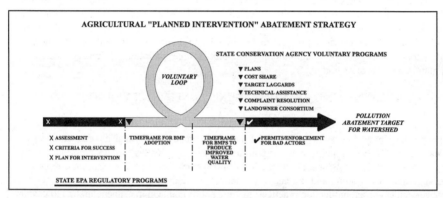

To function effectively, "planned intervention" programs require timely, proactive cooperation between conservation agencies and regulators to first recognize potential CAFO pollution problems in their early stages, and then to formally target agency efforts to efficiently address the most acute problem sources. Under "planned intervention," conservation agencies help provide CAFO operators the financial, technical and educational means with which to achieve environmental compliance, while environmental regulators, in cooperation with conservation agencies and CAFO operators, establish a watershed pollution-abatement target and a realistic time frame for achieving that goal. Absent all of these elements, "planned intervention" can easily dissolve into "unplanned intervention," with all of the attendant negative aspects inherent therein.

"Unplanned intervention" occurs when a CAFO pollution problem rises to the level of public attention but environmental policymakers fail to mobilize to address the problem in a timely manner, either through pure command-and-control measures or in cooperation with conservation agencies. Eventually, the pollution problem becomes critical and regulators are forced by public and media pressure to intervene on an *ad hoc* basis, often imposing harsh penalties on a few producers to make a public statement and awaken the regulated community. "Unplanned intervention" is often characterized by a purely reactive agency approach to CAFO pollution, relying on public complaints to initiate agency action, with little or no permitting of CAFO facilities. Such was the situation in Texas in September 1989, when the Texas Water Commission (TWC)[4], the state water quality regulatory agency, imposed heavy fines against nine milk producers in Erath and surrounding counties. Erath County is the number one milk-producing county in Texas, with approximately 70,000 cows.

While "unplanned intervention" may have a role in the overall regulatory scheme, particularly in addressing large influxes of CAFOs to an area over a short period of time, the regulated community would be better served by a more predictable system of lower, uniformly imposed penalties. Moreover, the long-term costs of "unplanned intervention" are likely higher than the combination of cost sharing and "bad actor" enforcement under "planned intervention," since a virtual army of inspectors would be necessary to police CAFO lands during manure application and storm events. TWC eventually instituted a Dairy Outreach inspection program in Erath County which somewhat remedied the former "unplanned intervention" approach. However, the Dairy Outreach program does not formally mesh regulatory efforts with programs run by state and federal conservation agencies, thereby failing to take full advantage of potential long-term cost reductions and other benefits provided by "planned intervention."

The opposite approach to "unplanned intervention" in CAFO pollution problems is "planned non-intervention." This term describes a situation in which policymakers proceed with a wholly voluntary approach to pollution abatement once a CAFO pollution problem is recognized, relying on the educational, financial and technical assistance provided by conservation agencies. This approach benefits from a long-established system of conservation agencies that is positively perceived by most CAFO operators. The greatest shortcoming of "planned non-intervention" is incomplete pollution abatement due to the conduct of "bad actors" who refuse to adopt BMPs absent the threat of penalties.

The salient characteristics of "planned intervention," "unplanned intervention" and "planned non-intervention" approaches to CAFO pollution abatement are listed in Table 1.

Table 1 - Characteristics of Three Approaches to CAFO Pollution Abatement

Approach	Planned Non-Intervention	Planned Intervention	Unplanned Intervention
Strategy	Voluntary approach using conservation agencies for cost share, organizational, technical and educational assistance	Combination of voluntary BMP adoption and regulation to insure most efficient environmental compliance	Reliance on ad hoc regulatory approach, characterized by selective enforcement through high fines
Problem Recognition	Early recognition of potential CAFO pollution since conservation agencies have dealt closely with farmers for decades	Early recognition of potential CAFO pollution problem and targeted response through proactive inspection and water quality monitoring by regulators and input from conservation agencies	Untimely recognition of potential CAFO pollution generally through reactive response to complaints; frequently coming after great influx of CAFOs to an area over short period of time
Agency Cooperation	Little cooperation and communication between regulatory and conservation agencies	Close cooperation and communication between regulatory and conservation agencies from the moment potential CAFO pollution problems arise	Little cooperation and communication between regulatory and conservation agencies
Producer Risk	Low risk, high level of predictability for CAFO operators due to absence of regulatory program	Low risk, high level of predictability for CAFO operators due to articulation by regulators of expected water quality improvements after BMP implementation and the time frames for BMP implementation and water quality improvements	High risk, low level of predictability for CAFO operators due to high penalties and selective enforcement
Public Sector Cost	Moderate, foreseeable, one-time costs in the form of cost-share financing	Moderate, foreseeable, one-time costs in the form of cost-share financing; some enforcement costs for "bad actors"	High, continual enforcement costs
Extent of Pollution Abatement	Incomplete pollution abatement due to presence of "bad actors"	Complete pollution abatement, initially through voluntary compliance, backed up by regulation of "bad actors"	Incomplete absent exorbitant regulatory personnel costs

FORMALIZATION OF "PLANNED INTERVENTION" IN TOP DAIRY STATES

The mix of regulatory and voluntary agricultural pollution abatement strategies in existence today is as diverse as the states presently addressing CAFO pollution problems. In an attempt to illuminate the interrelationships among regulatory and voluntary CAFO pollution abatement programs around the country, the Texas Institute for Applied Environmental Research (TIAER) at Tarleton State University surveyed state regulators and state conservation agency (SCA) personnel in the nine states that, together with Texas, constitute the top ten milk-producing states in the country: Wisconsin, California, New York, Minnesota, Pennsylvania, Michigan, Ohio, Washington, and Iowa. The top dairy states provide a sampling of the extent to which jurisdictions with significant numbers of CAFOs have adopted "planned intervention" strategies to address CAFO pollution.

The activities of SCAs and local conservation districts (LCDs) provide the best indication of the type and level of CAFO-related voluntary pollution abatement activity in most states. A review of 136 Areawide Waste Treatment Management Plans developed under section 208 of the federal Clean Water Act revealed that LCDs are the most prevalent choice among the states as institutions for implementing voluntary agricultural pollution-abatement programs.[5] For example, since 1985, section 201.026 of the Texas Agriculture Code has provided that TSSWCB "shall plan, implement, and manage programs and practices for abating agricultural and silvicultural nonpoint source pollution."

Table 2 displays responses by personnel from SCAs to a variety of queries. Representatives were asked whether their agencies are currently involved in any pollution-abatement programs designed specifically for CAFOs. Respondents from Wisconsin, New York, Pennsylvania, Michigan, Ohio, Washington and Iowa indicated that their agencies now are involved in such programs. Respondents from California and Minnesota indicated no direct involvement in pollution abatement programs targeting CAFOs. Of the seven respondents indicating that their agencies currently target CAFO pollution through specific initiatives, respondents from four states--Wisconsin, New York, Ohio and Iowa--indicated that CAFO pollution abatement programs are watershed-based. Respondents from Pennsylvania and Washington indicated that their programs target individual producers who are the subject of pollution complaints.

Table 2. "Planned Intervention" in Dairy States

State	Environmental Regulatory Agency	SCA	Involvement in CAFO Programs	Watershed Based	CAFO Based	Formal Regulatory/Statute Authorization
Wisconsin	Dept. of Natural Resources	State Soil and Water Conservation Office	Yes	Yes	No	Yes
California	Water Quality Control Board	Dept. of Conservation	Advisory	NA	NA	NA
New York	State Dept. of Environ. Conservation	State Soil and Water Conservation Comm.	Yes	Yes	Yes	Yes
Minnesota	Pollution Control Agency	Assoc. of Soil and Water Conservation Dist.	Advisory	No	Yes	No/Voluntary
Pennsylvania	Dept. of Environ. Resources	Bureau of Land and Water Conservation	Yes	No	Yes	Yes
Michigan	Dept. of Natural Resources	Dept. of Agriculture	Yes	No	Yes	Yes
Ohio	Environ. Protection Agency	Dept. of Natural Resources	Yes	Yes	Yes	Yes
Iowa	Dept. of Natural Resources	Soil and Water Office	Yes/Voluntary	Yes	Yes	No/Voluntary
Washington	Dept. of Ecology	Conservation Comm.	Yes	No	Yes	Yes

Table 2 also displays responses concerning SCA authority for NPS pollution abatement programs, or, more specifically, CAFO pollution-abatement programs. Authorization through state statute or agency regulation indicates a more formal role in state NPS and CAFO pollution-abatement efforts for SCAs than initiatives based solely on informal agency policy. Respondents from Wisconsin, New York, Pennsylvania, Michigan, Ohio and Washington indicated formal statutory or regulatory authorization for nonpoint source or CAFO pollution-abatement programs. Respondents from California, Minnesota, and Iowa indicated no such authority.

Personnel from both state environmental regulation agencies and SCAs were asked three similar questions regarding agency interaction between regulatory and voluntary pollution-abatement programs: 1. Do state statutes or agency regulations provide a grace period after a CAFO is found in violation of environmental regulations to allow that CAFO to voluntarily implement pollution-abatement BMPs prior to the imposition of a penalty?; 2. Does your agency have formal or informal agreements with [state regulatory]/[natural resource] agencies providing for a grace period after a CAFO is found in violation of environmental regulations to allow that CAFO to voluntarily implement pollution-abatement BMPs prior to the imposition of a penalty?; and 3. Do state or local statutes or agency rules, or informal policy, provide for a specific waiting period to observe water quality improvement subsequent to voluntary BMP adoption by CAFOs? Table 3 displays responses to the three queries.

Table 3. Timeframes for "Planned Intervention" in Dairy States

State	Statutory Grace Period		Grace Period by Agreement		Timeframe for Successful Abatement	
	SCA	Regulatory Agency	SCA	Regulatory Agency	SCA	Regulatory Agency
Wisconsin	Yes	Yes	Yes	Yes	Yes a	No
California	NA	Dependent on violation	NA	No	NA	Yes/Informal
New York	No	No	Informal	No	Yes/Informal	No
Minnesota	Yes	Yes	No	No	Yes/Informal	No
Pennsylvania	No	Yes/Informal	Informal	Yes/Informal	Yes a	No
Michigan	Yes	Yes	Yes	Yes	No	No
Ohio	No	Dependent on violation	No	Yes	No	No
Iowa	No	No	No	No	NA	No
Washington	Dependent on violation	Yes	Dependent on violation	Yes	Yes/Informal	No

a BMP implementation satisfies all regulatory requirements.

The disparity between some responses by regulators and state conservation personnel within the same state is informative. Such differences of opinion indicate the need for greater clarity in agency rules and memoranda of understanding between regulatory agencies and SCAs. Effective "planned intervention" depends on clear communication between regulators, conservation agency personnel and the regulated community.

TEXAS: FROM "UNPLANNED INTERVENTION TOWARD "PLANNED INTERVENTION"

Texas provides an excellent example of a two-step shift from "unplanned intervention" in CAFO pollution problems toward a "planned intervention" scheme. Prior to the mid-1980s, Erath County was the state's second largest milk-producing region, characterized by small-to-medium size dairies. However, by 1991, Erath County had surpassed Hopkins County as the state's number one milk-producing county. Erath County's ascension to the top is evidenced by a 148-percent increase in cow numbers between 1980 and 1989,[6] as well as an impressive increase in average dairy size.

While TWC long had authority to regulate all sources of water pollution in the state, active regulation of CAFOs did not begin until 1987. In April of that year, TWC promulgated regulations requiring dairies with 250 milking head or more to obtain a waste water discharge permit. However, those regulations were only sporadically enforced. Meanwhile, the population of dairy cows in Erath County continued to climb as a result of an influx of new operations as well as increases in existing county herds. Nonetheless, between 1987 and 1989, TWC did not enforce its regulations uniformly, and no formal pollution abatement strategy was formulated between TWC and natural resource agencies.

September 1989 marked both the zenith of "unplanned intervention" in CAFO pollution in Erath County and the impetus for a "planned intervention" strategy. In that month, TWC levied a total of over $490,000 in fines against nine area dairy operators. The fines shocked local milk producers because of the amount and perceived selective enforcement involved, i.e., many observers felt that the fined operations were singled out for no apparent reason.

The first step in a "planned intervention" strategy for Texas dairies began shortly thereafter. This initial phase targeted facilities with 250 milking head or more through a reinvigorated permit process. In December 1989, TWC published a formal resolution to enforce the 1987 permit and enforcement regulations against existing dairies. Further, TWC divided existing dairies into three priority categories, requiring permit applications from all catagories no later than June 1, 1991.

The large fines imposed in September 1989, coupled with a clear time frame for dairy permit procurement and compliance, resulted in rapid compliance with TWC regulations requiring construction of waste containment structures. According to the results of a TWC inspection of all Erath County dairies conducted in late 1992, only 10 permitted dairies from a total of 88 displayed "major deficiencies," while 55 permitted dairies exhibited "minor deficiencies."[7] Twenty-nine permitted dairies were in total compliance with TWC regulations for structural BMPS.[8]

While vigorous TWC inspection and fines resulted in substantial compliance by permitted dairies with TWC regulations requiring waste water containment structures, implementation of structural BMPs alone does not satisfy TNRCC's "no-discharge" rule. Anaerobic lagoons require careful management to insure proper operation and appropriate dewatering. Otherwise, illegal discharge from lagoons can occur during heavy rains. Moreover, apart from the implementation of a vegetative filter strip, control of runoff from manure application fields depends primarily on proper management. A legion of inspectors would be needed to consistently monitor dairy operations to insure manure is not applied during rain events, on frozen ground or beyond agronomic rates. TNRCC only recently received funding from EPA to station two full-time inspectors in Erath County. That number is sorely inadequate to inspect the proper implementation of managerial BMPs on approximately 225 county dairies.

Surveys conducted by TIAER in Erath County in 1991-92 to determine the extent of producer compliance with TWC regulations revealed the difficulty entailed in detecting noncompliance with managerial regulations. In most cases, trained surveyors were unable to detect the time and rate of application of solid manure and lagoon effluent after application.[9] Systematic soil sampling may indicate nutrient buildup due to over-application, but gives no indication of antecedent nutrient runoff. Depending on sample location and time of year, soil samples may provide little or no information on dangers to surface water from over-application. Like managerial BMPs, effective soil sampling depends on producer diligence.

Further, compliance with TWC structural regulations by Erath County dairies came at considerable cost. Estimates run approximately $100 per cow to install a waste water lagoon and associated structural BMPs.[10] A local constituency committee convened by TIAER as part of an EPA-funded section 319 project concluded that unpermitted dairies contribute significantly to environmental degradation but that many of these producers cannot implement necessary pollution-abatement BMPs without substantial cost sharing. Sixty-one percent of Erath County dairies are unpermitted; TWC's recent inspection program found that 94 of 137 unpermitted dairies exhibited "major deficiencies."[11] Thus, the results of both TIAER's compliance survey and TWC's inspection of county dairies support the need for a complete "planned intervention" program for Erath County dairies, integrating voluntary BMP adoption with back-up regulation. "Planned intervention" provides a realistic way to control the implementation of managerial BMPs on all Erath County dairies, permitted and unpermitted, and to bring unpermitted dairies into compliance within a reasonable time frame without undue economic hardship.

As a result of TIAER's section 319 program, TIAER's constituency committee recommended an alternative compliance program for unpermitted dairies. The Texas Clean Water Council, appointed by TNRCC Chairman John Hall, and the Texas Association of Conservation Districts both endorsed the committee's recommendation. Thus, TSSWCB drafted proposed legislation to shift oversight of small dairy environmental compliance from TNRCC to TSSWCB. TSSWCB was already empowered to coordinate agricultural NPS pollution abatement programs for Texas under section 201.026 of the state Agriculture Code. While no point source/NPS dichotomy exists in the Texas Water Code, TSSWCB proceeded under the assumption that dairies with fewer than 250 milking head are nonpoint pollution sources within the agency's NPS jurisdiction, in the same way EPA considers dairy operations with less than 200 head nonpoint pollution sources.

TSSWCB's proposed legislation was embodied in Texas Senate Bills 502 and 503 and House Bills 1230 and 1231. That legislation passed both houses by unanimous vote and was signed into law by the governor. Senate Bill 502 took effect on September 1, 1993; Senate Bill 503 became effective April 29, 1993.

The new law provides for a direct link between voluntary BMP adoption under TSSWCB and TNRCC regulatory programs where voluntary pollution-abatement efforts break down. Amended section 201.026(c) directs TSSWCB to establish a "water quality management plan certification program" in areas identified as "having or having the potential to develop agricultural or silvicultural nonpoint source water quality problems." TSSWCB, together with LCDs, will handle all complaints concerning agricultural nonpoint source pollution, and, where a problem is verified, "develop and implement a corrective action plan to address the complaint." The alternative dispute resolution mechanism inherent in the new scheme is important for suspected polluters and complainants alike. TIAER's constituency committee had recommended such an alternative to avoid the high costs of traditional litigation. In the event a pollution problem persists, "the state board shall refer the complaint to the Texas Natural Resource Conservation Commission." Thus, amended section 201.026 clearly outlines a "planned intervention" strategy for agricultural NPS pollution abatement.

TSSWCB is currently exploring a watershed/micro-watershed approach to implement the "planned intervention" required under section 201.026. TSSWCB and TNRCC have the opportunity, through rules and memoranda of understanding, to establish a comprehensive "planned intervention" program for all CAFOs, not just those falling under TSSWCB's agricultural NPS jurisdiction. In a watershed exhibiting actual or potential agricultural NPS pollution, TSSWCB could assist LCDs in organizing landowners within the watershed into a consortium. These stakeholders would then meet under the direction of LCD officials to identify pollution sources and recommend abatement strategies. Local district officials would communicate this information to TSSWCB to coordinate development of necessary water quality plans by SCS, cost-share financing from state or federal sources, educational activities by the Extension Service and any other assistance that can be brought to bear. Although permitted CAFOs fall under TNRCC jurisdiction, permitted CAFOs within the watershed could take part in the consortium together with all other agricultural nonpoint pollution sources. Should TSSWCB adopt this approach, the agency would direct a holistic, watershed-based agricultural NPS pollution-abatement program. By building on the direct link between voluntary BMP implementation and regulation outlined under section 201.026, TSSWCB's micro-watershed abatement program could provide a complete "planned intervention" program for CAFO pollution abatement.

Senate Bills 502 and 503 represent a revolutionary role for LCDs and TSSWCB. For years, LCDs and TSSWCB have played a relatively low-key role in organizing conservation plans to control erosion. Responsibility for agricultural NPS pollution will significantly increase TSSWCB's workload and thrust TSSWCB and local districts into an often contentious arena. Local district elections may now become more contested than in the past, since agricultural NPS pollution, and particularly pollution from Erath County dairies, is an emotional issue for many rural landowners. Thus, TSSWCB's managerial capacity may be tested as never before.

The success of a "planned intervention" program can be gauged at the mouth of the watershed or micro-watershed where the program is implemented. This fact should relieve the current strain on regulatory resources, and provide a feasible method through which runoff from manure application fields can be monitored. However, success can be determined only after research establishes appropriate criteria and time frames for NPS abatement. Prior research indicates that water quality improvement subsequent to BMP implementation may become evident only after many years.[12]

REFERENCES

1. Rasmussen, *Enforcement in the U.S. Environmental Protection Agency: Balancing the Carrots and the Sticks*, 22 Environmental Law 333, 336 (1991).

2. Logan, *Agricultural Best Management Practices and Groundwater Protection*, Journal of Soil and Water Conservation, March-April 1990, at 201, 205.

3. Epp & Shortle, *Agricultural Nonpoint Pollution Control: Voluntary or Mandatory?*, Journal of Soil and Water Conservation, Jan.-Feb. 1985, at 111, 111.

4. On September 1, 1993, the Texas Water Commission combined with the Texas Air Control Board and other agencies to form the Texas Natural Resource Conservation Commission.

5. Davidson, *Commentary: Using Special Water Districts to Control Nonpoint Sources of Water Pollution*, 65 Chi.-Kent L. Rev. 503, 514 (1989) (citing Beck, *Agricultural Water Pollution Control*, in 2 Agricultural Law 223 n.362 (J. Davidson ed. 1985, Supp. 1988)).

6. TIAER, Livestock and the Environment: Interim Report to the Joint Interim Committee on the Environment 72nd Texas Legislature 55 (1992).

7. TWC, Confined Animal Feeding Operations, Erath County Dairy Outreach Program 1 (1993).

8. *Id.*

9. TIAER, Final Report on Section 319 Nonpoint Source Management Program for the North Bosque Watershed 43, 44 (1992).

10. EPA, National Pollutant Discharge Elimination System General Permit and Reporting Requirements for Discharges from Concentrated Animal Feeding Operations, 58 Fed. Reg. 7610 (1993).

11. TWC, *supra* note 7, at 1.

12. Clausen, Meals & Cassell, *Estimation of Lag Time for Water Quality Response to BMPs*, in The National Rural Clean Water Program Symposium 173, 177 (1992).

POLICY AND SCIENCE ADDRESSING CAFO ODOR IN TEXAS

Larry C. Frarey *

ABSTRACT

Odor from concentrated animal feeding operations (CAFOs) recently has garnered considerable attention from government institutions and researchers in Texas. In April 1993, the Texas Supreme Court decided the case of *F/R Cattle Company, Inc. v. The State of Texas*.[1] The central issue in that case was whether the Texas Air Control Board[2] (TACB) had jurisdiction to control odor produced at a 6,000-calf feeding operation in Erath County, Texas, the state's top milk-producing county. In the wake of the *F/R Cattle* opinion, the 73rd Texas Legislature attempted to modify the Texas Clean Air Act[3] to provide greater regulatory predictability concerning CAFO odor. On a third front, late in 1992, TACB chairman Kirk Watson appointed an odor task force to recommend modifications to TACB procedures and regulations to provide greater efficiency and effectiveness. Finally, in 1992, TACB contracted with the Texas Institute for Applied Environmental Research (TIAER) and the Texas Agricultural Extension Service (TAEX) to conduct a study of odors produced by Erath County dairies. **Keywords**. Livestock odor, odor regulation, odor research.

CURRENT REGULATORY FRAMEWORK

Section 381.002 of the Texas Health and Safety Code provides that "[t]he [Texas Natural Resource Conservation Commission (TNRCC)] is the state air pollution control agency. The [TNRCC] is the principal authority in this state on matters relating to the quality of the state's air resources and for setting standards, criteria, levels, and emission limits for air content and pollution control."[4] Section 382.002 of the Texas Clean Air Act provides that the state's policy and purpose are to safeguard air resources "by controlling or abating air pollution and emissions of air contaminants." Subsection 382.003(2) of the Act defines "air contaminant" as "particulate, matter, radioactive material, dust, fumes, gas, mist, smoke, vapor, or odor, including any combination of those items, produced by processes other than natural." Section 382.025 provides that TNRCC may take "any action indicated by the circumstances" to control air pollution.

Texas has codified nuisance odor law in the subsection 382.003(3) definition of "air pollution." Moreover, TNRCC regulations expressly prohibit the release of nuisance odor:

> No person shall discharge from any source whatsoever one or more air contaminants or combinations thereof, in such concentration and of such duration as are or may tend to be injurious to or to adversely affect human health or welfare, animal life, vegetation, or property, or as to interfere with the normal use and enjoyment of animal life, vegetation, or property.[5]

TNRCC's enforcement of the nuisance odor regulation is exclusively complaint driven. Complainants calling TNRCC are asked a standard series of questions, including their identities. TNRCC responds to every complaint received by dispatching a field investigator to the site as soon as feasible. Nonetheless, because of the considerable distance often separating TNRCC's regional offices from complaint sites, the number of complaints verified from the total number received is

---

* Policy Analyst, Texas Institute for Applied Environmental Research, Tarleton State University, Stephenville, Texas.

very low. For example, an average of one complaint in 50 received by TNRCC from Erath County is verified.[6]

The TNRCC field verification process relies solely on the subjective judgment and experience of field inspectors. Inspectors travel to allegedly offending facilities and sniff the air at various points near the property, both upwind and downwind. Based on this "sniff test," inspectors determine whether the facility constitutes a nuisance. In doing so, inspectors attempt to place themselves in the shoes of complainants in deciding whether the odor would significantly interfere with complainants' use and enjoyment of their property. Despite its inherent subjectivity, TNRCC air quality inspectors are generally satisfied with the present inspection protocol, but believe that the process tends to favor odor producers rather than complainants.[7]

Sections 382.081-382.085 of the Act empower TNRCC to institute legal enforcement actions and provide penalties for violations of the Act. Further, the Act preserves a private cause of action for nuisance by an aggrieved party. Section 382.004 states that "[t]his chapter does not affect the right of a private person to pursue a common law remedy available to abate or recover damages for a condition of pollution or other nuisance, or for both abatement and recovery of damages." A Texas appellate decision has upheld the right to a private cause of action for nuisance in the face of a defense argument that TNRCC regulations occupy the field.[8] The same court held that possession of a TNRCC permit does not shield a defendant from liability for damages in a private nuisance suit.

During the 1980s virtually all states, including Texas, enacted "right to farm" laws as a counterweight to private nuisance actions against agricultural operations. Section 251.001 of the Texas Agriculture Code provides that "[i]t is the purpose of this chapter to reduce the loss to the state of its agricultural resources by limiting the circumstances under which agricultural operations may be regulated or considered to be a nuisance." Subsection 251.004(a) provides a one-year statute of limitations for private nuisance actions. However, that provision has no effect on actions brought by TNRCC to enforce the Clean Air Act.[9]

In addition to enforcement of the nuisance regulation, TNRCC controls CAFO odor through its section 382.051 permitting authority. However, section 116.6 of TNRCC regulations provides that the agency may exempt certain facilities from the permitting process if the agency determines that "such facilities will not make a significant contribution of air contaminants to the atmosphere." All TNRCC Standard Exemptions are filed in the office of the Secretary of State. Standard Exemption 62 excludes a variety of CAFOs from obtaining an individual permit, including "[l]ivestock animal feedlots designed to feed less than 1,000 animals."[10]

JUDICIAL ACTIVITY ADDRESSING CAFO ODOR

The Texas Supreme Court's April 21, 1993 opinion in the case of *F/R Cattle Company, Inc. v. The State of Texas* created great uncertainty concerning TNRCC's permitting and enforcement jurisdiction over CAFO odor. In *F/R Cattle*, the ultimate issue for resolution by the Supreme Court was the meaning of "natural processes" as used in subsection 382.003(2) of the Clean Air Act. Section 382.003(2) defines "air contaminant" to include "odor ... produced by processes other than natural." The Supreme Court observed that "[t]he crux of the dispute between the parties is over the meaning of this phrase."[11]

Section 382.011 of the Act provides that "[t]he board shall seek to accomplish the purposes of this chapter through the control of air contaminants by all practical and economically feasible methods." Thus, if the odor produced by the F/R Cattle Company facility were deemed to fall within the "natural processes" exception to the Act's definition of "air contaminant," TNRCC would have no jurisdiction to regulate the facility since the agency is charged with regulating air pollution and contaminants. The Supreme Court held that the odor emanating from F/R Cattle Company likely falls within the "natural processes" exception, and remanded the case to the court of appeals for a complete review of the trial court's factual findings. However, the State's motion for rehearing remains pending before the Supreme Court.[12]

In reaching its decision, a majority of the Supreme Court overturned a decision by the Eleventh District Court of Appeals, Eastland,[13] adopting instead the reasoning of two earlier Texas appellate decisions: *Europak, Inc. v. County of Hunt*, 507 S.W.2d 884 (Tex.Civ.App.--Dallas 1974, no writ) and *Southwest Livestock and Trucking Company v. Texas Air Control Board*, 579 S.W.2d 549 (Tex.Civ.App.--Tyler 1979, writ ref'd n.r.e.). *Europak* involved an action by the State and Hunt County to enjoin construction of a horse slaughter and packing facility without first obtaining a TACB permit. The trial court enjoined construction and the appeals court affirmed. The appeals court established the following two-prong test for determining a "natural process" exempt from TACB jurisdiction: 1) a process which occurs in nature, and 2) a process which is affected or controlled by human devices only to the extent normal and usual for the locale. The court held that while horse slaughter and processing may represent a process occurring in nature, "the evidence is sufficient to support a finding that concentration of such a large number of animals into such a small area would not be normal or usual in this vicinity."[14] The court in *Southwest Livestock* applied the *Europak* test in favor of TACB jurisdiction in a case involving a livestock loading facility, holding that "[i]t should not be considered normal, usual or natural to find odoriferous livestock pens situated in such close proximity to urban land uses such as homes and small commercial establishments which are particularly susceptible to strong odors."[15]

In its appeal to the Texas Supreme Court, F/R Cattle Company urged the Court to overturn the holding of the appeals court for Eastland and follow the two-prong test from *Europak* and *Southwest Livestock*. The Eastland court had held that

> it is abnormal and unusual, without regard to location to concentrate approximately 6,000 baby calves in 1,500 small hutches and in weaning pens.... The odor at the defendant's calf feeding facility was not produced by natural processes. The trial court erred in dismissing the State's petition. The Texas Clean Air Act is applicable, and the Texas Air Control Board has jurisdiction.[16]

An important distinction between the *F/R Cattle* case and both *Europak* and *Southwest Livestock* is that in *F/R Cattle* the appeals court rejected the trial court's factual determination in favor of F/R Cattle Company and decided the "natural processes" issue as a matter of law, while in the latter cases the appeals courts upheld trial court findings that the offensive facilities were abnormal for their locales. In *F/R Cattle*, the Supreme Court held that a determination of "natural processes" is inherently factual:

> The State's alternative argument, that the test in *Europak* and *Southwest Feedyards* [sic] is a question of law for the court is also problematic. The State offers no suggestion what standards the court might apply. As the court in *Europak* observed, a criterion based on the number of cattle would not suffice. At one end of the spectrum is a remotely located ranch with human involvement limited to fencing, feeding, and other typical ranching activities. At the other end of the spectrum is a feedlot located within city limits that clearly interferes with a [sic] neighbors' use of their property. In between, the circumstances would run the gamut, so that making an evaluation of whether emissions are produced by natural processes is an inquiry best left to the finder of fact.[17]

The dissent[18] rejected the two-prong test for jurisdictional purposes, arguing that the first prong--whether the process under scrutiny occurs in nature--should determine jurisdiction as a matter of law. The dissent noted that TACB had unilaterally circumscribed the broad jurisdiction provided under a "human intervention" interpretation of "natural processes" by requiring a permit from only those livestock operations housing 1,000 head or more. Further, the dissent underscored the illogical result obtained through adoption of the "locational normality" prong of the *Europak* test, a point made by the State's counsel to the Court during oral argument:

> Significantly, F/R Cattle Co. makes no argument that its facility would be outside the Board's authority if it were the only livestock facility in the region. The crux of its argument is that *Europak* and *Southwest Livestock* require consideration of "locational normality." By adopting this view, the majority sanctions a nonsensical approach to protecting the environment: as the air quality deteriorates, so does the Board's jurisdiction.[19]

LEGISLATIVE ACTIVITY ADDRESSING CAFO ODOR

Opponents of CAFO odor regulation introduced Senate Bill 684 and House Bill 2633 during the 73rd Texas Legislature in the spring of 1993. As originally filed, both bills called for an exemption from TNRCC odor regulation for agricultural operations. "Agricultural operation" included those facilities involved in agricultural production but not agricultural processing. Immediately following the *F/R Cattle* decision, proponents of the bills appeared to enjoy sufficient support for passage. However, both opponents and proponents of the legislation eventually introduced several amendments which significantly modified the original bills. Ultimately, the legislative session ended before compromise legislation could be adopted.

Initial modifications to both bills called for TNRCC jurisdiction over agricultural operations where public health was affected by emissions. The House Committee on Environmental Regulation then amended H.B. 2633 to require TNRCC to satisfy a rigorous two-prong test prior to asserting jurisdiction over an agricultural operation for odor emissions: 1) that the operation was not complying with appropriate production practices established by TAEX; and 2) that the odor emissions caused bodily injury or adverse public health effects. Thereafter, the full Senate rejected the House bill in favor of broad TNRCC jurisdiction where public safety is affected by odor emissions. Another Senate amendment broadened TNRCC jurisdiction further in cases involving "substantial adverse effect to quality of life."[20] A final Senate amendment sought to establish a "first in time, first in right" TNRCC jurisdictional provision in cases where "the board finds a party was located in the immediate vicinity of the agricultural operation prior to the establishment of the agricultural operation."[21]

Given the divergent positions on TNRCC jurisdiction over CAFO odor prevailing in the House and Senate, a Conference Committee was appointed to seek a compromise. However, the 73rd Legislature adjourned prior to resolution of outstanding issues.

AGENCY ACTIVITY ADDRESSING CAFO ODOR

In November 1992, TACB Chairman Kirk Watson announced the formation of a 28-member nuisance odor task force to provide TACB recommendations for more efficient and effective nuisance odor regulation. Observers attributed formation of the task force to two well-publicized controversies: 1) organized opposition to emissions emanating from the East Austin petroleum tank farm, and 2) the *F/R Cattle* case. While task force members initially intended to produce recommendations in time to inform debate during the 73rd Texas Legislature, the complexity and divisiveness of the issues under discussion necessitated greater time. The task force eventually met a total of eleven times between December 1992 and August 1993.

The task force's agricultural subcommittee was charged with evaluating "the regulation of emissions from agricultural operations to determine whether enforcement relating to nuisance odors should continue in its present form or be changed."[22] Based on deliberations by that subcommittee, the task force reported four recommendations specifically addressing agricultural issues. One recommendation received the unanimous support of task force members; three recommendations were supported by ten task force members and opposed by 13 members. The task force unanimously recommended that TACB/TNRCC should not attempt to formulate agency rules concerning agency jurisdiction over agricultural operations due to the complexity of the issues involved in that determination.[23]

One split recommendation stated that "Texas Air Control Board Rule 101.4 should be amended to state: 'This provision shall not apply to any emission of air contaminants produced by natural processes.'"[24] Those members supporting the recommendation argued that agency rules should be brought in line with section 382.003(2) of the Texas Clean Air Act in the wake of the Texas Supreme Court's *F/R Cattle* opinion. Those members opposing the recommendation stated that the amendment to agency rules would be premature given the pending motion for rehearing before the Supreme Court in *F/R Cattle*, and that the amendment would be superfluous since the statutory language applies to all agency rules.

A second split recommendation stated that the following definition of "natural processes" should be added to TNRCC rules: "'Natural processes' means those processes which occur in nature and are affected or controlled by human devices only to an extent normal and usual for the particular area involved."[25] Supporters of the recommendation claimed that the new definition was warranted now that the Texas Supreme Court and two appellate courts have adopted the same two-prong test represented in the proposed definition. Opponents of the recommendation noted that any such recommendation was premature while the motion for rehearing remained pending before the Supreme Court.

The final split recommendation provided that TNRCC should determine whether an agricultural operation is complying with "good agricultural practices" and applicable air permit provisions at the time the agency responds to a complaint, and that no penalty should be assessed against an operation in compliance. Supporters of this recommendation urged that farmers operating within the parameters of controlling permits and generally accepted agricultural practices should not be penalized simply because agricultural operations inherently create odor. Opponents of the recommendation countered that the recommendation represents the same policy debated during the 73rd Texas Legislature, and which failed to garner majority approval. Moreover, opponents argued that agricultural operations--particularly large agricultural operations--must expect the same odor regulatory treatment as any other economic sector.

RESEARCH ACTIVITY ADDRESSING CAFO ODOR

In 1992, TIAER and TAEX initiated a study of odor produced by dairies in Erath County, Texas pursuant to a contract with TACB. The final report of that research was presented to TNRCC in January 1994.[26]

The measurement component of the project required TIAER, in cooperation with TAEX and TACB, to select three "state-of-the-art" dairies where odor intensity from a variety of potential sources could be measured. TIAER erected one 10-meter meteorological tower at an unobstructed location on each of the three dairies. The towers provided a wealth of data at five-minute intervals while odor measurements were taken at the dairies, including wind speed, wind direction, temperature and relative humidity. Those weather data, along with additional information provided by the Weather Service Meteorological Observatory, Stephenville, Texas, were essential for odor dispersion modeling efforts undertaken as part of the project.

Odor intensity measurements were taken at 15 to 20 sites at each dairy on six different occasions between October 1992 and August 1993. A variety of weather conditions and times of day were encompassed in the measurement schedule. Each visit to a single dairy required approximately four hours to complete all measurements. Six trained panelists provided odor intensity data at each site using either a Barnaby & Sutcliffe scentometer or a butanol olfactometer. Measurements were taken near obvious odor sources, such as anaerobic lagoons, solids settling basins and animal confinement areas, and at other locations both near to and distant from production areas. Where possible, several property-line measurements were taken.

The scentometer and meteorological data collected at the three dairies were used to calibrate a dispersion model. While dispersion models have been used successfully for urban and industrial applications, relatively few studies in the United States have employed odor dispersion modeling in an agricultural context.[27] Modeling results from the Erath County study appear promising. However, considerable work must be done before odor dispersion modeling proves sufficiently reliable to provide the basis for dairy siting decisions or to isolate individual odor sources among several nearby agricultural operations.

IMPLICATIONS OF RECENT JUDICIAL, LEGISLATIVE, REGULATORY AND RESEARCH ACTIVITY

Substantial resources were expended in 1992-93 in Texas to address the problem of CAFO odor. That heightened level of activity notwithstanding, relatively little progress was made toward resolving many difficult issues. Efforts by both the TACB odor task force and the 73rd Texas Legislature to improve CAFO odor regulation proved inadequate. Moreover, the Texas Supreme Court's *F/R Cattle* decision actually increased regulatory uncertainty which may require considerable time and effort to resolve.

Prior to the *F/R Cattle* opinion, TACB enforced the Rule 101.4 nuisance provision against all sources of odor, including CAFOs and other agricultural operations. Further, TNRCC Standard Exemption 62 required only CAFOs with 1,000 head or more to obtain an air quality permit. However, in *F/R Cattle*, the Supreme Court ruled that agricultural operations located in an area encompassing substantially similar agricultural activity involve "natural processes" and are thus exempt from TNRCC permitting and enforcement jurisdiction. Thus, subsequent to *F/R Cattle*, TNRCC must expend significant resources attempting to determine whether the agency has either permitting or enforcement jurisdiction over a particular CAFO prior to attempting to assert that jurisdiction. Because the Supreme Court provided no guidance to aid TNRCC in establishing the parameters of agency jurisdiction, that process could prove long and expensive in terms of agency and judicial resources. Those parameters include the degree of similarity between individual livestock operations necessary to invoke the "natural processes" exemption, as well as the threshold number of such operations per given geographic area. In *F/R Cattle*, 21 dairy operations within a three mile radius were sufficient to extend the exemption to a calf-raising facility. However, myriad factual scenarios exist involving different types and densities of livestock operations. Ultimately, TNRCC simply may decline to pursue enforcement actions against livestock operations under the nuisance rule to avoid expending limited agency resources. That stance could result in undue hardship for rural landowners complaining of livestock odor. Those complainants would have little recourse beyond initiating private lawsuits.

The recently filed case of *Wilson et al. v. F/R Cattle*, No. 93-05-21472-CV (266th Judicial District, Erath County, Tex., 1993) provides an example of the private actions that may result from TNRCC's diminished jurisdiction to regulate livestock odors. The plaintiff Wilsons are owners of property adjacent to the F/R facility, and frequent complainants to TNRCC concerning odor emanating from that facility. The Wilsons refrained from filing a private action against F/R Cattle Company until the Supreme Court ruled against TNRCC jurisdiction. Thus, one apparent result of the *F/R Cattle* decision is the transfer of enforcement responsibility for CAFO nuisance odor--and associated litigation costs--from the public to the private sector. Interestingly, the complaint in *Wilson* does not state a cause of action for nuisance; rather, plaintiffs rely on breach of contract, negligence, gross negligence and trespass theories. Plaintiffs may have excluded the nuisance cause of action to avoid the harsh result provided under the Texas right-to-farm provision in the event the nuisance claim were dismissed. Right-to-farm provisions generally apply only to nuisance claims.

The issue of TNRCC air permitting jurisdiction over livestock operations soon may be moot with the recent consolidation of TACB and the Texas Water Commission (TWC) under TNRCC. Since 1987, TWC has addressed odor from CAFO manure application fields in agency regulations.[28] Some of the construction criteria for odor control currently required in former TACB permits could be incorporated into a unitary TNRCC waste discharge/air permit. However, legislative action may prove necessary to modify the "natural processes" exemption for permitting purposes where permit provisions patently address livestock odor. For example, if, under current law, TNRCC were to make permit issuance contingent on satisfactory odor dispersion modeling results, an applicant could argue that under *F/R Cattle* the agency has no jurisdiction over potential odor produced at the facility, particularly if the facility is located within close proximity to several other livestock operations.

Once proven reliable, odor dispersion modeling will likely represent a standard CAFO permit requirement. Odor dispersion modeling can improve dairy siting decisions by predicting the movement of livestock odor from a proposed location to insure sufficient buffer area between the dairy and neighboring property. Such predictions should help reduce the incidence of private nuisance odor litigation. Moreover, field odor monitoring data, a key input for accurate dispersion modeling, will likely improve in the near future as more triangular, dynamic, forced-choice olfactometers are utilized for field measurements. Late in 1993, TIAER ordered a triangular, dynamic, forced-choice olfactometer to improve its odor measurement capability.

In 1995, the 74th Texas Legislature should make every effort to resolve the "natural processes" issue to avoid the needless expenditure of judicial resources, and to provide greater predictability to TNRCC, the regulated community, and rural landowners. The case-by-case factual determination of "locational normality" required by the majority in *F/R Cattle* must be circumscribed so that TNRCC's permitting and enforcement jurisdiction over CAFOs depends once again on a bright line test, and new techniques such as odor dispersion modeling can be employed for more effective permitting decisions.

REFERENCES

1. *F/R Cattle Company, Inc. v. The State of Texas*, ____ S.W.2d ____, No. D-2481 (opinion issued April 21, 1993, motion for rehearing pending).

2. On September 1, 1993, the Texas Air Control Board merged with the Texas Water Commission and other agencies to form the Texas Natural Resource Conservation Commission.

3. V.T.C.A., Health & Safety Code §§ 382.001-382.141 (1992 & 1993 Supp.).

4. V.T.C.A., Health & Safety Code § 381.002 (1992).

5. 31 TAC § 101.4 (1989).

6. Presentation by Mark Gibbs, TNRCC Air Quality Division, Dec. 14, 1992, Texas A&M.

7. Conversation with Bob Kramer, TNRCC field investigator, Dec. 21, 1992, Fort Worth, Texas.

8. *Manchester Terminal Corp. v. Texas TX TX Marine Transportation, Inc.*, 781 S.W.2d 646, 650 (Tex.App.--Houston [1st Dist.] 1989).

9. C. Szopa, Texas Air Control Board Draft Memorandum, Dec. 16, 1992, at 1.

10. Texas Air Control Board Standard Exemption List, Exemption 62, Revised June 18, 1992.

11. *F/R Cattle*, No. D-2481 at 5.

12. On November 24, 1993, the Texas Supreme Court withdrew its April 21 opinion in *F/R Cattle* and substituted a new opinion. The bulk of the majority opinion remained unchanged from the April 21 version. However, the State's motion for rehearing was denied as part of the November 24 opinion.

13. *State of Texas v. F/R Cattle Company, Inc.*, 828 S.W.2d 303 (Tex.App.--Eastland 1992).

14. *Europak*, 507 S.W.2d at 891.

15. *Southwest Livestock*, 579 S.W.2d at 552.

16. *F/R Cattle*, 828 S.W.2d at 307.

16. *F/R Cattle,*, No. D-2481 at 8-9.

18. *F/R Cattle,* No. D-2481 (Spector, J. and Gammage, J., dissenting).

19. *Id.* at 3.

20. Nuisance Odor Task Force, Report to the Texas Air Control Board, Report of the Agricultural Issues Subcommittee (Appendix B) 14 (1993).

21. *Id.*

22. *Id.* at 2 (1993).

23. *Id.* at 18.

24. *Id.*

25. *Id.* at 19.

26. TIAER & TAEX, Final Report: Preliminary Research Concerning the Character, Sources and Intensity of Odors from Dairy Operations in Erath County, Texas (1994).

27. *See generally* P. Gassman, Simulation of Odor Transport: A Review (1992) (Paper No. 92-4517 *presented at* ASAE Winter meeting, Dec. 15-18, 1992).

28. 31 TAC § 321.37 (1989) ("If land application is utilized for disposal of waste and/or wastewater, the following requirements shall apply: (1)(C) Disposal of waste and wastewater shall be done in such a manner as to prevent nuisance conditions such as odors and flies.").

FARMER DECISION ALTERNATIVES UNDER ENVIRONMENTAL REGULATIONS

M. Rudstrom, R. A. Pfeifer, S. N. Mitchell, O. C. Doering III[*]

ABSTRACT

We investigated farm decisions likely to result from environmental regulations restricting chemical use in the White River Basin (WRB) in Indiana. Specific environmental regulations examined were limiting atrazine to post-emergent applications, banning atrazine, banning all triazines, and reducing soil erosion to "T".

A flowchart was developed showing the decision-making process a farmer uses when a crop production change is necessary due to regulations. Selected options from the flowchart were then modelled to investigate the economic impact of the change.

Representative farms were developed using land resource endowment based on the National Resource Inventory (NRI) information for the White River Basin (WRB). The National Agricultural Statistics Service (NASS) 1991 Area Studies Survey furnished information on the current cropping practices and operator characteristics in this area. This data was analyzed to determine operator characteristics that most affect decisions regarding cropping practices.

PCLP, a microcomputer based farm planning model, was used to evaluate the effects of herbicide use restrictions on the tillage system employed, crops grown, and profitability. With no additional restrictions, chisel plow systems have higher returns to resources than no-till or moldboard systems, given our representative resource base. As atrazine and triazine restrictions are tightened, returns to resources decrease, regardless of farm size or tillage system. Under any of the proposed herbicide use restrictions, no-till became a more economically preferable option. The extent to which it was more profitable relative to the other systems was dependent upon the restriction imposed. Under all restrictions, farms with smaller acreages were more adversely affected. Also, the different farm and operator characteristics strongly influenced the decision path farmers are likely to take to meet environmental restrictions.
KEYWORDS. Pesticide restriction, Decision-making, Soil erosion.

THE DECISION-MAKING PROCESS

We believe that over the next decade farming will be influenced more by policies like environmental regulations, trade agreements, and safety and health concerns than by traditional farm legislation. How farmers are likely to react and adapt to such pressures, and what the actual impact will be on agriculture become critical questions for policy makers.

When a farmer is faced with a new policy which will force him to alter his cropping practices, how does he respond? Many efforts have been made to model the results of a farmer's decision, or to develop a process by which to choose the "optimal" decision. Our goal, however, was to take a closer look at the options as a farmer sees them and explain why some of those alternatives may or may not be realistic.

---

[*]M. Rudstrom, R.A. Pfeifer, S.N. Mitchell, Research Assistants, and O.C. Doering III, Professor, Department of Agricultural Economics, Purdue University, West Lafayette, IN. This work completed under USDA/ERS Cooperative Agreement Nos. 43-3AEM-0-8002 and 43-3AEM-2-80098.

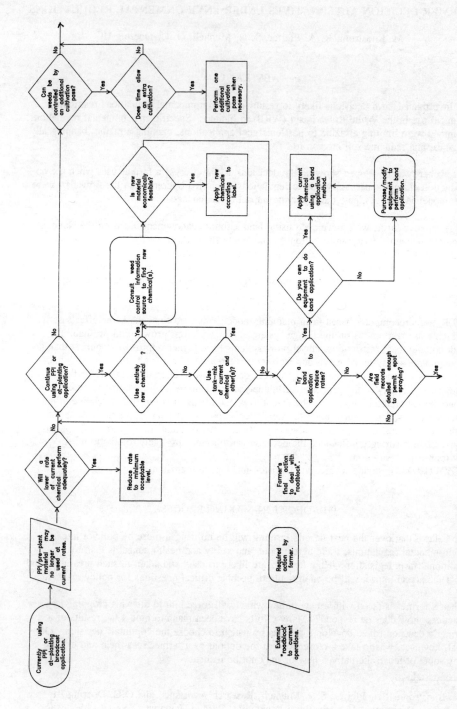

Figure 1. Decision-Making Flowchart for Farmers Faced with Pesticide Use Regulation

To better illustrate these options and 1) how they may impact a farmer's decision, and 2) why his final decision may not be the most obvious, or most cost-effective, or just not make the most sense at first glance, we have developed flowcharts as a representation of a farmer's thought process when faced with making a change. (A portion of this chart is presented as Figure 1.) Unlike an expert system, this flowchart does not necessarily guide a farmer to the "right" technical decision, but the one which makes the most sense to him. In effect, the flowchart simply illustrates the chain of decisions as they may occur.

This particular chart assumes a farmer starts off moldboard or chisel plowing in a corn/soybean rotation. This is only one of a myriad of a flowcharts which could have been developed and it is certainly not an exhaustive list of options. This is a generic chart based on our interviews, experience, and input from numerous professionals who work directly with farmers. Although it is presented as a one-way decision-making process, one should keep in mind that as a farmer reaches the 'bottom' or end of a branch of the flowchart, he may decide that he wants to pursue another alternative, making the route to a final decision rather complex. Once a decision has been made, however, the cost of going backwards may be sizable. As one moves further from the status quo (further to the right and down on the flowchart), the costs of a change go up and more alterations must be made to the cropping system.

DEVELOPMENT OF MODEL FARMS

Once a farmer has chosen a possible avenue of change, it is prudent to evaluate the impact of the change before actually beginning to use it. The modelling program used in this study to model the effect of change is PCLP, the Purdue Crop/Livestock Linear Program (Dobbins et al., 1992). PC-LP generates a dollar value for returns to resources given inputs such as yield levels, field operations, soil resources, labor, and machinery.

The White River Basin, located in south central Indiana and occupying an area of more than 11,000 square miles, was chosen as the geographic region for this study because of the data available from the National Resources Inventory developed by SCS (1987) and the Area Study Survey performed by ERS and NASS (1991), under the direction of the USDA, for the President's Water Quality Initiative. The survey was conducted to learn more about farm operators and their farm practices with regard to tillage and cropping systems and fertilizer and pesticide use.

Nine model farms were developed that reflected the physical and farmer characteristics found in the White River Basin identified by examing a number of variables in the NASS survey data and the NRI data. The data indicated that time spent working off the farm influences how farmers make their decisions concerning changes to their systems. Days worked off farm also correlates well with changes in farm size, i.e. the farm sizes will be 400, 500, and 1200 acres which correspond to farmers working weekends, part-time, and full-time on the farm.

Each of the farm sizes was given the same land quality resource base. Because the White River Basin does vary considerably with regards to topography and soil resource base, the model farms were assigned soil resources which reflect the distribution of soils found within the White River Basin. Tillage systems chosen for use in the model were moldboard plow, chisel plow, and no-till. Hence, we had three 400 acre farms - one 400 acre farm using a moldboard plow, one 400 acre farm using a chisel plow, and one 400 acre farm with no-till as the tillage system - and so on for the 500 and 1200 acre farms resulting in nine model farms.

As is typical of the White River Basin, our model farms produce corn alternating with narrow-row soybeans in a two year rotation. Machinery was sized so that field operations could be performed in a timely manner, standard rates of fertilizer and pesticides were used and yields reflected those possible on the soil complement and tillage system combination chosen.

ECONOMICS OF CHANGE

A base run was established for each sized farm in each tillage system in which there were no restrictions on pesticide use or soil loss. Returns to labor and management were then calculated (Table 1) by subtracting fixed costs (machinery and land costs) from returns to resources. Returns to labor and management were used to evaluate, by comparison, what happened when we ran the models under the different policy regimes of pesticide and soil loss restrictions.

Table 1: Per Acre Returns to Labor and Management for the Base Farms

Tillage System	1200 acre	500 acre	400 acre
Moldboard	$51.23	$34.14	$19.26
Chisel	$55.61	$43.97	$24.64
No-till	$56.72	$41.15	$22.49

The reduced tillage system (either chisel plow or no-till) does have an economic advantage when compared to the conventional moldboard system. In the case of the 1200 acre farm there also seems to be a further economic advantage to being in the no-till system as compared to the chisel plow system. This is due to a further reduction of machinery costs in the no-till system. The high initial cost of obtaining the new no-till equipment prevents the smaller farms from receiving as large a decrease in machinery costs per acre as the larger farms.

Impact of Triazine Herbicide Use Restrictions

What will farmers do if triazine herbicides are further regulated? Without exception, when the question "What would you do if atrazine or another material was banned?" was posed to leading farmers, they responded, in effect, "I would find a substitute material". Chemical dealers also indicated that this would be the route farmers would take, and, indeed, have chosen in the past. For the model farms examined in the PCLP, we selected alternative herbicides as well. The inputs for the base farms were adjusted to reflect higher herbicide costs for each of three pesticide restriction scenarios. The PCLP output indicates what impact those changes have on returns to resources.

Atrazine Post Only

The first restriction that we examined was in limiting atrazine to post applied applications only. Even choosing the least cost alternative the herbicide costs for the moldboard and chisel systems went up $7.88 an acre. The no-till system had already been using atrazine in a POST application and so no changes due to the restrictions were noted (Table 2).

For both the moldboard and the chisel systems the atrazine post only restriction caused a decrease in returns to labor and management. The smaller farms seemed to be penalized

Table 2: Per Acre Returns to Labor and Management (Atrazine Post Only) and Percentage Decrease From the Base

Tillage System	1200 acre		500 acre		400 acre	
Moldboard	$48.47	5.4%	$30.19	11.6%	$15.31	20.5%
Chisel	$52.31	5.9%	$35.95	18.2%	$20.69	16.0%
No-till	$56.72	0.0%	$41.15	0.0%	$22.49	0.0%

more heavily by this restriction. The reason for this is again one of scale; the smaller farmers had smaller per acre returns to labor and management to begin with and an increase in production costs of even a few thousand dollars would cause a larger percentage decrease in the returns to labor and management compared to the larger farms. The no-till farms would see no change under this scenario because they were in compliance with the new regulation before it was imposed. Given our model, the no-till system is now the most financially rewarding tillage system on all sized farms investigated when atrazine is restricted to post applications.

Atrazine Ban

A complete atrazine ban again prompts substitutions of herbicides to take place. The previous restriction (atrazine used post only) had prompted the moldboard and chisel plow farms to move away from atrazine use and so banning atrazine would have only the same affect as restricting its use to post only. However, the no-till farms were forced to use the next best alternative increasing in the per acre cost of herbicide by $1.90.

Table 3: Per Acre Returns to Labor and Management (Atrazine Ban) and Percentage Decrease From the Base

Tillage System	1200 acre		500 acre		400 acre	
Moldboard	$48.47	5.4%	$30.19	11.6%	$15.31	20.5%
Chisel	$52.31	5.9%	$35.95	18.2%	$20.69	16.0%
No-till	$55.90	1.4%	$40.28	2.1%	$21.70	3.5%

Again the smaller farms are more heavily penalized by this new restriction when we look at their per acre returns to labor and management (Table 3).

Triazine Ban

Continuing to increase the severity of the pesticide restrictions, the next logical restriction would be a complete triazine ban. The no-till farms already made the switch away from triazines at the last restriction. They are now followed by the moldboard and chisel systems whose pesticide bill has now climbed $10.70 an acre over the base.

From the results presented in Table 4 again we see a familiar pattern in the smaller farms losing a much larger share of their returns to labor and management to the increased restrictions. The range in decrease in returns from the banning of all triazines is from 1.4% to 28%.

Table 4: Per Acre Returns to Labor and Management (Triazine Ban) and Percentage Decrease From the Base

Tillage System	1200 acre		500 acre		400 acre	
Moldboard	$46.96	8.3%	$28.68	16.0%	$13.80	28.3%
Chisel	$50.80	8.6%	$34.49	21.6%	$19.18	22.1%
No-till	$55.90	1.4%	$40.28	2.1%	$21.70	3.5%

These results also show that the SEQUENCE of the restrictions is important. Depending upon the relative economic advantage of one system over another when a restriction is imposed, a farmer may or may not wish to change systems. Additionally, the financial status of the farm would determine if a moldboard or chisel plow system could be converted to a no-till system in a manner which would take advantage of the higher per acre returns to labor and management. Depending on the severity of the restriction, it may convince a farmer to get out of farming altogether, particularly for some of the smaller farms.

Impacts on Yield and Weed Control

An underlying assumption in these models is that a farmer can make substitutions in pesticides away from atrazine that provide equivalent weed control. But atrazine is a very good broad spectrum herbicide; it has a great deal of efficacy on both grass and broadleaf weed species and is especially good at controlling some rather difficult weeds to control like quackgrass. In our model we assumed that the farms did not have any particular noxious and hard to control weeds, this allowed the substitutions to take place rather easily in that all that changed was herbicide price. Unfortunately, this assumption in reality is probably flawed.

The other area where we may have been optimistic is in the assumption that crop tolerance to our substituted herbicides was good. However, some of the substitute herbicides that we selected have been known to cause some crop injury on some soil types under some conditions. Because of the dynamic nature of a growing crop it is difficult to really know how much yield potential may be lost do to herbicide crop injury. (New developments in herbicide resistant hybrids may mitigate this though.)

If we assumed that using a different herbicide caused increased crop injury and/or an increase in weed pressure that decreased yields by corresponding 5% yield reduction how much would we reduce the returns to management and labor? What if that reduction in yield was 10%? The models were run again to investigate the impact of this factor. A triazine ban was assumed and yield penalties of 5% and 10% were put into the model.
A 5% yield reduction cuts directly into returns to labor and management (Table 5). The impacts of a 10% yield reduction are even more drastic. The 400 acre farm is not sustainable economically. Small farms may be least able to make the major changes necessary to adjust to new restrictions and the accompanying yield impacts and may be forced out of business.

Other Alternatives

The preceding changes to the model reflected what the leading farmers said they would do to accommodate a regulation - use a new material. The no-till farmers, even under a complete triazine ban should only incur a $1.90 an acre increase in herbicide costs based on our herbicide choices.

Table 5: Per Acre Returns to Labor and Management and Percentage Decrease from the Base (Triazines Banned and a Yield Reduction Incurred)

Tillage System	Farm Size (acres)					
	1200		500		400	
	Yield Reduction 5%					
MB	$36.69	28.38%	$18.65	45.37%	$3.73	80.63%
CH	$40.12	27.85%	$24.49	44.30%	$9.13	62.95%
NT	$41.36	27.08%	$26.44	35.75%	$12.79	43.13%
	Yield Reduction 10%					
MB	$27.48	46.36%	$9.60	71.88%	-$5.40	128.04%
CH	$30.88	44.47%	$15.35	65.09%	-$0.10	100.07%
NT	$35.94	36.64%	$21.05	48.84%	$6.39	71.59%

The moldboard and chisel plow farms (more representative of WRB farms) face more substantial herbicide price increases. When atrazine was limited to a post application only, it became, in effect, an atrazine ban because the next best alternative herbicides did not contain atrazine. Limiting atrazine to post applications increased herbicide on rotation corn $7.88 an acre from a total of $13.70 to $21.58 an acre. Upon enacting the atrazine ban, the total per acre herbicide costs are $24.40 for rotation corn. These large overall increases in cost may send the farmer "back to the flowchart".

For the models presented, the farmer was assumed to continue a blanket treatment. That is, covering the entire field with the material. Some farmers may prefer to switch to a band, or over the row only application, as suggested in column B of the flowchart. This would cut pesticide costs about 40%, but would require a capital outlay to modify the equipment. In addition, an extra cultivation pass may be required to clear the row middles of weeds, increasing the susceptibility of the soil to erosion.

Under a triazine ban, even with a 40% reduction possible in pesticide costs made possible with band application, the herbicide costs would still be $14.64 an acre. This is 7% higher than the status quo. Additionally, an extra cultivation pass may be required at an approximate cost of about $5 an acre and equipment would need to be modified to spray in a band over the row. This illustrates that upon closer examination, what at first seemed a fine and economical choice is not clearly superior to using the full rate of the more expensive alternative herbicide.

Soil Loss Limitations

Soil erosion is another phenomenon which causes problems environmentally: soil particles carry attached herbicide molecules to waterways, soil which has been moved by erosive forces clogs streams and rivers causing expensive cleanup, and, as the soil is lost from a field the field is less productive. For these reasons and many more it is desirable to limit soil loss from agricultural land. Accomplishing this does not come without a price, of course. Tillage increases the susceptibility of the soil to erosion, so reducing tillage should also reduce erosion. But, a major benefit of tillage is weed control. If a farmer reduces his tillage, in most cases he must replace this with another form of weed control, usually overall herbicides. One must be clear about the advantages and disadvantages and consider the picture, being careful not to focus on simply reducing soil loss by any means.

Using the universal soil loss equation (USLE), the average rate of soil loss was calculated for each of the tillage systems used in our model farms. (The rate of soil loss would be the same no matter what the size because each farm size has the same soil resource base.) All of the tillage systems would be acceptable if soil loss was limited to 2T, or two times the rate at which the soil is regenerated, in this case T is 4.67 tons/acre/year given the soil resource base of our model farms. When the restriction is tightened to 1T, the moldboard system is no longer acceptable and the chisel plow system is only marginally acceptable. Therefore, the farmer using a chisel plow system may not have the option of increasing his tillage with additional cultivation instead of using a substitute herbicide under a herbicide use restriction.

CONCLUSIONS

We believe this analysis indicates the important questions that need to be asked as the nation imposes more restrictions on agriculture to achieve environmental quality or health/safety goals. What options really are available to manipulate existing cropping systems? Will completely different systems be required? What regulatory paths put farmers in a severe economic bind? Do the expected environmental regulations more severely impact small and moderately sized farms? So far, our work indicates there are very definite restrictions on changing current systems, the sequencing of restrictions can be critically important, and there are severe penalties for smaller farms trying to comply.

REFERENCES

1. 1987 National Resources Inventory, Soil Conservation Service.

2. 1991 Area Study Survey, conducted by the USDA-Economic Research Service and the National Agricultural Statistics Service.

3. Purdue Crop/Livestock Linear Program (PC-LP). Version 2.0. Developed by C.L. Dobbins, Y. Han, P. Preckel, and D.H. Doster. Copyright 1992 by Purdue Research Foundation, West Lafayette, IN 47907.

COMPOSTING OF YARD TRIMMINGS -- PROCESSES AND PRODUCTS

R. A. Nordstedt and W. H. Smith[*]

ABSTRACT

Yard wastes or trimmings have been or are in the process of being banned from lined landfills in many states. As a result, many programs have been set up around the country and in Florida to process yard trimmings into useful products. The products which are produced in these facilities vary in characteristics and usefulness. This paper will summarize the processes which are in use or available for producing composts, mulches, topsoil, fuel, and other products from yard trimmings. The characteristics of these materials will be described. The influence of other factors, such as local vegetation and method of collection will also be discussed. In some cases, the composting of yard trimmings can be improved by the addition of other materials, such as animal manures, food wastes, and sewage sludges (biosolids). The resulting product may also be of higher quality. Limitations of this approach will be discussed.
KEYWORDS: Yard wastes, Yard trimmings, Compost, Mulch, Municipal solid waste.

OVERVIEW OF YARD TRIMMINGS PROBLEM

Introduction

The quantities of municipal solid waste (MSW) which are produced have been increasing. Nearly four kilograms per day (over eight pounds per day) of MSW are produced per Florida resident. The seriousness of the growing amounts of MSW is magnified by the scarcity of environmentally desirable and economically available landfill sites. Because of groundwater contamination problems with old sanitary landfills, the costs of constructing and operating new landfills to prevent environmental problems have increased dramatically. Although there is more than adequate land area available to construct new landfills, it is becoming very difficult to find sites which are acceptable to nearby residents.

Florida responded to this problem with the passage of the Solid Waste Management Act of 1988 (Chapter 88-130, Florida Statutes). This legislation set recycling goals to reduce landfilling (Earle et al., 1991). It also included a ban on landfilling of yard trimmings after January 1, 1992. Yard trash is defined in the Solid Waste Management Act as vegetative matter resulting from landscape maintenance and land clearing operations. It includes tree and shrub trimmings, grass, palm fronds, leaves, trees, and stumps. Yard trimmings is used synonymously with yard trash or yard waste, since it may remove some of the negative connotation associated with the words trash or waste.

Quantities and Characteristics of Yard Trimmings

In Florida, approximately twenty percent of the solid waste stream is yard trimmings. However, this fraction may increase to over fifty percent in some localities during some seasons of the year. Also, counties with small populations typically have a very small percentage of yard trimmings in their municipal solid waste stream.

---

[*]R. A. Nordstedt, Professor, Agricultural Engineering Department, and W.H. Smith, Professor and Director, Center for Biomass Energy Systems, Institute of Food and Agricultural Sciences, University of Florida, Gainesville 32611.

The characteristics of yard trimmings in Florida are different from other areas in the country in several respects. First, the yard trimmings are produced throughout the year compared to northern states where they occur mainly as grass and leaves in the summer and fall seasons. Also, the nature of many Florida plants is quite different from more temperate plants. For example, live oak leaves, magnolia leaves, pine needles, and palm fronds found in Florida are more difficult to degrade than leaves of tree species like maple, oak, and poplar. In general, the yard trimmings in Florida contain more woody and brushy material. Also, yard trimmings in Florida usually contain significant amounts of sand. An average analysis of yard trimmings from a processing site in Alachua County, Florida, is given in Table 1 (Barkdoll and Nordstedt, 1991).

Table 1. Average Analysis of Yard Trimmings from Alachua County, Florida.[a]

%N	%P	%K	%Ca	%Mg	%S	%OM	C/N	pH	Soluble Salts mmho/cm
0.76	0.10	0.24	1.28	0.14	0.08	71	93	5.7	4.13

[a] Average values on a dry weight basis for samples taken over a six month period.

The materials can vary from only leaves and/or grass clippings to very large, woody material. The type of plant vegetation collected as yard trimmings will affect the moisture content and odor potential of the material. Yard trimmings with a large fraction of grass clippings may cause odor problems if they are not managed properly. Yard trimmings may vary from about 30% moisture (mixed wood, leaves, brush, and grass) to about 75% moisture for fresh grass clippings (Barkdoll and Nordstedt, unpublished data). In Alachua County, Florida, yard trimmings which are collected separately from tree surgeon debris may contain 15-35% branches greater than 2.5 cm (1 inch) in diameter, 28-56% leaves and debris, 4-11% evergreen branches, 0-4% grass, 1-5% palm fronds, and 1-5% pine needles as the primary constituents (Shiralipour and McConnell, personal communication).

In some communities wood waste (debris from tree surgeons, land clearing, etc.) will be combined with home yard trimmings, but in other cases the two types of waste will be collected and handled separately. Thus, a large scale processing facility may have several material flow strategies to reduce costs and to produce a greater number and higher value of desired end products.

Yard trimmings represent a fraction of MSW which can be processed into useful products. Although the processes are usually referred to as composting, the end products may be compost, mulch, topsoil, etc. If mulch is the end product, it is sometimes put through a composting or heating process to destroy weed seeds and plant pathogens. It is important to define the end product in the planning stages, since this will influence all aspects of the yard trimmings processing system. Also, the presence of foreign material, such as plastic, metal, and glass, will have an influence on the processing and acceptance of the finished product.

Yard trimmings and land clearing debris can also be used as fuel, particularly where the yard trimmings contain a large amount of woody material. This choice will be influenced by local markets and transportation costs. In some cases, state or local law may prohibit or discourage the use of yard trimmings as fuel.

HOME YARD COMPOSTING

A successful yard trimmings management program should consider reductions in the quantities which are placed at curbside in addition to establishment of a centralized processing site for yard trimmings. This can be accomplished by reducing the quantities which are generated and by encouraging the use of yard trimmings in the home landscape or garden. Educational programs to encourage homeowners to not collect grass clippings are very common. If proper mowing practices are utilized, leaving grass clippings on the lawn can be beneficial and does not cause thatching or other problems with the lawn.

Reductions in the quantities of yard trimmings which are generated can also be accomplished by landscape design and the selection of plant species. The planting of shrubbery which does not require frequent pruning will reduce the quantities of yard trimmings which are generated. Many of the educational programs which are in place to encourage homeowners to reduce the use of water in the landscape are also effective in reducing the quantities of yard trimmings which are generated.

When it is not possible to further reduce the quantities of yard trimmings which are generated or to use them in the landscape, composting of the remaining trimmings can convert the materials into a materials which can be utilized in gardens, flower beds, and other places in the landscape. Numerous home composting units can be purchased, or the homeowner can construct his own compost bins.

LARGE SCALE COMPOSTING SYSTEMS

In spite of efforts to reduce the quantities of yard trimmings which are generated and placed at curbside for collection, there will probably always be a yard trimmings stream which must be collected and managed at a centralized processing site. These facilities can be operated by government entities or by private businesses. This is a decision which should be made early in the planning process by local governments. Visits to successful governmentally operated and privately operated facilities can be very informative.

<u>Collection</u>

Yard trimmings can be collected from the home at curbside or from a drop off site, or they can be brought to the central processing site by homeowners and landscapers. When yard trimmings are collected from the home, they may be loose or in some type of container. If the material is loose and consists of grass and/or leaves, then vacuum collection systems can be used. However, vacuum collection is not common in Florida where the yard trimmings tend to contain woody and brushy material.

In some communities, the use of paper or plastic bags for yard trimmings is required or permitted. Paper bags will usually degrade in the composting system at the central processing site. However, the use of plastic bags may require special equipment for breaking bags or screening the plastic from the raw material or the finished product. Degradable plastic bags are also available, although their efficacy has not been firmly established (Cole and Leonas, 1991). If plastic or paper bags are used, care must be taken at the central processing site to make sure that the wind does not blow them into neighboring areas. Visual aesthetics of the site and of the finished products may also be a factor. Pieces of plastic or other foreign material in the finished compost or mulch is not attractive to prospective users of the materials. Also, plastic and paper bags may hide contaminants which can reduce the capacity or damage the processing equipment.

Another option for collection is for the homeowner to place yard trimmings at curbside in a rigid container (plastic or metal) or in bags. The containers or bags are then emptied into the collection vehicle. A major advantage of this method is that the yard trimmings can be visually inspected for foreign material which may damage processing equipment or impair the quality of the finished products. Examples of potential foreign material include used lawn mower blades, glass or aluminum beverage containers, pesticide containers, or other MSW which should be place in a sanitary landfill or otherwise properly disposed.

Processing Site Selection and Layout

Although there is always a tendency to select sites which are not valuable or useful for other purposes, the site conditions and proximity of the surrounding population must be critically evaluated. Potential problems include inadequate drainage and storm water control, odor and/or noise complaints, and inadequate planning for handling and storage of materials during periods of high waste generation.

Every yard trimmings processing site should have a set of scales to weigh incoming and outgoing material, even if local or state regulatory agencies do not require them. All material should be inspected as it enters the facility to determine its characteristics and the presence of any contaminants or foreign material. A variable fee schedule can be imposed to reflect the subsequent cost of processing various material and to discourage the inclusion of contaminants and foreign materials. Yard trimmings which contain foreign material can also be refused at the site. Most facilities can benefit by separating yard trimmings from tree surgeon and land clearing debris (Schroeder, 1990). Other separations of incoming raw material may be desirable to produced specialized products or to decrease processing and handling costs.

Material flow should be carefully considered in site layout to eliminate interference between receiving and materials preparation, size reduction, composting, screening, and product shipping. Vehicle traffic patterns should be carefully planned, and the land areas required for each activity should be designated. Future changes in the tonnage of material handled or the types of products produced should also be considered.

Visual appearance of the facility is very important. A clean, well-landscaped facility will imply that a good product is being produced.

Size Reduction

Size reduction can reduce the volume of the yard trimmings, increase surface area for more rapid decomposition or composting, and produce a visually aesthetic or functionally efficient particle size distribution in the finished product. The equipment for accomplishing size reduction is usually referred to as shredders, grinders, or chippers. Most of this equipment has been adapted from wood processing or animal feed processing equipment.

Selection of size reduction or grinding equipment should take several factors into account. The capacity of the processing equipment should be adequate for the quantity of material which is delivered to the site. If the quantity of material which is generated does not justify the purchase of grinding equipment, then contracting with an independent equipment owner/operator to grind on a periodic basis may be the best solution. Secondly, the capabilities of the grinding equipment must match the characteristics of the yard trimmings and the intended products which are produced. Thirdly, the presence of foreign material, such as large metal objects, should be taken into consideration when selecting the type of grinding equipment.

The most common type of size reduction equipment used for yard trimmings is the tub grinder. The tub grinder is a horizontal hammer mill with a rotating cylindrical feed hopper. The particle size is regulated by screens or bars which the material must pass through before it can escape the hammers.

The tub grinder has been found to produce a material which is satisfactory as a mulch. It is more tolerant of foreign material than some other types of size reduction equipment, and it is capable of handling large volumes of loose, brushy material. It is available in many sizes, ranging from small units which are powered by a tractor power-take-off shaft to very large units with diesel engines over 750 kw (1000 hp). However, tub grinders require a high degree of maintenance because of wear on the hammers.

Wood chippers range in sizerom small tractor power-take-off driven units to very large stationary or portable units with an integral power unit. A common type consists of a large, heavy flywheel with knives mounted on the sides. This machine is probably the most useful for processing clean wood into wood chips for use as mulch. Frequent sharpening of the knives is required, especially if foreign material or sand are present in the raw materials.

At processing sites which handle land clearing debris and other large material, specialized equipment may be required. In most cases, this may be required only periodically and contracting with an independent owner/operator for this operation may be the best choice. One type of equipment utilized in this type of operation consists of a large, rotating disk with teeth on one face of the disk. A hydraulic ram pushes material into the rotating disk. Although this equipment will also handle mixed yard trimmings, other types of equipment such as the tub grinder are better suited for grinding of yard trimmings.

Screening

Screening equipment can be used at various places in a yard trimmings processing facility. Screens can be used to remove plastic or soil from incoming raw material. They can also be used following composting to produce a uniform high quality compost or mulch. In many facilities, a screen is considered to be a luxury which will be purchased at a later date. However, screens are becoming much more common as their usefulness in increasing the quality and value of the end products is recognized. Most screens used in yard trimmings processing are rotating or trommel screens. However, vibrating screens are also in use.

Composting Systems

Yard trimmings are usually subjected to a composting process to kill weed seeds and plant pathogens. This has been shown to be the result of both the temperatures and the compounds produced in the composting process. Depending upon the size reduction, screening, and other processing parameters, the end product may or may not be compost. Other products include mulch, topsoil, soil amendment, landscape wood chips, or fuel. This will be discussed more later.

The composting systems that are currently in common use for composting of yard trimmings are static piles, turned windrows, and bin systems. Aerated static piles and other types of vessel systems are generally used when yard trimmings are used as a bulking agent for materials such as sewage sludge (biosolids) or animal manures.

Static piles are large stationary windrows which are turned infrequently. The least equipment is used in managing this system. Piles are generally turned with a front end loader or crawler tractor equipped with a bucket, root rake, or blade. Most effective turning of these piles occurs when the pile is moved bucket by bucket from one location to an adjacent area.

The rake may be used to aerate the pile in place. The least efficient aeration and mixing occurs when the pile is simply pushed from one site to an adjacent site.

Turned windrows are similar in configuration to static piles but are usually smaller in size. Their size is determined by the type of windrow turner in use. These windrows may be turned as frequently as daily. A windrow turner may handle a windrow as large as 2 meters (6 feet) high and 5 meters (15 feet) wide, and there are many different types currently on the market. Two main distinctions between the various windrow turning machines are the use of the elevating face versus the drum types. Windrow turners with elevating faces have teeth on the face which lift and aerate the compost. Windrow turners that utilize rotating drums have teeth or flails attached to the drum to aid in mixing and aeration. Swinging flails attached to the drums seem to aid in particle size reduction as well as aeration.

Bin composting systems are usually used when yard trimmings are mixed with another material for co-composting. In these cases, the yard trimmings are usually reduced to a smaller size than if a mulch product is being produced. The objective is usually to produce a high value end product. The bin system is also advantageous when a highly reactive material such as food waste, animal manures, or sewage sludge (biosolids) is being composted with the yard trimmings, since the bin system results in more controlled conditions in the composting process.

ALTERNATIVE SYSTEMS AND END PRODUCTS

Potential end products from processing of yard trimmings include mulch, compost, topsoil, landscape wood chips, and fuel. Many of the early processing facilities were successful in grinding yard trimmings and giving them back to local residents for use as mulch in the home yard. Problems with weed seeds were sometimes encountered. This problem was solved by allowing the raw materials which had undergone size reduction to go through a composting or heating cycle to kill weed seeds and plant pathogens (Shiralipour and McConnell, 1991; Hoitinck et al., 1976). This phenomenon was found to be a combination of temperature rise, exposure time, and compounds produced in the composting process.

If compost is one of the end products, it is usually anticipated that it will have a higher value than other potential end products. In most cases the compost will have to meet state or other quality criteria. This will probably require additional expense in processing and testing which will have to be recovered in the sale of the product. Composting of yard trimmings and compost quality criteria are regulated by the Florida Department of Environmental Protection in Rule 17-709 of the Florida Administrative Code. In the case of yard trimmings which contain large amounts of woody material, the energy and expense which is required to produce compost may not be justified unless it is co-composted with animal manures, sewage sludge (biosolids) or other high nutrient wastes to reduce the composting time and to enhance the value of the compost.

In Florida, the yard trimmings usually contain varying amounts of soil and sand. If the raw material is screened prior to or following grinding, the mixture of sand and readily degradable organic material will produce a topsoil product which can be used in landscaping projects.

If tree surgeon debris or other large woody material with a minimum amount of leaves and small branches can be separated from other yard trimmings, then it may be possible to produce wood chips which can be marketed as a landscape chip for use as a substitute for conventional mulch materials. This is becoming a popular product in some areas, and it may become more popular as the availability of some other mulch materials becomes limited.

Another product from tree surgeon debris, land clearing debris and yard trimmings is fuel wood. Yard trimmings can be ground for fuel depending upon its characteristics. If foreign material or large amounts of grass clippings are present in the material, it is not suitable for fuel. Woody yard trimmings typically have a moisture content of 35% or less.

A generalized flow diagram showing options for processing yard trimmings into various products is shown in Figure 1. Although the most common procedure is to grind, compost, and produce mulch, other options are available.

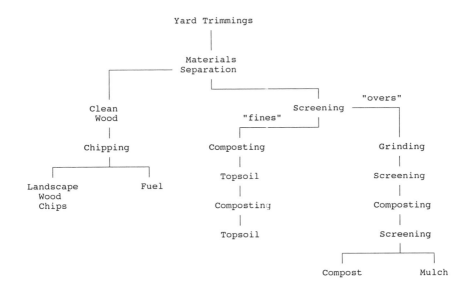

Figure 1. Generalized Flow Diagram for Processing of Yard Trimmings.

SUMMARY

The quantities of yard trimmings which are placed curbside for collection can be reduced through homeowner education. Grass clippings can be left on the lawn with good mowing practices, and other yard debris can be utilized in the landscape. Home yard composting of trimmings can produce materials which can be used in gardens and flower beds.

Yard trimmings which are placed curbside for collection separate from other municipal solid waste can be processed at centralized facilities for conversion in mulch, compost, topsoil, landscape wood chips and fuel. Following the passage of the Solid Waste Management Act of 1988 (Chapter 88-130, Florida Statutes), numerous facilities have been established for processing of yard trimmings. It has been a learning process, but the facilities are becoming more successful.

Careful planning, design and operation of yard trimmings processing facilities is necessary for them to be successful. Furthermore, markets must be developed to generate the economic return which will justify the expenditures for the required equipment. Inadequate or undersized equipment will increase chances of failure.

REFERENCES

1. Barkdoll, A.W. and R.A. Nordstedt. 1991. Strategies for yard waste composting. Biocycle 32(5):60-65.

2. Cole, M.A. and K.K. Leonas. 1991. Degradability of yard waste collection bags. Biocycle 32(3):56-63.

3. Earle, J.F.K, R.A. Nordstedt and M.S. Hammer. 1991. Overview of the Florida solid waste management act of 1988. Cooperative Extension Service Bulletin 272, University of Florida, Gainesville. 8p.

4. Hoitinck, H.A.J., L.J. Herr and A.F. Schmitthenner. 1976. Survival of some plant pathogens during composting of hardwood tree bark. Phytopathology 66:1369-1372.

5. Schroeder, R.M. 1990. Operating a wood waste recycling facility. Biocycle 31(12):56-58.

6. Shiralipour, A., and D.B. McConnell. 1991. Effects of compost heat and phytotoxins on germination of certain Florida weed seeds. Soil and Crop Sci. Soc. Fla. Proc. 50:154-157.

BENEFICIAL USES OF COMPOSTS IN FLORIDA

Wayne H. Smith[1]

ABSTRACT

With public and private sector support, the University of Florida's Institute of Food and Agricultural Sciences launched a statewide program to evaluate the effects of composts from various municipal solid waste fractions--yard trimmings, waste papers, RDF fractions--sometimes composted with sewage sludge and/or animal manures. These have been tested in land uses involving the most intensively and extensively grown crops in Florida such as citrus, tomatoes, peppers, watermelon, container-grown ornamentals, turfgrass, sod, landscape trees and shrubs, and slash pine. Benefits associated with compost utilization in addition to helping solve solid waste problems include plant growth and yield, probably through improved moisture relations; pest suppression; and nutrient retention and release. Recycling carbon from waste resources to its origin is an environmentally sound way of sustaining ecosystem function.

INTRODUCTION

Composting is being encouraged as an alternative to landfilling and incinerating of municipal solid waste because of environmental problems associated with these practices. Similarly, composting of animal manures, crop residues and food processing residuals is also seen as an environmentally benign way of recycling these biodegradable organics. Impetus in Florida is from the Solid Waste Management Act of 1988 which set goals of one-third recycling, one-third landfilling and one-third incinerating by 1994 to ameliorate environmental problems. In 1987 about 3% was recycled and almost 80% landfilled.

In Florida the solid waste stream has grown to about 20.3 million tons annually in 1992 (DEP, 1993). This represents over 8 pounds per person per day or about 1.5 tons per person per year, twice the national average of about 4 pounds per person per day. The total exceeds the 18.4 million tons projected in 1988 for 1994. This has occurred in spite of increasing recycling to almost 27%. Biodegradable organics that could be composted comprise about 60% of the total MSW or about 12 million tons annually. Yard waste (landscape maintenance and land-clearing debris) makes up 15.5 percent or 3 million tons annually, the largest single fraction other than mixed papers in Florida MSW. Animal manures represent about 530,000 dry tons of solids annually (Smith, 1990). Over half is from dairy operations, with horses being the second highest generator. Since litter and bedding materials are often included with manure solids, the tonnages of biodegradable organics from these animal operations could be greater.

Composting all of the biodegradable fraction of MSW could yield about 6 million tons of compost annually. Should all the yard waste be composted, about 1.6 million tons could be produced annually; if mulched, larger quantities would be available for use. Since yard waste can no longer be placed in a Class I landfill and these materials are clean and easy to collect separately, a high priority has been assigned to this fraction.

Composting is a natural process that occurs in forest or swamp floors, wherever adequate organics accumulate. In the forest the process is aerobic (occurs with oxygen); while in the aqueous swamp floor, it is anaerobic (oxygen absent). While organics from waste resources

---

[1]Director, Center for Biomass Programs, University of Florida, Gainesville, Florida.

can be composted by either process, we are referring to aerobic composting products. Whatever quantity is composted, agriculture, horticulture, silviculture, road and parks management and land reclamation will be looked to as the sites to recycle these materials. Cropping continually removes organic materials and the nutrients they contain from the site of production. Without returning these materials, their benefits are lost and the production system cannot be sustained. In the last few decades, soil nutrients, biological population balance, and water supplies have been maintained by chemical and other energy intensive means made possible by large supplies of inexpensive petroleum. This approach has worked well for increasing crop yields, but not without serious environmental costs. Returning organics, usually back to their site of production, can help restore soil organic levels that can re-establish ecological balance, reduce soilborne pests, improve water and nutrient retention, and serve as a modest supply of slow-release fertilizer.

POTENTIAL USES FOR COMPOST

Total acreage of potential land uses for composts in Florida are tabulated in Table 1. Depending upon assumptions, one can easily show that these uses can assimilate the annual production of composts and mulches in Florida. For example, up to 200 tons per acre MSW composted with sewage sludge were applied to a pine plantation in 1970 with positive growth responses and no adverse effects (Jokela et al., 1990). If only 50 tons were applied to each of the 170,000 acres reforested annually in Florida, 8.5 million tons of composts could be recycled each year. Should just 20 tons of compost be applied each year to just one-fifth of the citrus, vegetable and field crop acres, and less than one-tenth of the pasture and turfgrass areas, this would consume all of the total potential annual compost production in Florida.

Table 1. Potential Uses for Compost in Florida.*

Pasture	2,000,000 acres
All citrus	791,290
All Vegetables	418,000
Field Crops (not including sugar cane)	581,000
Annual Forest Plantings	170,000
Sod Farms	47,518
Greenhouse Crops	7,200
Lawns	1,000,000
Mined Lands	248,000
State, County and City Roads	134,825 miles
Recreation Areas (parks, playing fields, trails, golf courses)	not estimated

*Compiled from data provided by Florida Departments of Agriculture and Consumer Services, Environmental Protection, and Transportation.

Should half of the yard waste (1.5 million tons) be prepared as mulch and put through the composting process just to pasteurize, little weight loss would occur, so most of the mulch would be available for use. There are one million homes, uncounted commercial landscapes, and many acres of parks and other recreation areas that can easily consume this volume of mulch from yard waste resources.

RESEARCH AND DEMONSTRATION PROGRAM

To obtain additional research results to those in the literature (Shiralipour et al., 1992) that could guide use, a comprehensive compost utilization program was implemented in Florida. The cooperative program addressing water conservation/compost utilization involves support from the Department of Environmental Protection, the Department of Agricultural and Consumer Services, Procter and Gamble, St. Johns, South Florida and Southwest Florida Water Management Districts, Keep Florida Beautiful, Reuter Recycling, Amerecycle, Enviro-Comp and Bedminster.[2] The national Compost Council played a significant facilitating role in organizing the program. Support for other projects has come from the Florida Center for Solid and Hazardous Waste, the American Paper and Forest Products Association, the Palm Beach County Solid Waste Authority, and in-house funds of the University of Florida's Institute of Food and Agricultural Sciences.

Crops chosen for experimentation ranged from high value, low volume to low value, high volume (crop value/compost utilization potentials). Contributing projects addressing the different land uses are located at University of Florida/IFAS research centers, Gainesville campus departments or on cooperator lands (Table 2). Most of the experiments have completed one annual cropping cycle or growing season. For specific information, the individual investigators may be contacted at their University of Florida locations. In field applications, compost rates usually were applied once and the cropping cycles imposed. In Florida, lands are often cropped twice (e.g., tomatoes, either preceded or followed by peppers, squash, cucumber or watermelon or corn following a green manure crop). In the 7-year-old slash pine, a surface application was made, while in the new plantation the compost was incorporated into the soil and then planted. The latter procedure was followed for citrus. The potato crop was rotated with fallow. The sod was produced in a compost layer on a plastic sheet. Turfgrass was first tested in pots in mixes from 0 to 100% compost or soil. The optimum rate was then imposed in a landscape for establishing turfgrass. In the potting media, compost was evaluated in treatments ranging from 100% compost to 100% compost replacement of just the peat portion of the container media.

RESULTS

The emerging pattern of responses of crops receiving the compost treatments are encouraging (Table 3). Early results revealed that immature compost consistently caused N-rob and inhibited growth, especially of annual crops such as peppers and tomatoes. In one tomato planting, the investigator reported that compost suppressed early growth, but the "crop exploded" late in the season such that there was only a modest yield reduction. A side-by-side planting of tomatoes with mature and immature compost showed the immediate benefit of compost when mature. In rotations, the second crop showed superior growth where the compost was applied. For example, when watermelons following tomatoes, the yields were improved over 30% where the composts were applied. A tomato crop following peppers yielded larger fruit and significantly more marketable tomatoes. With sod and turfgrass production, 30% compost resulted in better growth than with other treatments.

---

[2]Funds for this project were administered by the Florida Department of Agriculture and Consumer Services, Bob Crawford, Commissioner, as provided by public and private entities for the purpose of demonstrating yard/solid waste compost utilization and water conservation.

Table 2. University of Florida/IFAS Project Leaders, their Locations and Targeted Crops.

Project Leader(s)	Location	Crops
Drs. Larry Parsons and Adaire Wheaton	Citrus Research and Education Center, Lake Alfred	Citrus
Drs. Herb Bryan, Bruce Schaffer and Jonathan Crane	Tropical REC, Homestead	Tomatoes, Mango, Squash
Dr. Thomas Obreza	Southwest Florida REC, Immokalee	Tomatoes, Watermelons
Drs. Gary Clark, Craig Stanley and Donald Maynard	Gulf Coast REC, Bradenton	Bell Peppers, Tomatoes
Dr. George Fitzpatrick	Ft. Lauderdale REC	Philodendron, Lantana, Schefflera, Oleander, Jessamine, West Indian Mahogany, Live Oak
Dr. Richard Beeson	Central Florida REC, Sanford	Viburnum, Ligustrum, Azalea
Drs. John Cisar and George Snyder	Ft. Lauderdale REC West Palm Beach	Turfgrass/Landscape St. Augustinegrass
Dr. Hans Riekerk	Forestry, Gainesville	Slash Pine Trees
Dr. Don Graetz	Soil and Water Science, Gainesville	Compost characterization. No crops involved.
Dr. A. E. Dudeck	Environmental Horticulture, Gainesville	Turfgrass/Sod St. Augustinegrass
Dr. Ray N. Gallaher	Agronomy, Gainesville	Field Corn, Squash, Sweet Corn, Okra, Cowpea
Dr. James H. Graham	Citrus REC, Lake Alfred	Citrus
Dr. Dale R. Hensel	Hastings REC	Potatoes
Dr. Dennis McConnell	Environmental Horticulture, Gainesville	Dracaena, Azalea, Peperomia, Shumard Oaks, Red Maples, Pickerelweed, Schefflera, Anthuriums, Begonias, Boston Fern, Burford Holly, Chlorophytum, Crape-Myrtle, Daylillies, Dieffenbachias, Ficus Benjamina, Golden Pothos, Juniper, Orchids, Yaupon Holly

(Table 2 continued on next page)

(Table 2 continued)

Project Leader(s)	Location	Crops
Drs. James Stephens and Stephen Kostewicz	Horticultural Sciences, Gainesville	Vegetable Gardening--Broccoli, Carrots, Green Beans, Kale, Okra, Pole Beans, Southern Peas and Greenhouse Vegetables--Lettuce, Tomatoes, Watermelon, Cucumbers. Vegetable Gardening, Organic culture--Cabbage, Cucumbers, Mustard, Okra, Peppers, Tomatoes, Southern Peas
Dr. Peter Stoffella	Ft. Pierce REC Boynton Beach	Tomatoes, Peppers, Cucumbers
Dr. George Fitzpatrick	Ft. Lauderdale REC	Ficus, Aveca Palm, Jasmine, Dwarf Oleander, Hibiscus, Philodendron, Schefflera
Dr. Don Graetz	Soil and Water Science, Gainesville	Chemical, physical and biological analyses

Pesticide and nutrient leaching in compost treated potted media appears to be reduced, according to preliminary results. The potato crop data showed that immature compost, incorporated in the soil sufficiently in advance of planting so that it could stabilize in situ, gave positive yield increases (ca 300 cwts per acre) in the first crop, pointing out the importance of having mature, stable compost or being able to allow it to stabilize in the soil. Phytoparasitic nematode assays in vegetable plots treated with compost showed a reduced population in comparison to plots that were only fertilized.

Responses by citrus have been mixed thus far; root length and density and trunk diameters were increased at one test site, but not at another site. The responses by pine, the other perennial tree crop, have not been noted except in survival of seedlings which was reduced by severe weed competition where compost was applied. In organic gardening projects, it was learned that second year responses were usually superior to first year results.

Laboratory analyses of soluble elements and total metals have not revealed concerns for toxic levels in composts. Carbon mineralization studies show that immature compost mixed with soil becomes stable within 100 days. Incidentally, about 100 days passed between application of yard trimming compost and the planting of potatoes; and, not only was growth not suppressed, yields were increased by 300 cwts of potatoes per acre.

Table 3. Compost Benefits Emerging from Florida Research

Crops	Major Results
Citrus	Increased length and concentration of roots in site 1; enlarged trunk diameter at site 2.
Container Grown Ornamentals	Greater plant biomass in 100% compost or as peat replacement in media. Response--species and compost specific.
Turfgrass	Enhanced growth and quality with compost in soil--optimum 30%; improved water-holding capacity; delayed plant wilting; increased nutrient absorption; and reduced pesticide leaching.
Vegetables	
Tomatoes and Squash	Mature compost increased yield of larger, marketable tomatoes; and boosted plant height.
Tomatoes and Watermelon	Compost increased soil organic content, water-holding capacity, mineral concentrations and pH levels. Immature compost did not increase tomato yield, however watermelon yields up 54% in rotation crop.
Bell Peppers and Tomatoes	Mature compost produced extra-large tomatoes and increased the total fruit yield from 2100 to 2600 boxes per acre. Immature compost decreased the pepper yield up to 14%.
Corn and Vegetables	Immature yard trimmings compost decreased corn yield in year 1; yield increased in year 2. Three of four plant parasitic nematodes were suppressed.
Potatoes	Yard trimmings compost increased potato yields up to 39%.
Landscape Trees and Shrubs	Top-dressing of compost reduced transplant stress and lowered irrigation requirements.
Reforestation	Abundant weed growth where compost was applied coupled with dry spring possibly caused seedling survival to decrease. Growth was not affected with 7 year-old trees. Massive weed growth may enhance species diversity and wildlife habitat.

CONCLUSION

In the aggregate these several projects frequently have demonstrated growth benefits from compost when mature and properly applied. Other benefits, identified but more difficult to ascertain, are associated with water use efficiency, pest suppression, and prevention of leaching of certain inorganic and organic materials. Because of the encouraging results, opportunities are being sought to expand the program of research and demonstration to further define the effects of composts and understand how they affect ecosystem processes and structure. Recycling waste resources in this way not only has these values, but also helps solve waste management problems in communities. Organics recycling as compost is environmentally sound and can contribute to a sustainable agriculture.

REFERENCES

1. Department of Environmental Protection (DEP). 1993. Annual Report. Florida Department of Environmental Protection, Tallahassee, Florida.

2. Jokela, E.J., W.H. Smith and S.R. Colbert. 1990. Growth and elemental content of slash pine 16 years after treatment with garbage composted with sewage sludge. J. Environ. Qual. 19:146-150.

3. Shiralipour, A., D. B. McConnell and W. H. Smith. 1992. Uses and benefits of MSW compost: A review and an assessment. Biomass and Bioenergy 3(3-4):267-279.

4. Smith, W.H. 1990. Organic waste management in Florida. BioCycle 31(4):52-55.

AGRICULTURAL SURFACE WATER MANAGEMENT (ASWM) IN SOUTHWEST FLORIDA

Sandra L. Means and Joseph T. Mustion [*]

ABSTRACT

The Agricultural Surface Water Management (ASWM) program was created by the Southwest Florida Water Management District (SWFWMD) to provide a permit exemption to farmers. Soil Conservation Service (SCS) has been funded by SWFWMD to assist the public with these exemptions. The program, in its third year, is scheduled to run for three years for a total cost of $750,000.

SCS uses ecosystem planning to develop a Resource Management System (RMS) to qualify agricultural projects for an exemption from permitting. An RMS uses a two-pronged approach which combines agronomic practices and engineered structural elements to address water quality, water quantity and environmental concerns. Agronomic practices include the management of nutrients and pesticides, irrigation water management, conservation cropping sequence, crop residue use, and other appropriate improved management practices. Structural elements include downstream filter areas, grade stabilization structures, and structures for water control to protect wetlands and water quality by reducing the potential for downstream impacts.

Implementation of a RMS plan serves to develop sustainable agriculture while not adversely effecting the existing ecosystem.

Keywords: Sustainable agriculture, ecosystem planning.

PROGRAM BACKGROUND

Letter of Exemption

Prior to 1991, the agricultural community was concerned that the standards for surface water management permitting for agriculture from the Southwest Water Management District (SWFWMD), which were the same as were used in urban situations and were inappropriate. Public workshops were held by the Soil Conservation Service (SCS) and SWFWMD addressing these concerns. After numerous meetings, SWFWMD decided to make special provisions for agricultural surface water management.

In 1991, SWFWMD developed criteria to exempt new agricultural projects from a Management and Storage of Surface Water (MSSW) permit. The exemption criteria for a permanent agricultural operation includes implementation of a Resource Management System (RMS), which consists of agronomic and engineering practices. If an agriculturalist agrees to meet the criteria for exemption, SWFWMD provides an letter of exemption to the farmer allowing them to proceed with production.

In 1992, SWFWMD developed two more types of exemptions to an MSSW permit. These are temporary and ordinary agricultural exemptions. A temporary agricultural exemption is obtainable for vegetable farming on a 7 year rotation, 2 years farming and 5 years site abandonment. Since farming is of short duration, the criteria for the temporary exemption is more flexible than the permanent agricultural exemption. The ordinary agricultural exemption is obtainable for well drained sites where deep beds are unnecessary, such as "ridge-type" citrus.

---

[*] Sandra L. Means, Agricultural Engineer, and Joseph T. Mustion, Area Resource Conservationist, U.S. Department of Agriculture, Soil Conservation Service, Palmetto, Florida.

The criteria for all three types of exemption address three areas of concern: water quantity, water quality and environmental protection. A Resource Management System (RMS) developed for the project must address these three areas of concern. An agriculturalist must have a surface water management plan, either by a letter of exemption or an MSSW permit, to meet water use permitting requirements.

SCS role

To facilitate the implementation of the new exemption process, USDA-SCS and SWFWMD on October 1, 1991 entered into a reimbursable agreement (ASWM) to provide technical assistance to owners and lessees of vegetable and citrus operations with surface water management plans. Through this agreement, SCS helps the agriculturalist to proceed with production by meeting permitting exemption criteria through implementation of a Resource Management System (RMS). This agreement in its third year, is scheduled to last three years for a total cost of $750,000.

For this agreement, SCS maintains a team of 2 conservationists, an engineer, and a technician to provide surface water management planning for agriculturalists. Their engineering assistance is provided through the local SCS field offices.

The Process

The farmer has an idea about a proposed project and contacts the local SCS District Conservationist. If the project can be assisted through the exemption process, the District Conservationist involves in the SCS-ASWM team. The process takes on a three meeting format.

The SCS-ASWM team meets with the farmer on-site for a contact meeting. At this meeting, the needs of the farmer are determined and his site is evaluated for environmental and engineering concerns. A SWFWMD environmental scientist and engineer (SWFWMD Ag. Team) are usually also present to stake wetland boundaries and assess their condition.

After the first meeting, a topographic survey is done if one is needed for the site and an engineering plan is developed. A second meeting with the farmer is arranged to discuss the proposed project. The farmer reviews the engineering plan and any changes to meet the farmer's needs within the exemption criteria are made at this time.

After the second meeting the engineering plan is finalized and a resource management system is developed. A third meeting is arranged to sign-off on the letter of exemption. The farmer, SCS-ASWM team, and SWFWMD Ag. team meet in a local office to finalize the process. The Resource Management System complete with engineering design is proposed to SWFWMD. If the SWFWMD Ag. team approves, .the letter of exemption is signed. By signing the letter of exemption the farmer agrees to construct the project as designed and can proceed with SWFWMD approval.

The SCS-ASWM team provides assistance in the layout of the project at construction. It also provides follow-up on the system after operation has begun.

From contact meeting to signoff, it takes about a month, barring complications. The product is a plan which is sensitive the environmental needs and promotes sustainable crop production with improved management techniques.

RESOURCE MANAGEMENT SYSTEM (RMS)

A Resource Management System (RMS) provides a record of farming practices that will be used on the project. SCS has offered this service to farmers nationwide for over 50 years, but a written record of farm activities is not required by state law. To meet the requirements of the letter of exemption for permanent agriculture, a Resource Management System (RMS) is required. This document helps both the client and the public. The client is able to draw upon the expertise of SCS and other governmental agencies to improve the farm operation and makes his operation more environmentally sound. SCS standards call for certain improved management practices (IMPs) for vegetableor citrus operations. Application of these minimum IMPs assure sustainable agricultural production while minimizing environmental impact. The agriculturalist can use this document to keep records of how and when practices were implemented. This record can be used as a guide to see if any problems occurred in the past year, and subsequently avoid having the same problems the next year. This document can also record improvements made by using the plan which can benefit other clients.

An RMS includes agronomic practices specific to the cropping needs to provide sustainable crop yields for the project area. These may include but are not limited to nutrient and pesticide management, irrigation water management, conservation cropping sequence, and crop residue use. An RMS consists of six parts. An attempt is made to put the parts into a logical, understandable order.

General Site Information

The first section includes an index to list the contents of the entire RMS and in which section of a six-part folder information appears. The first section also includes general maps that are used to locate the site (county map), confirm land ownership (plat book), and determine general topography (USGS topographic map).

Environmental Information

In this section an environmental assessment is made, a list of threatened and endangered species that could potentially be found on site is provided, and an environmental evaluation is made. In addition, documentation related to the topic above (e.g. maps, correspondences) is included.

The policy of the USDA is that an environmental assessment be performed on projects that receive SCS assistance. General assessments are made concerning cropland, grazing land and woodlands. Addition topics include existence of designated natural areas, prime, unique or locally important farmlands, pipeline or utility easements, and floodplains in the project area. Documentation is provided as needed to show the existing conditions. Measures will be taken to prevent adverse effects if any of these conditions exists.

One part of this environmental assessment is consideration of the effects of the project on threatened and endangered species. Typically written requests are made to the Florida Game and Freshwater Fish Commission (state) and the Fish and Wildlife Service (federal) inquiring into known records of threatened and endangered species on the proposed project and the surrounding area. Records of this correspondence are also included.

Wetland areas within the project are described in this section. The terms of the permanent letter of exemption allow no encroachment within 15.2 m (50 ft) of wetlands. This action helps eliminate direct impacts to wetlands. The area where two habitats meet is called an ecotone. A 15.2 m (50 ft) wetlands buffer helps preserve the ecotone between the wetland and upland habitat. If there are wetlands in the project area, a copy of the U.S. Department of the Interior National Wetland Inventory Map is included. Interpretation of the map symbolling is also included.

Federal law requires a cultural resource investigation be done on projects receiving federal assistance. SCS employees have been trained to assess sites for potential cultural resources. If the project has a historic or archaeological site, and the site is considered significant, further measures may be necessary. This is considered part of the environmental assessment.

Conservation Plan Map and Soil Information

A conservation plan map is a diagram of the proposed project boundaries and all associated enduring features. Typically, enduring features include: irrigation pipeline layout, structures for water control, and drainage orientation. Details such as pipe, structure, and ditch sizes are provided separately in the engineering plan.

A copy of the soils map from the county soil survey is included. The project boundary is outlined on this map. Non-technical soil descriptions are provided for the soils that areonthe project. These soils descriptions are specific for the land use (i.e. pasture, citrus, vegetable crops).

Record of Decisions

Discussion with the client concerning his farm operation provides information on which the selection of practices to implement is based. The determination of which practices to apply is based on the minimum required practices as designated by SCS standards. Engineering decisions will also influence the selection of practices. Confirmation of the use of these practices is made by the client, since they are the ultimate decision maker.

The "Record of Decisions" is a document which contains this information and is included in this section of the RMS. It is used in conjunction with the conservation plan map, which is facing in the previous section. Brief information is given about the specific practices which occur in the fields as numbered on the conservation plan map. The planned acreage and date for which a given practice is planned is also included. Additional information for practices is cross-referenced to the last section of the RMS.

Nutrient and Pest Management

Application of nutrients and pest control are necessary components of modern agriculture. A section of the RMS is dedicated to nutrient and pest management. Using good management techniques, these practices can be implemented to sustain economical production with consideration to minimizing environmental effects.

The following is a list of documentation that is included in the RMS concerning nutrient and pest management.

Soil test results. The ASWM program requires that the land owner obtain a double acid soil test. The results of this soil test and consultation with the Institute of Food and Agricultural Sciences (IFAS) is used to determine the rate of nutrient application.

Nutrient and Pest Management Support Data Sheet. The client designates what IMPs will be used by checking off the practices on this sheet. Nutrient IMPs listed include tissue testing, irrigation water management, applicator calibration, and 14 other practices. Pesticide IMPs include scouting, equipment calibration, proper chemical container disposal and 17 other practices.

Pesticide and Nutrient Management Jobsheet. This information on this job sheet is based on the leaching and runoff potential of soils in the project area. Recomendations on nutrient and pesticide application in light of these factors are made. SCS advises that the client obtain a copy of IFAS publication "Managing Pesticides for Crop Production and Water Quality Protection" for the particular crop that will be grown in the project. This publication helps the user select the appropriate pesticides by their potential for leaching and runoff.

Nutrient Budget Worksheet. The nutrient budget worksheet can be used to account for the use of alternate nutrient sources such as legume crops and animal or municipal waste application.

Operation and Maintenance Plans for Nutrient and Pest Management. These sheets list practices that contribute to the safety and efficiency of nutrient and pesticide application. Practices such as annual equipment calibration and use of protective clothing are specified.

Job sheets and Other Related Material

This section contains job sheets and work sheets for conservation practices, irrigation water management information, and operation and maintenance plans. These are referenced in the "Record of Decisions". Job sheets provide specific cover types, seeding, and fertilizer rates for different practices. Operation and maintenance plans recommend measures to protect and maintain the life of structures.

ENGINEERED DESIGN

Each RMS contains an engineered design to provide the crop with the needed drainage, irrigation, and proper management of surface water runoff in addition to water quality treatment and environmental protection. A three to four page engineering design detailing the site plan with all necessary cross sections and details are provided. The system is designed according to SCS standards and specifications and must meet the following criteria:

Environmental protection

Environmental protection is provided by a 15.2 m (50 ft) minimum undisturbed upland buffer around all wetlands. Agricultural operations, including access roads, must be at least 15.2 m (50 ft) from the wetland edge. This buffer limits the potential for direct impacts by equipment or sedimentation to the wetlands. Watershed to wetlands and other outlets must be maintained. Diversion of water away from wetlands or other outlets is avoided. Discharge to wetlands must imitate the pre-project state. If runoff sheet flowed into a wetland prior to the project, it should continue to do so after development. Limited conversion from sheet flow to multiple point discharge is allowed when adverse impacts to wetlands and adjacent property are avoided.

The condition of downstream wetlands dictates the amount of water quality treatment required by SWFWMD. The condition of the point of discharge is evaluated by a SWFWMD environmental scientist during a initial site examination. The environmental scientist determines the required travel time (Tt). Travel time varies from 15 to 30 minutes dependent on the proximity of the point of discharge to state waters.

Water Quantity

Cropping systems are designed for gravity drainage; No discharge pumps are used. Outlets are evaluated for capacity to provide positive drainage. Row arrangement is laid out in accordance with existing watershed breaks. Field and pickup ditches are provided to convey surface water to the outlets and are sized according to the crop needs. Ditch depth shall not exceed 46 cm (18 in.) depth on average, and provide no more than 15 cm (6 in.) lowering of the SHWT. Water quality treatment must take place prior to discharge into an outlet. Ditches outleting into wetlands must daylight prior to the 15.2 m (50 ft) upland buffer. Cropping systems must be designed in such a way that cause no upstream or downstream impacts. Hydrology of each site is evaluated for potential impacts. *Technical Release 55: Urban Hydrology for Small Watersheds* from the Soil Conservation Service is used to evaluate pre and post project peak discharge. When pre and post peak discharge differ by more than 10%, attenuation of the peak discharge is required. Natural attenuation by wetlands downstream of the project can be accounted for to attenuate peak discharge. SCS accounts for 2.54 cm (1 in.) of storage per 0.4 ha (1 ac) of wetland. Generally, this works out to dedicating 0.4 ha (1 ac) of downstream wetland acreage per 0.4 ha (1 ac) of cropland. Downstream wetland acres used for attenuation must be owned or under the control of the client.

Pipe structures are included in the design to safely convey water from the field to filter strips and outlet ditches.

Water Quality

Water Quality Treatment is provided by Overland or Grassed Channelized filter strips. All surface water must be treated prior to discharge off the project area. Filter strips are designed for a specific travel time (Tt).

Grass channelized filter strips (GCFS) are vegetated trapezoidal channels directly downstream from a discharge area where discharge from the project flows thru prior to leaving the site. It is assumed that low flow velocities and shallow depth of flow provide adequate water quality treatment. Flow in the channels is restricted to for 15 cm (6 in) depth or less. Bottom depth below land surface shall not exceed 0.6 m (2 ft). Since GCFS's are to be maintained by farm equipment, a minimum 3 m (10 ft) bottom width and 4:1 side slopes are used. A drainage rate of 0.012 (m^3/s)/ha (0.17 cfs/ac) is used for citrus and 0.009 (m^3/s)/ha (0.13 cfs/ac) is used for vegetables.

Overland Filter Strips (OFS) are grassed or wooded area directly downstream from the project where discharge is restored to sheet flow. Ditch bottom is brought out at a flat slope till it intersects with natural ground. This is referred to as "daylighting" the ditch. It is assumed that sheet flow occurs in the distance it takes to daylight. Travel time is calculated as the time it takes water to sheet flow across the filter strip area. OFS lengths range from 15 to 107 m (50 to 350 ft), dependent upon slope and travel time necessary.

CONCLUSIONS

The ASWM program provides agriculturalists in Southwest Florida with an alternative. By implementing a Resource Managment System (RMS) that meets the criteria established by SWFWMD, a farmer can proceed legally with a new project faster and with less initally invested than an MSSW permit. The project must meet limitations of the exemption criteria, and not all sites qualify. The RMS provides project planning for both agronomic and engineering practices. It documents how and when practices will be implemented. The use of improved management practices planned to the cropping needs of the site provides sustanable crop production.

Participation in the the ASWM program, is not farming "fence row to fence row", as has been done in the past. It is farming the available land which is best suited for the proposed crop, making the least amount of impacts to the environment.

ACKNOWLEDGMENTS

This project is funded by the Southwest Florida Water Management District through a reimbursable agreement with the U.S. Department of Agriculture, Soil Conservation Service.

SCREENING TOOL TO PREDICT THE POTENTIAL FOR GROUND WATER OR SURFACE WATER CONTAMINATION FROM AGRICULTURAL NUTRIENTS

W.F. Kuenstler, D. Ernstrom, and E. Seely[1]

ABSTRACT

The Soil Conservation Service is addressing a wider array of natural resource management concerns by developing better tools to help field personnel identify and solve the problems they encounter. We are developing a series of tools for our field offices that will help landusers become aware of and solve water quality problems from agricultural nutrients.

Phase I of this project is a screening tool that will be used to predict the potential for nitrogen and/or phosphorus losses form a field to surface or ground water. The procedure uses two sets of attributes, one for inherent site characteristics, the other for management activities. Site characteristics include field location in relation to water bodies, land slope, and various soil physical and chemical properties. Management activities include the type of crop being grown, application of organic nutrient sources (manure, sludge), irrigation, and others. The evaluation process uses attribute criteria to determine if that attribute has a high, medium, or low potential for contributing to nutrient losses. The attribute rankings are summarized for each field, giving an overall rating of the potential of nutrient losses from that field. This rating procedure will help our field personnel set priorities for providing assistance to landusers in dealing with water quality problems.

Keywords. Natural resource management, Nitrogen losses, Phosphorus losses, Nutrient management.

Introduction

For almost 60 years, the Soil Conservation Service (SCS) has been helping farmers manage their natural resources. For many years, the emphasis was on erosion control on sloping cropland, water management (irrigation and drainage), and flood control. Tools were developed that helped field offices work with landusers to plan and apply the components of the systems that solved these resource concerns. In recent years, there has been increasing emphasis on the water quality aspects of agricultural operations. Field offices have been hampered, though, by a lack of tools to help them in this effort. Intuitively, they knew they were having a positive effect on water quality by the application of existing conservation practices, but there were still many unanswered questions. How much effect are we having? Are we applying the correct practices? Are we concentrating our efforts in the right places? Within SCS, there was no good way to answer these questions.

In 1989, SCS initiated a water quality technology development effort. Part of this effort was to identify what was needed at the field level to help landusers address water quality concerns. The Field Office Water Quality (FOWQ) Tools project was begun as a result of this effort. The FOWQ Tools project has three parts:

---

[1] W.F. Kuenstler, Conservation Agronomist, D. Ernstrom, Soil Scientist, Technical Support Team, E. Seely, Project Manager, Decision Support Model Develpoment Team, USDA-SCS, Technology Information Systems Division, Fort Collins, CO.

1. A nutrient screening tool, to help determine the potential for nutrient losses
2. A nutrient management planning tool, to help develop sound nutrient management plans
3. A farmstead assessment tool (Farm-A-Syst), to help identify water quality problems originating on the farmstead itself.

This paper will discuss the first of these three tools.

The nutrient screening tool was developed to run as an integrated component of the SCS Field Office Computing System (FOCS), a proprietary software package. FOCS contains the databases and applications on which all the decision support tools developed by SCS are based. The screening tool is an automated procedure which identifies individual fields that have a potential for nitrogen or phosphorus losses to surface or ground water using (1) the inherent characteristics of the field (site sensitivity) and (2) the effects of management practices used on those fields (potential vulnerability).

This tool will be used by a conservation planner to conduct a water quality evaluation for site sensitivity and potential vulnerability, enabling him/her to work with agricultural producers (1) to evaluate management alternatives and (2) to develop practical and feasible plans to manage soil and water resources.

Nutrient Screening Processes & Features

The user starts by identifying the client being assisted, and all the tracts and fields that belong to this client. The client information and the list of tracts and fields are passed to the Nutrient Screening tool and displayed. By default all the tracts and fields are tagged for evaluation. The user can deselect one or more fields, to develop a list of those fields (Fig. 1) on which the screening will be done.

```
┌──────────────────────── Select Tract/Fields ────────────────────────┐
│                                                                     │
│     Tract      Field     Description              Mapunits          │
│  >  1000123    1230001   Sherwood-Corn-Field      Aab, BcA          │
│  >  1000123    1230002   Sherwood-Orchard         BaB, BcA          │
│  >  1000123    1230003   Sherwood-beans-field     AaB, BaB, BcA     │
│  >  1000123    1230024   Salisbury-Circle         BcA               │
│  >  1000123    1230025   Salisbury-Square         AaB, BaB, BcA, CaA│
│  >  1000123    1230026   Salisbury-Triangle       CaA               │
│  >  1000145    1450007   Sandiscove-Plot          BcA, CaA          │
│  >  1000145    1450057   Winchester-South         BcA               │
│  >  1000145    1450607   Bakersfield-West         BaB, BcA, CaA     │
│                                                                     │
└─────────────────────────────────────────────────────────────────────┘
     F5=Evaluate Fields        F6=View Saved Results        F7=Select All
     F8=Unselect All           F9=Edit Site Attrib          Shft+F9=Edit Soils
Move with ARROW keys.  Select/Unselect field with the SPACEBAR, F7 and F8.
```

Figure 1. Field Selection Screen from Screening Tool

For each field, all soils (components) are listed (Fig. 2). This information is available from the field soils table in FOCS, which would have been previously populated by the user using available soil maps. The user can select from this list the soils to be evaluated, and can add soils that were not identified on the soil map, using his/her knowledge of the site.

```
┌──────────────────────── Edit Soils ────────────────────────┐
│                Estimated                                    │
│   S/V  MUSYM   %MU   %LU   Component                        │
│    S    AaB    15     9    Windsor Loamy Sand               │
│    S    AaB    80    48    Campersfield Sandy Silt          │
│    S    BaB    91    40    Salisbury-Sherwood Sandy Loam    │
│                                                             │
│                                                             │
│                                                             │
│                                                             │
└─────────────────────────────────────────────────────────────┘

F3=Next Field        F5=Save          F6=Delete Soil        F8=Add Soil
F4=Prev Field        F9=Edit Site Attributes    Shft+F9=Edit Soil Attributes
Select F7 to view soils and F8 to view site attributes.
```

Figure 2. Example of Default Soils Data Available for Editing

To evaluate the effect of nitrogen and phosphorus on surface or ground water, the site sensitivity and potential vulnerability attributes for each field must be identified and rated. The site sensitivity attributes (Table 1) are those over which the farmer has little or no control, such as soils or field location. The vulnerability attributes (Table 2) are ones that are affected by the farmer's management activities.

All attribute names and values are displayed, and the user selects the attribute values that apply to the field and soil map unit component being evaluated. Since the soil attribute values displayed are based on map unit information from the soil survey, they can be edited at this time to ensure that they accurately reflect the characteristics of the soils at the site being evaluated.

After all field attribute values have been verified, each field is evaluated for sensitivity and potential vulnerability. A field can have both sensitivity and potential vulnerability rankings for each resource and nutrient combination (e.g. water resource = surface, ground water; nutrient = nitrogen, phosphorus).

The sensitivity evaluation process uses attribute criteria to categorize an attribute as being High, Medium, or Low sensitive. A CRITERIA encompasses all the algorithms for a single attribute that determine the risk rating. Every algorithm determines one risk rating and is called an EXPRESSION. For example:

Attribute	Expression	GWN	SWN	GWP	SWP
Slope	> 8%		3		3
	3 - 8%		2		2
	< 3%		1		1

The attributes used for the sensitivity evaluation are given 1, 2 or 3 risk ratings as the result of putting their values through these sensitivity criteria. Each attribute has a numeric value depending on the evaluation (1 = Low, 2 = Medium, 3 = High). The numeric values of all the sensitivity attributes for a field are then averaged to determine the field sensitivity index (SI) for every resource of concern. Depending on the SI value, the field receives a Low, Medium, or High

TABLE 1. SITE SENSITIVITY CRITERIA

PARAMETER	GWN *	SWN *	GWP *	SWP *
DEPTH TO IMPERMEABLE LAYER, cm.[1]				
<101		HIGH		MED
101-203		MED		MED
>203		LOW		LOW
FLOODING[2]				
FREQUENT		HIGH		HIGH
OCCASIONAL		MED		MED
RARE, NONE		LOW		LOW
RUNOFF CLASS[3]				
HIGH, VERY HIGH		HIGH		HIGH
MEDIUM		MED		MED
NEGLIGIBLE, VERY LOW, LOW		LOW		LOW
K FACTOR[4]				
>0.32		LOW		LOW
0.17 - 0.32		MED		MED
<0.17		HIGH		HIGH
ADJACENT TO SURFACE WATER BODY[5]				
YES		HIGH		HIGH
NO		LOW		LOW
SOIL ORGANIC MATTER PERCENT[6]				
<1	LOW		LOW	
1-4	MED		MED	
>4	HIGH		HIGH	
PERMEABILITY CLASS cm/hr[7]				
>5.1	HIGH		HIGH	
1.5 - 5.1	MED		MED	
<1.5	LOW		LOW	
AQUIFER RECHARGE AREA[8]				
YES	HIGH		HIGH	
NO	LOW		LOW	
SOIL TEXTURE[9]				
LFS OR COARSER			HIGH	
ALL OTHER TEXTURES			MED	
SOIL pH[10]				
> or = 7.5			LOW	
5.5-7.5			MED	
< or = 5.5			LOW	
SOIL ORDER[11]				
SPODOSOL OR ANDISOL			LOW	
ALL OTHER ORDERS			MED	

* **GWN** = ground water nitrogen; **SWN** = surface water nitrogen; **GWP** = ground water phosphorus; **SWP** = surface water phosphorus.

NOTES ON N-P SITE SENSITIVITY RATING CRITERIA

[1] Potential to generate interflow on sloping soils, with infiltrated soil water + nutrients returning to surface water as subsurface storm flow (Freeze and Cherry, pg. 218, 1979). Impermeable layer is any layer with a slow or very slow permeability class (< 0.51 cm/hr) and weathered bedrock (WB) or unweathered bedrock (UWB). Highly fractured bedrock should not be considered an impermeable layer.

[2] Potential for pickup by surface water of nutrients during overbank flow. Flooding classes from Soil Survey Staff, p. 163, (1991).

[3] Potential for nutrients in runoff to enter surface water. Runoff class as per Soil Survey Manual (Soil Survey Staff, draft version, 1991), as modified for use in the P Index, ver. 1.0.

[4] Potential for loss of sediment-bound nutrients from a site to impact surface water.

[5] Potential for surface water contamination due to site being in proximity to surface water body. This criteria replaces the criteria that gives distance variables and ratings for proximity to a surface water body. The tool should only rate the immediate field. The outputs from the field (water and nutrients) should be inputs to any field between the field being rated and a surface water body.

[6] Potential for increased N mineralization and subsequent conversion to nitrate because of more abundant organic N (Schepers and Mosier, 1991) and potential for a greater amount of mobile organic P (Alexander, 1977).

[7] Potential for soil to transmit water via saturated flow, including impact of macropores (Beven and Germann, 1982; Thomas and Phillips, 1979). Unsaturated flow is not considered directly because it is a function of matric potential, which varies depending on temporal changes in soil water content, and soil texture (Hillel, p. 108, 1971).

[8] Potential for ground water contamination due to site being within an aquifer recharge area. Nitrogen is considered potentially more mobile than phosphorus. Depth to an unconfined aquifer is not suitable criteria unless the material between the bottom of the root zone and the aquifer is characterized and rated.

[9] Potential for P sorption due to significant mineral surface area in the root zone. Texture class is that of the finest layer > 7.6 cm thick in the upper 152 cm of the soil profile. All other textures refers to a texture of loamy very fine sand or finer. Soil texture is being used in the current version as a proxy for surface area. P sorption correlates well with oxalate-extractable Fe and Al in soils (Breeuwsma, et al., 1986), so oxalate-extractable Fe and Al may be a better parameter to use if data is available.

[10] Potential for P adsorption or precipitation by carbonates (pH >= 7.5) and Fe and Al (pH <=5.5) (Brady, p. 462, 1974; Lindsay and Moreno, 1960). The pH is that of any layer > 7.6 cm thick within the upper 152 cm of the soil profile.

[11] Potential for P adsorption or fixation by soils with amorphous organic-mineral colloids (Spodosols) and allophanic material (Andisols).

TABLE 2. VULNERABILITY CRITERIA

PARAMETER	GWN *	SWN *	GWP *	SWP *
CROP GROWN[1]				
ALFALFA	LOW	LOW	LOW	LOW
CITRUS	HIGH	HIGH	LOW	LOW
CORN	HIGH	HIGH	HIGH	HIGH
COTTON	HIGH	HIGH	HIGH	HIGH
ONIONS	HIGH	HIGH	LOW	LOW
POTATOES	HIGH	HIGH	HIGH	HIGH
SORGHUM	HIGH	HIGH	HIGH	HIGH
STRAWBERRIES	HIGH	HIGH	HIGH	HIGH
SUGARBEETS	HIGH	HIGH	HIGH	HIGH
TOBACCO	HIGH	HIGH	HIGH	HIGH
WHEAT	LOW	LOW	LOW	LOW
TYPE OF IRRIGATION[2]				
DRIP	LOW	LOW	LOW	LOW
SPRINKLER	HIGH	LOW	HIGH	LOW
SURFACE	HIGH	HIGH	HIGH	HIGH
TIME OF NITROGEN APPLICATION[3]				
FALL	HIGH	HIGH		
SPRING (PRE-PLANT)	HIGH	HIGH		
SIDE DRESS	LOW	MED		
EROSION RATE > T[4]				
YES		HIGH		HIGH
NO		LOW		LOW
MANURE/ORGANIC WASTE APPLIED				
YES	HIGH	HIGH	HIGH	HIGH
NO	LOW	LOW	LOW	LOW
FERTILIZER APPLICATION METHOD[5]				
SURFACE APPLIED	HIGH	HIGH	LOW	HIGH
INJECTED/INCORPORATED	HIGH	LOW	HIGH	LOW

* **GWN** = ground water nitrogen; **SWN** = surface water nitrogen; **GWP** = ground water phosphorus; **SWP** = surface water phosphorus.

[1] Crops with high nutrient requirements increase the potential for nutrient losses.

[2] Systems that use greater volumes of water increase the potential for carrying soluble nutrients.

[3] Nitrogen applied closer to time of crop uptake has less potential for loss.

[4] As erosion rate increases, the amount of nutrients moving offsite attached to sediment increases.

[5] Surface applied nutrients have a greater potential for loss.

sensitivity rank (e.g. Low = 1.0 - 1.5, Medium = 1.6 - 2.5, High = 2.6 - 3.0) for each resource of concern. For soils the numeric values of the soil attributes are averaged to come up with a soil sensitivity ranking. When a field consists of multiple soils, only the numeric value of the soil with the worst sensitivity ranking is used to determine the field SI value, not the average of all soils. When the results of the sensitivity screening are displayed on screen only the H, M or L indicators will be visible, not the SI numbers.

The attributes used in the potential vulnerability criteria are given either a High or Low probability ranking, depending on the evaluation of the vulnerability attributes against the potential vulnerability criteria. For example:

Attribute	Expression	GWN	SWN	GWP	GWN
Erosion	= >T	H	H	H	H
Rate	= <T	L	L	L	L

If one or more attributes are marked with a High ranking, the field is ranked High. The degree of potential vulnerability is indicated by the number of attributes that received a High ranking. For example, if 5 attributes got high rankings for a particular field, the results of the evaluation for that field would show "5H". The number would provide the user with some information as to the degree of potential vulnerability for each field.

After the field sensitivity and potential vulnerability ratings have been computed, the information is displayed to the screen (Fig. 3), at which point the results can be saved by field to the "Nutrient Screening Results" database. The user will have the ability to sort the fields in ascending or descending order by sensitivity (SI), or potential vulnerability (#H) results, for the impact of nitrogen and/or phosphorus on ground and surface water. When the information is formatted the way the user wants it he can generate a report to give to the farmer or to add to the farmer's case file.

```
                       ─── Field Evaluation Summary ───
                                   Sensitivity         Pot. Vulnerability
                    - Evaluated -  Surface   Ground     Surface    Ground
     Tract    Field  By    Date    N    P   N    P     N    P    N    P
  R 1000123 1230001 PLC  10/04/93  M    M   M    H    1H   1H   1H    L
  R 1000123 1230002 RWK  05/14/91  H    L   L    L     L   4H    L   2H
  R 1000123 1230003 RWK  05/14/91  H    L   M    L     L   1H    L    L
  R 1000123 1230024 DKO  02/07/92  M    H   M    L    2H   3H    L    L
  R 1000123 1230025 RWK  05/14/91  M    M   L    H    1H    L   2H    L
  R 1000123 1230026 RWK  05/14/91  L    L   L    H     L    L   2H    L
  R 1000145 1450007 DKO  02/07/92  M    M   M    L    1H   1H   1H    L
  R 1000145 1450057 DKO  02/07/92  L    M   L    M     L    L   1H   1H
  R 1000145 1450607 DKO  02/07/92  M    L   L    H    2H   1H    L    L


  F5=Save       F6=Site Eval Summary       F7=Select All        F9=Sort
  F8=Unselect All    Shft+F6=Soil Eval Summary    Shft+F7=Acres
Use the SPACEBAR,F7 and F8 to select/unselect the fields to (R)eport on.
```

Figure 3. Summary of Screening Tool Evaluation by Field

The user will also be able to view which attributes contributed most to a field's ranking and why. This capability is achieved by showing in detail how we got to the nutrient screening results. The first level of detail shows a list of all the identified fields, each with their respective ratings for all resource and nutrient combinations for both sensitivity and potential vulnerability. The order in which the fields are shown depends on the criteria selected for sorting the list of fields. The second level of detail looks at all the attribute results for one single field. It shows all the attribute ratings for all resource and nutrient combinations for both sensitivity and vulnerability. The last level of detail provides information at the individual attribute level. It shows the attribute name, value and narrative explaining the criteria used for this attribute, some background information and guidance for interpreting the risk ranking.

Summary

The Nutrient Screening Tool gives SCS field offices a simple, technically sound procedure to help farmers determine the potential for nitrogen or phosphorus losses to surface or ground water from individual fields. This information can be used by the farmer to adjust management practices to minimize the potential for nutrient losses. It can also be used by an SCS field office to help them set priorities in providing technical assistance for nutrient management planning. The results from multiple runs of the screening tool can be summarized to show which cropping systems or which areas of the county have the greatest potential for nutrient losses. Technical assistance in nutrient management planning can then be targeted to these areas, insuring that in a time of shrinking budgets, the highest priorities are being addressed..

REFERENCES

Alexander, M. 1977. Soil microbiology. 2nd ed. John Wiley & Sons, Inc. New York, NY.

Beven, K. and P. Germann. Macropores and water flow in soils. Water Resources Res. 18:1311-1325.

Brady, N. C. 1974. The nature and properties of soils. 8th ed. Macmillan Publ. Co., Inc., New York, NY.

Breeuwsma, A., J. H. M. Wosten, J. J. Vleeshouwer, A. M. van Slobbe, and J. Bouma. 1986. Derivation of land qualities to assess environmental problems from soil surveys. Soil Sci. Soc. Am. J. 50:186-190.

Freeze, R. A., and J. A. Cherry. 1979. Ground water. Prentice-Hall, Inc., Englewood Cliffs, NJ.

Hillel, D. 1971. Soil and water. Physical principles and processes. Academic Press, New York, NY.

Lindsay, W. L. and E. C. Moreno. 1960. Phosphate equilibria in soils. Soil Sci. Soc. Am. Proc. 24:177-182.

Schepers, J. S. and A. R. Mosier. 1991. Accounting for nitrogen in nonequilibrium soil-crop systems. p. 125-138. In Follet, R. F., D. R. Keeney, and R. M. Cruse (eds.). Managing Nitrogen for Ground water Quality and Farm Profitability. Soil Science Society of America, Inc., Madison, WI.

Soil Survey Staff. 1991. Soil survey manual. Draft version. National Soil Survey Center, Lincoln, NE.

Thomas, G. W. and R. E. Phillips. 1979. Consequences of water movement in macropores. J. Environ. Qual. 8:149-152.

Wischmeier, W. H., and D. D. Smith. 1978. Predicting rainfall erosion losses-a guide to conservation planning. Agric. Handbook No. 537, United States Department of Agriculture, U. S. Government Printing Office, Washington, D. C.

SWINE WASTEWATER TREATMENT IN CONSTRUCTED WETLANDS

P. G. Hunt, F. J. Humenik, A. A. Szögi, J. M. Rice, K. C. Stone, and E. J. Sadler[*]

ABSTRACT

Swine production, a major enterprise in the Eastern Coastal Plain, has substantial waste generation and disposal considerations. Constructed wetlands have received considerable interest as a method of swine wastewater treatment that could decrease the amount of land required for application of wastewater. This study was undertaken to investigate the capacity of constructed wetlands that contained either natural wetland plants or water tolerant agronomic plants to treat swine wastewater. Six wetland cells (4 x 30 m) were constructed in Duplin Co., NC, in 1992. One set of two cells contained rush and bulrushes, and another set of two cells contained bur-reed and cattails. The third set of two cells contained flooded rice and soybean grown in saturated-soil culture. A nitrogen loading rate of 3 kg ha^{-1} day^{-1} was used during the initial four months; plant growth was excellent. The redox conditions of the wetland soils during this start-up period were highly reducing. These reducing conditions may inhibit N loss by nitrification and denitrification as well as decrease the long-term phosphorus removal efficiency. However, during this initial period, treated effluent concentrations of nitrogen and phosphorus were low and could have met discharge requirements in some areas. Rice yield was 2.8 Mg ha^{-1}, and group V and VI soybean yielded 1.9 and 3.3 Mg ha^{-1}, respectively. Long-term data over annual cycles for varying crop and hydraulic conditions are needed to assess the treatment sustainability.

KEYWORDS. Nitrogen, Phosphorus, Redox potential, Rice, Saturated culture soybean

INTRODUCTION

Constructed wetlands with either surface or subsurface flow and various vegetation (e.g. cattails, rushes, or reeds) have received considerable interest as methods of wastewater treatment (Hammer, 1989 and Reed, 1993). Since wetlands can effectively treat hydraulic loads of a few or several thousand m^3/d, they represent a method of advanced wastewater treatment for municipalities, industries, agriculture, parks, and homes. Properly designed and operated wetlands have the potential to reduce land requirements relative to other methods of natural treatment of wastewater, such as crop land irrigation. However, a consensus does not exist on the proper design and operation of wetlands for municipal or domestic wastewaters (Reed, 1993). This lack of consensus is even more apparent for the wide range of agriculture wastewaters, but pressing needs for wastewater treatment, popular press discussion of wetland treatment, and producer demands for cost effective methods have emphasized the need for more definitive design and performance data.

Swine production has become a major enterprise in the Eastern Coastal Plain. In 1990, Sampson Co., NC, and contiguous Duplin Co. were the largest and third-largest swine

---

[*] P. G. Hunt, Research Leader and Soil Scientist, A. A. Szögi, Soil Scientist, K. C. Stone, Agricultural Engineer, and E. J. Sadler, Soil Scientist, US Department of Agriculture, Agricultural Research Service, Coastal Plains Soil, Water and Plant Research Center, Florence, SC; and F. J. Humenik, Professor, and J. M. Rice, Agricultural Engineer, Agricultural Engineering Dept., NC State University, Raleigh, NC.

producing counties in the USA, respectively (NC Agric. Stat. Div., 1990). Swine wastewaters are highly concentrated in nutrients (C, N, and P), and their liquid nature makes the cost of transporting or pumping very expensive. Thus, disposal can be a difficult problem for producers that have limited land near their swine operations. Wastewater disposal must be done in a reliable and sustainable manner to avoid significant environmental damage to shallow ground waters and nutrient-sensitive streams of the coastal environment (Evans et al., 1984; Hubbard and Sheridan, 1989; and Stone et al. 1989 and 1992).

Questions exist about the long-term efficiency of constructed wetlands for swine wastewater treatment: specifically, questions exist about loading rates, oxidative/reductive conditions, denitrification potential, phosphorous removal efficiency, and ammonia toxicity to wetland plants. This study was undertaken to investigate the capacity of constructed wetlands that contained either natural wetland plants or water tolerant agronomic plants to treat swine wastewater.

MATERIALS AND METHODS

Wetland Cell Layout

Six, 4 x 30-m, wetland cells were constructed in Duplin Co., NC, in 1992 for swine wastewater treatment. The six cells were divided into three parallel sets of two end-on-end cells (Fig. 1). The cell bottoms and side walls were lined with clay, which was covered with 20 to 30 cm of loamy sand soil.

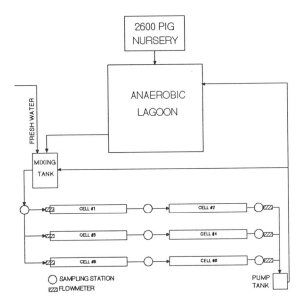

Figure 1. A schematic of the pig nursery, lagoon and constructed wetlands.

Monitoring Equipment[**]

Six V-notch weirs and six PDS-350 ultrasonic open-channel flowmeters (Control Electronics, Morgantown, PA) were installed at the inlet and outlet of each set of cells. A CR7X data logger (Campbell Scientific, Logan, UT) with three multiplexers was installed for hourly data acquisition of flow, weather parameters, and soil redox potential. Seven ISCO 2700 (ISCO, Lincoln, NE) samplers were installed. One sampler collected samples of the wastewater inflow, and the other six sampled the water at the end of each single cell. The water sampler combined hourly samples into daily composites.

Plant Materials

Four cells were planted to natural wetland vegetation in 1992. One set of two cells (two cells end-on-end) contained rush (*Juncus effusus*) and bulrushes (*Scirpus americanus*, *Scirpus cyperinus* and *Scirpus validus*), and another set of two cells contained bur-reed (*Sparganium americanum*) and cattails *(Typha angustifolia* and *Typha latifolia)*. The third set of two cells contained agronomic crops. One cell contained soybean (*Glycine max*) grown in saturated-soil culture on 1-m-wide beds that were surrounded by ditches of approximately 10-cm depth (Cooper et al., 1992 and Nathanson et al., 1984). Water level in the ditches was held at about 5 cm below the surface. Group V (cvs. Essex, Holladay, and Hutcheson) and group VI (cvs. Brim, Centennial, and Young) soybean cultivars were planted in 18-cm wide rows. The other agronomic cell contained flooded rice (*Oryza sativa* cv. Maybelle). Both agronomic crops were planted in May 1993, and plant densities were 750,000 plants ha^{-1} in both cases.

Physico-Chemical Analysis

Water samples were analyzed for electrical conductivity (EC) and pH by electrometric methods. Nitrate-nitrogen (NO_3-N), ammonia-nitrogen (NH_3-N), total Kjeldahl nitrogen (TKN), ortho-phosphate (*o*-PO_4), total phosphorus (TP), total solids (TS), total organic carbon (TOC), and volatile solids (VS) analyses were done in accordance with the USEPA recommended methodology by use of a TRAACS 800 Auto-Analyzer (Kopp and McKee, 1983). Soluble organic carbon (SOC) was analyzed with a Dhorman DC-190 carbon (C) analyzer (Rosemount, Santa Clara, CA), and chemical oxygen demand (COD) was analyzed by use of the Hach method (Gibbs, 1979). Soil redox potential was monitored continuously with a total of ninety Pt electrodes arranged in three clusters of five electrodes per cell with one Ag/AgCl reference electrode per cluster (Faulkner et al., 1989). Redox potential readings were taken every five minutes, averaged every hour and stored in the CR7X data logger. Redox readings were adjusted to the standard hydrogen electrode (Eh) by adding 200 mV.

Nutrient Loads

In order to prevent the potential of damage to wetland plants it was desirable to start wastewater application to the cells with low NH_3-N loading rates. Therefore, wastewater (Table 1) was diluted to a 1:15 ratio with fresh water and applied at an N rate of 3 kg ha^{-1} day^{-1}. This low daily N application rate required a high dilution in order to maintain the hydraulic conditions of a wetland (i.e. wastewater with 25 mg L^{-1} N provides 3 kg ha^{-1} day^{-1} of N with a loading depth of only 12 mm day $^{-1}$). Since 6-mm hydraulic loading

---

[**] Mention of trademark, proprietary product, or vendor does not constitute a guarantee or warranty of the product by the U.S. Dept. of Agr. and does not imply its approval to the exclusion of other products or vendors that may also be suitable.

would not meet evaporation transpiration demands during the summer months, the dilution and hydraulic loading were increased as needed to maintain the wetland and the 3 kg ha^{-1} day^{-1} N application rate.

Table 1. Characteristics of non-diluted wastewater from the anaerobic lagoon.

PARAMETERS	UNITS	MEAN	STD. DEV.
pH		7.53	0.14
TS	(g kg^{-1})	1.86	0.47
VS	(g kg^{-1})	0.73	0.32
TOC	(mg L^{-1})	235	124
COD	(mg L^{-1})	737	237
BOD$_5$	(mg L^{-1})	287	92
TKN	(mg L^{-1})	365	41
NH$_3$-N	(mg L^{-1})	347	52
NO$_3$-N	(mg L^{-1})	0.04	0.03
TP	(mg L^{-1})	93	11
o-PO$_4$	(mg L^{-1})	80	9

RESULTS

All equipment used for pumping, diluting, and measuring flow and distribution has worked well and required little maintenance during this start-up period. The continuous flow loading was automated with float control valves in the mixing tank, which provided automated loading of the desired proportion of lagoon liquid and dilution water. Outflow from the mixing tank for loading to the wetland cells was controlled by valves, which were periodically manually adjusted. Effluent from all three wetland series was pumped back to the lagoon in order to avoid any problems with discharge requirements. Since the wetlands could treat only a small portion of the lagoon wastewater, most of the wastewater was pumped onto crop land as needed for storage space during the summer. Neither odors nor mosquitos were a problem.

During periods of high temperatures and low rainfall, standing water (4 to 8 cm) was not maintained throughout cells 2 and 4. Vegetation planted in cells 1 and 2 remained predominant, but some intrusion of voluntary species occurred in cells 3 and 4. As temperatures lowered, standing water was maintained in all wetland cells.

A half section of 6-inch PVC pipe was used to distribute flow across the width of the constructed wetland cells. However, streaming occurred near the discharge of the second cell in each series during periods when standing water was not maintained throughout that wetland cell.

Vegetative growth in the four natural vegetation wetland cells was very good. The estimates of above ground dry matter production for cells 1, 2, 3, and 4 were 23.4, 17.9, 10.2, and 8.4 Mg ha^{-1}, respectively. The rice yield was 2.8 Mg ha^{-1} which is an acceptable production yield. Groups V and VI soybean yielded 1.9 and 3.3 Mg ha^{-1}, respectively. Soybean yields were not as high as those reported by Cooper (1992), but there were substantial cultivar differences including some high yields. Young soybean yielded 3.8 Mg ha^{-1}. Thus, it appears that there is genetic variability for high yield potential of soybean in saturated culture when swine wastewater is used for irrigation.

The electrical conductivity (EC) of the applied wastewater was significantly lowered by flow through all of the wetland cells (data not shown). This decreased EC indicated that there was a lowering of total electrolytes - probably by several mechanisms including precipitation, soil fixation, plant uptake, incorporation into soil organic matter, ammonia volatilization, and denitrification.

Average daily pH values ranged from 7.5 to 8.1. However, higher pH values may have occurred as a result of diurnal algal activity. These higher short-term pH values together with high summer temperatures could have induced some NH_3-N volatilization.

Soluble organic C concentrations indicated that energy was available to establish strong anaerobic conditions; the inflow mean was 12 mg L^{-1}, and the outflow ranged from 7 to 17 mg L^{-1}. Wide temporal ranges were observed because the anaerobic lagoon effluent and wetland waters were sometimes significantly diluted by rainfall. Redox potentials indicated strong reducing conditions in all wetland cells (Table 2). These conditions were unfavorable for nitrification. Thus, NH_3-N remained the prevalent nitrogen form. Limited nitrification also prevented significant subsequent loss of nitrogen by denitrification. The reducing conditions, lack of nitrification and denitrification, and high ammonia-N have been reported to be significant problems for treatment of municipal wastewater in constructed wetlands throughout the USA (Reed, 1993).

Ammonia-N concentrations in the wastewater decreased from 21 mg L^{-1} to < 4 mg L^{-1} after treatment in the rush/bulrushes (cells 1 and 2) and bur-reed/cattails (cells 3 and 4), and it decreased to < 1 mg L^{-1} after treatment in the soybean and rice cells (Table 3). The mass removal of ammonia-N by wetlands with all three vegetative communities was over 99% in all cells. The substantial decrease was probably from plant absorption and NH_3-N volatilization. However, some of the ammonia-N may have been nitrified especially in the soybean and rice cells.

Nitrate-N concentrations were low in the inflow wastewater because of the anaerobic conditions of the lagoon. Very little nitrate-N was accumulated in the treated wastewaters; inflow and outflow concentrations were < 1.0 and < 0.1, respectively. These low nitrate-N concentrations along with the low redox conditions and the presence of ammonia-N suggest that very little nitrification/denitrification occurred. However, in wetlands, nitrification and denitrification occurs at the interface of anaerobic and anaerobic zones and more detailed measurements are needed before a firm conclusion can be made about the extent of nitrate-N loss by denitrification.

It appeared that even the highly diluted swine wastewater in the free-water-surface wetlands caused sufficient oxygen demand and diffusion reduction to prevent the aerobic process of nitrification. Thus, an oxidative step seems necessary. This oxidation could be accomplished by treatment of the wastewater via overland flow. In overland flow the water film is only a few mm thick and in close contact with the nitrifying population of the soil surface. Thus, nitrification should occur rapidly, particularly if overland flow was done after treatment in the wetland cell had reduced the organic content of the wastewater.

In the wastewater, phosphorus was present mostly in the form of orthophosphate (Table 4). We believe that its effective removal was predominately by plant uptake, precipitation, and sorption to the soil substrate. Mass removal by wetland with all three vegetative communities was over 99%.

Table 2. Soil redox potential (Eh) ranges for every wetland cell (June - September 1993).

CELL	Eh RANGE (mV)[*]
1	150 to -240
2	240 to -240
3	100 to -240
4	280 to -220
5	400 to -250
6	165 to -235

[*] Nitrate is normally absent when soil Eh is below 300 mV.

Table 3. Mean, standard deviation, and range for NH_3-N of daily composite wastewater samples (June - September 1993).

PLANTS	SAMPLER	MEAN (mg L^{-1})	STD.DEV. (mg L^{-1})	RANGE (mg L^{-1})
	INFLOW	21	6	6.0 - 31
J/S[a]	CELL 1	4	4	0.3 - 13
	CELL 2	2	4	0.3 - 11
S/T[b]	CELL 3	3	3	0.1 - 10
	CELL 4	2	2	0.1 - 4
SOYBEAN	CELL 5	8	6	1.0 - 14
RICE	CELL 6	0.2	0.1	0.1 - 0.3

[a] J/S = *Juncus* sp. and *Scirpus* sp.
[b] S/T = *Sparganium* sp. and *Typha* sp.

Table 4. Mean, standard deviation and range for o-PO_4 of daily composite wastewater samples (June - September 1993).

PLANTS	SAMPLER	MEAN (mg L^{-1})	STD. DEV. (mg L^{-1})	RANGE (mg L^{-1})
	INFLOW	4.0	0.8	3.0 - 5.0
J/S[a]	CELL 1	1.0	0.6	0.2 - 2.0
	CELL 2	0.2	0.1	0.1 - 0.2
S/T[b]	CELL 3	0.7	0.6	0.2 - 2.0
	CELL 4	0.1	0.1	0.1 - 0.2
SOYBEAN	CELL 5	2.0	0.4	2.0 - 3.0
RICE	CELL 6	0.3	0.2	0.2 - 0.5

[a] J/S = *Juncus* sp. and *Scirpus* sp.
[b] S/T = *Sparganium* sp. and *Typha* sp.

SUMMARY

Growth of the rushes, bulrushes, bur-reed, and cattails was very good. Rice yield was 2.8 Mg ha^{-1}, and groups V and VI soybean yielded 1.9 and 3.3 Mg ha^{-1}, respectively.

The low N loading rate of 3 kg ha^{-1} day^{-1} was chosen because it is a currently recommended level and wetlands with higher loading rates have not been producing an effluent that could be stream discharged. The preliminary results presented are for this loading rate during a four-month period with excellent plant growth. The treated effluent concentrations of nitrogen and phosphorus were low and could have met discharge requirements in some areas; however, longer term research including dormant plant periods is still needed. Phosphorus is of concern on a long-term basis since the highly reducing conditions may lower the removal efficiency.

The redox conditions of wetland soil during this start-up period were highly reduced. The presence of ammonia-N in the discharge effluent and the very low concentrations of nitrate-N throughout the wetlands suggest that the cell did not support nitrification which must occur before removal of nitrogen by denitrification. However, nitrification and denitrification in wetlands occur at the interface of anaerobic and anaerobic zones, and more detailed measurements are needed before a firm conclusion can be made about the extent of nitrate-N loss by denitrification.

In any case, an oxidative component and a reductive component in sequence will be necessary for nitrogen removal. If it does not occur in the natural wastewater-plant-soil interface of the wetland cell, it could be produced by construction of an oxidative component and recycling of the effluent. Long-term data over annual cycles for varying crop and hydraulic conditions will be necessary to determine if constructed wetlands can produce a dischargeable or significantly weakened swine wastewater effluent so that less land will be required for wastewater management.

REFERENCES

1. Cooper, R. L., N. R. Fausey, and J. G. Streeter. 1992. Crop management to maximize the yield response of soybeans to a subirrigation/drainage system. pp. 466-473. IN Drainage and Watertable Control. Proceedings of the Sixth International Drainage Symposium. Am. Soc. Agric. Eng., St. Joseph, MI.

2. Evans, R. O., P. W. Westerman, and M. R. Overcash. 1984. Subsurface drainage water quality from land application of swine lagoon effluent. Trans. ASAE 27:473-480.

3. Faulkner, S. P., W. H. Patrick, Jr. and R. P. Gambrell. 1989. Field techniques for measuring wetland soil parameters. Soil Sci. Soc. Am. J. 53:883-890.

4. Gibbs, C. R. 1979. Introduction to Chemical Oxygen Demand. Technical Information Series -- Booklet No. 8. Hach Technical Center for Applied Analytical Chemistry. pp. 1-16. Hach Company. Loveland, Colorado (USEPA-Approved for NPDES reporting. Federal Register, Vol. 45, No. 78, Monday April 30, 1980, page 26811).

5. Hammer, D. A. (Ed.). 1989. Constructed wetlands for wastewater treatment: Municipal, industrial, and agricultural. Lewis Publishers, Chelsea, MI. 831 p.

6. Hubbard, R. K. and J. M. Sheridan. 1989. Nitrate movement in groundwater in the Southeastern Coastal Plain. J. Soil and Water Cons. 44:20-27.

7. Kopp, J. F. and G. D. McKee. 1983. Methods for chemical analysis of water and wastes. USEPA Report No. EPA-600/4-79020. Environmental Monitoring and Support Lab., Office of Research and Development, U.S. Environmental Protection Agency, Cincinnati, OH. 521 p.

8. Nathanson, K., R. J. Lawn, P. L. M. DeJabrun, and D. E. Byth. 1984. Growth, nodulation, and nitrogen accumulation by soybean in saturated soil culture. Field Crops Res. 8:73-92.

9. North Carolina Agriculture Statistics Div. 1990. North Carolina agricultural statistics, 1990. Raleigh, NC. 76 p.

10. Reed, S.C. 1993. Subsurface flow constructed wetlands for wastewater treatment: Technology assessment. EPA-832-R-93-001. Office of Water, USEPA, Wash., DC.

11. Stone, K. C., K. L. Campbell, and L. B. Baldwin. 1989. A microcomputer model for design of agricultural stormwater management systems in Florida's flatwoods. Trans. ASAE 32:545-550.

12. Stone, K. C., R. C. Sommers, G. H. Williams, and D. E. Hawkins. 1992. Implementation of water table management in the Eastern Coastal Plain. J. Soil and Water Conservation 47:47-51.

ACKNOWLEDGMENT

We would like to thank the pig nursery owners, Gerald and Paulette Knowles, for permission to work on their farm and for their cooperative attitude. We would also like to thank Murphy farms, particularly Mr. Gary Scaff, for assistance with construction and operational issues of the wetlands.

NUTRIENT AND WATER MANAGEMENT PRACTICES FOR SUSTAINABLE VEGETABLE PRODUCTION IN THE LAKE APOPKA BASIN

C.A. Neal, E.A. Hanlon, J.M. White, S. Cox and A. Ferrer[1]

ABSTRACT

Lake Apopka was identified in the 1980's as a priority water body in need of restoration and preservation by the Florida legislature (Conrow, 1989). Controlling nutrient loading from agricultural lands is a key component of the restoration plan. It has been estimated that 76 billion L of water are discharged annually from the agricultural lands in the basin, principally from 5260 ha of Histosols used for vegetable production. The quality of discharge water is poor due to high concentrations of nutrients. Phosphorus (P) is the nutrient of primary concern.

A cooperative project between Florida Cooperative Extension Service, Soil Conservation Service and Agricultural Stabilization and Conservation Service was initiated in 1991. The objectives were to improve nutrient and water management practices for vegetable production within the Lake Apopka drainage basin, resulting in reduced nutrient loading to the lake. This is being accomplished through water and nutrient management practices including water table control, land leveling, calibrated soil testing, installation of water control structures, and other practices.

Changes documented by a 1993 survey of farmers show that all ten targeted farms are participating in the effort. Phosphorus application to vegetable crops has been reduced by 155400 kg, or 55%, compared to 1991. Dramatic improvements in water management have also been made.

KEYWORDS. Fertilizer, Histosols, Phosphorus, Water Quality.

BACKGROUND

Lake Apopka is the fourth largest lake in Florida and is the headwater lake for the Oklawaha Chain of Lakes. It is located at latitude 28 37' N and longitude 81 38' W. The nearest major metropolitan area is Orlando, located about 48 km to the southeast, outside of the Lake Apopka drainage basin.

The hypereutrophic condition of this lake is a reflection of past waste discharges from municipalities, industry, and drainage water from agricultural lands. Excess water resulting from rainfall or intentional flooding of Histosols for pest management is still pumped into Lake Apopka from the farmland. This water may contain elevated levels of P originating from mineralization and/or from P fertilization of the Histosols. Phosphorus loading of the

---

[1]C.A. Neal, Extension Agent III, Florida Cooperative Extension Service, 30205 SR 19, Tavares, FL 32778; E.A. Hanlon, Assoc. Professor, Dept. of Soil and Water Science, University of Florida, Gainesville, FL 32611; J.M. White, Assoc. Professor, Central Florida Research and Education Center, 2700 E. Celery Ave., Sanford, FL 32771; S. Cox, District Conservationist, Soil Conservation Service, 2012 E. Michigan St., Orlando, FL 32806; A. Ferrer, Sr. Agricultural Assistant, Florida Cooperative Extension Service, 30205 SR 19, Tavares, FL 32778.

discharge water is a function of many factors, some of which can be controlled by growers using this productive land (Anderson et al., 1994).

The Lake Apopka Hydrologic Unit Area Project (LAHUAP) was initiated in 1991 to assist growers in developing and adopting sound and efficient water and fertilizer management programs for vegetable production. Recommended best management practices for Histosols in the Everglades Agricultural Area have been described by Bottcher and Izuno (1993) and Anderson and Howell (1993). Selected practices which fit the LAHUA situation and objectives are being emphasized through educational and technical assistance to growers. The Florida Cooperative Extension Service (FCES), the Soil Conservation Service (SCS) and the Agricultural Stabilization and Conservation Service (ASCS) are cooperating agencies involved in the LAHUAP effort.

EDUCATIONAL ACTIVITIES

At the beginning of the project, the ten vegetable farms within the LAHUA were surveyed to collect baseline data on fertilizer and water management. Two years later, the survey was repeated to document what changes had been made.

A number of educational activities were implemented to encourage growers to adopt recommended changes in nutrient and water management. Concurrently, replicated field trials were conducted to improve our database on crop nutrient requirements for the major vegetable crops and soils in the LAHUA.

<u>Information Packages</u> Each grower received a specially prepared 3-ring notebook containing the most current information from the FCES and SCS regarding crop production fertilizer use, water control, and pest management. Sections within the books included individual tissue and soil test results for each farm. Additional sections were designed to contain current and future communications regarding the LAHUAP.

<u>Newsletter</u> A quarterly newsletter, "Soil and Water", is prepared by the LAHUAP team. The newsletter consists of technical articles addressing soil, water and fertilizer management; cost-share assistance; progress reports and a who's who of project members. The newsletter is designed to be inserted into a special section of the notebook.

<u>Soil Testing</u> In an effort to promote the use of calibrated soil testing, FCES provides free soil testing with a one-week turnaround time for farms in the LAHUA. Over 200 samples per year are analyzed for nitrogen (N), P, potassium (K), magnesium, calcium, zinc, copper, manganese and pH under this program. Eighty percent of the samples to date tested high to very high in P, thus resulting in a recommendation that no P fertilizer be added to the field.

<u>Tissue Testing</u> Tissue tests have been used to monitor selected fields for crop nutrient status where growers had reduced fertilizer application rates. The results of the laboratory tests were also compared with in-field sap tests. No deficiencies of N, P, or K have been detected.

<u>Group Meetings</u> LAHUA personnel conducted several instructional sessions aimed at informing the growers about the physical and chemical properties unique to Histosols. Emphasis was placed on how different management strategies affected P loading in drainage water from fields. The importance of controlling water movement to control P movement was stressed.

Field Research/Demonstration Experiments Replicated trials were conducted in farm fields to check the calibrated soil test P recommendation against the grower rate and other P rates. Two studies on carrots and three studies on sweet corn have been completed. As predicted, the data have shown no significant yield response due to P fertilization where soil test levels were already high to very high (Crnko et al., 1993).

Field Days Two field days have been conducted at the sites of the experiments. Demonstrations of tissue sap testing and water table monitoring methods, as well as reports concerning the P trials, were included at the field days.

Resource Management Plans SCS prepared plans on 5160 ha within the LAHUA. The plans recommended management practices such as land leveling, water table control, irrigation water management, and nutrient and pesticide management. Practices are currently being implemented on all of the farms. One of the largest farms (1000 ha) constructed a tailwater recovery system to allow for on-farm reuse of water.

RESULTS

Nutrient Management Figure 1 shows the average amount of P applied to each of the major crops in the LAHUA in 1990-91 compared with 1992-93. All ten targeted farms reduced P application rates on some or all of their crops, for a net reduction of 155400 kg of P. This represents a 55% reduction in the amount of P applied within the project area, which totaled 284450 kg in 1990-91 and 129000 kg in 1992-93 (Fig. 2). Because of the widespread adoption of calibrated soil testing by the farms, nitrogen and potassium applications have been reduced concurrently, by 186429 and 320390 kg, respectively.

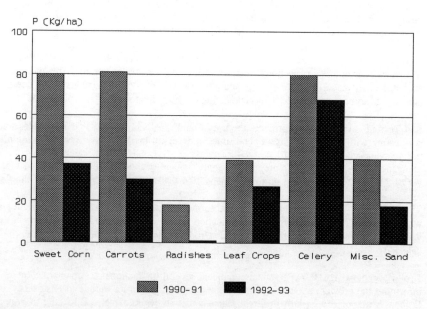

Figure 1. Average amount of P (kg/ha) applied to each of the major vegetable crops produced in the LAHUA, 1990-91 and 1992-93 seasons.

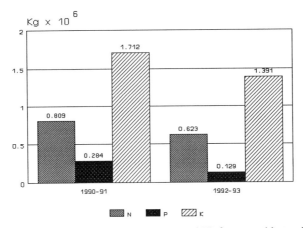

Figure 2. Total application of macronutrients (N, P, and K) for vegetable production in the LAHUA, 1990-91 and 1992-93 seasons.

Water Management Measures to promote water recycling and reduce discharges have included installation of 14 internal pumps, 164 water control structures, and 3 floating staff gauges. In addition, 3240 ha were land leveled and 2835 ha were mole drained to improve water use irrigation efficiency and drainage.

The tailwater recovery system on one farm has reduced water intake from Lake Apopka by over 80%, and has reduced the amount of water and P discharged from the farm to the Lake by 65%.

The combined efforts of vegetable producers in the LAHUA have reduced the amount of P which is potentially discharged into Lake Apopka. These changes are expected to reduce the environmental impact and contribute to the sustainability of agriculture within the LAHUA.

REFERENCES

1. Anderson, D.L, E.A. Hanlon, A.B. Bottcher, and A. Ceric. 1994. Phosphorus in Drainage Waters from Organic Soils (Histosols) of Florida. Univ. of Fl., Gainesville. Circular (in press), 3 pp.

2. Anderson, D.L. and M.D. Howell. 1993. Implementation of the Everglades Surface Water Improvement and Management Plan in the EAA. Univ. of Fl., Fl Exp. Sta. Abstract J. Series no. A-00248, 1 p.

3. Bottcher, Del and Forrest Izuno. 1993. Procedural Guide for the Development of Farm-Level Best Management Practices Plans for Phosphorus Control in the Everglades Agricultural Area. Fla. Coop. Extn. Service, Univ. of Fl., Gainesville, FL, 43 pp.

4. Conrow, R.C. 1989. Draft SWIM Plan for Lake Apopka. St. Johns River Water Management District, Palatka, FL.

5. Crnko, G.S., A. Ferrer, E.A. Hanlon, C.A. Neal, and J.M. White. 1993. Carrot and Sweet Corn Yields When Fertilized According to Soil Test Results. Proc. Fla. State Hort. Soc. 1993: (in press).

IMPACT OF BMP'S ON STREAM AND GROUND WATER QUALITY IN A USDA DEMONSTRATION WATERSHED IN THE EASTERN COASTAL PLAIN

K. C. Stone, P. G. Hunt, J. M. Novak, and T. A. Matheny[*]

ABSTRACT

A USDA Water Quality Demonstration project in the Herrings Marsh Run (HMR) watershed in Duplin County, North Carolina was initiated in 1990. The HMR watershed is representative of an eastern Coastal Plain watershed with intensive agricultural practices (*e.g.*, crop, swine, poultry, and cattle production). Stream sampling stations were established at four locations in the watershed to evaluate the influences of agricultural practices on stream water quality. Ninety-two monitoring wells were installed on 21 farms throughout the watershed to evaluate the influences of agricultural practices on ground water quality. Stream water at a HMR tributary was consistently of lower quality than water from the HMR main channel. Nitrate-nitrogen and ortho-phosphate levels consistently exceeded 5 and 1 mg/L, respectively, in stream water at the HMR tributary. Mean nitrate-nitrogen in ground water at five farms exceeded 10 mg/L, possibly caused by animal waste water application and/or excessive use of commercial fertilizers. Both triazines and chloroacentanilides were detected in approximately 20% of the monitoring wells during some sampling months. The majority of these pesticide detects had concentrations well below the maximum contaminant level for safe drinking water. These data indicate that current and past agricultural management practices have degraded water quality in both stream and ground waters.

INTRODUCTION

Even though significant progress has been made in the development and implementation of agricultural best management practices (BMP's), nonpoint pollution of surface and ground water by agriculture is a major water quality concern (Bjerke, 1989; Bouwer, 1987; Hurlburt, 1988; Hubbard et al, 1986; and Hubbard and Sheridan, 1989). A five-year water quality demonstration project involving federal, state, and local agencies, private industry, and local land owners was initiated in 1990 on a watershed located on the Cape Fear River Basin in Duplin County, North Carolina. The 2044 ha demonstration watershed, Herrings Marsh Run (HMR), is one of the eight original demonstration projects funded as part of the USDA's Presidential Water Quality Initiative, and it is located within the Goshen Swamp Hydrologic Unit Area one of the 37 original hydrologic unit area projects (United States Department of Agriculture and Cooperating State Agencies, 1989). Duplin County has many of the characteristics of an intensive agricultural county in the eastern Coastal Plain of the USA. Duplin County, NC, has the highest agricultural revenue of any county in North Carolina. In 1990, it had the highest population of turkeys and the fourth highest population of swine of any county in the United States (North Carolina Dept. of Agriculture, 1990).

Agricultural management practices on the watershed are typical for the southeastern Coastal Plain and include 1093 ha of cropland, 708 ha of woodlands, and 212 ha of farmsteads, poultry facilities, and swine facilities. The major agricultural crops on the watershed include tobacco (131 ha), corn (415 ha), soybeans (273 ha), wheat (121 ha), and vegetables (162 ha). The predominant soil series in the watershed is Autryville (Loamy, siliceous, thermic Arenic Paleudults); secondary soil series are Norfolk (Fine-loamy, siliceous, thermic Typic

---

[*] Agricultural Engineer, and Soil Scientists, respectively, USDA-ARS, Coastal Plain Soil, Water, and Plant Research Center, Florence, SC.

Kandiudults), Marvyn-Gritney (Clayey, mixed, thermic Typic Hapludults), and Blanton (Loamy siliceous, thermic Grossarenic Paleudults).

Current annual nutrient usage for crop production on the watershed is estimated at 145 metric tons of nitrogen, 64 metric tons of phosphorus, and 243 metric tons of potassium. Although swine and poultry operations produce sufficient quantities of waste to supply over half of the needed nutrients, 90% of the nutrients applied to cropland are supplied by commercial fertilizers. The application of large quantities of commercial fertilizers coupled with the production of large quantities of animal waste provides a potential for nitrogen and phosphorus contamination of surface and ground water. The objective of the initial phase of the project has been to evaluate the effect of current agricultural management practices on stream and ground water quality within the watershed.

METHODS

Surface water sampling stations were established in August 1990, at three locations within the watershed (Fig. 1). Station 1, Red Hill, was located at the stream outlet from the watershed. Station 2, Herrings Marsh Run Tributary, was located along a tributary downstream from intensive swine and poultry operations. Station 3, Herrings Marsh Run Main, was located along the main stream run flowing through woodlands. The woodlands located above Station 3 has a substantial riparian buffer compared with the riparian buffers for Station 2. Station 3 was chosen to represent background conditions due to the large riparian buffer and because no animal production facilities were in the subwatershed. Station 4, Red Hill tributary, was installed in August 1991, to provide additional information about the eastern portion of the watershed. The U. S. Geological Survey in Raleigh, NC, installed gaging stations at the initial three stations in April 1991. A gaging station was installed at Station 4 in August 1991. Isco 2700 automated water samplers were installed at each station. Sample collection was continual from October 1990 to the present time. The water samplers combine hourly samples into a daily composite. The samples were collected weekly and transported to the laboratory for analysis. The gaging stations measure flow at 15-min intervals using automated water level recorders.

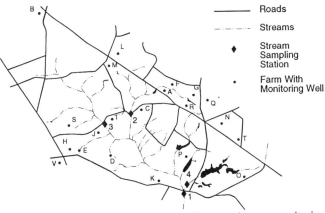

Figure 1. Location of stream gaging stations and farms with ground water monitoring wells on Herrings Marsh Run watershed.

Ground water monitoring wells were established on 21 farms in the HMR watershed (Fig. 1) beginning in August 1991 and continuing through March 1993. These farms exemplify the agricultural practices used in the watershed. The farms were selected to cover the watershed both on a geographical basis and to represent the farming practices on the watershed. The majority of the farms with monitoring wells are in row crops with and without implemented nutrient management plans. Two of the row crop farms have their main source of nitrogen from poultry litter and poultry compost. The other farms have practices which include the application of swine lagoon effluent to pasture (Farm A) and a farm with pasture for hay production. Monitoring wells at Farm F are located around a poultry compost facility.

Ground water monitoring wells were installed using a SIMCO 2800** trailer-mounted drill rig equipped with 108-mm inside diameter hollow stem augers. The well casings and screen were 50-mm threaded schedule 40 PVC, and well screens were 1.5 m long. Well bottoms were placed on an impermeable layer or to a depth of 7.6 m if the impermeable layer could not be located above that depth. Water table depths in the watershed were generally 1.5 to 3 m below the soil surface. Monitoring wells were constructed according to N. C. Dept. of Environmental Management regulations. A filter pack of coarse sand was placed around well screens. An annular seal of bentonite was placed above the filter sand. Concrete grout was then placed above the bentonite to the soil surface to prevent contamination from the surface. Locking well covers were installed to prevent unauthorized access. WaTerra foot valves (model D-25) and high density polyethylene tubing were installed in each well to provide dedicated samplers.

At Farm A, ground water sampling transects were established in the swine lagoon effluent application field. Both transects had wells in the spray field (in-field), at the edge of the field (field-edge), and at the edge of a stream (stream-edge) that flows through the farm. Wells were placed at two depths, 4.6 and 7.6 m in each location, and one well was placed at 15 m. The 4.6- and 7.6-m sampling depths were immediately below and approximately 3 m below the water table, respectively. The area between the field-edge and stream-edge contained a small riparian buffer. Stream grab samples were taken adjacent to the stream-edge samples for two of the transects. At the other farms monitoring wells were placed around fields to a depth of 7.6 m. These wells were located both up and down slope of the various practices to monitor ground water flowing from the fields.

Before samples were collected, the static well water depths were measured, and three well volumes were purged. Glass sample collection bottles were rinsed with the well water before sample collection, filled with sample, packed in ice, and transported to the laboratory. Wells were sampled monthly.

All water samples were transported to the USDA-ARS, Soil, Water, and Plant Research Center in Florence, SC, for analysis. Water samples were analyzed using a TRAACS 800 Auto-Analyzer for nitrate-nitrogen, ammonium-nitrogen, total Kjeldahl nitrogen, ortho-phosphorus, and total phosphorus using EPA Methods 353.2, 350.1, 351.2, 365.1, and 365.4, respectively (U.S. EPA, 1983). EPA-certified quality control samples were routinely analyzed to verify results. Ground water samples were initially screened for triazines and chloroacentanilides using immunoassay techniques and detects were confirmed by gas chromatography/mass spectroscopy (GC/MS). All statistical analysis of the data was accomplished using SAS version 6.07 (SAS, 1990).

**Mention of trademark, proprietary product, or vendor does not constitute a guarantee or warranty of the product by the U. S. Dept. of Agr. and does not imply its approval to the exclusion of other products or vendors that may also be suitable.

RESULTS AND DISCUSSION

Mean daily nitrate-nitrogen concentrations (Table 1) of water leaving the watershed outlet, Red Hill (Station 1), and at HMR tributary (Station 2) were two and four times higher, respectively, than background concentrations as represented by HMR main (Station 3). Daily mean nitrate-nitrogen concentrations at the HMR tributary periodically exceeded 10 mg/L. Overapplication of swine lagoon effluent and undersized, overloaded lagoons are likely contributors to the elevated nitrate-nitrogen concentrations in the HMR tributary.

Mean daily ammonium-nitrogen concentrations of water at the watershed outlet and at the HMR tributary were two and seventeen times higher, respectively, than background concentrations (Table 1). Ammonium-nitrogen concentrations at the watershed outlet and at HMR tributary exceeded limits considered harmful to humans (0.5 mg/L) and fish (2.5 mg/L) (U.S. EPA, 1973). During the first month of the sampling period, daily ammonium-nitrogen concentrations at HMR tributary ranged from 6 to 12 mg/L. These high concentrations of ammonium-nitrogen indicate that a significant discharge of animal waste products into the waterway had occurred. After the first month, daily mean ammonium-nitrogen concentrations did not exceed 4 mg/L.

Table 1. Mean daily nutrient concentrations and mass fluxes over the sampline period for four stream monitoring stations in the Herrings Marsh Run watershed in Duplin County, NC.

	Stations*			
	1	2	3	4
Concentration	mg/L			
NO_3-N	2.01	5.34	1.18	1.26
PO_4-P	0.14	0.54	0.06	0.07
NH_4-N	0.15	0.42	0.08	0.18
Mass Flux	kg/day			
NO_3-N	22.17	19.61	3.56	2.18
PO_4-P	2.24	2.04	0.17	0.13
NH_4-N	2.08	1.34	0.28	0.37
	m^3/s			
Stream Flow	0.147	0.041	0.034	0.025

* Station 1 is located at the watershed outlet. Station 2 is the Herrings Marsh Run tributary. Station 3 is the Herrings Marsh Run main and is used as a background reference. Station 4 is the Red Hill tributary.

Mean daily ortho-phosphorus concentrations of water at the watershed outlet and at HMR tributary were five and ten times higher, respectively, than background concentrations (Table 1). On ten daily occurrences, ortho-phosphorus concentrations of water at the exit of the watershed were significantly higher than at HMR tributary. The origin of these peaks is unclear.

Stream flow data from the USGS Gaging stations (USGS, 1992) began in April 1991 for three stations (Red Hill, HMR tributary, and HMR main) and in August 1991 for the fourth station (Red Hill tributary). The stream flow data were integrated with the stream monitoring data to calculate the mass loading of nitrate-nitrogen and ammonium-nitrogen. The mass nitrate-nitrogen leaving the watershed (Red Hill) averages approximately 30 kg/day. The HMR tributary monitoring station has approximately 22 kg/day leaving that sub-watershed. The mass ammonium-nitrogen at Red Hill and HMR tributary frequently exceed 5 and 3.5 kg/day, respectively. Ortho-phosphorus mass flux at the watershed outlet averages 7 kg/day while the HMR tributary and HMR Main averages 3.5 and 0.7 kg/day, respectively.

Mean of monthly ground water nitrate-nitrogen and ammonium-nitrogen concentrations for the monitored farms are presented in Table 2. Both nitrate-nitrogen and ammonium-nitrogen concentrations in ground waters at Farm A consistently exceeded 10 mg/L. In addition, the mean nitrate-nitrogen and ammonium-nitrogen concentrations in stream water at Farm A were 8 and 4 mg/L, respectively. At Farm A, the elevated nitrate-nitrogen concentrations was believed to be directly related to the land application of swine wastewater that has been an on-going operation since 1986. The spray field for the waste application was undersized due to expansion of the swine operation since its original design. Prior to 1991, the spray field had no permanent grass cover; row crop or weed fallow served as the ground cover. Additionally, it was suspected that the overloading may be degrading the performance and efficiency of nutrient removal in the lagoon.

Table 2. Mean of monthly nitrate-N and ammonium-N concentrations in ground water monitoring wells located within the demonstration watershed.

Farm	Nitrate-N		Ammonium-N	
	(mg/L)	Std. Dev.	(mg/L)	Std. Dev.
A	72.7	64.9	16.5	16.2
B	10.9	6.4	0.25	0.3
C	16.8	4.13	0.27	0.28
D	5.62	5.49	0.29	0.42
E	6.48	1.24	0.28	0.34
F	11.1	8.71	0.44	0.97
G	7.67	2.1	0.2	0.41
H	10.3	8.55	0.36	0.52
I	8.23	1.4	0.24	0.39
J	0.73	1.92	0.18	0.22
K	1.65	1.33	0.18	0.23
L	4.28	6.39	0.26	0.71
M	8.18	7.15	0.26	0.24
N	1.49	1.9	0.25	0.2
O	7.41	4.1	0.2	0.28
P	7.7	6.44	0.23	0.16
Q	7.96	7.48	0.24	0.33
R	17.1	24.8	0.23	0.26
S	6.02	1.91	0.19	0.11
T	5.11	2.12	0.15	0.16
U	2.23	2.35	0.22	0.07
V	2.47	0.16	0.17	0.07

Elevated nitrate-nitrogen concentrations at Farm F are possibly due to pre-existing contamination from contiguous poultry houses. The elevated nitrate-nitrogen concentrations at other farms in the watershed are likely related to nonpoint sources of nitrogen due to overapplication of fertilizer. It appears that improved nutrient management will be helpful on these farms. The majority of other farms appear to have appropriate nutrient management budgets since the nitrate-nitrogen concentrations were less than 10 mg/L.

Over a 10-month period, immunoassay analyses of ground water showed that triazines and chloroacentanilides were detected in less than 20% of the 92 monitoring wells. Detects in monitoring wells varied monthly. Both atrazine and alachlor were confirmed (with GC/MS) in a number of ground water samples. The majority of detected concentrations were well below the maximum contaminate level for safe drinking water.

SUMMARY

Results from the initial phase of the five-year project indicate that most of the streams and ground waters of the watershed have acceptable water quality. The stream water at the watershed outlet and station 2 had elevated nitrate-nitrogen and ammonium-nitrogen concentrations. The elevated concentration levels at station 2 are believed to be directly related to the high concentration of swine production facilities in that subwatershed along with the reduced riparian buffers. The mass flux from this subwatershed accounts for over two-thirds of the nitrate-nitrogen leaving the watershed at the watershed outlet.

Ground water on the farms sampled in the watershed has indicated that over eighty percent have nitrate-nitrogen concentrations below 10 mg/L. Four farms had elevated nitrate-nitrogen concentrations between 10 and 20 mg/L. Only one farm had nitrate-nitrogen concentrations in ground water that exceeded 20 mg/L. Triazines and chloroacentanilides were periodically detected in some wells but the majority of the concentrations were below safe drinking water standards. It appears that traditional agricultural management practices on the watershed have had an adverse impact on the quality of surface and ground water at specific locations.

REFERENCES

1. Bjerke, K. D. 1989. Pesticides, common sense and groundwater. J. Soil and Water Cons. 44(4):262.

2. Bouwer, H. 1987. Effect of irrigated agriculture on groundwater. J. Irrig. and Drainage Eng. 113(1):4-29.

3. Hubbard, R. K., G. J. Gascho, J. E. Hook, and W. G. Knisel. 1986. Nitrate movement into shallow groundwater through a Coastal Plain sand. Trans. ASAE 29:1564-1571.

4. Hubbard, R. K. and J. M. Sheridan. 1989. Nitrate movement to groundwater in the southeastern Coastal Plain. J. Soil and Water Cons. 44:20-27

5. Hurlburt, S. 1988. The problem with nitrates. Water Well J. 42(8):37-42.

6. North Carolina Department of Agriculture. 1990. Annual agricultural survey. Raleigh, NC.

7. SAS. 1990. SAS version 6.07. SAS Institute, Cary, NC.

8. U. S. Geological Survey. 1992. Provisional stream flow data. USGS, Raleigh, NC.

9. United States Department of Agriculture and Cooperating State Agencies. 1989. Water quality program plan to support the President's water quality initiative. USDA, Washington, D.C. 29 pp.

10. U.S. EPA. 1973. Water quality criteria. U.S. Government Printing Office. Washington, DC.

11. U.S. EPA. 1983. Methods for chemical analysis of water and wastes. USEPA-600/4-79-020. J. F. Kopp and G. D. McKee. Environmental Monitoring and Support Lab. Cincinnati, OH. Office of Research and Development. Cincinnati, OH.

Dairy Park: An Environmentally Sound Alternative to Current Dairy Production Practices

A. K. Butcher and R. Neal[*]

ABSTRACT

Dairies and other confined animal feeding operations (CAFOs) have undergone tremendous changes in the last several decades. Today, fewer dairies produce more milk due to advances in production and dairy technologies which concentrate more cows on less acreage. As an example of the trend in the modern dairy industry, in only eight years, the number of dairy cows in Erath County, Texas, increased 148% while the number of dairies increased only 3%. The growth and concentration of the industry strained the area's water and air quality, quality of life, and the industry's traditionally strong relations with the local community.

The continuing industrialization and concentration of CAFOs challenges the traditional structure, regulation, and location of dairy operations. As large dairy operations face public opposition, even in rural areas such as Erath County, the problem of where to locate new dairies arises. Negative public pressure also confronts dairies wishing to expand. Clearly, dairies do not belong in urban areas, but, if rural areas begin to oppose dairies, agriculture and regulators must examine exactly where these facilities will (or should) locate and how they will operate in the future. One proposal to alleviate CAFO pollution concerns is the creation of an industrial style "dairy park" utilizing free stall dairies. The park would provide a suitable area where individual dairy operations, adhering to rigidly specified technologies, would collocate to take advantage of centralized environmental systems, including fresh water, animal waste treatment, permitting, and resource recovery. Sufficiently sized buffer zones would surround the park to decrease nuisance odor, preserve water quality, and minimize complaints from neighbors.

Key Words: Dairy Park, Agricultural Park, Centralized Animal Waste Treatment

Background

Erath County is not the only example of environmental problems stemming from large dairy operations. In Washington's Snohomish River area, "urbanization is swallowing up farms." (Nexis, 1993.) "As more city folk seek the country life, tougher regulations for managing waste follow. Farmers need enough land to spread out their cow manure and waste so there won't be too high a concentration of nitrates in one place." (Id.) In the words of one Washington State dairy operator, "I'm afraid if we came back 20 years from now there won't be a dairy industry left." (Id.)

Twenty-five years ago, P. H. Jones stated that

> [the Agribusiness Park] represents an entirely new suggestion for locating intensive livestock operations together. It is basically like an industrial park but located in the rural countryside. It presupposes the managed location of mutually

---

[*] Allan K. Butcher, J.D., is a Policy Analyst and Robert Neal, M.S. is a Research Economist at the Texas Institute for Applied Environmental Research, Tarleton State University, Stephenville, Texas 76402. The Institute will issue a detailed report on the dairy park concept in the Spring of 1994.

supportive agribusiness activities. Conceptually, there would be a core of intensive livestock operations with different livestock as neighbors. Adjacent to this central area would be support services such as tractor sales and service, farm equipment, pumps, feeders, etc. Government extension offices might also be located somewhere in the centre, as would professional services.

***

If the agribusiness park is then surrounded by a large area of crop-producing land, not only would buffering be achieved but animal wastes would provide the necessary fertilizer to produce the feed necessary for the livestock. In addition, by instituting a centrally managed handling system, costs could be substantially reduced and benefits substantially increased. (Jones, 1977, at 47.)

Variations on a dairy park project have appeared in the past. Previously proposed or constructed agricultural facilities with centralized waste treatment include:

- Tillamook, Oregon's, proposed central digester which would accept the wastes from 10,000 cows on forty existing local dairies; (Mattocks, 1993.)
- A Hawaiian digester project that consumes dairy manure along with organic waste from grocery stores and other local industry; (Mattocks, 1993, and Wagner, 1993.)
- Centralized treatment of pig wastes in Malaysia, instituted to counter pollution and social problems; (Teoh, et al, 1988, at 96.)
- A mammoth 20,000 cow dairy park in Mexico with a 100 acre-foot lagoon and an average herd size of 200 head per dairy operation; (Cortez, 1993.)
- Various proposed strategies for the Chino, California, area including massive treatment plants and electrical co-generation; (Morse, 1993.)
- "Wagon wheel" and "cow pool" designs in the United States that required extensive cooperation among the dairies and used centralized milking parlors; (Roy, 1963, at 244-263.)
- Isreali "Moshave" facilities that provide common waste treatment, feeding, milking, and security; (Patterson, 1993.)
- And, finally, an ambitious plan in Singapore to relocate that nation's pig industry into a single area with standardized barns and centralized waste treatment. The Singapore government developed a comprehensive plan for the financial success of the project while protecting the area's environmental resources. (Taiganides, 1992 at ix and 1993.)

However, these previously proposed or constructed projects are distinguishable from this proposal which emphasizes centralized environmental systems for collocated dairies while preserving the individual dairy operator's autonomy and independence. All of the component dairies at the park would be distinct, independently housed operations, responsible for maintaining individual herds according to each operator's management style. For a modern dairy park to operate successfully, the project must not only have strong financial backing and highly developed and engineered systems, but must also preserve an acceptable level of independence for the individual dairy operations. Each operator in the dairy park would continue to enjoy control over the herd, feeding, and milking.

Ownership and Management of the Park and Dairies

Various options exist for the development and ownership of the park itself and of the individual component dairy facilities. Each of these options affects the initial cost of the park, the range of services provided, the number of park staff required, and how the park would obtain the necessary operating permits. The park's owners would, at a minimum, need to procure sufficient land for the facilities and buffer zone as well as design, construct, and maintain the park's centralized environmental systems.

A private, investor-backed entity, such as a for-profit joint venture, partnership, or a corporation could build and operate the park. The entity would have a built-in profit motive to encourage efficiency, responsiveness, and a high quality of management. However, the need to return a profit to the investors would also mandate higher costs for services to be passed along to the component dairies. A private effort would require proof of sufficient capital, perhaps through a bond or other performance device, to assure regulators of the owner's long term commitment to the project.

Cooperative ownership of the park and facilities by the dairy components represents another ownership possibility. A cooperative arrangement, long a mainstay of agricultural operations, would provide the component dairies with direct management authority and would promote accountability and efficiency of operations. However, the park concept promises dairy operators a vehicle by which to escape the regulatory and waste disposal issues that the owner and manager of the park would have to deal with on a daily basis. Additionally, cooperative investment would require significant financial commitments which would limit the potential dairy participants to operators with substantial assets. As in a traditional dairy setting, the dairy operators might best reserve their finances and management efforts for the success of their individual dairy and herd.

Under either of the ownership scenarios, the developer of the park could be a party other than the eventual owner. For example, a public or quasi-public entity, such as an agricultural or industrial development authority, or even a chamber of commerce, could assist private parties interested in the project or provide the initial spark for private development. The public entity could advance the concept, recruit potential dairies, begin the planning and permitting, and search for an appropriate private entity to handle the park's actual financing, development, and operation. Such an arrangement might prove beneficial to a locality in which the citizenry desires to attract the dairy industry and associated businesses and professionals yet maintain environmental quality.

The day-to-day operation of the facility would be the responsibility of a professional manager and staff reporting to the owners. Existing agricultural investment and management companies with significant experience in large CAFO operations could operate the park for the owners. Hiring a manager would insulate the owners from day to day decisions and would ensure that the management possessed expertise in dairy operations and in waste handling. He or she would hire and coordinate a staff to oversee and maintain specific operational areas of the park. Depending on the park's ownership of, and level of responsibility to, the individual dairies under the various options described, the park staff would also maintain the individual dairy facilities.

Ownership and management options for the component dairies range from a fully developed park with pre-built, turn-key dairy facilities available for lease to a more basic park providing only the centralized environmental systems in which each dairy buys land and builds its own facilities. Any facility built by an individual dairy would have to utilize the park's environmental systems. The fully developed park would require a greater initial investment and higher operating and maintenance expenses which would be passed along to the dairies. Although at some sacrifice to the operators' independence, with all dairy facilities standardized and under park ownership and maintenance, the fully developed park could exercise greater control over the environmental compliance of each dairy.

At the other extreme, a more basic park, providing fewer services, would require a lower level of investment from the park. Because each operation would need to buy land and build its own facility, this option would require more management and investment input from each dairy. With fewer facilities and services provided by the park whose costs are passed on to the dairies, this option would allow a greater financial return to the dairies. Additionally, the sale of land to the dairies would generate income to the park at an early stage of operations. Under a variation of this option, the park could lease, rather than sell, land to an operator who would then build a dairy facility.

Other options, between these two extremes, also exist. For example, the park would sell undeveloped parcels to incoming dairies to build their own facilities, but, would also pre-build some complete dairy facilities for lease. These pre-built facilities might be attractive to beginning dairy operators or established operators wanting to expand to a second herd with a minimum of investment.

Depending on local regulation, the options in which the dairy operator leased land or facilities from the park might ease permitting considerations for the park. By preserving privity, or direct contractual links through the lease to the permit holder, the land owner, and the dairy operator, regulators could pursue enforcement actions against any individual dairy operator at the park. Otherwise, if an operator owned the dairy and the land, but the park held the permit, the regulator might be impeded in taking action against an individual operator. However, a regulatory body could consider the park as a point source and issue a permit to the park's waste treatment plant rather than permit the park as a dairy operation. Under this scenario, ownership of the individual dairies, their land, and privity to the permit would not matter. The regulator would take enforcement actions against the waste treatment plant, not the dairies.

Siting, Technologies, and Systems

Environmental and economic advantages of the free stall dairy design make it the best choice for a dairy park. The free stall's roof, concrete floors, and drainage system allow for implementation of an efficient, centralized waste management system. Although the free stall, as with a traditional open lot dairy, must retain all contaminated rainfall runoff, the area subject to manure contamination is very small compared to the open lot and the animal waste is collected and processed several times per day by automated systems. Apart from having a smaller area subject to adverse environmental impact, the free stall also allows the dairy operator greater control over the herds' living environment than a comparable open lot. These benefits allow for implementation of the park's centralized environmental systems.

Proper siting of the dairy park also represents an integral part of the park's pollution control measures. Proposed sites should provide geologic conditions that protect both surface water and groundwater. Additionally, the site's topography should promote efficient operation of the park's waste treatment system. A surrounding buffer zone, sized according to demographic, meteorological, and odor dispersion criteria, would assist in mediating odors and land use strategies would protect adjacent areas from nuisance sensitive development. These strategies could range from simple ownership of the surrounding land by the park to legally defined land use controls such as covenants, deed restrictions, and zoning. Additionally, a local governmental entity with authority to contract could indemnify the park against nuisance suits in exchange for taxes and the park's economic development potential.

The dairy park's location should be functional, with direct access to transportation, electrical service, fresh water, milk collection services, and dairy support industries. However, such a well supported location could also place the facility squarely in the public's eye. To receive and maintain the requisite permits without insurmountable public opposition, the park must not only operate efficiently, but must also work to dispel any public fears that the park is not operating in an environmentally sensitive manner. Landscaping and design should present the image of a carefully operated and maintained facility with environmental concerns taken seriously. A poorly maintained or designed exterior might cause the public to lose confidence in the dairy park's overall operational and environmental capabilities. (Azevedo, et al, 1974, at 37.)

Placing any dairy under centralized environmental systems would reduce the work load on the regulatory agencies. With more dairies under a blanket permit, the regulators would deal with fewer individual operations. Operations locating outside the park, in various locations around the countryside, would have to obtain their own permits, perhaps a multiple year process, and deal individually with regulatory officials. These traditional dairies would create new, individual, waste treatment systems managed by that dairy operator, rather than participating in a centralized environmental system managed by a waste treatment specialist. As opposed to centralized systems, monitoring of individual on-farm waste treatment plants is cumbersome, while operation can be a burden to the farm operator more concerned with herd management. (Taiganides, 1983, at 209.)

The park's environmental system must manage, control, and use the by-products of agricultural production in a manner that sustains or enhances the quality of air, water, soil, plant, and animal resources. (Peterson, et al, 1993, at 54.) The system, cost prohibitive for most individual dairies, would emphasize wise use of waste as a resource, and, employing recycling, water and soil conservation, composting, and reuse of materials, would provide solutions to all identified waste problems. The park's system approach should address all aspects of the waste, including waste production, collection, storage, treatment, transfer, and utilization. (Id., at 57-59.)

The concrete alleys of the park's free stall barns would collect waste from the herds. Flushing systems in each alley would transfer the waste to a central treatment plant which would include a pretreatment stage, such as a solids separation device. Further treatment would utilize biological, physical, and chemical means to reduce the waste to an inert, usable form. The treated liquid from the waste system would be disinfected to prevent the possibility of disease transfer between herds and would then be recycled to the various dairies as flush water. The park would compost all solids removed during the waste treatment process for use as bedding in the free stalls, animal feed, mulch, organic matter, or plant nutrients. Properly treated and certified as pathogen and weed free, these waste products could be marketable.

The park's system approach to waste treatment would include full utilization of recommended Best Management Practices in all phases of construction and operations along with a sophisticated environmental system. After mechanical solids separation, the park could rely on technologies adapted from municipal sewage treatment plants to treat the waste. Treatment designs such as an aerobic treatment basin, an aerobic ditch, and trickling filters, have all been successfully proven for large scale agricultural use. (Taiganides, 1992 and 1993.)

Additionally, other technologies show promise for large scale agricultural waste treatment. Ultrafiltration and reverse osmosis technologies employed after hydrocyclone pretreatment may offer a cost effective treatment alternative that would produce chemically pure water along with dry, compostable matter. (Clay, 1993.) Another alternative, although significantly more expensive, is large scale anaerobic digestion, as used in Hawaii and as proposed for a dairy waste treatment plant in Washington. (Mattocks, 1993.) However, because the digester requires a drier manure input than the other proposed systems, a digester based system might require additional labor, scrapers, and trucks to transport the waste as opposed to using an automated recirculating flush system.

The Permit

With centralized environmental systems, the dairy park would operate under "blanket" permits from all applicable regulatory agencies. The park's permits, covering a geographically defined area and specifying a maximum number of total cows, would authorize operation of all individual component dairies as long as those individual dairies abided by predetermined, technologically based practices. The public would benefit by having dairies incorporate environmental safeguards and locate in an area with minimal residential impact. For the park's dairy operators, the blanket permits would provide a measure of predictability enabling operators to focus their time and energies on milk production, rather than attempting to meet regulations on an individual basis.

Concurrent with the environmental and permitting benefits of the dairy park to component operators, the blanket permit would also benefit the regulatory agencies' enforcement efforts. With the park's management holding centralized responsibility for all of the component operators' environmental compliance, regulatory agencies would only have to deal with one entity knowledgeable about, and in control of, the environmental compliance aspects of several CAFOs.

The park would initially house a "start up" number of cows in free stall barns. The park would have the ability to expand and house a maximum number of cows according to conditions of the blanket permit as long as the park meets the permit's requirements. The permit would ideally specify the maximum number of cows for the entire facility, not the number or size of the component dairies or barns. The permit would not control how the park apportions specific numbers of animals among the component dairies. The size of the individual free stall components would be negotiated between the entering dairy operators and the park's management.

A master plan for the park, included within the blanket permit, would guide the allocation of the dairies and animals to ensure available space for mid-sized as well as large operations. Various sized "slots" available for incoming dairies would allow the park to accommodate the widest possible range of dairies. By reserving slots for smaller dairies, the park would preserve space for operators just entering the business or those expanding to a second herd. Even though large dairy operations are more cost effective and efficient than are smaller operations, (Matulich, 1978, at 642-647.) the park should encourage some smaller operations. These mid-sized dairies, with 500 to 700 cows, would not only provide good public relations to a large segment of the dairy community, but would also provide a means for the park to fine tune the total number of cows for the park's greatest operating efficiency.

Conclusion

The dairy park offers an outstanding model for areas of the world with CAFO related environmental problems, or, for areas recruiting agriculture but attempting to avoid such problems. The dairy park would form a prototype, through technology and thorough planning, of how modern agriculture may co-exist with the increasing pressures of urbanization. Various opportunities to study these issues could conceivably lead to funding assistance for the park from universities, foundations, and governmental entities.

Distinct from other CAFO operations, management and operators at the dairy park would have several powerful incentives to promote environmental compliance. At a very basic level, only dairy operators interested in the dairy park concept and in environmental compliance would want to participate in the project. Additionally, peer pressure among the component dairies, the importance of maintaining the blanket permit for the collective good, and pressure to succeed from the regulators, the public, and the investors would promote permit compliance. One large park, geared to environmental compliance and only able to remain in operation by consistent maintenance of high environmental standards, could compete with the level of compliance of individual operations housing far fewer numbers of animals.

The park would provide an agricultural-environmental showplace, and, with an on-site analytical laboratory to monitor and certify the waste treatment and compost quality, would offer an opportunity for continuing research into odor, buffer zones, and agricultural pollution control technologies and practices. Study areas could include water use, wastewater handling, solid waste, composting, and odor control measures. Other park research could include dairy nutrition studies with an emphasis on nutrient reduction in the manure.

References

1. Azevedo, J., and P.R. Stout, 1974, Farm Animal Manures: An Overview of Their Role in the Agricultural Environment, California Agricultural Experiment Extension Service, Manual 44.

2. Clay, Gaylen, 1993, Environmental Water Purification Co., Plano, Texas, personal communication, October 26, 1993.

3. Cortez, Sabino, 1993, AgKone, Incorporated, Stephenville, Texas, personal communication, October 6, 1993.

4. Jones, Philip H., 1977, "Criteria and Guidelines for the Selection of Animal Feedlot Sites," Animal Wastes, Taiganides, E. Paul, editor, Applied Science Publishers, London.

5. Mattocks, Mr. Rick, 1993, Unisyn Biowaste Technology, Seattle, Washington, personal communication, October 6, 1993, brochures, and undated newsclippings of his company's projects.

6. Matulich, Scott C., 1978, "Efficiencies in Large Scale Dairying: Incentives for Future Structural Change," American Journal of Agricultural Economics, 60:4.

7. Morse, Deanne, 1993, Animal Waste Specialist, University of California at Davis, personal communication, October 5, 1993.

8. Nexis, 1993, The Seattle Times, "Living in Holstein Heaven," July 16, 1992, Final Edition, at F1.

9. Patterson, Allan, 1993, Consulting Engineer, personal communication.

10. Peterson, Don C., and William H. Boyd, 1993, "The SCS Approach to Planing Animal Waste Management Systems for Total Resource Protection," Proceedings of the Integrated Resource Management for Landscape Modification for Environmental Protection Conference, American Society of Agricultural Engineers, Chicago, Illinois.

11. Roy, Ewell Paul, 1963, Contract Farming U.S.A., Interstate Printers and Publishers, Danville, Illinois. "Cow pools are now in operation in Florida, California, Michigan, Virginia, Iowa, Kansas, Ohio, Texas, and Missouri, among other states."

12. Taiganides, E. Paul, 1983, "Animal Waste Management and Recycling," <u>New Strategies for Improving Animal Production for Human Welfare</u>, Proceedings of the Fifth World Conference on Animal Production, Japanese Society of Zootechnical Science.

13. Taiganides, E. Paul, 1992, <u>Pig Waste Management and Recycling, The Singapore Experience</u>, International Development Research Center, Ottawa, Ontario.

14. Taiganides, E. Paul, 1993, personal communication, November 1, 1993.

15. Teoh, Soo See, E. Paul Taiganides, and Teow Chong Yap, 1988, "Engineering design parameters of Wastes from Pig Farms in Malaysia," <u>Biological Wastes</u> 24:95.

16. Wagner, 1993, The Honolulu <u>Star-Bulletin</u>, "Firm makes Food Waste a Cash Crop," June 21, 1993, supplied by Unisyn Biowaste Technology, Seattle, Washington.

DAIRY PROTECTS ENVIRONMENT THROUGH IMPLEMENTATION OF BEST MANAGEMENT PRACTICES

Elwyn O. Cooper[1]

ABSTRACT

BMP's selected by Gustafson Dairy were carefully integrated into a workable plan that would allow the dairy to function in harmony with the environment. The environmental concern was the protection of ground and surface water resources of Green Cove Springs, FL, located to the north and east, the St. Johns River to the east, and Governors Creek which flows through the property.

OVERVIEW OF PLAN SELECTION PROCESS

The Gustafson Dairy is the largest family owned and operated dairy in the U.S. The dairy is located in the southern part of Green Cove Springs, FL, about 1.6 km (1 mile) west of the St. Johns River. Governors Creek flows through the property adjacent to the areas receiving the nutrients from the dairy operation. The dairy itself lies in the NE corner of a 4,047 ha (10,000 acre) parcel of land owned by the dairy.

The objective of Gustafson Dairy was to install an environmentally sound waste management system that would allow the milking of 4,000 head of cows and a dry cow herd of 1,500.

In 1990, Gustafson Dairy contacted the Soil Conservation Service (SCS) for technical assistance in the planning of a waste management system. SCS offered the following planning and design approach:

1) Waste system selection must be done by team approach with the regulatory agencies as a team member.
2) System must meet SCS standards and specifications.
3) System selected must be completely acceptable to the owner and solve the resource concern.

After several conferences between Gustafson Dairy and SCS relative to different waste management alternatives, the Best Management Practices (BMP's) were selected. The BMP's were integrated into a waste management plan that would meet the objectives listed above. Once a preliminary plan was developed, the Florida Department of Environmental Protection (FDEP) was given the opportunity to review it on site and provide input. This step greatly facilitated the permitting process that followed.

BEST MANAGEMENT PRACTICES SELECTED

The best management practices BMP's selected consisted of the following:

---

[1]Project Engineer for the USDA Soil Conservation Service, Gainesville, Florida.

Dike
Fencing
Irrigation System, Sprinkler
Irrigation Water Conveyance Pipeline, High Pressure
Pipeline
Pumping Plant for Water Control
Structure for Water Control
Surface Drainage Field Ditch
Surface Drainage Main or Lateral
Waste Storage Pond

DESCRIPTION OF WASTE MANAGEMENT SYSTEM SELECTED

Figure No. 1 shows the layout of the waste management system selected.

The selected alternative required confining the cows to an area adjacent to the milking facility. This was necessary to minimize the walking distance for the cows and minimize cow stress. Using pastures would have required drainage improvements and the cows to walk long distances. The high intensity area (HIA) on which the cows are confined consists of 80 ha (198 acres). The area is broken up into two different HIA's with HIA No. 1 for the milking herd and HIA No. 2 for the dry cow herd. The HIA's are located to the east and south of the dairy with a 9 m (30 foot) wide access road on top of a dike that isolates the area from non-cow use areas. The dike is designed to contain within the HIA's the runoff from the 25 year 24 hour storm event from the dairy complex. Parallel to the dike and on the HIA side of the dike is a perimeter ditch discussed below. The HIA's are fenced to provide an exercise and loafing area for each herd.

The dry cows were re-located from an area about 1 mile away to HIA No. 2. The dry cows were given their own feed barn and calving barn. The feed barn is located on one side of the HIA but adjacent to an elevated and fenced access lane that is used to convey the dry cows to and from the dairy milking herd. The calving barn is located on the access lane so that all the dry cows have to pass through it as they either enter the dry cow HIA or leave it immediately following calving.

Field ditches are planned within the HIA's to provide drainage. The field ditches outlet into the perimeter ditch via pipe overfall structures. The field ditches are designed for a removal rate of 51 mm (2 in) in 24 hours.

The perimeter ditch was constructed 1.2 m (4 feet) deep, 3 m (10 feet) bottom width, with 3:1 sideslopes on the dike side and 4:1 on the HIA side. The storm runoff from the HIA's enters the perimeter ditch and flows from each end to a central location where an earthen sump is located. The flow enters the sump by means of aluminum water control structures.

Solids are removed from the dairy waste from the milking and feeding complex prior to its entering the waste storage pond. Solids removal takes place in two ditches which alternately convey the waste water to the sump. The ditches are designed with a skimmer structure at the outlet to allow the solids to drop out. When the ditch becomes full of solids, they are removed and hauled away for processing into commercial

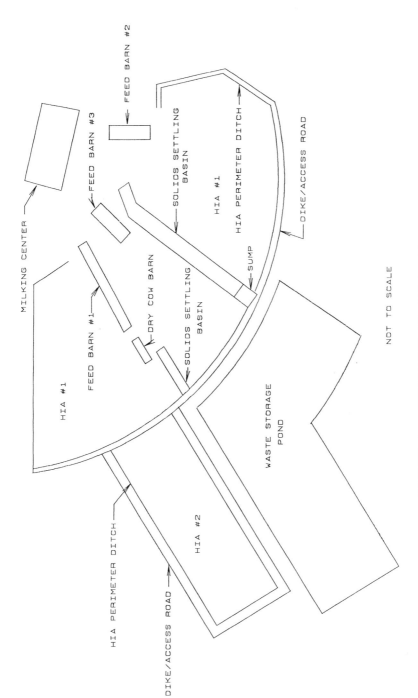

FIGURE 1. SCHEMATIC OF BMP's AT DAIRY COMPLEX.

fertilizer. A structure at the upper end of the ditches allow the operator the ability to control which ditch receives the waste water while providing time for solids to dry in the adjacent ditch prior to removal. The ditches are about 519 m (1,700 feet) in length, 6.1 m (20 foot) bottom width, 2:1 sideslopes, and with a 6.1 m (20 feet) wide access road between the ditches as well as on the HIA side of each.

The earthen sump which collects all runoff from the dairy complex and HIA was lined with concrete riprap with a geotextile placed under the riprap. The geotextile prevents seepage failures to the sideslopes of the sump. Two axial flow pumps with electric motors that are float switch controlled provide an outlet from the sump to the waste storage pond. One pump is capable of 15,142 L/min (4,000 gpm) and is set to operate under normal inflow conditions. The other pump is capable of 44,955 L/min (11,876 gpm) and is set to operate in conjunction with the smaller pump during and immediately following heavy rainfall events. This provides a total outlet capacity of 60,097 L/min (15,876 gpm) or 102 mm (4 in) in 24 hours from the HIA. Both pumps discharge into a waste storage pond.

The waste storage pond is 34 ha (90 acres) in surface area with a volume adequate to store 30 days inflow. The embankment is constructed of material taken from a borrow area within the reservoir and parallel to the embankment. The emergency overflow structure was installed through the embankment. The structure is sized to discharge the pump inflow of 60,097 L/min (35.3 cfs) at 0.15 m (0.5 feet) of head. To increase the potential for wildlife (ducks and wading birds) five islands were constructed within the reservoir. These 0.04 ha (0.1 acre) islands provide a nesting area for the ducks as well as roosting areas.

The nutrient laden wastewater in the waste storage pond is used as irrigation/fertilizer for cropland and for recycling to flush the feed barn floors. Two solids handling pumps are located at the NW corner of the waste storage pond. One is the irrigation pump capable of 155 L/s (2,460 gpm) for distributing the wastewater over approximately 259 ha (640 acres) via centerpivots and traveling volume guns. The other pump, an existing 63 L/s (1,000 gpm) pump, was relocated for recycling the wastewater. Recycling involves the refilling of the 37,854 L (10,000 gallon) flush tanks for the feed barns and for discharge onto the one non-flushable feed barn floor during the scraping operation.

The irrigated area of approximately 259 ha (640 acres) required about 17,700 m (11 miles) of pvc pipeline which ranged in diameter from 152 mm to 305 mm (6 inch to 12 inch). Four booster pumps were required at key locations within the pipeline network to maintain adequate pressure for the distribution system.

The old existing lagoon system will be cleaned out and used for non-waste water purposes.

WASTE SYSTEM UPDATE

Installation of the waste management system is approximately 98% complete. The dairy has decided to modify their operation to include total confinement for the milking herd. The modification will also include converting the HIA's to pastures for the dry cows. These modifications are being implemented to reduce cow stress which increases milk production, reduce cow health problems associated with HIA's, and reduce maintenance cost of the HIA's.

Seven (7) monitoring wells have been installed to provide data as to how the system is functioning. The following parameters are being monitored quarterly:

        Total Nitrogen   (as N)
        Nitrate Nitrogen (as N)
        Total Phosphorus (as P)
        Ortho Phosphorus (as P)

Initial data indicates the system is addressing the environmental concerns. As additional data is obtained, it will be analyzed and the system operation modified as needed.

The United States Department of Agriculture (USDA) prohibits discrimination in its programs on the basis of race, color, national origin, sex, religion, age, disability, political beliefs and marital or familial status. (Not all prohibited bases apply to all programs). Persons with disabilities who require alternative means for communication of program information (Braille, large print, audio tape, etc.) should contact the USDA Office of Communications at (202) 720-5881 (voice) or (202) 720-7808 (TDD). To file a complaint, write the Secretary of Agriculture, U.S. Department of Agriculture, Washington, D.C. 20250, or call (202) 720-7327 (voice) or (202) 720-1127 (TDD). USDA is an equal employment opportunity employer.

DESIGN OF A ROTATIONALLY GRAZED DAIRY IN NORTH FLORIDA

M.P. Holloway, A.B. Bottcher, Ron St.John[*]

ABSTRACT

This paper describes the concept and recent studies of Intensive Rotational Grazing (IRG) for use on dairies. Detailed design criteria and general layouts are provided for a IRG dairy to be constructed in North Florida. The dairy will use a 14 day rotation system with paddocks located under center pivots to provide irrigation water, cooling water, shade for the cows, and a means of fertigation. The cows will spend approximately 85% of their time in the paddocks and, for the majority of the year, will receive supplemental feed only while being milked.

Keywords: Intensive Rotational Grazing, Dairy, Pasture, Florida

INTRODUCTION

The latest trend in the Florida dairy farm has been to move the cows off pastures into concrete floored cooling or confinement barns. These structures provide shade and cooling for the cows, resulting in an increase in milk production, as well as to facilitate the collection of manure for an environmentally sound waste management system. However, these buildings require a large capital investment, and have the possibility of increased odors due to the short holding times of the manure in waste storage ponds and the large sprayfields associated with the spreading of the collected manure. Since most of the feed is supplied to the cows via silage and other imported commodities, these dairies also have large open feed areas that provide fly breeding sites and additional sources for odor. Some of the large diaries recently constructed in North Florida have received numerous complaints from neighbors regarding flies and odor, and several counties in North Florida have adopted special zoning ordinances (in addition to required state regulations) for new dairies.

For many years, the goal of agriculture has been to maximize production. Recently the surge has been to maximize profits, which does not necessarily correspond to maximized production. This is true for the dairy industry in a form of management called Intensive Rotational Grazing (IRG). This concept takes the cows out of the expensive cooling barns and puts them back on the pasture in a rotation system that allows for maximum forage production and nutrient intake of the pasture grasses. Consequently, this system eliminates the large sprayfields and open feed storage areas, thus eliminating the major fly and odor sources found on the large confinement type dairies. The only manure that must be handled is generated in the milking barn and can be treated in an economically feasible anaerobic lagoon and applied to a small sprayfield. An IRG dairy may not produce the amount of milk that could be produced using cooling or confinement barns, but it has a much lower capital and operational costs. Although there have not been any published economic studies done at the time of writing this paper, initial data from established IRG dairies in Texas seems to indicate that an IRG may be much more profitable than the confinement type dairies. This paper will explain the design concepts for a proposed IRG dairy in North Florida.

---

[*]M.P. Holloway and A.B. Bottcher are vice president and president, respectively, of Soil and Water Engineering Technology, 3448 N.W. 12th Ave., Gainesville, FL 32605 (904) 378-7372. Ron St.John is Managing Partner of Alliance Dairies.

LITERATURE SEARCH

Forms of rotational grazing have long been practiced in the United States, but true IRG has not been widespread. In normal rotation scenarios, each individual field or paddock is large, animal densities in the paddock are low, and the length of time that animals stay in the paddock is long (Murphey et al., 1986)(usually months at a time). The benefits of rotational grazing are not realized under this type of system because desirable plants are grazed continuously for a large period of time, giving undesirable plants a competitive advantage, the desirable vegetation is not harvested at peak periods of production, and rest periods are usually not at appropriate times to allow the desirable vegetation a chance to re-establish. For these reasons, traditional rotational grazing (with extended grazing periods) does not offer many advantages over continuous grazing, and should be placed in the same category. Intensive rotational grazing (also refereed to as Voisin grazing management)(Voisin, 1959) has two main rules that overcome the problems of continuous or traditional rotational grazing. First, the rest period (between grazings) of the paddock should vary with the climatic conditions and the growth rate of the pasture vegetation (Murphy et al., 1986) and should be based on producing an optimum crop height with a balance of protein and roughage. Second, the grazing period of each paddock should be less than 6 days for normal animals and less than 2 days for lactating or fattening animals (Murphy et al., 1986).

Mayne et al., (1987) conducted studies in the United Kingdom comparing vegetative uptake (calculated by measuring vegetative height after a grazing cycle in the paddock) to milk production. They determined that the critical vegetative height below which milk production per cow declines may vary with the production potential of the animal, but in general, the higher the residual ward heights (maximum of 3 inches) the higher the milk production. All of their experiments were conducted on the same type of vegetation, but it is obvious that vegetative height for maximum production also varies tremendously with the type of vegetation. Farmers in North Florida commonly report producing over 7 tons/ac (dry matter) from non-irrigated bermudagrass fields. Limited research from South Florida showed that two week grazing periods result in a higher protein content of the grazed grass, but significantly reduced dry matter yields (Adjei, 1989). The increased protein content demonstrates that large amounts of nitrogen are still being used by the grazed grass. However, the decrease in dry matter yields indicates the need for irrigation and intensive management of the pasture grasses on an IRG dairy.

Mayne et al., also suggested that time of calving plays an important role in maximizing milk production from grazing cows due to the growing seasons of the vegetation. Shoemaker et al., (1992) conducted studies using a variety of breeding technologies to determine if seasonal grazing was feasible in Ohio. Although it does not appear that seasonal grazing is cost effective, they were able to establish and maintain cows that calved in early spring to effectively utilize the maximum pasture production during the summer.

Intensive Rotational Grazing also appears to produce better quality milk and has no negative impact on udder health. Goldberg et al., (1992) found that mean standard plate counts from monthly bulk tank samples taken for one year on Vermont dairy farms were lower on IRG dairies during the grazing season than on continuous grazing or confinement dairies. Mean bulk tank counts of streptococci, other than *Steptococcus agalactiae*, during the growing season were also lowest for the IRG dairies.

It appears that little research has been done on the environmental impacts of IRG dairies. No published scientific reports were found on odor, fly control, groundwater quality, or surface water quality associated with these type of dairies.

NORTH FLORIDA DAIRY DESIGN

This section contains detailed design considerations for a 1400 cow facility that is to be constructed in Gilchrist County Florida (40 miles west of Gainesville, Florida). The dairy facilities will occupy slightly over 500 acres of a proposed 1925 acre site. The additional acreage of the permitted site is forested and will serve as an extensive buffer around the dairy facilities. The site contains two center pivots, each about 200 acres in size, that have been used to grow commercial crops such as corn and cotton. Pasture will be established under these pivots and will serve as the IRG pastures for the 1200 head of lactating animals (the milking herd).

Basic Concept

Figure 1 shows the overall farm layout and an expanded view of the paddocks under a pivot. These paddocks are based on a 14 day rotation, with two herds of cows (herd A and herd B) under each pivot. The cows will spend approximately 4 hours a day either in the milk barn, or walking to and from the milk barn. The cows will be milked twice a day and will be switched to a new paddock after every morning milking. In order to maximize the rotational effect, the cows will only be let into the front half of the pasture after the morning milking. An electric tape will be used to separate the front and rear parts of the paddock and will be dropped after the afternoon milking to allow the cows access to the full paddock. This results in providing the cows with fresh grass after each milking. This daily cycle results in the cows spending about 20 hours in each paddock, and since there are 14 paddocks, each paddock will rest for 13 days between grazings. This scenario could very easily be switched to a 28 day cycle during periods of low growth, or for other reasons, by simply keeping the cows in a paddock for four milkings, or 40 hours.

Each paddock is about six to seven acres in size and will contain a watering trough centered in the paddock. Supplemental feed will be supplied to the cows in the paddocks only during periods of minimal vegetative growth, and when supplied will be provided on portable feed wagons. The majority of the animals feed will be provided by the highly nutritious pasture grasses. These grasses are expected to contain large amounts of protein due to the optimal growth stage that will be maintained by the 14 day cycle. Additional feed will be provided to the cows in the milk barn while the cows are being milked to assure a complete feed ration. The paddocks are placed under the center pivots for a number of reasons. It is vital that the pasture grass maintain an optimal growth rate, therefore requiring supplemental irrigation at times. The pivot provides an easy means of irrigation, as well as fertilization when required. It is anticipated that the pivots will also serve as a structural support for a mister line and possibly shade cloth. It should be noted from Fig. 1 that the paddocks are laid out in such a fashion that both herd A and B can be under a pivot at the same time. The misting and shade may be very crucial during the hot summer months in an attempt to keep milk production as high as possible.

Except for the short section near the milk barn, the travel lanes will be crowned dirt and grass paths that will be scraped when necessary. The short section of lanes between the edge of the pivots and the milk barn will be concrete. The travel lanes within the pivot circles (see Fig. 1) are positioned so that they receive a maximum of 5 days of traffic during the 14 day rotation. Experience on other diaries has indicated that a stabilizing fabric may need to be installed on portions of the lane to maintain the integrity of the crown and provide adequate drainage, but with proper maintenance the lanes can provide reliable alleys of transportation for the cows.

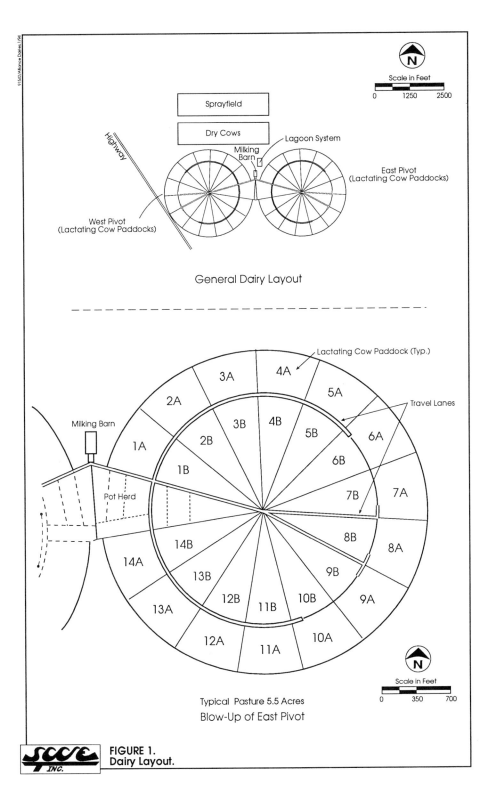

FIGURE 1.
Dairy Layout.

As shown in Fig. 1, the milk barn will be constructed between the two pivots to minimize the travel distance for the cows. As previously mentioned, the barn will be specially constructed to allow the cows to feed on a purchased concentrated blend while being milked. The barn will be flushed after each herd is milked, and the effluent will travel by gravity to a two cell lined lagoon system.

The section of the pivot closest to the milk barn is reserved for the pot herd (sick cows, cows that are about to calve, and cows that have just calved). As shown in Fig. 1, this area is subdivided into smaller paddocks to allow rotation of this small herd also. The dry cows will be confined on about 70 acres of pasture that will be irrigated with a traveling gun and divided into paddocks so that it is also rotationally grazed. The sprayfield will be located just north of the dry cow pasture and will consist of planted pines irrigated with the same traveling gun system. Although the dry cow pastures will normally be irrigated with fresh water, it the portion without any cows can be used as an alternative sprayfield. Heifers and calves will be raised offsite.

Nutrient Balance

A detailed average daily nitrogen balance for the entire dairy is presented in Fig. 2. As with any nutrient cycle, the numbers presented in Fig. 2 are approximate and are expected to vary considerably throughout the year. It should be noted that all values are given in total pounds of nitrogen per day, not per acre. Phosphorous retention on the soils present onsite is expected to be very high, and groundwater samples on other diaries in the area indicated that phosphorous contamination of groundwater is not a concern. Nitrogen production estimates are based on values form ASAE (1989) and personal experience on other dairies in the area. Loses are based on a variety of sources and experience, including Moore (1989). Plant uptake rates are based on the Institute of Food and Agricultural Science, University of Florida (IFAS) recommendations.

It should be noted that a large amount of nutrient needs for the rotational pastures will be supplied by commercial fertilizer. The reason for such a heavy reliance on the commercial fertilizer is to allow a great deal of flexibility within the nitrogen budget. Due to the ease of spreading fertilizer through the center pivot, it is possible to give small boosts of nitrogen during peak growing times, while not overloading the system during periods of slow growth. Continuous monitoring of crop uptake and nutrient application will provide a gauge for commercial nitrogen fertilization, and an internal monitoring well will verify that excessive fertilization has not occurred.

Benefits of Intensive Rotational Grazing

As well as the expected economic benefit, several other aesthetic benefits are expected on this dairy. First, only a high concentration feed will be imported to the dairy for most of the year and will be stored in closed silos. Therefore the only exposed feed on the dairy will be in the feed troughs located in the milk barn. This should significantly lower the fly problems associated with most dairies. It also eliminates the large silage pits which can be both a source of odor and flies.

Odor will also be decreased by the fact that a majority of the manure will be spread by the cows on the grassed paddocks. This portion of the manure (85%) should have minimal odor due to the aerobic conditions in the open fields. Because there is such a small percentage of manure to be handled through a treatment system, it is economically affordable to construct a true anaerobic lagoon for the waste, and in this particular diary, a two stage lined anaerobic/aerated lagoon system will be installed.

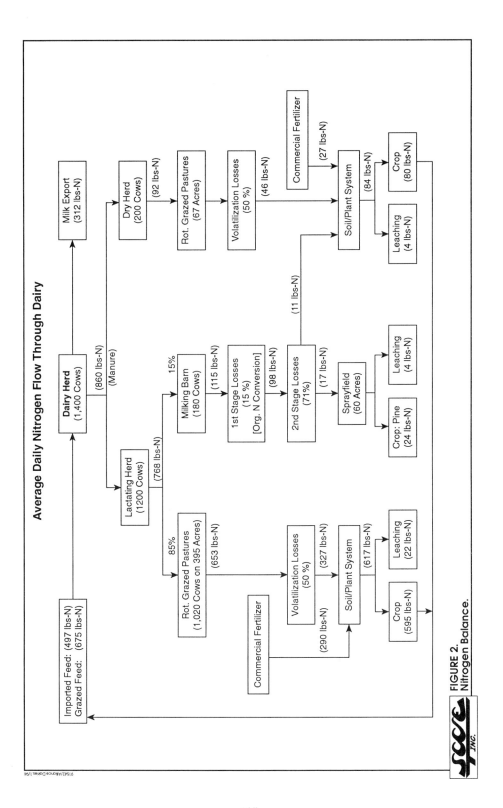

FIGURE 2. Nitrogen Balance.

Because the paddocks are separated using only single strand electric wire, the visual appearance of the dairy will be extremely appealing. It will give the impression of cows grazing on a large lush pasture when viewed from a distance. On this particular diary, the milk barn will be barely visible from the road, and the dry cows, sprayfield, and lagoon system will not be visible form the highway.

CONCLUSION

Due to wide spread citizen opposition against large confinement type diaries, as well as the expected economic benefit of an IRG dairy, an IRG dairy seems to be a very attractive alternative for a new dairy for the North Florida area. These type of dairies consist of considerably less capital investment, and appear to be very profitable in other areas of the country. It is estimated that local opposition to a confinement type dairy at the proposed location would have been so great that it would not have been granted a required special exemption from the county, however it appears that this IRG plan will pass and the dairy will be under construction by the time this paper is presented.

REFERENCES

Adjei, M.B., P. Mislevy, R.S. Kalmbacher, and P. Busey. 1989. "Production, Quality, and Persistence of Tropical Grasses as Influenced by Grazing Frequency". Soil and Crop Science Society of Florida Proceedings. Volume 18.

ASAE. 1989. American Society of Agricultural Engineers STANDARDS. R.H. Hahn and E.E. Rosentreter, Editors. pp 437-438.

Goldberg, J.J., E.E. Wildman, J.W. Pankey, J.R. Kunkel, D.B. Howard, and B.M. Murphy. 1992. "The influence of Intensively Managed Rotational Grazing, Traditional Continuous Grazing, and Confinement housing on Bulk Tank Milk Quality and Udder Health". Journal of Dairy Science. 75:1. pp 96-104.

Mayne, C.S., R.D. Newberry, S.C.F. Woodcock, and R. J. Wilkins. 1987. "Effect of Grazing Severity on Grass Utilization and Milk Production of Rotationally Grazed Dairy Cows". Crop and Forage Science. 42:1 pp 59-72.

Moore, J.A., and M.J. Gamroth. 1989, "Calculating the Fertilizer Value of Manure from Livestock Operations". Oregon State University Extension Service, Corvallis, Oregon.

Murphy, W.M., J.R. Rice, and D.T. Dugdale. 1986. "Dairy Farm Feeding and Income Effects using Voisin Grazing Management of Permanent Pastures". American Journal of Alternative Agriculture. volume I, number 4. pp 147-152.

Shoemaker, D.E., S.R. Shoemaker, and D.L. Zartman. 1992. "Seasonal Milking with Intensive Grazing: The Ohio Experience". Presented at the American Dairy Science Association 1992 Annual Meeting, Ohio State University, Columbus Ohio.

Voisin, A. 1959. Grass Productivity. Philosophical Library Inc., New York. 353 p.

DAIRY LOAFING AREAS AS SOURCES OF NITRATE IN WELLS

D. E. Radcliffe, D. E. Brune, D. J. Drommerhausen, and H. D. Gunther[*]

ABSTRACT

Nine dairies in north Georgia were surveyed using a ground electromagnetic (EM) conductivity meter to determine the source of high nitrate in dairy wells in the region. Ground EM conductivity increases with water and soluble salt content and has been used to map contaminant plumes from landfills and animal waste lagoons. The pattern was the same on all nine dairies so this report focuses on one dairy where EM readings were supplemented with soil samples and groundwater observation wells. There was little evidence of seepage from the lagoon, but EM conductivity was highest in the unpaved loafing area adjacent to the milking barn. Soil samples were taken to a depth of 2.1 m in 0.3 m intervals at 45 locations in the loafing area. The distribution of mean soil profile nitrate nitrogen (NO_3-N) content followed a pattern similar to the EM readings. The highest nitrate contents were observed at 0.3-0.6 m (maximum of 215 µg g^{-1}) but high readings extended to the deepest depth sampled (maximum of 80 µg g^{-1} at 1.8-2.1 m). Groundwater observation wells were installed at five locations: one next to the dairy barn, three in the loafing area, and one beyond the loafing area. Nitrate contents in the three wells in the loafing area in January, 1994 were 96, 105, and 120 mg L^{-1} NO_3-N. Concentrations in the well next to the barn and beyond the loafing area were 36 and 14 mg L^{-1}, respectively. We conclude that the loafing areas at this dairy and the other dairies we surveyed are a source of groundwater contamination by nitrate because of the high animal density and consequent high waste deposition rate. **KEYWORDS.** Nitrate, Animal Waste, Dairy, Electromagnetic Conductivity.

INTRODUCTION

Several studies have found evidence of subsurface losses of nitrate from dairies and confined cattle operations. In Florida, the Department of Environmental Regulation installed monitoring wells on nine dairies and of the 34 wells installed, 16 had nitrate concentrations above 10 mg L^{-1} NO_3-N (Darling, 1992). In a Wisconsin study, 48% of the diary wells sampled had nitrate concentrations above what was considered background level (Goodrich et al., 1991). Goodman (1985) reported that livestock confinement areas were responsible for much of the nitrate contamination in an aquifer in South Dakota. Manure from barnyards and corrals in northeastern Nebraska was the source of most of the nitrate in domestic and stock wells (Exner et al., 1985). Barry et al. (1993) developed a nitrogen budget for a dairy farm in Ontario and found a surplus that resulted in an estimated groundwater recharge concentration of 58 mg L^{-1} NO_3-N.

Ground EM conductivity surveys have been used to detect contaminant plumes from landfills (Jansen et al., 1992), high salinity areas on western farms (Hendrickx et al., 1993), and swine lagoons (Brune and Doolittle, 1990). Ground conductivity is a function of the soluble salt content, water content, and dielectric properties of the soil and parent material. If water contents and parent material are reasonably uniform at a site, then the EM readings can be interpreted as a

---

[*]D. E. Radcliffe, Associate Professor, Crop and Soil Sciences Department, University of Georgia, Athens, GA; D. E. Brune, Professor, and H. D. Gunther, Graduate Assistant, Agricultural and Biological Engineering Department, Clemson University, Clemson, SC; and D. J. Drommerhausen, Graduate Assistant, Geology Department, University of Georgia, Athens, GA. This research was supported in part by a grant from the Georgia Water Resources Program and a Cooperative Agreement with the USDA Soil Conservation Service Area 2 Office, Watkinsville, GA.

reflection of changes in soluble salt content. In assessing swine lagoon for seepage using EM surveys, Brune and Doolittle (1990) found that the dominate anions detected in the contaminant plumes were nitrate, chloride, sulfate, and carbonate

Approximately 150 of Georgia's 600 dairies are located in the five county area that comprises the Little River / Rooty Creek Hydrologic Unit in north Georgia. The University of Georgia Cooperative Extension Service in cooperation with the USDA Soil Conservation Service sampled 138 rural wells within the Hydrologic Unit from June 1991 to October, 1992. Thirteen percent of the wells sampled had NO_3-N concentrations above 10 mg L^{-1}. These wells had concentrations that ranged from 10.5 to 54.7 mg L^{-1} NO_3-N. Among the wells that were located on farms (96), 17% of the wells had above 10 mg L^{-1} compared to 5% for the nonfarm wells (Gould, 1993).

In the summer of 1993 we ran EM surveys on nine dairies in the Little River / Root Creek Hydrologic Unit to try and determine the source of nitrate in wells in the region. The pattern of EM readings was the same on all nine dairies so this report focuses on one dairy where EM measurements were supplemented with soil samples and groundwater observation wells.

MATERIALS AND METHODS

We used an EM 34-L3 conductivity meter (Geonics Ltd., Mississisauga, Ontario) which generates a ground EM field with a transmitter coil held at the soil surface. A second receiver coil 10 m away from the transmitter detected the strength of the EM field and allowed a calculation of the effective ground electrical conductivity. Readings were taken in a horizontal and vertical dipole orientation which measured shallow (field strength greatest at 0-2 m) and deep (field strength greatest at 4.5 m depth) conductivity, respectively.

The dairy described in this report (designated MO-1 in our survey) was located in Morgan County. The unconfined operation consists of about 45 milking cows with waste from the milking barn draining through a pipe to a lagoon (Fig. 1). The soils at the site are Cecil series and the parent material is granite gneiss, granite, and schist (Soil Survey Staff, 1965). The 0.8 ha field between the lagoon and the milking barn had been used as the loafing area from approximately 1952 to 1980 and is currently used as a pasture and feeding area. The term "loafing area" is used for the unpaved field, usually at the entrance to the milking barn, where milking cows are kept when they are not in the barn or pasture. Cows remain in these areas for 12-14 h per day and there is usually very little vegetation present, especially in the winter months. The field that is currently being

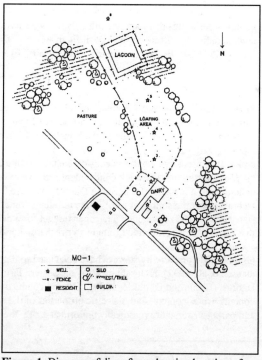

Figure 1. Diagram of diary farm showing location of EM readings (dots) and wells (stars).

used as a loafing area at MO-1 is located beyond the trees to the southwest of the old loafing area (upper right corner of Fig. 1).

We took EM readings in a transect around the lagoon, in four transects in the loafing area that separated the lagoon and dairy barn, in a transect radiating from the dairy barn SE into a pasture with a side spur into an area where round bales had been fed, and in a transect located just N of the dairy barn and running NE-SW. In August, 1993, we returned to the site and took soil samples in 0.3 m intervals to a depth of 2.1 m at 45 locations along the four transects in the loafing area where EM readings were made. We also took soil samples at 15 locations along the transect that ran SE into a pasture (Fig. 1). The soil samples were refrigerated immediately and later ground and sieved and extracted with 2M KCl. Soil nitrate was measured using an autoanalyzer (Lachat Instruments, Milwaukee, WI). In November, 1993, we installed five groundwater observation wells in a rough transect starting near the existing supply well beside the barn (well #1) and running out through the loafing area to the pasture beyond the lagoon (Fig. 1). The wells were installed to a depth approximately 1.5 m below the water table with a 3 m section of screen installed to extend 1.5 m above and below the water table. A sand pack was poured to 0.6 m above the screen followed by a 0.6 m bentonite seal and grouting to the surface. A manhole cover and cement pad was added at the surface. The wells were developed by pumping for about 2 h with a purging pump. Samples were taken on 12/7/93 and 1/4/94 with a bladder pump.

RESULTS AND DISCUSSION

The shallow EM readings did not indicate any evidence of a contaminant plume from the lagoon. Assuming that groundwater flow followed the topography, as is usually the case in the Piedmont region (LeGrand, 1988), a plume should have resulted in the highest readings to the SW of the lagoon (Fig. 2). Instead the highest readings occurred in the loafing area. Almost all the readings within the loafing area were above 11 mS m^{-1} and the highest readings (16 mS m^{-1}) were observed very close to the point where cows entered and exited the barn. The transect that ran out through the pasture to the SE of the barn showed that EM readings decreased to the 5-6 mS m^{-1} normally observed as background in this region (Brune et al., 1992) as we moved away from the barn. There were high readings (13 mS m^{-1}) in the area where round bales had been fed. The transect to the N of the barn, which was up-slope of the dairy (Fig. 1), resulted in near background levels, indicating that there was no up-gradient source of soluble salts

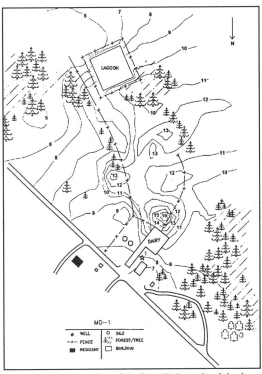

Figure 2. Contour plot of shallow EM conductivity in mS m^{-1}.

(Fig. 2).

The deep EM readings (Fig. 3) showed a pattern that was similar to that observed in the shallow readings in that there was no evidence of a plume to the SW (down-gradient) of the lagoon and the highest readings were observed in the loafing area where electrical conductivity was generally above 10 mS m^{-1}. Unlike the shallow EM readings (Fig. 2), the highest readings (13 and 14 mS m^{-1}) occurred in the middle and southern end of the loafing area away and downslope from the barn. There was another area of high deep EM readings (13 mS m^{-1}) just north of the two silos next to the dairy barn (Fig. 3).

Soil NO$_3$-N contents in the loafing area, averaged over the 0 - 2.1 m sampling depths, are shown in Fig. 4. Like the shallow EM readings (Fig. 2), the highest soil nitrate contents (44 to 106 µg g^{-1}) occurred close to the barn. Most of the sample locations within the loafing area had a mean content above 10 µg g^{-1}. There were some high readings (~50 µg g^{-1}) at the southern end of the loafing area near the lagoon (Fig. 4) that did not correspond with the shallow EM readings (Fig. 2).

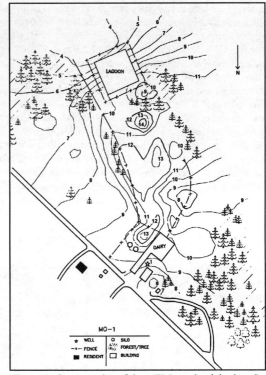

Figure 3. Contour plot of deep EM conductivity in mS m^{-1}.

In Fig. 5, the average NO$_3$-N content distribution with depth of the samples from the loafing area is shown. Mean concentration decreased from the surface down to the 0.6 - 0.9 m sampling interval and then increased to the deepest interval (1.8 - 2.1 m). Nitrates were most variable (as indicated by the standard deviations) at the 0.3 - 0.6 m depth interval and this was the depth where the highest reading (217 µg g^{-1}) was observed. It is clear from this graph that nitrate had moved below the root zone (0 to approximately 1.5 m) in this soil.

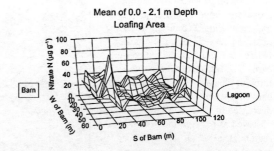

Figure 4. 3D plot of mean soil nitrate N for the 0.0 - 2.1 m sampling depth within the loafing area as a function of distance west and south of the milking barn.

Soil NO$_3$-N contents at each sampling depth as a function of distance along the transect through the pasture SE of the barn (Fig.1) are shown in Fig. 6. Contents at all depths decreased below 30 µg g^{-1} at a distance of about 120 m from the barn. There was an area with high readings below a depth of 1.2 m between 60 and 100 m from the barn, clearly indicating that nitrate was moving below the rootzone. This pattern followed that observed in the shallow EM readings in this area which started at 7-8 mS m^{-1} near the barn, increased to 9-10 mS m^{-1}, and then decreased to background levels.

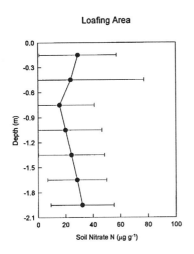

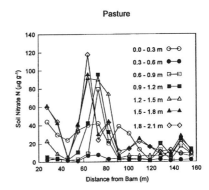

Figure 6. Soil nitrate N as a function of distance from the milking barn in the pasture to the southeast for various soil depth increments.

Figure 5. Mean soil nitrate N as a function of depth within the loafing area. Error bars represent ± one standard deviation.

Table 1. Water table depths, elevations, and nitrate concentrations from wells.

Well Number	Type of Well	Water Table Depth 1/4/94	Water Table Relative Elevation 1/4/94[a]	NO$_3$-N Concentration	
				12/7/93	1/4/94
		(m)	(m)	(mg L^{-1})	
1	Supply	-6.32	-6.32	7	7
2	Observation	-4.82	-6.28	12	36
3	Observation	-3.79	-6.82	95	96
4	Observation	-2.72	-7.17	93	105
5	Observation	-3.48	-7.60	97	120
6	Observation	-8.43	-9.97	15	14

[a]Elevation relative to top of well # 1.

Nitrate concentrations in the supply well at the barn and in the observation wells installed along a rough transect starting near the barn and going out through and beyond the loafing area (Fig. 1) are shown in Table 1 for two sampling dates. Concentrations were much higher in the three wells located in the loafing area than in the other wells on both dates. The relative water table elevations measured on 1/4/94 indicated that, along the transect, the component of groundwater flow was away from the barn toward the loafing area and lagoon, as we expected. This indicates that the high readings in the loafing area could not have been due to seepage from the lagoon. The slightly lower elevation at the supply well was probably caused by the frequent pumping that occurs with this well. The nitrate concentration in the supply well was above 3 mg $L^{-1}NO_3$-N, the level considered to be background (Madison and Brunett, 1985), indicating that some contamination had occurred. Nitrate concentrations in wells 2-5 increased between sampling dates which may indicate that nitrates were flushed from the soil profile with the onset of winter rains.

These results (and similar patterns of EM readings from the other 8 dairies we surveyed) have convinced us that the loafing area is a source of groundwater nitrate contamination at this and other dairies in the area. We believe that the large number of cows in these relatively small areas has resulted in waste deposition rates that have caused subsurface losses to groundwater of nitrate. However, it's not certain that the elevated nitrate levels in the supply well are due to leaching from the loafing area since the general direction of groundwater movement is away from the barn toward the loafing area. Contamination could occur if the cone of depression caused by pumping the supply well extends to the northern end of the loafing area. The elevated deep EM conductivity (13 mS m^{-1}) just above the silos in Fig. 3 may be evidence of a plume that is being drawn toward the barn by pumping.

The sustainable solution to the contamination problem posed by the loafing areas is likely a rotational grazing system which would more evenly distribute dairy waste.

REFERENCES

1. Barry, D.A.J., D. Gorahoo, and M.J. Goss. 1993. Estimation of nitrate concentration in groundwater using a whole farm nitrogen budget. J. Environ. Qual. 22:767-775.

2. Brune, D.E. and J. Doolittle. 1990. Locating lagoon seepage with radar and electromagnetic survey. Environ Geol. Water Sci. 16:195-207.

3. Brune, D.E., P.W. Westerman, and R.L. Huffman. 1992. Terrain conductivity to quantify seepage from animal waste lagoons. Annual report to the South Carolina Soil Conservation Service June 1991 - June 1992. Dep. of Ag. and Biol. Eng. Clemson University. Clemson SC.

4. Darling, W.A. 1992. Status of Florida regulations of dairy farm waste management. p. 67-70. In J. Blake, J. Donald and W. Magette (ed.). National Livestock, Poultry and Aquaculture Waste Management. Am. Soc. Ag. Eng. St. Joseph, MI.

5. Exner, M.E., C.W. Lindau, and F.R. Spalding. 1985. Groundwater contamination and well construction in southeast Nebraska. Ground Water 23:26-34.

6. Goodrich, J.A., B.W. Lykins, and R.M. Clark. 1991. Drinking water from agriculturally contaminated groundwater. J. Environ. Qual. 20:707-717.

7. Goodman, J. 1985. Agricultural sources of nitrate contamination in a shallow sand and gravel aquifer of eastern South Dakota. p. 264-267. In Perspectives on nonpoint source pollution. Proc. of a National Conference Kansas City, MO. 19-22 May 1985. Iowa Dep. of Natural Resour., Des Moines, IA.

8. Gould, M.C. 1993. A summary of nitrate levels in well water samples. Water Quality

Courier: A Newsletter on Water Quality Issues. 2:3-4. University of Georgia Cooperative Extension Service. Athens, GA.

9. Hendrickx, J.M.H., B. Baerends, Z.I. Raza, M. Sadig, and M.A. Chaudhry. 1993. Soil salinity assessment by electromagnetic induction of irrigated land. Soil Sci. Soc. Am. J. 56:1933-1941.

10. LeGrand, H.E. 1988. Piedmont and Blue Ridge. *in* W. Back, J.S. Rosenshein, and P.R. Seaber (ed.). Hydrogeology. Geological Society of America. Boulder, CO.

11. Madison, R.J. and J.O. Brunett. 1985. Overview of the occurrence of nitrate in ground water of the Unites States. U.S. Geol. Surv. Water Supply Pap. 2275.

12. Soil Survey Staff. 1965. Soil survey Morgan County Georgia. Series 1962, No. 6. U.S. Gov. Print. Office. Washington, DC.

13. Tranel, L. 1991. Dairy pasture economics. *In* Managing the Farm. Vol 24, No 4. Department of Agricultural Economics, University of Wisconsin, Madison WI.

MEASUREMENT OF SEEPAGE FROM WASTE HOLDING PONDS AND LAGOONS

G. G. Demmy and R. A. Nordstedt[*]

ABSTRACT

Concern about the impact of animal waste handling facilities upon groundwater quality has become more acute during the past several years. Of particular interest are unlined waste holding structures, such as farm ponds and lagoons. These structures are known to seep, but average annual seepage volumes remain largely unquantified. However, these structures have been cited as significant contributors to groundwater degradation (Andrews, 1992). Without an understanding of the relative contribution of a pond to the total nutrient groundwater loading of animal operation as a whole, cost effective groundwater protection best management practices for these operations cannot be designed.

A simple device was developed and demonstrated to measure evaporation corrected changes in water level between 6 and 60 mm, with a resolution of 1 mm under laboratory conditions. The device may be implemented in the field to make rapid measurements of pond water mass balances. Seepage may be directly quantified measuring changes in pond water level during periods when the only water mass flows are seepage and evaporation. Short term average seepage rates from an unlined North Florida dairy farm waste holding pond in clayey sand were measured to be on the order of 10^{-7} m/s. While near real time seepage rates may be measured, several separate measurements at various pond stages should be made before average annual seepage is estimated.

Keywords: Leakage, Infiltration, Wastewater.

INTRODUCTION

Investigation of groundwater quality in the vicinity of animal waste holding ponds and lagoons has revealed the possibility of appreciable seepage from such structures. Traditional methods of seepage characterization have generally relied upon indirect techniques, such as leachate detection, nutrient retention, and mathematical modeling rather than a rigorous quantification of seepage rates through direct measurement. An obvious problem with attributing all groundwater quality deterioration in the vicinity of a wastewater holding structure to that structure is the possible exclusion of other dubious management practices from consideration in a comprehensive groundwater quality protection scheme. Clearly, a technique which can inexpensively and accurately make rapid direct measurements of the average seepage rate of a pond at a given stage could significantly aid in the development of cost effective best management practices (BMPs) for farmsteads with existing earthen storage structures.

TRADITIONAL METHODOLOGIES

Methods for the detection and evaluation of seepage and its impact upon groundwater quality have been widely documented in the literature (Reese and Loudon, 1983, Bouwer and Rice, 1968, Brockway and Worstell, 1968).

---

[*] G. G. Demmy, Graduate Student, R. A. Nordstedt, Professor. Agricultural Engineering Department, University of Florida, Frazier Rogers Hall, Gainesville, Florida.

SEEPAGE EFFECTS

Sewell (1978) measured rapid increases in groundwater nitrate, chloride, fecal coliform, and fecal streptococci concentrations in the vicinity of an anaerobic dairy waste lagoon on a silt loam underlain by sand and clay layers after it was initially loaded. The concentrations were reported to decrease some years after the initial loading (Sewell, 1978). Ritter et al. (1981) monitored groundwater in the vicinity of a two-stage anaerobic swine waste lagoon on soil described as sandy loam to loamy sand with clay lenses, and reported sustained elevated concentrations of chloride, chemical oxygen demand, and ammonia in one well. However, the investigators concluded the seepage was localized and the pond had a minimal impact on groundwater quality, based on the lower measured concentrations of nutrients found in other monitoring wells (Ritter, et al., 1981). Feng, et al. (1992) found that the seepage rate from a swine wastewater lagoon was sufficient to cause mounding of the groundwater, but no quantification of the seepage rate was made.

SEEPAGE QUANTITIES

Davis et al. (1973) measured the reduction in seepage rate from 5.8 cm/d to 0.5 cm/d (6.7×10^{-7} m/s to 5.9×10^{-8} m/s) in the span of four months in a dairy farm lagoon using ring infiltrometers placed into the structure. Similarly, Robinson (1973) recorded a reduction in seepage rate from 11.2 cm/d to 0.3 cm/d (1.3×10^{-6} m/s to 3.5×10^{-8} m/s) in six months in an unlined dairy farm pond. Performing rigorous mass balances on pilot scale ponds, Hills (1976) observed considerable and consistent sealing with time, concluding that the infiltrate contains approximately 0.1 percent of the influent potential pollutants. Barrington et al. (1983) measured pond infiltration rates using field leveling equipment, reporting rates from 0.3 cm/d to 1.2 cm/d (3.5×10^{-8} m/s to 1.4×10^{-7} m/s) from unlined animal waste storages. In a comparison of several infiltrometer measurements to mass balance calculations, McCullough-Sanden and Grimser (1988) found the two methods yield similar results in quantifying seepage from drainwater evaporation ponds. Seepage was found to be highly dependent upon properties of the water, such as salinity (McCullough-Sanden and Grimser, 1988).

GENERAL CONSIDERATIONS

An obvious approach to quantifying the seepage from a storage structure is to perform a long term mass balance on the storage structure system. In principle, such a mass balance approach is sound. However, closer examination of the practical aspects of measurement of the various pond inputs and outputs reveals that this approach may not be sensitive enough to accurately quantify, or even detect, significant seepage. Consider the average annual seepage rate from a circular 30 meter (100 ft) diameter pond, typical of smaller North Florida dairies, necessary to cause groundwater standards to be exceeded 30 meters (100 ft) from the pond edge. Assume an average ammonium concentration in the pond water to be 150 mg/l. Assume all ammonium is converted to nitrate, and no denitrification takes place within the 30 meter zone of interest. Assume an annual average recharge of 23 cm (9 in) pure water and complete mixing of the pond leachate and recharge water occurs. A seepage rate greater than 5 cm/yr (2 in/yr) would be sufficient for the groundwater nitrate concentration to exceed the 10 mg/l primary standard in this conservative model. If the dairy flushed an average of 75 m^3 (20000 gal) per day into the pond, the total flush volume alone would have to be measured to within 1 m^3 (260 gal) or 1.5% in order not to mask a seepage rate of this magnitude. Obviously, even if the allowable seepage rate is an order of magnitude greater than that predicted by the model, accurate measurement of pond inputs and outputs (i. e. pumping, evaporation, and precipitation) may remain problematic. Of course, this model does not consider the mixing of leachate with the existing groundwater, denitrification, and a host of other processes which would tend to reduce the measurable concentrations of nitrate from a pond source, but it does indicate that a minute sustained seepage, sufficient to cause serious deterioration of groundwater quality, may go undetected by traditional mass balance techniques. Moreover, it also indicates that any seepage measurement method must be able to measure rates on the order of 10^{-9} m/s.

In principle, integrated pond seepage may be measured by measuring the change in water level during periods with little or no significant pond mass fluxes other than seepage. Such periods occur on calm nights when the farm is idle and the evaporation rate is minimal. Choosing six hours as the maximum amount of idle time which may reasonably be expected, a viable measurement technique must be able to measure water level changes on the order of 30 mm.

Minute changes in liquid level have been measured in a variety of ways, some of which exhibit potential for application to measurement of seepage from waste holding facilities. While many of these techniques do not explicitly correct for evaporation, the devices could certainly be incorporated into a measurement scheme which does. Wang et al. (1991) demonstrated a relatively simple and inexpensive non-contact technique using ultrasonic interferometry to measure changes in liquid level elevations with a precision of 10 mm. The specifications for the Validyne P305D differential pressure transducer claims an accuracy of 0.25% of a full scale of 5.6 cm H_2O, an accuracy of 140 mm (Validyne Engineering Corp.)[*]. This transducer is relatively expensive (>500 $US) and requires additional data collection equipment, not to mention that its accuracy falls outside of the target 30 mm range. Other more sensitive pressure transducers may be found in the trade literature, but their cost may reasonably be expected to equal or exceed that of the P305D. Additionally, very low pressure differential pressure transducers are quite delicate, and not well suited to the rugged conditions found in the field. The specifications for the Aromat Corporation LM series of analog laser sensors claim a resolution of 5, 2, and 0.2 mm for the LM 100, LM 200, and the LM 300 series sensor respectively (Aromat Corporation). However the relatively high cost of the sensors (list price 1800 $US, 2100 $US, and 7000 $US, respectively), may be prohibitive. The literature describing the MTS Temposonics II linear displacement transducer claims the device has a resolution of 3 mm (MTS Systems Corporation). All of these devices require external power and some sort of device for recording or displaying data. While many of these devices could have been incorporated into a measurement scheme, none were implemented due to high cost or insufficient ruggedness for general field use.

DESCRIPTION OF APPARATUS

The apparatus as constructed is comprised of two open topped square aluminum boxes, or pans, 0.91 m (3.0 ft) on a side and 0.15 m (0.50 ft) deep (see Figure 1). At the corners are fastened 3.8 cm (1.5 in) mounting tubes 30 cm (12.0 in) in length. These tubes are oriented vertically, and flush with the bottom of box. The tubes are fitted with 6 mm (0.25 in) wing bolts for clamping the pans into position on 2.5 cm (1 in) diameter poles which have been rigidly placed in the liquid containment structure of interest. The pans are hydraulically connected by a 1.3 cm (0.5 in) diameter nylon tube 0.4 m (16 in) in length, and flow between the pans is controlled by a 1.3 cm (0.5 in). ball valve.

---

[*] Disclaimer: Mention of any product name is meant neither as an endorsement nor a proscription of that product or the company which produces it, but is included for the information of the reader only.

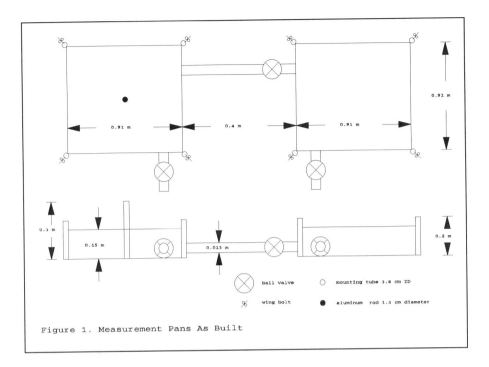

Figure 1. Measurement Pans As Built

Additional ball valves connected to the pans allow for equilibration with the body of liquid into which the pans are placed. The measurement pan has a 1.3 cm (0.5 in) diameter aluminum rod 30 cm (12 in) in length mounted vertically upward in the center of the pan. An aluminum collar with an ell shaped blade 2.5 cm (1.0 in) wide may be mounted to the rod in different locations. The ell points downward an serves as a fixed pivot (see Figure 2). A 0.5 m^2 (5.4 ft^2) styrofoam slab 5 cm (2 in) thick floats in the center of the measurement pan. A 2.5 cm (1.0 in) wide aluminum blade is mounted vertically upward on a small aluminum plate which rests on the float. Between the two blades rests a lever 2.5 cm by 5.1 cm by 0.3 cm (1.0 in by 2.0 in by 1/8 in). Two small grooves are machined into the lever 0.64 cm (0.25 in) apart, one on either side of the lever, to serve as a register for the blades, and to keep the float from moving due to surface tension, electrostatic, or other effects. A small back silvered mirror is mounted on the lever. The angle of the mirror may be adjusted by a set screw. The collar must be adjusted so the lever is as horizontal as possible. The relative angle of the mirror is unimportant, and may be adjusted to aid in preparing for measurement. A more complete description of the design, testing and implementation of the device may be found in Demmy (1993) and Demmy et al. (1993).

The apparatus works on the principle of positive displacement. When the water level changes, the float is displaced by the amount of change, which in turn displaces the lever (see Figure 3). A mirror placed upon the lever is moved through some angle, which may be measured by observation of the reflection of some graduated instrument (i. e. ruler or staff gage) placed at some distance from the mirror. By the principle of similar triangles, the change in water level, h, will be the change observed on the graduated instrument, H, times twice the distance of the mirror to the instrument, L, divided by the length of lever arm created between the float blade and the fulcrum, l. Thus the gain, or effective magnification, of the system may be altered by moving the graduated instrument.

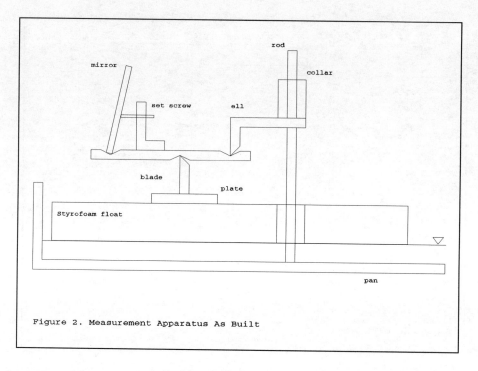

Figure 2. Measurement Apparatus As Built

In order to compensate for evaporative losses, two pans of equal area are hydraulically connected via a tube and a ball valve (see Figure 1). These pans are in turn hydraulically connected to the liquid which is to be measured. One pan, called the measurement pan, contains the measurement device (i. e. the float, lever, mirror, etc.), and the other pan, called the evaporation pan, contains free liquid, to provide correction for, or measurement of, evaporation. Since oasis effects are generally known to be quite severe, measurements should be taken when evaporation rates are very low, and any differences between the evaporation rate of the pan and the structure into which the pan was placed, will be minimized in an absolute sense. The evaporation pan is allowed to equilibrate with the liquid, then is isolated by shutting all valves which feed it. The time is noted (t_0). After some span of time, the measurement pan is isolated, and the time noted (t_1). The evaporation pan is covered to prevent further evaporation during the measurement. The graduated instrument is observed in the mirror, and a relative position on the instrument is noted (H_0). The pans are allowed to equilibrate, and the new position is noted (H_1). The change in water level is calculated as h = 2Hl/L. The average rate of change in liquid level during the measurement period is h divided by the time span over which the evaporation pan was isolated, $t_1 - t_0$.

FIELD IMPLEMENTATION

A dairy waste holding pond located in Lafayette County, Florida was selected for field testing of the seepage measurement device. Installation of monitoring wells in the vicinity of the pond indicates that the pond lies on a deep sand with a number of clay lenses. The water table has been observed to vary from 3 to 6 meters (10 to 18 ft) below the average ground surface. The average bottom of the pond lies between four to eight feet below the average ground surface. The pond is roughly circular in shape, approximately 30 m (100 ft) in diameter. Groundwater is believed to flow North towards the Suwannee

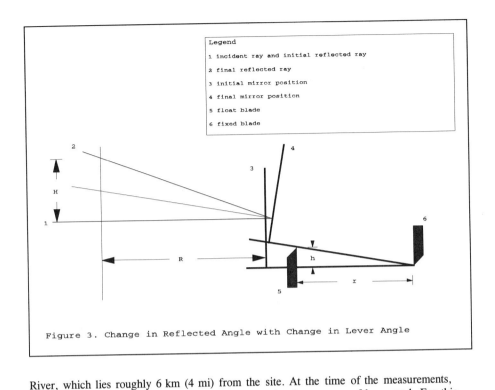

Figure 3. Change in Reflected Angle with Change in Lever Angle

River, which lies roughly 6 km (4 mi) from the site. At the time of the measurements, pond stage was quite high, with pond water covering large areas of bare sand. For this reason, it is believed that the seepage rates measured are not representative of a well sealed pond. However, it does suggest that proper management of ponds in coarser grained soils is essential to adequately protect groundwater quality.

The pans were set up on the North side of the pond in approximately 2 meters (6 ft) of water. Measurements were made at night, when evaporation rates were lower, and the weather was calmer. Several measurements of free water evaporation from the evaporation pan were made with in order to detect any problems, such as shifting of the pans. All of the evaporation measurements indicated either no detectable evaporation or slight evaporation. There was no mysterious accumulation of water which would indicate a differential shift of the pans. Evaporation measurements were carried out by first isolating the pans hydraulically from the pond. After allowing the two pans to equilibrate, the pans were isolated by shutting the ball valve which connects them. A period of 10 minutes was allowed to pass, after which, the valve was re-opened. The response of the measurement pan was monitored and recorded. Observation of the measurement pan response was facilitated by projecting a reflection from the mirror on to the staff gage with a small laser. While the use of the laser was strictly unnecessary, it made the measurement process significantly easier. The evaporation rate, ranging from 1×10^{-9} m/s to 1×10^{-8} m/s, varied considerably between measurements, but all were on the same order of magnitude. The average evaporation rate for all nine recorded evaporation measurements was 6×10^{-9} m/s, with a standard deviation of 4×10^{-9} m/s. Direct measurements of seepage, corrected for evaporation, were made by first allowing the evaporation pan to equilibrate with the pond. The pan was then hydraulically isolated from the pond and the measurement pan. After a period of time, the valve connection the two pans is re-opened, after isolating the equilibrated measurement pan from the pond. The response of the measurement pan is monitored and recorded. Three direct measurements were made. The seepage rates were measured to be 7.4×10^{-8} m/s (time interval = 112 min), 2.0×10^{-7} m/s (time interval = 118 min), and 5.0×10^{-7} m/s (time interval = 22 min), for an average of 2.6×10^{-7} m/s. This rate corresponds to a saturated infiltration rate into finer textured soils such as fine sands, silts, loess, and till (Domenico and Schwartz, 1990). The pond is located on sand, and thus these measurements indicate that some sealing has taken place.

In the time intervals in which the evaporation pan was isolated, the measurement pan was used to monitor the pond water level. Six indirect seepage measurements were made during these interim periods. Subtracting the average evaporation rate from the water level rate of change measured with the measurement pan yielded an indirect measurement of the seepage. The average seepage rate measured in this manner was 1.9×10^{-7} m/s with a standard deviation of 1.7×10^{-7} m/s, and was on the same order of magnitude as the direct measurements. These results indicate the pond at the time of measurement was seeping at a rate two orders of magnitude larger than the rate predicted to cause possible deterioration of groundwater quality.

GROUNDWATER RECONNAISSANCE

Examination of groundwater quality in the vicinity of the pond further indicate that some sustained seepage is taking place, with total nitrogen concentrations greater than 20 mg/l not uncommon for monitoring wells within 20 meters of pond (see Figures 4-8). Total nitrogen values were highest in wells immediately adjacent to the pond, with values consistently and rapidly decreasing with distance, indicating that natural attenuation is taking place. An anomaly was well A which often exhibited nitrate nitrogen concentrations well in excess of the total nitrogen values of the wells lying nearer to the pond. However, the pond may not be the primary source of nitrogen in this well, as a great deal of nitrogen fertilizer has been applied to crops in the vicinity of this well.

SUMMARY AND CONCLUSIONS

A simple and inexpensive measurement scheme was developed and demonstrated to detect and quantify seepage from a dairy farm waste holding pond, though the scheme may applicable to virtually any other type of liquid retention structure. Other processes, such as free water evaporation, may be measured in near real time. The seepage rate of an unlined dairy waste holding pond in a clayey sand was found to be on the order of 2×10^{-7} m/s. Groundwater quality in the vicinity of the pond indicate that appreciable seepage is taking place, but significant natural attenuation is also occuring. Night time evaporation rates were found to be quite variable, but on the order of 10^{-8} m/s.

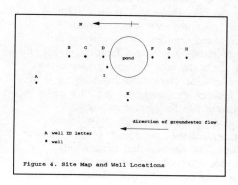

Figure 4. Site Map and Well Locations

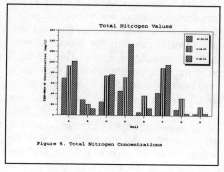

Figure 5. Total Nitrogen Concentrations

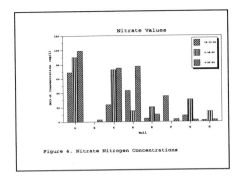

Figure 6. Nitrate Nitrogen Concentrations

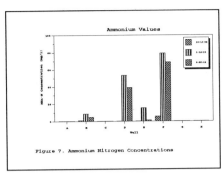

Figure 7. Ammonium Nitrogen Concentrations

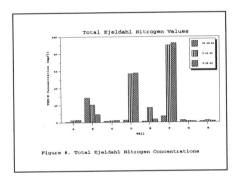

Figure 8. Total Kjeldahl Nitrogen Concentrations

REFERENCES

Andrews, W. J. 1992. Reconnaissance of Water Quality at Nine Dairy Farms in North Florida, 1990-91. USGS Water Resources Investigations Report 92-4058. USGS. Tallahassee, Florida. 39 p.

Aromat Corporation. Product specifications literature. New Providence, New Jersey.

Barrington, S. and P. J. Jutras. 1983. Soil sealing by manure in various soil types. Paper presented at the 1983 Winter Meeting of the ASAE. ASAE paper no. 83-4571 ASAE. St. Joseph, Michigan.

Bouwer, H. 1961. Variable head technique for seepage meters. Journal of the Irrigation and Drainage Division. ASCE. 87(1): 31-44.

Bouwer, H. and R. C. Rice. 1968. Salt penetration technique for seepage measurement. Journal of the Irrigation and Drainage Division. ASCE. 94(4): 482-92.

Brockway, C. E. and R. V. Worstell. 1968. Field evaluation of seepage measurement methods. Proceedings of the Second Seepage Symposium. USDA. Phoenix, Arizona. 150 p.

Chang, A. C., W. R. Olmstead, J. B. Johanson, and G. Yamashita. 1974. The sealing mechanism of wastewater ponds. Journal of Water Pollution Control Federation. 8(1): 126-130.

Collins, E. R., Jr., T. G. Ciravolo, D. L. Hallock, D. C. Martens, E. T. Kornegay, and H. R. Thomas. 1975. Effect of anaerobic swine lagoon on groundwater quality in high water table soils. Managing Livestock Wastes, Third International Symposium on Livestock Wastes. ASAE. pp. 303-305, 313.

Culley, J. L. B. and P. A. Phillips. 1982. Sealing of soils by liquid cattle manure. Canadian Agricultural Engineering. 24(2): 87-9.

Dalen, L. D., W. P. Anderson, and R. M. Rovang. 1983. Animal manure storage pond groundwater quality evaluation. Paper presented at the 1983 Winter Meeting of the ASAE. ASAE paper no. 83-4572. ASAE. St. Joseph, Michigan.

Davis, S., W. Fairbanks and H. Weisheit. 1973. Dairy waste ponds effectively self sealing. Trans. ASAE. 16(19): 69-71.

Demmy, G. G. 1993. Measurement of Seepage from Dairy Farm Waste Holding Ponds. Master's Thesis submitted to the University of Florida.

Demmy, G. G., A. B. Bottcher, and R. A. Nordstedt. 1993. Measurement of leakage from dairy waste holding ponds. Paper presented at the 1993 International Summer Meeting. ASAE paper no. 93-4017. ASAE. St. Joseph, Michigan.

Grimser, M. E. and B. L. McCullough-Sanden. 1989. Correlation of laboratory analyses of soil properties and infiltrometer seepage from drainwater evaporation ponds. Trans. ASAE. 32(1): 173-6, 180.

Hart, S. A. and M. E. Turner. 1965. Lagoons for Livestock Manure. Journal of Pollution Control Federation. 37(11): 1578-96.

Hills, D. J. 1976. Infiltration characteristics from anaerobic lagoons. Journal of Water Pollution Control Federation. 48(4):695-709.

Huffman, R. L. and P. W. Westerman. 1993. Tracking seepage with terrain conductivity and wells. Paper presented at the 1993 International Summer Meeting. ASAE paper no. 93-4016. ASAE. St. Joseph, Michigan.

McCullough-Sanden, B. L. and M. E. Grimser. 1988. Field analysis of seepage from drainwater evaporation ponds. Trans. ASAE. 31(6): 1710-4.

Miller, M. H., J. B. Robinson, and D. W. Gallagher. 1976. Accumulation of nutrients in soil beneath hog manure lagoons. Journal of Environmental Quality. 5(3): 279-82.

MTS Systems Corporation. Product specifications literature. Research Triangle Park, North Carolina.

Nordstedt, R. A., C. B. Baldwin, and C. Cc. Hortenstine. 1971. Multistage lagoon systems for treatment of dairy farm waste. Livestock Waste Management, Second International Symposium on Livestock Waste. ASAE. pp. 77-80.

Osterberg, C. G. 1972. Sealing of Anaerobic Dairy Waste Lagoons in Sandy, High Watertable Soils. Master's Thesis. University of Florida. Gainesville, Florida. 75 p.

Phillips, P. A. and J. L. B. Culley. 1985. Groundwater nutrient concentrations below small-scale earthen manure storages. Agricultural Waste Utilization and

Management, Fifth International Symposium on Livestock Waste. ASAE. pp. 672-9.

Reese, L., and T. Loudon. 1983. Seepage from earthen manure storages and lagoons: a literature review. Paper presented at the 1983 Winter Meeting of the ASAE. ASAE paper no. 83-4569. ASAE. St. Joseph, Michigan.

Ritter, W. F. and A. E. M. Chirnside. 1983. Influence of animal waste lagoons on ground-water quality. Paper presented at the 1983 Winter Meeting of the ASAE. ASAE paper number 83-4573. ASAE. St. Joseph, Michigan.

Ritter, W. F., E. W. Walpole, and R. P. Eastburn. 1981. An aerobic lagoon for swine manure and its effect on the groundwater quality in sandy-loam soils. Livestock Waste: A Renewable Resource, Fourth International Symposium on Livestock Waste. ASAE. pp. 244-6, 251.

Robinson, A. R. and C. Rohwer. 1959. Measuring seepage from irrigation channels. Technical Bulletin 1203. USDA Agricultural Research Service. Washington, D. C. p. 82.

Robinson, F. E. 1973. Changes in seepage rate from an unlined cattle waste digestion pond. Trans. ASAE, 16(1): 95-6.

Sewell, J. I. 1978. Dairy lagoon effects on groundwater quality. Trans. ASAE. 21(5): 948-52.

Sewell, J. I., J. A. Mullins, and H. O. Vaigneur. 1975. Dairy lagoon system and groundwater quality. Managing Livestock Wastes, Third International Symposium on Livestock Waste. ASAE, pp. 286-8.

Validyne Engineering Corp. Product specifications literature. Northridge, California.

Wang, Y., C. Mingotaud, and L. K. Patterson. 1991. Noncontact monitoring of liquid surface levels with a precision of 10 micrometers: a simple ultrasound device. Rev. Sci. Instr. 62(7): 1640-1.

THE AGSTAR PROGRAM: ENERGY FOR POLLUTION PREVENTION

K.F. Roos[1]
United States Environmental Protection Agency
Washington, DC

ABSTRACT

AgSTAR, a voluntary agricultural program sponsored by the Environmental Protection Agency, Department of Energy, and Department of Agriculture's Soil Conservation Service, shows manures are resources which offer financial returns to producers. AgSTAR, a key component of President Clinton's Climate Change Action Plan, encourages the widespread use of methane recovery technologies to increase livestock production profits and demonstrate that industry and environment can work together to create a cost-effective and environmentally aware America. By investing in methane recovery technologies, AgSTAR participants realize substantial returns through (1) reduced electric, gas, and oil bills, (2) revenues from high quality manure byproducts; and (3) savings on manure management operational costs. Partners also reduce pollution associated with water resources, odors, and global warming.

While methane recovery for on-farm energy production has been implemented to a limited extent in the U.S., a number of financial, technical, and informational barriers currently limit the full use of the technology. AgSTAR is strategically designed to overcome these barriers. As part of AgSTAR's strategy a series of products, information, and services are scheduled to become available to assist partners in evaluating, identifying, and installing technologies offering high rates of financial return.

KEYWORDS. Livestock manure, Anaerobic digestion, Cost-effectiveness, Greenhouse gas.

ATMOSPHERIC METHANE CONCENTRATIONS AND OPPORTUNITIES FOR PROFITABLE REDUCTIONS IN AGRICULTURE

AgSTAR is one of the fifty initiatives included in President Clinton's Climate Change Action Plan which takes measures in all sectors of the economy that emit greenhouse gases while guiding the U.S. economy toward environmentally sound economic growth.

AgSTAR will profitably reduce emissions of methane. Methane currently accounts for over 15 percent of expected warming from climate change. While methane's concentration in the earth's atmosphere is small, it has a sizeable contribution to potential future warming because it is a potent greenhouse gas and because methane's concentration has been increasing dramatically. Its global concentration has more than doubled over the last two centuries, after remaining fairly constant for the preceding 2,000 years, and continues to rise.

Methane's rising concentration is largely correlated with increasing populations, and currently about 70 percent of global methane emissions are associated with human activities such as energy production and use (coal mining, oil and natural gas systems, fossil fuel

---

[1] K.F. Roos is the AgSTAR Program director and is also co-chair in the AgSTAR inter-agency work group representing EPA. For more information regarding the program contact: EPA; 401 M St., SW; 6202-J; Washington, DC 20460.

combustion); waste management (landfills and wastewater treatment); livestock management (ruminant enteric fermentation and manure handling); biomass burning; and rice cultivation (US EPA, 1993a).

Reductions in methane emissions of 30 to 40 million metric tons per year, or about 10 percent of annual anthropogenic (human related) emissions, would halt the annual rise in methane concentrations (US EPA, 1993a).

Methane Emissions From Livestock Manure Management Systems Are Substantial

Livestock manure contains un-digested organic material. When handled under anaerobic conditions, microbial fermentation produces methane. The United States like many developed countries manages manure from large concentrations of cattle, swine, and poultry using liquid manure management systems that are conducive to anaerobic fermentation and methane production. The emissions of methane from livestock manure are driven by the quantity of manure produced, how it is handled, and the temperature at which it is handled. The manure management system employed in particular is very important, with liquid or slurry based systems (such as lagoons) converting large portions of the available carbon to methane and pasture systems converting fairly small portions. Emissions vary from system to system and throughout the year.

Methane emissions from livestock manure in the U.S. in 1990 are estimated to be in the range from 1.7 to 3.6 Tg/yr with a central estimate of 2.3 Tg/yr, or about 10 percent of total U.S. anthropogenic emissions (US EPA, 1993b). Of the total 1990 U.S. emissions of 2.3 Tg/yr, two livestock groups account for about 80 percent of total emissions:

- Swine account for about 1.1 Tg/yr or about 50 percent; and
- Dairy cattle account for about 0.7 Tg/yr or about 30 percent.

Furthermore, methane emissions from livestock manure are expected to increase significantly during the next decade. As the demand for animal products increase, the number of animals and the amount of manure produced will also increase. In addition, the use of livestock manure management systems that promote methane production (e.g., anaerobic lagoons) will likely increase substantially because of concerns over applying manure during times when crops cannot utilize the nutrient value of the manure. Emission estimates for the year 2000 range between 3.1 and 5.6 Tg/yr, and emission estimates for 2010 range between 3.3 and 6.0 Tg/yr (US EPA, 1993b).

Profitable Opportunities To Reduce These Emissions

Methane recovery systems can collect the methane produced by liquid manure management systems so that the methane can be used as a fuel for use as an on-farm energy source. Methane obtained from manure management systems is generally in the form of a medium BTU gas, commonly referred to as biogas. Biogas consists of 60-70% methane and 30-40% carbon dioxide, which can provide on-farm energy in the following areas:

- **Electricity Production**: Methane can be used to fuel a engine generator;
- **Heating**: Methane can be used to fire boilers and space heaters; and
- **Cooling**: Methane can be used to fire chillers and other refrigeration equipment necessary in the production, handling, and storage of food.

Technologies available for methane recovery applications include the following:

Covered Anaerobic Lagoon: Anaerobic lagoons are among the simplest manure storage and treatment systems in current use. By retro-fitting or re-configuring a existing lagoon with a cover made from a flexible gas tight material methane is recovered and can be used as an on-farm energy source (Safley and Westerman, 1990).

Plug Flow and Complete Mix Digesters: Digesters have been used for many years, particularly in China and to a limited extent in the U.S. and other countries, to produce energy from livestock manure. Generally, digester systems have: poured concrete floors and walls; a flexible gas tight cover to capture methane; heated to facilitate rapid decomposition of the manure and reduce system size; and configured as either tanks (complete mix) or trenches (plug flow).

The EPA has conducted a series of field evaluations and characterizations of representative livestock production facilities in intensive dairy and pork producing regions in the U.S. (RCM, 1990-91 a,b,c,d). Based on these studies national assessments of profitable methane reductions were developed and submitted to Congress (US EPA, 1993c) as a requirement under the Clean Air Act Amendments of 1990.

Results from this analysis (US EPA, 1993b) indicate that about 2000 dairy and 2000 pork producing facilities could profitably utilize methane recovery technology. These results are consistent with other economic studies on methane recovery (Lusk, 1991; NEOS, 1993). The utilization of this technology, where profitable, could reduce methane emissions by .5 to .8 Tg by the year 2000, and .6 to 1.0 Tg by the year 2010.

THE AGSTAR PROGRAM

AgSTAR, a voluntary agricultural program sponsored by the Environmental Protection Agency, Department of Energy, and Department of Agriculture's Soil Conservation Service, shows manures are resources which offer financial returns to producers. AgSTAR, a key component of President Clinton's Climate Change Action Plan, encourages the widespread use of methane recovery technologies to increase livestock production profits and demonstrate that industry and environment can work together to create a cost-effective and environmentally aware America. By investing in methane recovery technologies, AgSTAR participants realize substantial returns through (1) reduced electric, gas, and oil bills, (2) revenues from high quality manure byproducts; and (3) savings on manure management operational costs. Partners also reduce pollution associated with water resources, odors, and global warming.

While methane recovery for on-farm energy production has been implemented to a limited extent in the U.S., a number of barriers currently limit the full use of the technology including:

- Poor technical and economic perception;
- Lack of access to comprehensive technical information and expertise; and
- Difficulty in obtaining financing.

AgSTAR is strategically designed to overcome these barriers by making available a series of products, information, and services to assist partners in making informed technical and economic decisions to improve facility profits while also playing a vital role in promoting the program to other producers. These products and services include:

AgSTAR Decision Support Software System: A computer software package that enables program participants to survey facilities for lagoon retrofit and new digester construction alternatives through a series of user friendly input screens. The software has a report generation feature to develop clear and informative survey summary reports suitable for use by facility managers, corporate financial staff, and senior management comparing the financial performance of the existing system to the methane recovery option.

AgSTAR Upgrade Manual: A comprehensive methane recovery handbook and reference guide organized for specific livestock rearing methods and manure management strategies such as, recycle flush, single cell and double cell lagoons, liquid slurry storages, and mechanical scrape systems. The manual will include general manure management principles, odor control, nutrient management strategies, technical design, energy applications, economics, national and local financing availability, and case studies.

Public Recognition: AgSTAR will place public-service advertising in major magazines, newspapers, and trade journals; reports innovative applications by participants; and publishes a semi-annual newsletter on participants economic, technical, and environmental success. To promote their own AgSTAR activities, EPA will distribute ready-to-use promotional materials such as brochures, video tapes, and the AgSTAR logo shown in Fig. 1.

Figure 1. The AgSTAR Logo

High Visibility Demonstration Projects: Many of the methane recovery facilities constructed during the "Energy Crisis" of 1975 did not consider the importance of simple operation, servicing, and financial returns, causing a poor technical perception among livestock producers. AgSTAR is developing high visibility projects which demonstrate attractive financial returns with simple operational procedures to re-establish producer confidence in the technology.

Financing: Investments in on-farm energy systems require up-front capital. AgSTAR endeavors to identify and develop financial services for program participants exhibiting desirable "on-farm" economic performance.

AgSTAR Installation Directory: A comprehensive AgSTAR participant and installation directory to facilitate information exchange within the livestock community.

Research & Development: AgSTAR continually identifies and develops mechanisms which improve economic, technical, and managerial performance.

AgSTAR: How Does it Work

The Partner Program: To become an AgSTAR Partner, a livestock producer signs a Memorandum of Understanding (MOU) with EPA. In the MOU, AgSTAR participants agree to survey their facilities within one year of signing the MOU and to install AgSTAR technology within two years of the survey if profitable. Profitability is defined as those projects which provide an annualized rate of return that is at least equivalent to the prime interest rate plus six percentage points or the APR of the loan plus six percentage points, whichever is lower. An AgSTAR partner may, at their option, elect to accept a lower rate of return. Partners also agree to appoint an implementation manager who oversees participation in the program and is the primary link to EPA's program staff.

The Ally Program: AgSTAR Ally programs are comprised of members in the agriculture and energy industries, state energy and regulatory offices, universities, and electric utilities, and are designed to encourage the wide-spread use of methane recovery systems for the livestock sector. To become an AgSTAR Ally industries, organizations, universities, and utilities sign a MOU similar to those signed by partners, but agree to promote methane recovery systems to livestock producers, educate their industry, and implement programs to stimulate greater development and use of AgSTAR technologies. Through the Ally Program, EPA encourages greater investment in the development and marketing of innovative methane recovery and manure treatment technologies.

Other Benefits Of AgSTAR Technologies

Enhancing anaerobic activity to yield higher methane production and profitability also results in the following additional benefits:

Reduced Ground And Surface Water Pollution: Anaerobic treatment of manure is very effective at reducing the damaging effects on ground and surface water in two ways. First, manures high biological oxygen demand are indicative of high strength organic materials which contribute to eutrophication when introduced into surface waters. As illustrated in Figure 2, high reductions in the oxygen demand of manure are achieved as a result of anaerobic bacterial action on the volatile solids portion of manure (Gunnerson and Stuckey, 1986). Second, as illustrated in Figure 3 (NC State University, 1991), anaerobic treatment of livestock manure converts a large proportion of the organic nitrogen content of fresh manure into ammonia (NH_3), the primary constituent of commercial fertilizers (Loehr, 1984). Ammonia is readily available for plant uptake and used by plants during the crop year. Plant uptake of ammonia prevents conversion to nitrate and subsequent leaching of nitrate during non-crop growing periods. Applications of anaerobically digested material to crop land additionally reduce nitrate concentrations when compared to commercial fertilizers while also reducing sediment and nutrient runoff from rainfall events (Mostaghimi et al., 1988).

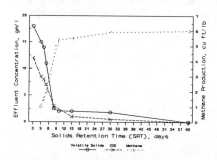

Figure 2. Anaerobic Stabilization of Manure

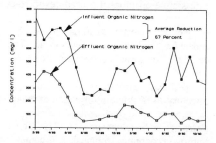

Figure 3. Organic Nitrogen Reduction

Public Health: It is estimated that more than one hundred diseases may be transmitted from animals to man through viruses, bacteria, protozoa, and helminths. Risks of disease transmission are determined by indicator organisms which include *Escherichia coli*, *Streptococcus faecalis*, *Salmonella typhimurium*, and coliform bacteria. Figure 4 illustrates the large indicator organism reductions achieved under anaerobic conditions (Edgar and Hashimoto, 1991).

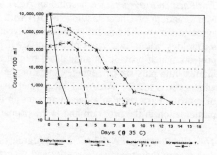

Figure 4. Pathogen Destruction

Reduce Manure Handling Expense: Anaerobically digested manure is managed as a liquid. If fields are nearby, pumping liquid and spray irrigating is less expensive than hauling.

Reduce Odors And Enhance Fly Control: Anaerobic digestion eliminates most noxious odors associated with livestock manure. Odor control results from the bacterial destruction of the volatile organic acid component of manure.

PROGRAM STATUS

Although still in its infancy, the AgSTAR Program has already initiated pilot activities in some high density livestock regions in the U.S.. Specific states and activities are discussed below.

North Carolina

North Carolina has almost doubled in swine production over the past two years, from about 2.8 million hogs in 1991 to 4.6 million in 1993, accounting for about $730 million in pork sales for the state. This expansion has created a need to re-assess a wide range of environmental issues as related to manure management including disposal, nutrient management, and odor control.

As a response to this expansion, the North Carolina Energy Division has recently developed two projects demonstrating the financial and environmental benefits of methane recovery technology. These demonstrations include:

Randleigh Dairy Farm: Through assistance from the Southeastern Regional Biomass Energy Program (SERBEP), N.C Agricultural Research Service, the N.C. Dairy Foundation, and technical support from North Carolina State University the first covered lagoon in the state was completed in 1988. This project was designed to evaluate the performance of methane recovery and on-farm energy production under low temperature (psychrophilic) digestion of manure and has averaged close to 4,000 ft^3/day of biogas production. The project demonstrated that an earthen psychrophilic digester has economic advantage over conventional mesophilic digesters because of lower capital cost and reduced operation and maintenance expenses (Lusk, 1991; Safley and Lusk, 1990). These findings are also consistent with the experience at Royal Farms, California, the first covered livestock manure lagoon system in the U.S. (Chandler et al., 1983).

Carroll's Foods, Inc.: More recently the N.C. Energy Division developed a second larger covered lagoon system in conjunction with the state's largest swine corporation. This methane recovery system is designed for a 1000 sow farrow to finish facility located in Sampson County, one of the most densely populated swine areas in the state. The system is currently recovering up to 30,000 ft^3/day of methane which will be used in an electric generator to provide a substantial portion of the facility's energy requirements.

Because of the Energy Division's strong involvement in methane recovery applications in the state and their overall strategy to promote the technology through private sector participation, the Energy Division is developing the broader AgSTAR implementation plan for the state through the AgSTAR Ally Program.

As a next step a roundtable meeting is being planned to bring together the major stakeholders concerned with energy and livestock issues in the state. Stakeholders include representatives of the pork producing trade groups, livestock producers, methane recovery experts, electrical utilities and cooperatives, and state and federal legislative and regulatory agencies. These stakeholders are being brought together to discuss other potential barriers and identify win/win solutions necessary to develop and successfully implement the AgSTAR Program.

Texas

Major expansion of the dairy industry has occurred in the last ten years in Erath County, Texas. Dairy operations in Erath County have been identified by the Texas Water Commission (TWC) as point and non-point sources of water pollution under Section 319 of the Federal Clean Water Act (Sweeton and Wolfe, 1990). Under the TWC regulatory statute, dairies with greater than 250 head in the milking herd must apply and submit manure management plans for approval of operating permits.

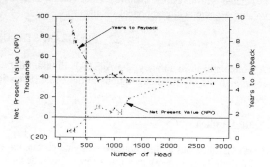

Figure 5. Economic Analysis of Covered Lagoons at Texas Dairies

The best identifiable practices under the TWC regulations are to collect and store corral run-off and milking parlor wash water in lagoon systems or holding ponds.

The AgSTAR Program is currently identifying dairy sites in the Erath area to demonstrate covered lagoon technology as a profitable regulatory compliance strategy for those affected herd sizes. As illustrated in Figure 5, profitable alternatives exist for dairymen across a range of herd sizes (Roos, 1992).

Pennsylvania

AgSTAR has initiated a pilot effort to identify profitable manure management opportunities for small dairy operations, in the 40 to 100 milking herd size range in Lancaster county. The pilot is a cooperative effort including the State of Pennsylvania, Department of Agriculture's Office of International Cooperation and Development, Lancaster Conservation District, and Pennsylvania State University.

International Component: A group of local dairy producers and state officials have recently completed a technical exchange tour in China to gather information on the multitude of methane recovery systems and nutrient management strategies successfully implemented in this region of the world. This exchange was hosted by the Chengdu Biogas Research Institute largely responsible for the many successful applications developed in China.

Domestic Component: A follow-up workshop was held in Lancaster recently to share the findings of the trip with local livestock producers and to identify the level of interest in the future adaptation of the technology to local needs. As a next step "on-farm" technical and economic feasibility studies coupled with demonstration projects and local design, construction, and maintenance trainings may be planned.

California and Florida

Demonstration sites are currently being identified in high visibility dairying areas. These demonstration sites may integrate algae or duckweed production into the overall manure treatment system to gain additional economic benefit by substituting these high protein yielding aquatic feed sources for commercial feed. AgSTAR considers these types of innovative approaches as examples of cutting edge nutrient management strategies in areas where cropping season, land availability, and soil characteristics cannot handle the large nutrient loads from concentrated livestock production facilities.

CONCLUSIONS

Methane is an important greenhouse gas and also a valuable energy source. Livestock manure contains un-digested organic material. When handled under anaerobic conditions, microbial fermentation produces methane. Methane emissions from livestock and poultry manure in the U.S. in 1990 are estimated to be in the range from 1.7 to 3.6 Tg/yr with a central estimate of 2.3 Tg/yr, or about 10 percent of total U.S. anthropogenic emissions. Emission estimates for the year 2000 range between 1.9 and 5.7 Tg/yr, and emission estimates for 2010 range between 1.9 and 6.0 Tg/yr.

AgSTAR, a voluntary program sponsored by the Environmental Protection Agency, Department of Energy, and Department of Agricultures's Soil Conservation Service, shows manures are resources which offer financial returns to producers. AgSTAR, a key component of President Clinton's Climate Change Action Plan, encourages the widespread use of methane recovery technologies to increase livestock production profits and demonstrate that industry and environment can work together to create a cost-effective and environmentally aware America. Recovery systems can collect the methane produced by liquid manure management systems so that the methane can be used as a fuel as an on-farm energy source. The EPA has conducted a series of field evaluations and characterizations of representative livestock production facilities in intensive dairy and swine producing regions in the U.S. Results from this analysis indicate that about 2000 dairy and 2000 swine producers could profitably utilize the technology and reduce methane emissions by .5 to .8 Tg by the year 2000, and .6 to 1.0 Tg by the year 2010.

REFERENCES

1. Chandler J.A., S.J. Hermes, and K.D. Smith, 1983, "A Low Cost 75kW Covered Lagoon Biogas System," California Energy Commission, in Energy from Biomass and Wastes VII.

2. Edgar T.G. and A.G. Hashimoto, 1991, <u>Feasibility Study For A Tillamook County Dairy Waste Treatment And Methane Generation Facility</u>, Department of Bioresource Engineering, Oregon State University.

3. Gunnerson C.G. and D.C. Stuckey, 1986, <u>Anaerobic Digestion: Principles and Practices for Biogas Systems</u>, World Bank.

4. Loehr R.C., 1984, <u>Pollution Control for Agriculture</u>, Second Edition, Academic Press Inc.

5. Lusk P.D., 1991, "Comparative Economic Analysis: Anaerobic Digester Case Study", Bioresource Technology, #36, pg. 223-228, Elsevier Science Publishers.

6. Mostaghimi S., M.M. Deizman, T.A. Dillaha, C.D. Heatwole, and J.V. Perumpral, 1988, <u>Tillage Effects On Runoff Water Quality From Sludge-Amended Soils</u>, Virginia Water Resources Research Center, Virginia Polytechnic Institute and State University, Bulletin 162.

7. N.C. State University, 1991, personal communication with L.M. Safley, long term monitoring data of Randleigh Dairy Farm.

8. NEOS, 1993, <u>Energy Conversion of Animal Manures in 13 Western States</u>, prepared by NEOS Corporation for the U.S. Department of Energy Regional Biomass Energy Program. Contract No. 65-90WA05637. NEOS Corporation, Lakewood, CO, Draft Report.

9. RCM (Resource Conservation Management), 1990a, "Estimating the Costs and Benefits of Methane Recovery from New Texas Dairy Waste Lagoons to Mitigate New Methane Sources." Draft report prepared by RCM Inc. under sub-contract to ICF, Inc., for the U.S. Environmental Protection Agency.

10. RCM (Resource Conservation Management), 1991b, "Potential Methane Generation and Mitigation Study of the South Valley of California." Draft report prepared by RCM Inc. under sub-contract to ICF, Inc., for the U.S. Environmental Protection Agency.

11. RCM (Resource Conservation Management), 1991c, "Potential Methane Generation and Mitigation from Hog Waste Storages in the States of Iowa and Illinois." Draft report prepared by RCM Inc. under sub-contract to ICF, Inc., for the U.S. Environmental Protection Agency.

12. RCM (Resource Conservation Management), 1991d, "Potential Methane Generation and Mitigation from Hog Waste Storages in the State of North Carolina." Draft report prepared by RCM Inc. under sub-contract to ICF, Inc., for the U.S. Environmental Protection Agency.

13. Roos K.F., 1992, "Profitable Alternatives for Regulatory Impacts on Livestock Waste Management," National Livestock, Poultry, and Aquaculture Waste Management, American Society of Agricultural Engineers, Publication # 03-92, pp. 89-99.

14. Safley L.M. and P.D. Lusk, 1990, <u>Low Temperature Anaerobic Digester</u>, Energy Division, North Carolina Department of Economic and Community Development.

15. Safley L.M. and P.W. Westerman, 1990, "Psychrophilic Anaerobic Digestion of Dairy cattle Manure," Proceedings of the Sixth International Symposium on Agricultural and Food Processing Wastes, American Society of Agricultural Engineers Publication # 05-90, pg. 250-258.

16. Sweeten J.M., and M.L. Wolfe, 1990, "Runoff and Wastewater Management Systems for Open Lot Dairies", Proceedings of the Sixth International Symposium on Agricultural and Food Processing Wastes, American Society of Agricultural Engineers, Publication # 05-90, pp. 361-375.

17. U.S. EPA, 1993a, <u>Options For Reducing Methane Emissions Internationally Volume I</u>, Report to Congress, Office of Air and Radiation, Washington, D.C.

18. U.S. EPA, 1993b, <u>Anthropogenic Methane Emissions in the United States: Estimates for 1990</u>, Report to Congress, Office of Air and Radiation, Washington, D.C.

19. U.S. EPA, 1993c, <u>Options for Reducing Methane Emissions in the United States</u>, Report to Congress, Office of Air and Radiation, Washington, D.C.

Potential Options for Poultry Waste Utilization: A Focus on the Delmarva Peninsula[1]

C. A. Narrod; R. Reynnells, H. Wells[2]

Abstract

The disposal of poultry and other livestock waste is often a negative externality to society. These wastes are considered a major non-point source of pollution and in some areas are causing a growing environmental threat. The potential environmental risks due to disposal of these wastes are magnified as a result of the dense confinement of poultry and livestock and the decreasing amount of land available for waste disposal. This resource is a valuable source of nitrogen and phosphorous; however, due to the nature of some soils, saturation with these nutrients may be a problem. Due to externalities of animal production, various regulations have been developed. The most recent are the Clean Water Act revisions and the Coastal Zone Act Reauthorization Amendments of 1990. Stricter restrictions have the potential for relocations, which would have a severe impact on the local economy. One way to alleviate some of the possible impacts is to look at potential markets for the waste.

Background

The Delmarva Peninsula is located in the mid-Atlantic coastal plain eco-region that includes most of Delaware and the Eastern Shore of Maryland and Virginia. The area covers 16,000 km^2. In this area the poultry industry is highly concentrated with eight integrated companies having an annual production of about 517 million chickens in approximately 6,000 active production houses. The broiler farms average 50,000 head per growout and range in size up to 400,000 head and produced approximately 5 tons of manure per 100 chickens, per year. From these figures Carr and Brodie, (1992) estimated the annual by-product of production is 646,250 tons of poultry litter, 24,816 tons of dead birds, 12,480 tons of hatchery waste, and 15,510,000 gallons of dissolved air floatation skimmings. In Delaware and Maryland broilers rank first in gross agricultural income and many of the farmers use the poultry waste as a low-cost fertilizer for their crops. In Sussex County, however, approximately 50% of the cropland is used for soybeans. Soybeans, being nitrogen fixing, do not require additional sources of nitrogen (Sims and Wolf, 1993).

A concern is that poultry production is increasing on marginal land that can not assimilate the waste produced from this increased production. On Delmarva, there is a mixture of best management practices (BMPs) and nutrient management plans (NMPs) that are used in correctly land applying manure. Cost-share money is available to aid farmers in their manure management through the Agricultural Stabilization Conservation Service and State cost-share money, State Revolving Funds, PL 566 land treatment funds, and ASCS special projects programs. Farmers are encouraged by USDA, to have NMPs as a part of their soil conservation and water quality plan. They are required by ASCS and SCS to have NMPs to obtain cost-share money for manure storage sheds and dead bird composters. Maryland has developed a program to certify individuals to prepare the plans. A voluntary manure clearing house has been established to coordinate distribution of manure from surplus farms to deficit farms. This program is administrated by the Delmarva Poultry Industry, Inc., however, it hasn't been used to its full potential. Currently there is no apparent excess of manure, due in part to manure being applied by a nitrogen standard

---

[1] Parts of this project was funded separately by USEPA and USDA-Extension Service. For a more detailed discussion of these issues see Narrod, et. al, 1993.

[2] C. A. Narrod, Center for Energy and the Environment, University of Pennsylvania; R. Reynnells, USDA- Extension Service, National Poultry Leader, and H. Wells, EPA - Office of Pollution Prevention and Toxics, Agricultural Coordinator.

rather than a phosphorous loading standard. The report "Poultry Manure Storage and Process Alternatives Evaluation" by CABE Associates (1991) projected that even if the manure is applied based on a nitrogen standard, that by the year 2000, there would be 315,000 tons of manure in excess of what could be utilized on croplands under BMPs.

Because of the potential excess of manure it is necessary to address the question of government intervention. In principle, regulatory agencies were created to remedy the perceived failure of the free market to allocate resources efficiently (Portney, 1990). Government may intervene to correct the societal risks of externalities from production of a good. When water quality is impaired by agricultural production, society bears a cost not fully reflected in the market price for the agricultural products and inputs used in producing the products. If government is unable to correct for these problems, society bears the full cost.

In the poultry industry there are three potential externalities affecting water quality in the production, processing, and supply of inputs that, without regulation, are unlikely to be borne by the producer. Manure and litter are positive externalities and are valuable sources of nitrogen, phosphorous and organic material. However, due to the nature of some soils and some management practices, saturation with nutrients may be a problem, and thus becomes a negative externality when excess nitrogen or phosphorous leaches into the groundwater or contaminates surface water. A second externality is the by-products from processing which, when discharged untreated into the water may cause a major point source of water pollution. A third externality is the release of ammonia or odors from manure into the air, causing air pollution.

Brief description of the USA industry

The major components of the poultry industry are the broiler, layer, and turkey industries. These industries all have vertically integrated segments, and to some degree are horizontally integrated[3]. The typical broiler company consists of a hatchery, feed mill, processing plant, field service and management staff. The company also contracts with 150-300 growers of breeder flocks and growout flocks, who are usually within 25 miles of the central facilities (Rogers, 1979). The egg and turkey industry are somewhat less integrated and consolidated than the broiler industry, though they also have a corresponding supporting network of facilities.

Vertically integrated firms are cost-efficient and therefore can sell their product at a favorable price, while guaranteeing a timely supply of a standardized product which is uniform and of high quality. These coordinated systems can induce more rapid adoption of improved production technology and enable development of new advances and quick responses to changing consumer needs. The contract farmer owns the land and buildings' and provides the necessary labor and electricity. These contract farms are supported by the company's hatchery, feed mill, and service department. The purpose of the contractual arrangement is to form a minimum guaranteed payment for the grower of a base price to be paid per pound or per bird produced, and it often is adjusted based on feed conversion, mortality, and other factors. Vertical integration enables individual producers to move from a simple credit financing and open-market production to a variety of contractual agreements. Contracts are attractive to many chicken producers because many are small farmers with limited financial resources who are unwilling and often unable to obtain the production capital through the traditional sources, and to bear the market risks associated with entering the broiler or egg markets. By turning to these input suppliers for their financing they have an alternative source of production capital and a means to shift a substantial part of the financial and market risks to the contractor (ibid). An estimated 97% of all broilers

---

[3] *Vertical integration is the linking together of various portions of the production, input-supply, and market segments of an industry. Horizontal integration refers to the replication of units at the same stage of production.*

are now produced under some form of formal vertical coordination; the remaining are raised by independent producers for niche markets (Rogers, 1979).

It is recognized that there are some concerns about aspects of vertical integration in the poultry industry. The primary issues are: 1) if only one company is in an area growers will lack bargaining power forcing them to accept contract terms which are uniformly but often unilaterally set by the company[4]; 2) growers undertake a long-term financial obligation when they build a poultry house but do not receive a long-term commitment from the poultry company; 3) growers may have little recourse against adverse company decisions which affect them; 4) the structure of the industry is becoming monopolistic; and 5) the presence of poultry houses or other animal production facilities can create risks to human health and the environment. Of particular concern was that contractual arrangements were unfair to growers. Some feel that if the integrators had to shoulder all the risks, they would be better off building their own laying and growout facilities, as has been seen in the egg industry. Most companies recognize that the contracts are not highly profitable, but are proportional to investment and risks associated with the production. It has been suggested that a profit-sharing system option should be developed so growers may take part in the successful marketing of what they "grow," as well as getting help with additional expenses they may incur, such as handling waste in an environmentally sound way.

The responsibility for the manure in all operations is the grower's, whether this grower is a contract grower, independent grower, or a company. In many cases the grower sells manure or gives it away, whereby it becomes the responsibility of the individual using the waste. Poultry are owned by the company from hatchery through the growout stage, but if the bird dies it usually is the responsibility of the grower to dispose of it, whether contact, independent, or company-owned. There is an exception to this case in which one company has spent the last several years researching the potential of recycling dead broilers into animal feed. They have started programs that distribute custom-built freezers to their growers. Their growers are responsible for pouring cement floors for the composters and supplying the electricity to run these freezers. They have company crews that collect the carcasses and take them to the rendering plants that reprocess the fats, carbohydrates, and protein removed from processing waters and convert them to secondary nutrients that can be used as a commodity feed protein. Processing plant sludge is the responsibility of the company. Discharge from these processing facilities in most cases is recognized as a point source of pollution. Several other companies compost their skimmings.

Current regulations

Because of the increasing pollution of many of our water sources, it has been necessary to issue regulations and guidelines. Originally these regulations were administered by the states, though, because of declining surface water quality, the Federal government has been authorized to take a supervisory role in some states. The Clean Water Act of 1972 was one of the first laws to have a major impact on the industry with its focus on controlling pollution from point sources. It required that EPA regulate "concentrated animal feeding operations" which, up to this time, had been considered non-point sources of pollution. Section 208 of the bill called for the development and implementation of "areawide" water-quality management programs to ensure adequate control of all sources of pollutants and states were responsible for implementing appropriate control, which generally took the form of BMPs (ibid). In 1977 a Clean Water Act made provision for EPA to issue National Pollutant Discharge Elimination System permits for point source discharges such as those from industrial and municipal facilities, feedlots, and other discrete significant sources. In 1987 the Clean Water Act was again revised and became known as the

---

[4] *According to a discussion at a recent meeting in Maryland (July, 1993), this is not the case for the Delmarva Peninsula where there are eight vertically integrated companies and a grower is only allied to that particular company for one growout period. There are optional long term contracts also for ten years.*

Water Quality Act which expired in 1992. Under this congress tried to directly control non-point source pollution. In 1990, the Coastal Zone Act Reauthorization Amendments were passed to address more closely the problems of non-point source agricultural pollution. Through these amendments, states in the coastal zone will be required to develop programs for management of these problems. Under this act, states are required to develop coastal non-point pollution programs in designated coastal zones. The programs must provide implementation of management measures and are required to contain enforceable policies. Both small and large facilities will be required to manage stored runoff and solids with proper waste utilization and disposal methods. This act will effect all of Delmarva.

In addition to the clean water acts there are regulations to aid in maintaining clean air. The Air Pollution Act of 1956 was revised in 1990, as the Clean Air Act Amendment was passed. It had provisions of importance to agriculturalists. With the great increase in concentrated areas of animal production and humans, the handling of their waste has produced higher levels of ammonia and in some cases, methane in the air. Some states are starting to request atmospheric ammonia test results on air taken at the property lines of animal operations. In the poultry industry, methane may appear in compost piles which do not have enough oxygen, creating an anaerobic condition.

Survey

EPA and USDA are interested in preventing pollution. In order to better understand the potential impacts of regulation and the industries views, a survey was sent out in the summer of 1993 to the top ten broiler, layer, and turkey companies. It solicited their views on the current roles in poultry waste management of members of the poultry industry. The responses indicated that many companies, growers, and people involved in the poultry industry felt that if the question of manure ownership were ever challenged, the integrated company might be deemed responsible for the waste, because the integrated company supplies the waste inputs. It was felt that such a decision could harm growers if it resulted in more restrictions written into the contracts or in a move towards a totally company-owned growout unit. This could effectively remove farming opportunities. Several survey respondents felt that it was unlikely that companies would move their operations to other states, because the benefits will be short lived since most states are in the process of establishing further restrictions in attempts to reduce water pollution in their area. Many of the companies are aware of the environmental problems associated with poultry production and some are actively working with the growers to find alternative methods of disposing of dead birds and animal manure in less environmentally hazardous ways. Several companies, however, are leaving the responsibility of meeting the new standards solely in the hands of their growers. Many respondents were unclear of the ramifications of the Clean Water Act and the Coastal Zone Management Act on the industry.

Responsibility[5]

Part of the problem in choosing an effective and efficient policy that will correct the environmental problems associated with the poultry industry is determining who is responsible for the pollution. The nature of non-point source pollution makes it difficult to define a specific source. This is in part due to the latency period between the actual act of pollution and the manifestation of that pollution. This is compounded by the fact that much of the poultry waste is produced on growout farms but contracted by a particular company which supplies many of the inputs needed to raise the birds.

---

[5] *Refer to paper by Manale, Narrod, and Trachtenberg in this volume for a more thorough discussion of this topic.*

One potential solution to alleviate the manure problem in highly concentrated areas is to market the waste as a value-added product. By doing so, the perception of the manure situation could change (with education) from a liability to being viewed as a resource. It is important that attention be paid to this issue now while the industry and producers can still voluntarily change the situation without regulations or changes in consumers attitude about poultry products. When sold as bulk fertilizer, feedstuff, packaged fertilizer, or chemical extract it has a positive value. Various entrepreneurs who recognized the potential value of manure have developed alternative ways to add value and improve its utilization, such as marketing it as an enhanced fertilizer, an animal feed, methane energy, or a soil filler (in road beds, mine reclamation, and construction projects). The success of the marketing options depends on the types of crops utilizing the manure, the distance from the farm that needs it, and the willingness of the individual to use the manure as a product.

Currently most of the manure from poultry operations, including those on Delmarva, is directly applied to land for its plant nutrient value and its organic matter which improves soil structure. It also serves as a practical way to alleviate the problems associated with the buildup of stored litter and manure. For some growers land application is the least cost practice. For other growers who may not own crop land for disposal and/or do not have contractual agreements with farmers to take the litter and manure, disposal of manure and litter is a serious concern. Several states as well as Canada, are considering legislation to control the quantities of litter that can be spread as fertilizer according to the amount of phosphate needed. If this legislation passed, highly concentrated manure production areas may not be able to land-apply manure. The value of this manure will be limited compared with other sources of nutrient sources in part because of the high transportation costs associated with poultry manure.

Although litter and manure has many potentially local valuable uses, areas with concentrated poultry production may not have adequate cropland and livestock for effective litter utilization. Therefore, exporting litter from concentrated areas to surrounding areas may be both environmentally and economically beneficial. Bosch and Napit (1991) found the value of litter based on the savings that could be achieved by substituting litter for commercial fertilizer would range from $20 to $29 per ton depending on its use and the distance that the litter had to be transported.

Composting is an alternative way to market poultry litter. The composting process is relatively inexpensive but a guaranteed supply of manure, and a suitable market must be available before it will become an attractive alternative. Important considerations in marketing compost are quality of the product; consistency; price; color/texture/moisture; information; and a reliable source of delivery. Potential buyers of compost are: landscapers, commercial nurseries, home and garden centers, greenhouses, homeowners, farmers (fruit, vegetable, field crops, organic), golf courses and cemeteries, public works departments, road and highway contractors, park departments, turf growers and developers (Rynk, et al., 1992).

The development of a compost market in the Delmarva Peninsula appears feasible and various entrepreneurs are attempting to establish one. In Delaware, the extension service has already looked into the feasibility of setting up such a facility. "Poultry Manure Storage and Process Alternatives Evaluation" report by the CABE Associates, identified that it would be difficult to assimilate all of the manure on the Delmarva Peninsula in the future. They felt that there was, however, an opportunity for utilizing the manure in value-added markets such as the horticultural market that would divert the waste water away from the highly sensitive areas such as the Inland Bays and Nanticoke River. Currently, one company, is doing a pilot project using a fifty percent organic blend of poultry manure and chemical fertilizers. The results for crops grown on this medium are very good. Another company, is already established in the area is interested in starting a composting facility that will also pelletize the manure. Several of the agencies have applied for special non-point source pollution funding grants to demonstrate the market opportunities for pelletized broiler litter[7], but have been denied because among other things it would take 2-3 years to show a measured reduction in pollution rates, which exceeds the one year limitation on pollution prevention grants. Perhaps EPA and USDA should be looking more closely at this potential, rather than addressing the need for setting up a central composting facility.

Because ruminants are able to use the non-protein nitrogen in poultry wastes nitrogen can be used as a **feed** for cattle. Fontenot and Jurubescu (1980) calculated that cows can use the nitrogen in animal waste about three times more efficiently than nitrogen from plants. The acceptability of the use of poultry litter as a feed for cattle has been slow. This is because it was initially discouraged when in 1967 FDA issued a policy statement discouraging the feeding of litter to animals. However, in 1980, after extensive tests by University and USDA researchers, the FDA rescinded its earlier policy statement and announced that the regulation of litter should be the responsibility of the state department of agriculture (Ruffin and McCaskey, nd). Currently 21 states have regulations pertaining to the marketing of litter and animal wastes as feed ingredients, and four additional states are considering adopting regulations.

Methane produced through anaerobic digestion is another potential market for manure. Barth and Hegg (1979) have indicated that gas utilization was generally inadequate except in the case of one commercial plant which sold the gas to a commercial pipeline company. Shih (1987), however have done research showing biogas produced from a 50,000 layer hen house could be produced in such a quantity to replace the natural or propane gas for brooding. According to Shih the generation of biogas has multiple benefits beside the generation of a combustible fuel that should be taken into account when looking at the benefits from methane production. The economic feasibility of methane production is increased when the utilization options of the residues are also taken into account. A problem is that after methane has been generated, the excess nitrogen and phosphorous still need to be removed.

The transportation of farm waste is a biosecurity concern. Contamination may be direct through dust particles shed from transport vehicles or indirect through disease agents brought back to the farm from central processing facilities or other locations. This concern has been magnified since the Avian Influenza crisis. It will be difficult to address the feasibility of transporting waste off the farm unless there is some actual proof that there is no problem with transport of unprocessed products off farm. Policies promoting the utilization of a central compost facility will be limited unless scientists are able to compensate for, and change people's perceptions of, biosecurity problems based on facts, and policy makers are able to discern the real risks.

One purpose of this project was to determine the potential for setting up a central compost facility on Delmarva with the help of government. Based on preliminary numbers in our report it would be possible to assimilate all the manure on the Delmarva Peninsula based on use of a nitrogen standard to apply the manure. If these states chose to turn to a phosphorous standard, it is very likely manure in saturated areas will need to develop a method to get the manure off the peninsula. If so, this would be an additional reason to consider composting or processing at a central facility. It seems that there already exist a number of entrepreneurs who are interested in this, which would suggest that government intervention should be minimal in the actual set up. It is likely that a central processing facility will be the key to getting waste off the peninsula after the waste has been composted on the farms. Another option would be for each integrator to handle its own central composting facility to minimize or get around this concern. Alternatively, the clean-out operators or other entrepreneurs could start their own business. Perhaps what government should be looking at is promoting a central processing facility to pelletize and bag the compost after the manure and dead birds have been composted on the farm.

---

[7] *They already have a company who is willing to do this an export a bagged product for lawn and garden markets in the Northeast.*

When property rights are not well defined, Coase (1960) has suggested that the agents can negotiate among themselves for appropriate compensation for incidental services. The agents here would be the growers, vertically-integrated firms, and society who should work together at alleviating the environmental problems associated with the waste. The inability to define responsibility and the resulting failure to met societal goals could in the future result in having the government define the equity issue concerning ownership of/responsibility for the manure. These designations of responsibility mandated by the government could undermine the contractual arrangements that are critical to the success of growers and companies. A regulation that results in making the cost of pollution greater than the value of a company or contract grower's production could make these individuals question whether they should produce at all.

If a government decides it is necessary to intervene, it then has to decide how much protection is necessary, who pays, and if the amount will allow for tradeoffs. Specifically, there has to be a decision whether a zero-risk scenario will be pursued or if there is a safe level of pollution. A policy that promotes zero-risk, prevents trade-offs from being made. In particular there is likely to be a trade-off between productivity and water quality. One result in the reduction in productivity could be price increases for consumers. This could result in a change in consumption patterns since the income elasticity of poultry products is such that it changes with incomes. The consumption patterns that are most likely to be affected are those of the poorer portion of the population who spend most of their disposable income on food and currently are benefiting from this relatively cheap source of protein.

In areas such as the Delmarva Peninsula, where the soil is very sandy, a zero-risk situation might mean no production whatsoever due to the current excess of nutrients in the ground water. Such a policy could have a significant socioeconomic impact on this area, whose number one agricultural commodity is poultry. Part of the success of the poultry industry in the Delmarva can be attributed to the benefits of its spatial concentration resulting in cheaper prices. A zero-risk policy that results in a wider spatial relationship between farms and the suppliers of inputs could cause increases in prices. An alternative approach is to use the best available technology. The idea behind this is that the only allowed pollution is that coming from the best available technology. There are three problems in this approach: (1) there is not always a clear "best" technology; (2) it assumes that the use of such technology is worth its cost; and (3) it lacks a dynamic setting, locking the producer of the problem into a specific means of control (Portney, 1990).

There are three types of policy options that can be followed: voluntary, command-and-control, and market mechanisms. Voluntary approaches rely on getting individuals to adopt environmentally sound practices. These types of programs can save society money in the long run compared to a program that also has enforcement costs. The types of voluntary approaches that have been used in the water quality area are: education, technical assistance, and cost sharing for BMPs. As farmers have become aware of the current problems, many have switched to or improved their BMPs and NMPs so that pollution may be less of a problem in the future. However, as the poultry or other animal industries (and human population) continue to expand in this ecologically sensitive area that is already concentrated with animals, waste disposal is likely to continue to be a problem. Some voluntary approaches have been successful in Delmarva. The manure clearinghouse system of the Delmarva Poultry Industries, Inc. attempts to move the manure from surplus areas to deficit areas, but it is being underused. In the dead bird composter area, the most successful efforts seem due to cost-share programs. With manure sheds a smaller step has been taken, due to the misuse of sheds. Unless these voluntary approaches work successfully there will be moves to look at instituting other types of regulations.

Part of the problem of the voluntary approach to waste management is the ability to hold individuals accountable and enforce policy. The established Water Quality Acts fail to provide sanctions for inadequate progress in implementing water quality programs. Currently most states are relying on voluntary type programs that minimize economic effects. Malik et al. (1992), suggest that this is done in spite of these programs not meeting water quality goals. An easier method from a societal viewpoint might be to let the vertically integrated firms be the ones responsible for getting their growers to meet the environmental restrictions. The problem is where several firms compete in restricted locality, no one will want to be the first to bear the expense. Such a problem is a classic case of what is called a prisoner's dilemma in economics and promotes a free-rider situation.[6] For instance, if one company mandated all its growers do a certain practice that would require an excess cost to them, is likely that if all the other integrators do not also mandate this, that the grower would contract with another integrator. The risk of this not happening is an incentive not to cooperate in a highly concentrated area, such as Delmarva. For a successful strategy, the industry would have to agree through collective action to follow a certain policy, or the government could break the prisoner's dilemma through regulation or changing the prices that everyone faces. Currently the poultry industries on the Delmarva are highly organized in an industry sponsored organization called the Delmarva Poultry Industry that could easily and fairly facilitate various actions.

A command-and-control approach declares that there is a source-specific pollution problem in which a limit will be set and backed up by the threat of enforcement actions. In the past, most environmental regulation used in the United States has relied on a command-and-control approach. Manufacturing and industries have been the most affected by this type of regulation. This type of regulation, however, has failed to include the small producer (such as contract growers) and other individuals whose type of discharge was less amenable to this sort of regulation. The water pollution problem in the poultry industry is generated with the improper disposal of manure and dead birds at the contract grower stage of production. It is possible that these small producers will not be able to afford to meet the mandates enforced on them. Those that can't will be forced ultimately to go out of business. Is this an equitable solution? This is a particular concern in the poultry industry where most of the production is done by contract growers who are forced to comply or get out of the business. It is possible that any policy that regulates the control of manure would ultimately affect the grower. This is because if the company is forced to comply, it will likely write the necessary changes for compliance into the contracts of growers. This may not however be acceptable from a social or political viewpoint.

Economic incentives are instruments that are used to influence, rather than dictate the actions to a targeted party and allow businesses and consumers to make their own choices by providing continuous inducements, financial or otherwise, for sources to make reductions in the environmental pollution they release. These policies attempt to correct market failures by adjusting the costs faced by the private decision-makers to reflect the full social costs of their actions.

The major advantage of economic incentives is that they allow information about scarcity to be transmitted across actors via prices and quantities demanded. However these mechanisms can be restricted by information costs, economies of scale, high transaction costs, joint impact goods that effect others (e.g., Prisoner's dilemma, free rider problem), variability in supply and demand, short versus long-term effects and outcomes that may be "efficient" but socially undesirable. Summarized in table 1 are the different types of economic incentives that have been used by regulatory agencies to reduce environmental problems (EPA, 1991).

---

[6] *A prisioner's dilemma might occur that would prevent one company from making the first move for fear of loosing growers if other companies chose not to follow. The free-rider situation occurs when one individual makes the choice to follow a certain action and all the other individuals chose not to change and ride while benefiting from the changes of another.*

Table I Market Base Measures

Marketable Permits - are government-issued permits that use a system of allowable ceilings on the amount of discharge of pollutants (or the use of scarce environmental resources); These permits can be tradable.

Monetary Incentives - are methods to change market incentives, including direct subsidies, and the reduction of subsidies that produce adverse environmental effects, fees, or taxes;

Deposit/refund Systems - are schemes to discourage the disposal and encourage central collection of specific products;

Information Disclosure - are actions to improve existing market operations by providing information to consumers;

Procurement Policies - are means by which the federal government uses its own buying power to stimulate development of markets -- e.g., for recycled products;

* The type of economic incentives most related to the poultry industry are the marketable permits, subsidies/monetary incentives, information disclosure and taxes.

Reasons why poultry industry should be concerned

It is recognized that the potential environmental risks due to the disposal of wastes are magnified in Delmarva as the result of the dense confinement practices and the geographical concentration of the poultry industry. The flip side of having to ship waste from concentrated areas to less concentrated areas is to promote a policy that causes the de-concentration of all the animal industries. It would do this by limiting the size of flocks to be no larger than the carrying capacity of the local available land base. Such a policy would affect the spatial cost of production. Aho (1989) determined that if plants had to relocate farther apart it would increase costs to nearly one cent per pound for deliveries from the feed mill, processing plant, and hatchery. An increase in the supply band to a distance of 35 miles away from the growout operation would double the spatial costs and increase the costs to $2 million per year. Such an increase in cost would result in increased prices for the consumer. Such a policy might cause firms to move to areas outside the Coastal Zone area where regulations weren't so strict. It is likely that the benefits of this will be short lived as many states are already in the process of adopting stricter regulations.

King et al., (1983) did a study looking at the economic impact of converting commercial egg production to less intensive production systems with regard to animal right concerns. They concluded that the investment to convert 45,360 hens from 4 to 3 per cage would be a $91,000 investment, costing the U.S. industry $240,740,500 and one forcing the industry to convert to an all floor-deep litter system for 230,000,000 layers would cost $1,120,000,000. A family owned operation with 45,360 layers would require $250,000 for additional investment. The supply would drop 10 to 20% until conversion could take place and the price would move 60% = 85 cents to $1.35 per dozen. It is possible that similar actions could be used to reduce the amount of manure being produced for a given amount of land. It can be seen that this would result in huge price increase for the industry and the consumers. If a law went into effect mandating less intensive systems, everyone would be affected.

Options

Market Manure as a Value Added Product by treating manure as a resource focusing on its nutrient and soil amendment value. Poultry manure has been successfully marketed as commercial or residential fertilizers (often pelletized), animal feeds, methane sources and feedstocks for chemical extracts and soil amendments. It has also been home gardeners, commercial growers, and others with applications including bulk soil amendments for mine reclamation and construction projects. The success of the marketing options depends on the types of crops utilizing the manure, the distance from the farms that needs it, and the willingness of the individual to use the manure as a product.

They could do this through the use of policies that would facilitate an economically sustainable operation that it would eliminate the unfair advantage any one compost company would have with the fertilizer industry, which is subsidized, or the compost from sewage facilities which is often given away. Until that is done most compost/processing facilities may have long-term economic problems.

REFERENCES

Aho, P. 1989. Spatial Costs and the Economics of Manure Management. Paper presented at the Symposium on the Clean Water Act and the Poultry Industries, Washington D.C.

Bosch, D.; and K. Napit. 1992. Economics of transporting poultry litter to achieve more effective use as fertilizer. Journal of Soil and Water Conservation, 47: 342-346.

CABE Report. 1991. Poultry Manure Storage and Process Alternatives Evaluation. Cabe Associates, Inc., Consulting Engineers, Dover, Del.

Carr, L. and H. Brodie. 1992. Composting Hatchery By-Products and DAF Skimmings. Presented at Midwest Poultry Federation Convention, Minneapolis, Minnesota.

Coase, R. 1960. The Problem of Social Cost. Journal of Law and Economics, 3: 1-44.

EPA. 1991. Economic Incentive Options for Environmental Protection. US-EPA.

Fontenot, J. and V. Jurubescu. 1980. Animal Waste Utilization. In: Digestive Physiology and Metabolism in Ruminants, AVI Publishing Co., Westport, Ct. 641-664.

King, D.; Lasley, Holleman, Bray. 1983. Discussion of the Economic Significance of Converting to Less Intensive Commercial Egg Production in the U.S. ARS.

Malik, A; B. Larson, M. Ribaudo. 1992. Agricultural Non-point Source Pollution nd Economic Incentive Policies: Issues in the Reauthorization of the Clean Water Act," Resources and Technology Division, ERS, USDA Staff Report, AGES #9229.

Manale, A; C. Narrod, E. Trachtenberg. 1994. Opportunities Available to the Vertical Integrator to Achieve Water Quality Benefits in the Dairy and Poultry Industries. This vol.

Narrod, C.; R. Reynnells, H. Wells. 1993. Potential Options for Poultry Waste Utilization: A Focus on the Delmarva Peninsula. White Paper for USEPA/ESUSDA.

Portney, P. 1990. Public Policies for Environmental Protection, Resources For The Future.

Rogers, G. 1979. Poultry and Eggs. In: Another Revolution in US Farming. USDA.

Ruffin, B. and T. McCaskey. nd. Feeding Broiler Litter to Beef Cattle. Alabama Extension.

Rynk, R. ed. 1992. On Farm Composting Handbook, NRAES-54.

Shih, J. 1987. Ecological Benefits of Anaerobic Digestion. Poultry Science, 66: 946-950.

Sims, J. and D. Wolf. 1993. Poultry Waste Management: Agricultural and Environmental Issues. Advances in Agronomy, vol 50, ed. Sparks, Academic Press, Inc., Orlando. Fl.

OPPORTUNITIES AVAILABLE TO THE VERTICAL INTEGRATOR OR FOOD PROCESSOR TO ACHIEVE WATER QUALITY BENEFITS IN THE DAIRY AND POULTRY INDUSTRIES

A. Manale, C. Narrod, E. Trachtenberg[1]

ABSTRACT

The paper examines how the current structure of the dairy and poultry industries leads to water quality problems, as well as opportunities to correct these problems. Contracting between small producers and large food processors and vertical integrators increasingly characterizes the production of dairy and poultry products. Decisions by the processors and integrators regarding the volume of their production directly affects the intensity of production of milk and poultry production and hence the volume of animal waste that must be disposed of. Intense competition among producers for these contracts reduces the ability to pass on the cost of environmental protection to the middleman. Since the producers do not sell directly to consumers, they are not able to capture the "green" premium that consumers may be willing to pay. Nevertheless, economies of scale for addressing animal waste problems do present themselves at the level of the processor. These are identified and discussed. Possible policies to encourage responsible waste management are suggested.

WATER QUALITY AND THE DAIRY AND POULTRY INDUSTRIES QUALITY

Agriculture represents the most important source of water pollution in the United States. Nitrogen and phosphorous from agricultural runoff and leaching can cause eutrophication, fish kills, and methemoglobinemia in babies and animals who drink the water contaminated with high levels of nitrates. Animal operations, particularly those relating to dairy and poultry operations, account for one-third of all agricultural non-point (diffuse) source pollution (USEPA, Office of Water, 1993). Nutrient runoff and leaching from animal waste has impaired fisheries in 60,000 stream miles, caused extensive fish-kills in California and Florida, and contaminated ground water in 17 states.

Until the development of chemical fertilizers in this century, animal waste served as the chief fertilizer for crop production. When applied at agronomic rates to growing crops, nutrients derived from manure rarely pollute ground or surface waters. However, where there is insufficient land upon which to apply animal waste or, in the case of the poultry industry, the remains of dead chickens, the excess phosphorous or nitrogen that is not taken up by crops can move to ground or surface waters. Large amounts of animal waste over immense areas can lead to ammonia emissions that can cause acid rain. Overapplication of wastes over extended periods of time can cause the build up of phosphorous in the soil to levels that seriously affect the productivity of the land. The concentration of large numbers of animals relative to the availability of cropland has led to the overproduction of nutrients in relation to the economic feasibility of its use as fertilizer.

---

[1] A. Manale and E. Trachtenberg, EPA, Agricultural Policy Branch, Washington; C. Narrod, Center for Energy and the Environment, University of Pennsylvania, Philadelphia.

We would like to thank Amy Pagano, Fred Woods, and Basil Eastwood for their helpful comments on drafts of this paper.

The preponderance of poultry products are produced under contract to food processors and vertical integrators, large companies that link together various portions of the production, input-supply, and market segments of an industry. The preponderance of milk products in the dairy industry however are produced by individual producers who sell to producer-run cooperatives. These cooperatives then sell to milk-procurement processors. These milk procurement processors are regulated by milk marketing orders. These marketing orders regulate the processors, not the producers. The producers, however are affected by the marketing order when they ship to the regulated processor.

Secondary products, such as manure or dead animals, which have no value to the wholesale buyer and limited value to the specialized producer, become waste to be disposed of as cheaply as possible, by the producers. Production contracts with many integrators neglect to explicitly state who owns the animal waste or define the responsibility relating to its environmentally sound management. This is a problem, because no one wants to take responsibility for it. The nature of the contracts for the dairy producer, however differs from the poultry industry because the milk cooperatives act as middlemen between the producer and the milk procurement processors while the producers rely on their cooperative to perform marketing functions and to represent their interest as in establishing prices and terms of trade with the procurement processors. In this situation it is clear that the producer is the one responsible for the waste. Nevertheless, neither type of contract facilitates the passing of the environmental costs to the consumer.

Policies to address problems of animal waste should take into account the structures of the industries, because it can play a role in the clarification of the property rights. For industries that are highly integrated, greater responsibility of the integrator reduces the enforcement cost of ensuring compliance. In situations where middle men exist, passing on the cost to the procurement processor becomes more difficult. It is unclear how the assumption of greater environmental responsibility by the integrator or milk procurement processor will affect the concentration of the industry or the competitiveness of the small producer.

THE PROBLEM OF MARKET FAILURE AND ANIMAL WASTE

The environmental problems associated with the management of animal waste from poultry and dairy operations are the consequence of what economists call "market failure." Failure of the market to assign a value to environmentally sound management of animal manure or dead animals results in what is called an externality in economic theory. Imperfect information contributes to market failure.

<u>Externalities</u>

An externality is a cost that is borne by a person other than the one who caused the cost to arise (Portney, 1990). Externalities arise when an individual, in the course of rendering some service for which payment is received, coincidentally renders services or disservices to other persons for which payment cannot be extracted. In the case of the poultry and dairy industries, excessive release of nutrients into the environment associated with the production of such goods as broiler meat, eggs, or milk, results in contamination of drinking water supplies or impairment of fisheries affecting the welfare of downstream users. Because the downstream cost of environmental pollution is not borne by the producer, the price of the product does not always reflect what society must sacrifice to enjoy its benefits.

Studies indicate that consumers are willing to pay for environmental protection, (e.g., Carson & Mitchell, 1986). In this case, there would be incentives for the producer to internalize the externality and assume the cost of environmental protection. But with regard to the production of dairy or poultry products, consumers, as well as producers, have imperfect information regarding the associated hazards. Not knowing that the environmental problems are brought about by the aggregate impact of all producers in a watershed, an individual grower may not be aware of the environmental liability associated with producing the poultry or dairy product. Furthermore, the price paid to the producer by the integrator or processor is the same regardless of whether or not animal wastes are handled properly. When sold in the marketplace, the good conveys no information about how it is produced. The high cost of collecting such information prevents consumers from making an informed choice affecting his or her demand. The consumer is unwilling to pay more for an item unless he or she has certainty that it combines the attribute of environmental protection. Thus, the marketplace is unable to convey signals to the producer regarding consumer preferences that could ultimately affect how animal waste is handled.

There are three major externalities in the production and processing of poultry and dairy products that are unlikely to be borne by the producer. The first externality relates to waste. When spread at agronomic rates on cropland reflecting the physical characteristics of the soil, they represent a positive externality and a valuable source of nitrogen, phosphorous and organic material. Application beyond what is taken up by plants creates a negative externality when excess nitrogen and/or phosphorous leaches into the groundwater or contaminates surface water. A second externality arises when the by-products from processing are discharged directly into surface waters. In the United States, most vertically-integrated companies in the poultry industry have developed rendering plants to alleviate this form of water pollution. A third externality is caused by the release of ammonia into the air leading to air pollution, such as acid rain, and odor problems--a problem in heavily concentrated areas, such as large feedlots.

ORGANIZATION OF THE INDUSTRIES

Poultry Industry

Many companies in the poultry industry are vertically integrated. Their size allows them to purchase inputs in bulk and to take advantage of economies of scale. The centralization of decision-making allows coordination of production capacity at each stage of production to guarantee a timely supply of a standardized product of uniform and high quality that can sell at a favorable price ensuring dominant market share. Such coordinated systems can also affect more rapid adoption of improved production technology, enable development of new advances, and respond more quickly to changing consumer needs (Rogers, 1979).

Large integrated firms contract out much of their production to farmers who live not more than 25-30 miles from their processing plant (Heffernan, 1992). The contract farmer owns the land and buildings of the grow-out operation and provides the necessary labor and electricity. These contract farms are served by a company's hatchery and feed mill. An estimate of 97% of all broilers are now produced under some form of formal vertical coordination, the remaining are raised by independent producers for niche markets (Rogers, 1979).

The purpose of the contractual arrangement is to form a minimum guaranteed payment for the grower, which often is adjusted based on feed conversion, mortality, and other factors. It enables individual producers to move from a simple credit financing and open-market production to a variety of contractual agreements. By doing so, the risk associated with production is reduced. Contracts can differ in four ways: 1) the degree to which both parties participate in management during the production process, 2) the method of payment for the product or service produced, 3) how risk and profits are shared, and 4) whether one or both parties supply resources used in production (Mighell and Jones, 1963). These contracts by nature are for a short period of time. The length of time depends on the individual contracts between company and grower. For a broiler producer the contract might specify a base price to be paid per pound or per bird produced. A minimum bird weight may be required. Often premiums are based on the grower's efficiency and are then added to the base rate the grower receives. With turkeys, the production tends to follow the broiler pattern with conventional lending rather than work under contract.

In the production of eggs, contracting is less frequent, though it still dominates total production, as sixty percent are produced under contractual arrangements (Haynes, 1982). The contracts specify that the pullets, feed, medicine and field services are furnished by the company; the producers furnish the housing facilities, electricity, and labor. The eggs belong to the company and the producers are paid on the basis of the number of eggs produced, with bonuses for efficient performance. Either party can terminate the agreement. These contracts are attractive to many chicken producers, many of whom are small farmers, for the following reasons: with limited financial resources they are unwilling and often unable to obtain the production capital through traditional sources and to bear the market risks associated with entering the broiler or egg markets (Benson & Witzing, 1971; Reimund et al., 1981). By turning to these input suppliers for their financing, they have an alternative source of production capital and a means to shift a substantial part of the financial and market risks to the contractor.

The contract also serves to reduce the risk and the amount of capital needed by the company since it does not have to invest in growout and breeder facilities. Contracting also provides a steady flow of birds for the company. Were integrators to shoulder all the risks, building their own laying and growout facilities, as has been seen in the egg industry, might be the preferred alternative. Because there is low investment and little financial risk to growers, the entry cost of providing the service to companies is low. With the pool of potential growers available to companies likely to be relatively large, companies can pay relatively low fees for the service.
A small number of companies do, however, offer risk/profit sharing which enables growers to cover the additional expense of disposing of poultry in an environmentally sound way.

The responsibility for waste management differs among the broiler, layer, and turkey industries; among contract, independents, and company owned farms; among companies themselves; and among companies' contracts with growers. Since many of the layer operations are company-owned, the company is responsible for the waste. Some broiler companies have written into their contracts that ownership of the dead birds and manure is the grower's responsibility. In this case, the grower generally sells the manure or gives it away. Some of these companies have chosen to share the responsibility of the dead birds with their growers by underwriting the expense of collecting dead animals. The responsibility for the manure in all operations is the grower's, be he a contract grower, independent grower, or a company.

<u>Dairy Industry</u>

In the dairy industry a similar consolidation of the industry has occurred. Fewer dairy farms produce less milk, accordingly farms have become larger and more specialized. However there are many different types of producers and considerable heterogeneity within and across regions. Traditionally many dairies have operated with sufficient acreage to raise animals' feed supply and dairy herd replacements. In 1983 the average herd size for 120,655 producers selling milk to plants regulated by Federal milk marketing orders was 63 cows per farm (OTA, 1986). However the industry is changing. By 1992 the average herd size in the dairy records program was 88 cows (USDA, 1993). There remain many regional differences.

The main form of dairy farm organization throughout the United States is of a single family proprietorship. These producers tend to grow their own feed and apply the manure that is produced as a crop and pasture fertilizer (Jones et al, 1993). Nevertheless, in the Corn Belt, South West and South East partnerships are growing in number and corporate specialized farms are common in the North East, Lake States, South West, and North and Central Regions. These large producers purchase much of their feed requirements and are growing pasture and crops merely as a by-product of waste application to their fields (Jones et. al, 1993 citing TIAER, 1992). An important factor in this consolidation is the significant economies of size and scale in dairy production and dairy manure management (Matulich, 1976).

The structure of the dairy industry differs however from the poultry industry in that most dairy farmers sell to producer cooperatives (middle men) who then sell directly to milk procurement agencies as opposed to contracting with an integrator. These cooperatives strengthen the market power of the producers who market the milk as a collective (Babb and Bohal, 1979). Most members are able to increase their relative strength through the collective. (Granted some large producers would have more power if they marketed the milk by themselves.) Often these cooperatives contract with milk procurement producers that control milk marketing not production technology. These market-specific agreements set prices and quantities of milk that is purchased. The advantage of these arrangements are that they reduce the risk to both the milk procurement agency and the cooperative while cutting transactions costs. The advantage to the producer comes in the form of ready markets, reduced marketing costs, and lower price uncertainty. At the same time, the farmer keeps control over the production process. The procurement agency gains from having risk and uncertainty reduced through stable supplies of raw materials bought at the predictable prices.

Milk marketing orders also affect regional prices. They are used to establish a uniform price for volume sold. They do this by stratifying the prices processors are required to pay based on the use of the milk they purchased. Furthermore these orders establish a system of classified pricing and rules for distributing the proceeds from the sale of milk among producers (Dunn et al., 1979). They are divided into three classes: fluid milk products, perishable manufactured products, and storable manufactured products. These orders contain provisions such as when the producers must be paid, butterfat differentials, location adjustments applicable to prices, shipping requirements that plants must satisfy to qualify for pooling, and optional deduction to promote the sale of dairy products, which subsequently affects the terms of trade (OTA, 1986).

Current Roles of waste management for members of the dairy industry

The milk procurement contracts of the dairy industry, unlike the vertical integration in the poultry industry, do not extend to the inputs and production technology for the small producer. A different relationship is found between the processors and the producers with regard to manure disposal. At the present time, manure disposal is the responsibility of the individual farmer. In most cases the farmers have adequate land for the disposal of their waste, with the exception of the large openlot producers and areas where there are a large number of animals of all types and limited crop land. There is considerable risk of ground water and surface water run-off if the application of waste doesn't correspond to crop needs of plants. This can create a problem because the capital costs of an environmentally acceptable manure storage system for a herd of 50 milk cows can range from $10,000 to $85,000 (Milne, 1990). This cost does not include the field application equipment. This need to store waste is necessary as of now because the alternative markets for waste, such as composting and feed are relatively nonexistent in many areas of the country.

There are economies of size and scale in waste management (Fallert et al., 1993). According to recent estimates from a survey done by the Texas Food Policy Center (1993), the cost per cow to comply with waste management rules would be $400 per cow from a 50 head farm, compared to $228 per cow from a 175 head farm. These figures do not include the cost of managing the system. This same study indicated that, "large scale dairies that were not already in financial trouble appeared able to amortize the extra capital investment cost associated with compliance of dairy manure management regulations" (Outlaw, et al., 1993). These results also indicated that there was a potential for a substantial restructuring of the dairy industry in conjunction with meeting the proposed EPA standards for region 6 (Fed Register, Feb. 8, 1993) for dairy management. Larger dairies are able to assume the extra capital costs. This same study showed that the cost to put in proper waste management was substantial. It indicated that it would take several years for small farmers to recover financially after this initial investment. This again would make it difficult for the small farmers.

Policy Implications

The discussion of the structure and organization of the poultry and dairy industries should explain in large part why there has been only limited success with the current voluntary approaches to addressing the externalities of water pollution from the dairy and poultry industries. Where there is concentration of large numbers of animals in a geographic area, usually around a large processing facility in the case of poultry or within the confines of a milk marketing order, there is a definite financial cost to handling animal waste in an environmentally sound manner. For the smaller producer, the cost tends to be more burdensome than for the larger, though this clearly depends upon the indebtedness of the producer. The individual producer generally cannot pass this cost onto the buyer of the goods, the vertical integrator or processor, unless all (or most) producers within the production region also choose to internalize the environmental costs--a situation that is analogous to the game theorist concept "prisoner's dilemma." The latter, however, who generally has a large pool of potential producers with which to contract, has the option of whether or not to purchase or contract with any individual producer. Hence, an individual producer who chooses to internalize the environmental cost puts himself at financial risk because he must absorb these costs himself--in lieu of government cost-sharing programs. Furthermore, should there be a premium that consumers will pay for poultry or dairy products that are produced in ways that are more environmentally sound, the individual producer cannot necessarily garner this premium without incurring the substantial transaction cost of advertising. Even so, without objective assurance that the producer really is "green," consumers may not be willing to pay an additional amount over the "non-green" product.

The customary solution to the prisoner's dilemma in policy analysis is government regulation whereby rules on the handling of animal waste are imposed upon all producers. The Dutch, for example, have rules requiring that all excess animal waste from more than a given number of animals relative to cropland, be deposited at a central manure depot--a manure bank. In addition, a tax, though small, is also imposed once the animal-land ratio has been exceeded. The Germans similarly have rules limiting the amount of waste that can be legally applied to cropland. In some areas of Germany, government helps to reduce the transaction cost of finding alternative uses for the manure through the establishment of brokerages. However, enforcement of both regulations has been very spotty.

The administrative cost of enforcement is very high--there are many individual producers, but few inspectors. Furthermore, the European experience suggests that, even where violations have been identified, they are rarely prosecuted. Local administrative bodies upon which the responsibility for enforcement has been delegated, are not likely to impose fines or force compliance where rules are seen as representing a cost to a community and benefits accrue primarily to people outside the community (Manale, 1991).

A recurrent criticism of regulation is that it generally falls more heavily upon smaller producers. This, however, may not be a significant problem in the contract poultry industry. In existing contractual relationships, the vertical integrator generally does assume responsibility for all major investment, short of the actual land and capital costs, necessary for entry into poultry production. Technology to meet environmental regulations should therefore represent an additional required investment. In cases where there are short term contracts with the integrator, there is little incentive to invest in environmental technology because of the greater risk of not being able to get the returns back.

We see however that when property rights were defined for point source pollution as in the Clean Water Act, subsequent fines placed on the polluter have resulted in technological innovation such as been seen with further rendering plants in the poultry industry. The far lesser integration of the dairy industry suggests that raising the cost of production to meet environmental regulations would cause the demise of many marginal producers whose total revenue is not sufficient to absorb extra costs. Forcing them to do so would accelerate current trends towards larger, better capitalized operations. It is possible that the same thing will happen to the small producer in the poultry industry if integrators are forced to ensure that all their small contract growers have proper waste management measures. Specifically they might find it more cost effective to further integrate as seen in the layer industry rather than assume the environmental responsibility of having many contractors.

Policies to address the externalities associated with animal wastes, such as regulation directed solely at producers, could therefore have very different consequences depending upon the degree of integration of the industries and how the industries choose to respond. In order for sound public policy to be made, the environmental policies should be tailored to the unique features of the industry and designed in such a way so as not to impede social objectives. It is possible that the social objective of environmentally sound management may result in the small producers being put out of business which would generate another social cost. For instance, the dairy and poultry sector could adopt policies analogous to the 1990 Farm Bill Conservation Compliance provisions that tied eligibility to commodity price support to farm plans for reducing soil erosion. In this case, dairy price support would depend upon farm plans for environmentally sound management of dairy waste.

However, we would expect that either direct regulation of what producers generate or the legal determination of vertical integrator ownership of animal waste would lead to the assumption by the integrator of greater responsibility for poultry waste management. It is possible that they would then pass the cost on to their contractors or the consumers. In both industries, the social cost of direct regulation may not be acceptable. Thus we suggest that a more novel approach be considered. Certification for environmentally sound waste management, the creation of a special "green" grade coupled with a higher minimum price support level, and the creation of special "green" dairy labels to allow differentiation in the marketplace that would allow for any consumer premium for "green" dairy and poultry products to be passed onto producers is an alternative one. Such green labeling helps to offset the costs of promoting the adoption of technologies that use green production techniques. This will enable integrators, who currently don't differentiate the price that is paid to producers who handle their waste properly, to be able to.

References

Babb, E. and R. Bohall. 1979. Marketing Orders and Farm Structure. In Structure Issues of American Agriculture. USDA Agricultural Economic Report 438.

Benson, V. and T. Witzig. 1977. The Chicken Broiler Industry: Structure, Practices, and Cost, USDA. ERS. Marketing Research Report. #930.

Carson, R. and R. Mitchell. 1986. The Value of Clean Water: The Public's Willingness to Pay for Boatable, Fishable, and Swimmable Quality of Water. Discussion Paper No. QE85-08 revised, (Washington, D.C., Resources for the Future).

Fallert, R.; M. Weimar, and T. Crawford. 1993. Here's Why Milk is Moving West. Hoard's Dairy Magazine. 139(1) p 7.

Haynes, R. 1982. Poultry Industry Unique in Contract Operations. The Poultry Times. Jan. 18.

Heffernan, W. and D. Constance. 1992. Restructuring of the U.S. Meat-, Poultry-, and Fish-Processing Industries in the Global Economy. Paper presented at New Factory Workers in Old Farming Communities: Costs and Consequences of Relocating Meat Industries. April 12-14.

Jones, R.; L. Frarey, A Bouzaher, S. Johnson, and S. Neibergs. 1993. Livestock and the Environment: A National Pilot Project Detailed Problem Statement.

Manale, A. 1991. European Community Programs to Control Nitrate Emissions form Agricultural Activities. An Evaluation of their State of Implementation and Effectiveness. A report to the German Marshall Fund of the US.

Matulich, S. 1978. Efficiencies in large-scale Dairying: Incentives for future Structural Change. American Journal of Agricultural Economics, 60(4): 642-647.

Mighell, R. and L. Jones. 1963. Vertical Integration in Agriculture. Economic Research Service. Washington. 90p.

Milne, R. 1989 Dairy Manure Systems-Can They be Both Economically and Environmentally Viable? in Dairy Manure Management Symposium Proceedings. Syracuse. New York. NRAES.

Outlaw, J; R. Schwart Jr., R. Knutson, A. Pagano, J Miller, and A. Gray. Impacts of Dairy Waste Management. Texas Agricultural Food Policy Center, Working Paper 93-4, Texas A&M.

OTA. 1986. Technology, Public Policy, and the Changing Structure of American Agriculture.

Portney, P. 1990. Public Policies for Environmental Protection. Resource for the Future.

Reimund, D.; J. Martin, C. Moore. 1981. Structural Change in Agriculture: The Experience for Broilers, Fed Cattle, and Processing Vegetables. USDA. Economic and Statistic Service, Technical Bulletin. 1648.

Rogers, G. 1979. Poultry and Eggs. Another Farming Revolution. USDA. Washington.

USDA. 1992. Dairy Situation and Outlook Report. DS-429, Commodity Economics Division, Economic Research Service. Washington.

US Department of Agriculture. 1993. National Agricultural Statistic Service.

Federal Register. 1993. NPDES General Permit and Operating Regulations for Discharges for CAFOS. Feb 8.

EVALUATION OF MANURE WATER IRRIGATIONS

D. Morse, L. Schwankl, T. Prichard, A. Van Eenennaam[*]

ABSTRACT

The goal of this project was to develop a method for dairy producers to comply with the nutrient and irrigation water management components of the Coastal Zone Act Reauthorization Amendments. Dairy manure water irrigations on a corn silage crop were evaluated. The objectives were to determine: 1) normal application rates of water and nutrients; 2) uniformity of water and nutrient application; 3) estimates of nutrient distribution and water infiltration. The 1.91 ha (4.71 ac) border check was flood irrigated. Soil samples were taken prior to and following the pre-seeding irrigation and post harvest. Alternate irrigations were monitored. The irrigation valve was fitted with flow meters. During the first irrigation, flow meter readings and water samples were taken (30 to 40 min intervals) at the headland and at quarter points in the field. The field was marked at 61 m (200 ft) intervals to measure water advance rate. During the initial irrigation, tail water, wind speed and pH values for water samples were quantified. Irrigations consisted of manure and irrigation water mixes. The ratio changed within and between irrigations depending on consistency of water entering the field and availability of canal water.

Keywords: nutrient budgeting, manure water irrigation, infiltration rate.

INTRODUCTION

The California dairy industry is home to some 1.2 million dairy cows and an equal number of replacement animals. The geography of the state, precipitation and availability of agricultural production lands vary tremendously. Differences in these parameters greatly impact manure collection and handling techniques. A study by Meadows and Butler (1989) evaluated manure handling techniques by region. In the Chino (Southern California area) herds are large (800 cows), with little acreage for waste disposal. Operators pay to have manure solids hauled away. Approximately two-thirds of the state's dairy herd resides in the Central Valley. This number is expected to increase as dairies relocate from Southern California. In the Central Valley manure is collected via a variety of systems, including flushing. Greater than 60% of Central Valley dairy herds have flush systems as a component of their manure system. In the Southern Central Valley 40% of herds sell manure solids from the farm. Dairies in Coastal areas used some flushing (40%). Of the herds surveyed, 16% sold manure solids from the farm.

Across the state, 50.3% of herds used some type of flush system for manure cleaning. Inherent with including flush systems in a manure collection system is the need to ultimately dispose of the effluent and handle the solids. Some 90% of producers surveyed included manure water irrigations as part of the liquid waste disposal process. In the areas where fewer crops were grown, or which received great amounts of rainfall, producers also included the importance of evaporation from holding ponds as a critical component of liquid disposal (17.2% coastal; 19.9% Southern California).

The legislative climate in California during the late 1960s indicated the importance of

---

[*] D. Morse, Livestock Waste Management Specialist, Department of Animal Science; L. Schwankl and T. Prichard, Irrigation Specialists, Land, Air and Water Resources, University of California, Davis; A. Van Eenennaam, Dairy Advisor, San Joaquin County.

understanding potential threats of nutrient leaching to the groundwater underlying manure water holding ponds. University of California researchers carried out studies to evaluate sealing capabilities of holding ponds (Meyer, 1973). Additional calculations were made to estimate nutrient excretion by animals. Losses of nutrients (particularly nitrogen (N)) during storage were not evaluated. Research results from the early 1970s were used by the State Water Resources Control Board to estimate animal carrying capacity per acre of farmed land. Until recently, little additional work was undertaken to evaluate actual application of nutrients from manure holding ponds.

The recent publishing of 6217 (g) Guidance (EPA, 1993) indicated the immediate need to understand nutrient and water application rates when manure waters are used during irrigations. Sections on irrigation water and nutrient management are included in 6217 (g) Guidance. The goal of this project was to evaluate regular irrigations with manure/fresh water mixes to a corn silage crop. The objectives were to determine: 1) normal application rates of water and nutrients; 2) uniformity of water and nutrient application; 3) estimates of nutrient distribution and infiltration.

MATERIALS AND METHODS

This experiment was conducted in one 1.91 ha. (4.71 ac) border check operated by a dairy producer in California's Central Valley, during the 1993 corn growing season. Check size was 30.5 m (100 ft) wide and 626 m (2053 ft) long. The field was laser leveled with a slope of 0.15% prior to irrigation 1. Soil type was Veritas, sandy loam (loamy, mixed, thermic, typic Haploxerolls). Depth to groundwater at the bottom of the field was 1 to 1.4 m (3.5 to 4.5 ft). Hardpan was reached at the bottom of the field at a depth of 1 m (3.5 ft). No precipitation occurred during the trial. Alternate irrigations (3 of 6) were monitored.

Field layout

The soil was disked and harrowed. For the pre-seeding irrigation (irrigation 1), temporary borders were formed 30.5 m (100 ft) apart to control water released from the single valve. The field was border irrigated during irrigation 1. Furrow corrugations, perpendicular to the direction of water flow, were used to ensure uniform spreading of water across the border check. Borders were flattened; the soil worked. The crop was planted flat on a 0.76 m (30 in) spacing. During seedling stage, the crop was cultivated with a rolling cultivator set in a fashion to form furrows. The head and tail of the check (18 rows) were planted and cultivated perpendicular to the field slope and water flow. The remainder of the field was planted and cultivated parallel to field slope and water flow.

Soil

Soil samples were collected at three locations: 30.5 m (100 ft) from the top and bottom of the field and in the middle of the field. At each location 5 holes were sampled. Samples were collected at 30.5 cm (1 ft) depth intervals. Samples were collected 3 days prior to irrigation 1 and 16 days after irrigation 1, just after seeding. The center samples were used to determine soil moisture and bulk density values. The final soil samples were collected at harvest (October 15). Samples were dried in a 40°C forced air oven. Samples from the top of the field were analyzed individually and by volumetric composites by depth. Samples from the middle and bottom of the field were evaluated individually for the center hole (5 samples) and by volumetric composites by depth (5 samples). Samples were analyzed for electrical conductivity, NO_3-N, NH_4-N, Total Nitrogen, Olsen-P, and exchangeable K.

Irrigation Monitorings

One irrigation valve was used to supply water to the check. It was fitted with four flow meters (3 20-cm (8-in); 1 30.5- cm (12-in)) calibrated prior to monitoring. Water entering the field included a mixture of manure and canal waters. Flow meter readings were taken at 30 to 40 minute intervals during each monitored irrigation. Meters were monitored and tended to regularly to minimize clogging by foreign materials. Flow readings were used as midpoint representation for flow of water before and after the reading.

One liter water samples were taken at the valve at 30 minute intervals. During irrigation 1, quarter point samples also were taken at 30 minute intervals, pH and wind odometer readings were recorded, and an estimate of solids settling from irrigation waters were made at quarter points. Water samples were stored on ice, frozen and transported to the analytical laboratory. Samples were analyzed for electrical conductivity, total dissolved solids, NO_3-N, NH_4-N, TKN, total P and total K. Samples results were used as midpoint values to represent nutrient concentration prior to and post sample time. Nutrient delivery to the field was calculated as the sum of flow multiplied by nutrient concentration for each irrigation event. Settled solids were estimated by tacking pre-weighed and identified cheese cloth pieces (645 cm^2) to the soil. Five sites were used: 30.5 m (100 ft) from the headland and tail and at quarter points.

Estimates of water volume and nutrient application for irrigation events 2, 4 and 6 were calculated using the average volume from measured irrigations before and after the one being estimated. Nutrient concentrations used were the average values from the subsequent irrigation.

Water advance/infiltration

Water advance and recession times were monitored at 61 m (200 ft) intervals during irrigation 1. Water advance measurements, used in conjunction with a volume balance irrigation model (Walker and Skogerboe, 1984) allowed the determination of field-wide soil infiltration characteristics. Infiltration opportunity times were determined from the difference between irrigation water recession and advance times. Uniformity of irrigation water application was determined from the infiltration characteristics and infiltration opportunity times.

Crop production

Corn was planted June 24. Between irrigations 2 and 3 156 kg/ha of UN32 were applied. At harvest, four rows of corn were identified for sampling. Side by side rows, one third of the way in from each side of the check were chopped into a tared truck, weighed and sampled for nutrient content. Samples of chopped corn were taken at harvest, frozen and later analyzed for TKN, total P, total K and moisture.

RESULTS AND DISCUSSION

Wind odometer and irrigation water (manure and canal waters mixed) pH values were recorded during irrigation 1 only. Average wind speed was 25 kmph (15.5 mph) and pH values ranged from 7.3 to 7.9.

Irrigation water

Uniformity of irrigation water application was determined from the infiltration characteristics and infiltration opportunity times (Table 1). The uniformity of water application was relatively high for a border irrigation event (distribution uniformity = 83%). A distribution uniformity of 70 - 85% is considered high for sloping border fields (San Joaquin Valley Drainage

Program, 1987). While water application uniformity was high, the 610 m length of the field required that a large average depth of water (16 cm / 6.3 in) needed to be applied to simply advance water to the end of the field. Soil and soil moisture characteristics varied within the field, but volumetric soil sampling to a depth of 1.5 m (5 ft), both pre- and post- irrigation, indicated a soil moisture deficit in the 1.5 m profile of approximately 8 cm (3.2 in) (Table 2). Irrigation water was applied in excess of that required to bring the 1.5 meter profile to field capacity.

Table 1. Distance, advance times and infiltrated water calculations for irrigation 1

Distance		Advance Time	Infiltrated water	
m	ft	(min)	cm	in
0	0	0	19.76	7.78
61	200	30	19.53	7.69
122	400	70	19.10	7.52
183	600	120	18.52	7.29
244	800	165	18.16	7.15
305	1000	210	17.98	7.08
366	1200	251	17.35	6.83
427	1400	293	16.61	6.54
488	1600	325	15.29	6.02
549	1800	355	14.86	5.85
549	2000	390	14.50	5.71

Seasonal, field-averaged, applied water for the six irrigations was 61 cm (24.1 in). The estimated crop water use for the corn crop was 50.8 cm (20 in) (Snyder et al., 1987). This would result in a seasonal over-irrigation of approximately 10 cm (4 in). Since there was no runoff from the field, water applied in excess of that required to meet plant water demands would be lost to deep percolation. Determining deep percolation loses on a seasonally- and field-average basis ignores contributions to deep percolation from water application nonuniformity and from non-optimal irrigation scheduling of irrigations.

Table 2. Determination of soil moisture (cm/30.5cm) in samples taken at the top (T), middle (M), and bottom (B) of the field. Five samples (1 - 5) taken at 30.5 cm increments where ρ_{dry} is bulk density (g/cm³), θ_g is percent water (g/g) and θ_v is percent water (v/v).

	Pre-irrigation 6/11/93				Post irrigation 7/1/93			Post - Pre		
Location	ρ_{dry}	θ_g	θ_v	cm	θ_g	θ_v	cm	θ_g	θ_v	cm
T - 1	1.63	8.50	13.86	4.22	18.00	29.34	8.94	9.50	15.49	4.72
T - 2	1.75	6.00	10.50	3.20	12.00	21.00	6.40	6.00	10.50	3.20
T - 3	1.73	11.50	19.90	6.07	11.80	20.41	6.22	0.30	0.52	0.15
T - 4	1.68	15.20	25.54	7.77	15.20	25.54	7.77	0.00	0.00	0.00
T - 5	1.58	17.00	26.86	8.18						
M - 1	1.55	9.60	14.88	4.55	14.40	22.32	6.81	4.80	7.44	2.26
M - 2	1.74	13.80	24.01	7.32	15.80	27.49	8.38	2.00	3.48	1.07
M - 3	1.86	15.10	28.09	8.56	16.00	29.76	9.07	0.90	1.67	0.51
M - 4	1.65	21.90	36.14	11.02	21.50	35.48	10.82	-0.40	-0.66	-0.20
M - 5	1.60	23.70	37.92	11.56	24.50	39.20	11.94	0.80	1.28	0.38
B - 1	1.68	6.00	10.08	3.07	10.20	17.14	5.23	4.20	7.06	2.16
B - 2	1.77	16.40	29.03	8.84	14.30	25.31	7.72	-2.10	-3.72	-1.14
B - 3	1.63	16.00	26.08	7.95	16.60	27.06	8.26	0.60	0.98	0.30
B - 4		no sample				no sample				
B - 5	1.51	25.10	37.90	11.56	26.10	39.41	12.01	1.00	1.51	0.46

Nutrient applications

Irrigation events 1, 3 and 5 were monitored (flow recorded and water samples taken). The concentrations of nutrients were variable in each of the irrigations. A summary of water application, ranges of nutrient concentrations, electrical conductivity and total dissolved solids are provided (Table 3). Additionally, the quantity of applied nutrients is presented. This value is the sum of the water volumes multiplied by nutrient concentration for the various times when data and samples were recorded and analyzed.

The range in nutrient concentration was greatest during irrigation 1 and least during irrigation 3. This was due to changes in quantity of canal water added to the manure water and may have been due to variation in nutrient concentration of water in the actual undiluted manure water. Data from Idaho (Ohlensehln et al., 1993) indicated up to 10 fold variability of nutrient concentration of dairy lagoon water when sampled at various points throughout the surface and strata of the lagoon. The first irrigation had the greatest volume of water and was diluted the least (producer management strategy). Direct monitoring of the quantity of canal water dilution to the manure water sampling of lagoon waters prior to dilution would have helped to determine if lagoon nutrient concentration was variable. The variation in lagoon nutrient concentration may explain, in part, the higher concentration of nutrients during irrigation 5 than during irrigation 3 as lower manure water levels were present in the lagoon.

Application rate and timing is an important component to preventing groundwater contamination. Irrigation events occurred June 14 (1), July 4 (2), August 4 (3), August 13 (4), August 26 (5), September 21 (6). Total monitored application rates in kg/ha for irrigations 1, 3 and 5 were: NH_3-N, 66.28; TKN, 131.20; P, 41.49; and K, 139.71 (Table 3). Commercial N was applied (156.29 kg/ha) prior to irrigation 2. The estimated amounts of nutrients applied during irrigations 2, 4 and 6 were: NH_3-H, 39.35; TKN, 101.00; P, 29.57; and K, 54.34. The average NO_3-N was less than 1 ppm for all monitored irrigations. Total dissolved solids averaged 2467 with a range of 466 go 5218 ppm during irrigation 1.

In addition to knowing the application rate of nutrients to the field, it is important to understand the nutrient distribution within the field. After irrigating land with manure water of high solids, it is common to see considerable amounts of solids which have settled out of the irrigation water. The day of irrigation 1, solids at the 60.5 m distance from the irrigation valve were between 20 and 30 cm above the soil. One week after irrigation 1 the depth of these solids had settled to 6 to 8 cm (top of field), with less than 1 cm on cloths removed from the middle and bottom of the field. Cloths were retrieved and dry matter content of solids was determined and averaged. The average quantity of dry solids settled at the five distances from the headlands were 263.2, 33.1, 1.7, 2.8, and 21.2 g /648 cm^2. A section of field 3.05 m by 30.5 m would have received 341 kg (750 lbs), 48 kg (105 lbs), 2.5 kg (5.4 lbs), 4.0 kg (8.9 lbs) and 30.6 kg (67 lbs) at each of the areas sampled.

The variability in quantity of settled solids makes one ponder if other nutrients are distributed uniformly within the field. Quarter point samples were taken during irrigation 1 at half-hour intervals. Although there are not sufficient data to preform an analysis. The data provided (Table 4) indicate that nutrients appeared to be distributed somewhat evenly throughout the field. Data indicate nutrient concentration of settleable solids within manure water is far less than the nutrient concentration of whole manure water (Hermanson, 1981). Although the solids did settle at variable rates, this probably had little affect on nutrient distribution. It will however affect soil chemical characteristics.

Total nutrients present in soil (Table 5) and nutrient application rate (Table 3) was greater than the quantity of nutrients remaining in the soil post harvest and those harvested in the corn crop for N (108 kg/ha). Of greatest concern was the large quantity of nutrients applied during the commercial N application.

Table 5. Soil nutrient concentrations: (1) 3 days pre-irrigation 1; (2) 10 16 days post-irrigation 1; (3) 1 day post harvest.

Location[a]	NH_3-N (ppm)			NO_3(ppm)			TKN (%)			X-K (ppm)		
	1	2	3	1	2	3	1	2	3	1	2	3
T - 3 - 1	5.4	8.2	3.9	31.7	61.7	39.7	.125	.163	.106	105	165	124
T - 3 - 2	3.3	4.2	2.0	9.0	11.9	7.6	.038	.054	.032	100	105	71
T - 3 - 3	3.4	2.3	2.3	11.6	11.0	4.3	.026	.021	.020	73	67	64
T - 3 - 4	2.5	2.7	1.9	18.2	20.9	3.3	.015	.017	.019	79	99	68
T - 3 - 5	2.0	2.7	2.0	25.5	18.3	3.9	.016	.017	.016	76	71	
T - C - 1	4.6	8.0	4.4	33.9	45.4	22.3	.113	.105	.095	133	128	118
T - C - 2	3.0	4.3	2.3	12.3	12.7	8.6	.048	.047	.035	88	69	75
T - C - 3	2.1	3.6	2.0	15.0	9.6	6.3	.020	.023	.019	63	50	57
T - C - 4	2.9	3.1	2.0	19.5	12.8	7.2	.014	.018	.014	83	87	66
T - C - 5	2.9	3.4	2.6	13.7	18.4	8.9	.012	.017	.014	79	85	76
M - 3 - 1	5.2	10.2	13.3	13.2	25.2	45.7	.060	.067	.051	137	151	126
M - 3 - 2	4.1	5.3	4.5	4.3	5.3	49.9	.027	.018	.026	133	141	118
M - 3 - 3	2.8	4.7	2.8	3.5	4.1	6.6	.014	.015	.014	117	119	111
M - 3 - 4	2.3	4.2	2.8	6.2	5.2	5.4	.013	.011	.017	87	88	89
M - 3 - 5	2.9	5.1	2.1	5.2	6.4	6.4	.012	.009	.012	125	88	81
M - C - 1	5.9	8.9	8.8	9.9	16.9	35.1	.062	.062	.068	123	130	131
M - C - 2	4.0	5.3	3.6	5.0	6.8	16.8	.028	.026	.022	137	122	113
M - C - 3	3.0	4.7	3.0	3.6	4.0	6.4	.014	.015	.020	116	113	114
M - C - 4	2.7	3.7	2.4	4.4	4.3	4.7	.011	.012	.016	93	93	68
M - C - 5	3.5	4.8	2.2	5.6	5.1	5.8	.011	.012	.011	127	90	92
B - 3 - 1	3.7	4.3	7.3	7.6	14.0	5.9	.044	.038	.036	85	72	95
B - 3 - 2	2.1	3.5	2.8	4.3	5.7	9.5	.014	.016	.016	129	120	92
B - 3 - 3[b]	2.1	3.3	4.8	5.1	5.7	15.6	.012	.012	.012	121	95	104
B - 3 - 4	2.6	3.5	3.1	6.9	7.0	10.3	.011	.012	.012	145	153	152
B - 3 - 5	2.6	3.5	2.4	7.1	4.1	6.3	.008	.009	.008	141	149	145
B - C - 1	3.2	4.9	9.7	6.7	14.1	21.6	.033	.035	.049	87	92	79
B - C - 2	2.5	3.3	2.9	4.2	5.2	10.5	.017	.016	.017	123	108	99
B - C - 3	2.6	3.2	3.2	4.4	5.0	9.6	.012	.014	.014	95	85	62
B - C - 4	2.8	3.5	2.4	6.9	4.4	6.1	.011	.013	.014	143	139	131
B - C - 5	2.9	3.7	2.2	6.8	3.4	5.1	.008	.009	.009	137	96	156

[a] Location = T (top), M (middle), B (bottom) of field - C (composite), 3 (center hole) - depth from surface (1 - 5)

CONCLUSION

Water Management

Two strategies could be used to apply a lesser amount of irrigation water to the field while irrigating the entire check uniformly. First, the inflow rate to the check could be increased. This would result in a faster advance rate of water down the check; thus making infiltration opportunity times and infiltrated water more equal between the top and bottom of the field. This strategy is constrained by the maximum flow capabilities of the supply system. For the field investigated, inflow rate to the check could be increased slightly but not enough to result in the level of improvement desired.

A second strategy which would likely to be more effective would be to reduce the length of the border check to be irrigated. Irrigating a check only 305 m (1000 ft) long vs. the current length of 610 m (2000 ft) would substantially reduce the differences in infiltration opportunity times and infiltrated water along the check. While effective at improving irrigation performance, this strategy would be difficult to implement because it would require significant and expensive changes in the irrigation water supply pipelines. Keeping border-irrigated field lengths short is best implemented when the irrigation system is being designed and originally installed.

Nutrient Application and distribution

The variability in nutrient concentration as water was applied combined with variation in infiltration rate can result in extremely high over applications of N to parts of the field. As an example, the average concentration of N in irrigation 1 was 45.5 ppm. An application of 18.5 cm (7.3 in), 16.0 cm (6.3 in), or 13.5 cm (5.3 in) would deposit 33.6 kg (74.3 lbs), 20 kg (64 lbs), or 24.6 kg (54 lbs) per .4 ha (acre). To best use nutrient sampling during irrigations to determine nutrient application, the dilution and flow rate of water to the field should be constant. Further research to determine efficacy of sampling the source of manure water prior to application is being conducted.

REFERENCES

Environmental Protection Agency. 1993. Guidance specifying management measures for sources of nonpoint pollution in coastal waters. US EPA, 840-B-92-002, January.

Hermanson, R.E. 1981. Characteristics of solids separated from dairy manure. Cooperative Extension Service, Washington State University.

Meyer, J.L. 1973. Manure waste ponding and field application rates. Part I. Study findings and recommendation. University of California Agricultural Extension.

Meadows, C. and L.J. Butler. 1989. Dairy Waste Management in California: A Survey. University of California Sustainable Agriculture Competitive Grant Program. (#240).

Ohlensehlen, R.M., D.E. Falk, and M.V. Boggess. 1993. Waste characterization of 21 Southern Idaho dairy lagoons. J. Dairy Sci. 76 (supp. 1): 257.

San Joaquin Valley Drainage Program, Agricultural Water Management Subcommittee. 1987. Farm water management options for drainage reduction. August. San Joaquin Valley Drainage Program.

Snyder, R.L., B.J. Lanini, D.A. Shaw, and W.O. Pruitt. 1987. Using reference evapotranspiration (ETo) and crop coefficients to estimate crop evaportraspiration (ETc) for agronomic crops, grasses, and vegetable crops. Cooperative Extension. University of California Division of Agriculture and Natural Resources, Leaflet 21427.

Walker, W.R. and G.V. Skogerboe. 1984. The theory and practice of surface irrigation. A guide for study in surface irrigation engineering. Utah State University, Logan.

Table 3. Water and nutrient application rates and ranges for each of the evaluated irrigations (1, 3, 5) and estimated values for irrigations 2, 4 and 6.

Irr. No	Water in/ac	Range of Nutrient Concentration (ppm)				EC Range mhos/cm	Dissolved Salts Range ppm	Quantity Applied (kg/ha)			
		NH₃-N	TKN	P	K			NH₃-N	TKN	P	K
Measured irrigations											
1	6.38	2.9 - 61.1	5.3 - 83.6	0.1 - 129	8 - 129	0.16 - 1.58	28 - 860	47.15	73.76	23.81	114.61
Canal Water		0.2	2.7	<0.1	1	0.11	4				
3	4.58	1.1 - 32.5	16 - 87	3 - 56	2 - 17	0.10 - 0.72	30-320	5.93	31.67	10.98	5.72
5	2.24	15.0 - 27.3	36 - 51	9 - 15	31 - 36	0.41 - 0.56	266 - 346	13.20	25.77	6.70	19.38
Application amount (kg/ha)								66.28	131.20	41.49	139.71
Estimated irrigations (average nutrient concentration from next irrigation)											
2[a]	5.48	3.7	19.5	6.78	3.53			7.09	37.81	13.15	6.84
4[a]	3.41	8.15	15.91	4.13	11.96			32.26	63.19	16.42	47.50
6[a]	2.0	canal water									
Commercial Nitrogen (kg/ha)								105.83	156.29		
Total nutrient application (kg/ha)									388.49	71.06	194.05
Nutrients harvested in Corn silage									251.11	44.12	242.77

Table 4. Nutrient concentration of waters (ppm) at quarter points (0, 1/4, 2/4, 3/4, 4/4) during half-hour sample intervals.

Time	NH₃-N					TKN					P					K				
	0	1/4	2/4	3/4	4/4	0	1/4	2/4	3/4	4/4	0	1/4	2/4	3/4	4/4	0	1/4	2/4	3/4	4/4
10:30	34.0					46.5					16.6					103				
11:00	54.9					83.6					17.6					127				
11:30	14.1					21.4					3.7					24				
noon	8.7	49.3				16.4	85.9				2.3	26.9				13	148			
12:30	2.8	13.6				16.0	17.4				3.0	3.2				7	24			
1:00	37.0	5.2				78.8	30.0				16.9	5.0				125	13			
1:30	49.0	38.3				82.6	75.2				14.3	22.6				124	115			
2:00	39.1	30.2	31.6			71.9	57.6	52.1			22.5	33.1	22.7			96	127	93		
2:30	61.1	31.6	40.8			73.5	74.5	67.4			40.3	42.2	32.9			129	109	127		
3:00	15.9	27.6	28.4			25.7	70.1	71.2			10.5	38.2	40.6			33	93	118		
3:30	2.9	24.5	26.7	52.5		5.2	38.9	56.3	80.7		1	22.1	29.8	38.3		8	65	82	127	
4:00	34.7	5.1	13.3	36.9		37.4	10.5	18.8	75.5		35.7	2.2	11.9	27.8		80	11	29	106	
4:30			6.3	23.9	32.2			10.2	36.5	70.4			4.2	16.7	37.0			15	67	109
5:00			6.1	11.4	22.3			9.4	14.5	38.8			1	0.6	15.1			12	18	62
5:30					18.5					22.8					12.6					55

GLEAMS Modeling of BMPs to Reduce Nitrate Leaching in Middle Suwannee River Area

William R. Reck[1]

ABSTRACT

This study was initiated to estimate the effectiveness of conservation practices to reduce nitrate leaching to the Floridan Aquifer through improved agricultural waste management systems. Ground water monitoring of the Middle Suwannee River Area (MSRA) has shown high concentrations of nitrate nitrogen near (down gradient from) intensive agricultural operations. Dairy and poultry farms have been intensively monitored by US Geological Survey and have been found to have high nitrate levels below these operations compared to nearby control wells.

During the summer of 1992, the SCS conducted interviews with eight dairy farmers and 16 poultry farmers in the MSRA. Questions were asked about pastures, high intensity areas (HIA), barns, manure storage facilities, herd size, flock size, flocks per year, soils, crops grown, manure spreading methods and cooling methods. The data from the interviews was used to develop "typical" values to input into the computer model Groundwater Loading Effects of Agricultural Management Systems (GLEAMS) in order to predict nitrogen leaching under various management schemes.

Several alternative management scenarios were modeled to determine the most effective method of reducing the leaching of nitrate nitrogen. These BMPs include determination of optimal loading rates for poultry litter, efficient crop rotations, number of harvests for hayland crops, irrigation of agricultural crops with lagoon effluent, reduction of commercial fertilizer used, and control of HIA.

This information was used to estimate an average reduction in nitrate leaching of 77 percent for participating poultry farms and 55 percent reduction in nitrate leaching for participating dairy farms with the use of BMPs. The overall effectiveness of implementing BMPs for the MSRA was estimated to be a 50 percent reduction in nitrate nitrogen leached for all dairy and poultry farms combined.

INVESTIGATION AND ANALYSIS

Ground Water Quality Monitoring

In 1989 the Florida Department of Environmental Regulation's (FDER) extensive statewide ground water monitoring network revealed numerous instances of high nitrate concentrations in the Floridan Aquifer in Suwannee and Lafayette Counties. Although none of the measured concentrations exceeded the drinking water standard for nitrate, exceedances were expected to occur in the immediate vicinity of sources of contamination. The elevated

---

[1] Agricultural Engineer, USDA Soil Conservation Service, Gainesville, Florida.
"All SCS programs and services are offered on a nondiscriminatory basis, without regard to race, color, national origin, sex, age, religion, marital status or handicap."

nitrate concentrations correlated closely with livestock and poultry operations.

These results prompted FDER to initiate further studies to better define the extent of the problem. In 1990, FDER funded a study by the U.S. Geological Survey (USGS) to monitor ground water directly under and adjacent to intensive dairy and poultry operations located in the Middle Suwannee River Area (See Figure 1).

The USGS drilled monitoring wells, on 7 dairies in Suwannee and Lafayette Counties, into the top of the Upper Floridan Aquifer. The monitoring wells were installed adjacent to various land uses on the dairies. The USGS also drilled monitoring wells, on 5 poultry farms in Suwannee and Lafayette Counties, into the upper level of the Floridan Aquifer. The monitoring wells were installed adjacent to various land uses on the poultry farms.

The results of these monitoring efforts are reported in the Middle Suwannee River Area Watershed Protection Plan and Environmental Assessment by the SCS. USGS has conducted periodic testing of the water quality of springs. A graph depicting the results of this monitoring is presented in this same document.

Ground Water Quality Modeling

During the summer of 1992, SCS conducted interviews with eight dairy producers and sixteen poultry producers in the MSRA. The purpose of the interviews was to gain detailed information on the waste management needs and practices of the operators. Questions were asked about pastures, HIA's, barns, manure storage facilities, herd size, flock size, number of flocks per year, soils, crops grown, manure spreading methods and cooling methods. The data from the interviews was used to develop "typical" or "representative" values to input into the computer model GLEAMS in order to predict nitrogen leaching under various management schemes. All the figures reported below were developed from these interviews. The GLEAMS program was used to predict nitrate nitrogen leaching in the "No Action Alternative" as well as the "Recommended Alternative". The difference between these two represents the effects of the project.

<u>Poultry Evaluation</u>
The "No Action Alternative" poultry nutrient budget was based upon interviews with 16 poultry producers and was verified by interviews with industry leaders. Poultry litter analyses were also performed to determine litter nutrient composition and production. It was determined that the average poultry producer has 3.65 houses with approximately 19,000 birds per house and an average of 5.5 cycles per year. The analysis of 18 samples of litter taken from two poultry houses operated by two different producers showed that the average producer produced 485 Mg (535 tons) of litter per year with a nutrient analysis of TKN = 3.0 percent, ammonia N = 0.4 percent, total P = 1.3 percent, and total K = 1.9 percent. The interviews indicated that, on average, litter is currently being spread over 50 acres of cropland, with 7 ha (18 acres) of land receiving heavy loading of litter and 13 ha (32 acres) of land receiving lighter loadings. The waste disposal fields are currently not being intensively managed. A bermudagrass field harvested once a year was modeled in GLEAMS to determine the current (No Action) nitrogen leaching.

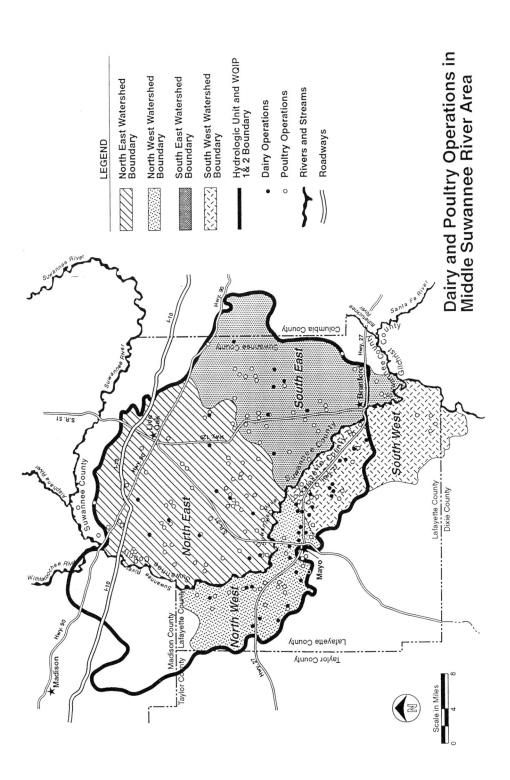

This scenario was found to leach an average of 430 kg N/year/ha (380 lbs N/year/acre).

Several alternative management scenarios were modeled in order to determine the most effective method of reducing the leaching of nitrate nitrogen for the "Recommended Alternative". The poultry litter was modeled as being spread over a bermudagrass/winter oats crop rotation for a varying number of acres with a varying number of applications of waste and a varying number of harvests. The results of this analysis can be seen in Figure 2 which represents a bermudagrass hayfield (overseeded with winter oats) harvested 5 times per year. It also shows the amount of nitrogen applied, the predicted amount of nitrogen taken up by bermudagrass/winter oats, and the predicted amount of leached nitrogen versus the amount of applied poultry litter in tons/acre (No commercial fertilizer applied). Nitrogen leached is graphed for the average litter produced per farm versus the number of acres of cropland (See Figure 3).

Figure 2

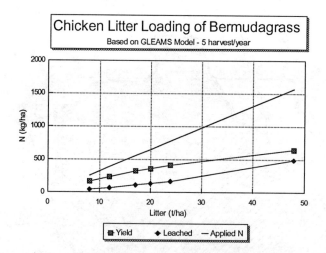

Figure 3

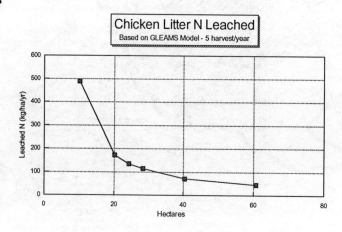

Using a realistic yield average and the data in the SCS Agricultural Waste Management Field Handbook (AWMFH) it was determined that a bermudagrass yield of 18 t/ha/yr (8 tons/ac/yr) of dry matter and an oats yield of 7 m^3/ha (80 bu/ac) would contain 340 kg/ha (300 lbs/ac) and 60 kg/ha (50 lbs/ac) of N/yr respectively. As Figure 1 indicates, a nitrogen uptake of 390 kg/ha/yr (350 lbs/ac/yr) can be achieved from a bermudagrass and winter oats rotation when litter is applied at 22 t/ha/yr (9.9 tons/ac/yr). This level of application and uptake will leach approximately 60 kg/ha/yr (50 lbs/ac/yr) of nitrogen. Implementation of this management scenario would reduce the total pounds of nitrogen leached from each participating poultry farm by 77 percent.

Dairy Evaluation
The "No Action" dairy nutrient budget was based upon a survey of 9 dairy producers which validated an earlier questionnaire administered by the Institute of Food and Agricultural Science (IFAS). It was determined that the average dairy producer has 310 cows. The cows spend 71 percent of their time in pastures at a density of 4.5 cows/ha (1.8 cows/ac). Dairies with waste storage pond effluent application areas have an average application area of 23 ha (56 ac) of cropland. These application areas consist mainly of bermudagrass and small grain rotations. Seventy-eight percent of the surveyed dairies have application areas. According to the survey, the average size of the holding area is 1.3 ha (3.3 ac), the average lane size is 0.8 ha (1.9 ac), and the average area of bare soil areas around watering tanks and shade is 0.7 ha (1.6 ac). Seventy-eight percent of those surveyed have a waste storage pond and 22 percent of them discharge directly into depressional areas. On average, waste storage ponds are pumped out to the application field every 14 days. Of the dairies surveyed, only one dairy had a lined waste storage pond.

The nutrient budget for the current average system was calculated with the following distributions; 71 percent of time in pasture, 4 percent of time in High Intensity Areas (HIA), and 25 percent of the time in the milking/feeding barn. Deposition of waste by dairy herd was taken to be a 1 to 1 ratio of excreted waste to percentage of time spent in the various locations. All of the waste from the milking/feeding barn is washed into the existing waste storage pond. The waste excreted in the HIAs was assumed to volatilize 50 percent of the total nitrogen with the remaining nitrogen being eventually leached. The waste excreted in the pasture was assumed to be evenly spread throughout the pasture.

The waste storage pond effluent application areas were modeled using GLEAMS with the current acreages, pumping rates, and crop management practices on bermudagrass and small grain rotations. Currently, the average dairy farm is fertilizing bermudagrass application areas with 45 kg/ha (40 lbs/ac) of N, 16 kg/ha (14 lbs/ac) of P_2O_5, and 45 kg/ha (40 lbs/ac) of K per year and is fertilizing small grain rotation with 64 kg/ha (57 lbs/ac) of N, 17 kg/ha (15 lbs/ac) of P_2O_5, and 64 kg/ha (57 lbs/ac) of K per year. Two different management practices, one representing bermudagrass and one for small grain rotations, were averaged together on a weighted basis based upon survey data. The nitrogen leached for the lagoon and application area of the average dairy was calculated to be 2,130 kg (4,700 lbs) of N per year. In the HIA (which consist of bare spots on pastures, holding areas, and lanes) the average amount of nitrogen leached was calculated to be 900 kg (2,000 lbs) of N per year.

Pasture leaching was calculated using an average of 71 percent of excreted manure occurring in pasture for varying numbers of cows per acre. The survey indicated that the average dairy producer fertilizes pastures at a rate of 53 kg/ha (47 lbs/ac) of N, 17 kg/ha (15 lbs/ac) of P_2O_5, and 53 kg/ha (47 lbs/ac) of K per year. Harvesting of pasture grass was based upon a percentage of dairy feed coming from pasture and based upon forage consumption estimation using methods developed by Heitschmidt and Stuth. Figure 4 illustrates the range of nitrogen application, nitrogen yield, and nitrogen leaching based upon cow density in the pasture (no commercial fertilizer added). The average nitrogen leached per farm from pastures is 3,400 kg (7,500 lbs) of N per year as calculated using the GLEAMS model.

Figure 4

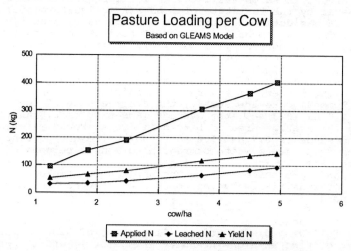

The nutrient budget for the "Recommended Alternative" was determined using procedures described in the AWMFH and upon modeling using GLEAMS. The waste collection system was changed to include the HIA with the milking/feeding area being collected and conveyed to the waste storage pond. The application area was modeled as a bermudagrass crop harvested 5 times yearly and overseeded with oats in the winter. The average yearly nitrogen leached from lagoons, application areas, and HIAs was reduced from 7,500 to 450 kg/ha/yr (6,700 to 400 lbs/ac/yr) for this scenario.

The Recommended Plan requires that pasture density be reduced from 4.5 cows/ha (1.8 cows/ac) to 3.5 cows/ha (1.4 cows/ac) and reduce the percentage of time cows spend on pastures from 71% to 60%. This reduces the estimated nitrogen leaching from pastures from 8,400 kg/ha/yr (7,500 lbs/ac/yr) to 6,300 kg/ha/yr (5,600 lbs/ac/yr). Implementation of this project will reduce the nitrogen leached from the average dairy by 56%.

GLEAMS

The GLEAMS model has recently included an animal waste application option to the model. This section was used for most of the GLEAMS modeling done for the Middle Suwannee River Project. Several problems were discovered during the modeling process, which have since been corrected by the authors of the model. The modeling process should not be considered an absolute answer, but rather a yardstick for alternative solutions. GLEAMS provided a method of analysis where research data was not available.

References

Andrews, W.J., 1992, Reconnaissance of

Bottcher, A.B., Holloway, M.P., and Campbell, K.L., 1990, BMPs for Mitigating Nitrate Contamination of Ground Water Under North Florida Daries, Institute of Food and Agricultural Science, University of Florida.

Heitschmidt, R.K., and Stuth, J.W., 1991, Grazing Management- An Ecological Perspective.

USDA, Soil Conservation Service, 1992, Agricultural Waste Management Field Handbook.

Quality of Well Water

On Tennessee Poultry Farms

H. Charles Goan, Paul H. Denton and Frances A. Draughon*

Abstract

What is "safe" drinking water for humans, poultry and livestock? Safe drinking water could be thought of as water that is free of harmful bacteria, viruses, parasites and chemicals. Since many disease-producing organisms are far to small to be seen with the naked eye, thousands of them may be in a single drop of water. This contaminated water can affect the health of farm families and the health and performance of their poultry flocks.

During a two-year period, a survey of well water quality was conducted on Tennessee poultry farms that grow meat type chickens (broilers). The survey included evaluations of the well site, well-head protection, manure handling and land application practices.

There are approximately 130 million broilers grown on more than 700 Tennessee poultry farms. These farms vary in size from as little as ten acres to several hundred acres. Each year a typical broiler house averages approximately six grow-outs with an average of 24,000 birds per grow-out.

The water sources for these birds are wells, springs and public utility water. On the farm, there can be one or more wells supplying water to the broiler house and/or the family residence.

Research on poultry farms in Georgia (French, et al., 1988) and Delaware (Ritter, 1989) found nitrate-nitrogen levels in water wells to be as high as 95 parts per million (ppm). Also, high bacterial levels were found in some wells. Nitrate-nitrogen levels exceeding 20 ppm may adversely affect broiler growing performance (Barton, et. al., 1989).

The objectives of this study were:

1. To determine the levels of nitrate-nitrogen, and bacteria in well water on Tennessee poultry farms.

2. To relate water quality data to manure handling and application practices, soil types, well location and conditions around the well head.

KEYWORDS. Water quality, Water wells, well protection, manure handling.

---

*H. Charles Goan, Professor and Poultry Specialist, University of Tennessee Agricultural Extension Service. Paul H. Denton, Professor and Soils Specialist, University of Tennessee Agricultural Extension Service. Frances A. Draughon, Professor, Food Technology and Science Department, University of Tennessee.

Geographic Location

Four geographic regions of Tennessee with heavy broiler production and distinctly different geology were selected for study. The Ridge and Valley region included McMinn, Polk, Bradley and Hamilton counties. This region is underlain by folded and faulted limestone, shale and sandstone. The Cumberland Plateau region includes Morgan, Fentress and Grundy counties. The region is underlain by relatively level bedded sandstone, shale and coal. The Highland Rim region includes Coffee, Franklin, Grundy and Lincoln counties. This region is underlain by level bedded limestone. The Central Basin includes Bedford and Lincoln counties. It is underlain by level bedded limestone. Karst topography is found in parts of the Ridge and Valley, Highland Rim and Central Basin.

Well Selection

The first step in the ground water study was to identify broiler growers in each of the four test regions using private water supplies (wells) who were willing to participate in the survey. Seventy-one broiler growers using well water for broiler production were identified by the county Agricultural Extension agents with the assistance of four integrated broiler firms operating in these regions.

A survey form was developed to assist in obtaining information about the well water usage, well location and physical characteristics, and the manure handling and application practices on the farm.

Sample Collection

During the first year of the study, water samples were collected from 105 wells on 71 broiler farms in the four test regions previously specified. The sampling procedure was as follows:

- Locate a water faucet near the wellhead.
- Allow water to run through the faucet at least 5 minutes to purge the expansion tank and associated plumbing.
- Flame sterilize the faucet.
- Collect 250 milliliters of water from the faucet in a pre-sterilized container.
- Store sample in an ice bath and return to the water quality laboratory within 12 hours of the sampling time.

Laboratory Analysis

The water samples were assayed for total coliform, fecal coliform and aerobic bacteria (APHA, 1975). <u>Salmonella</u> was analyzed according to the Conventional Procedure outlined in the Bacteriological Analytical Manual (FDA, 1992). Nitrate-nitrogen was determined using ion chromatography (APHA, 1975).

DISCUSSION

The laboratory results (Table 1) indicated that 62 percent of the water samples had coliform bacteria present and 97 percent of the water samples had aerobic plate counts of one cell or higher per milliliter. Fecal coliform bacteria were found in 50 of 105 samples (48 percent) which indicated contamination from human and/or animal sources. In addition, nine water samples tested positive for Salmonella. Five of the samples that tested positive for Salmonella were negative for the presence of fecal coliform bacteria. There was very little variation in the percentage of wells contaminated with fecal coliform bacteria between geographic regions.

Table 1. Summary of Laboratory Analysis for Bacteria in 105 Well Water Samples

Levels	Total Coliform CFU/100 ml	Fecal Coliform CFU/100 ml	Aerobic Plate Count CFU/ml
	Number of Samples		
0	40	55	3
1 to 10	21	19	58
11 to 100	15	16	20
> 100	29	15	24

Water samples from eight wells were found to exceed 10 ppm nitrate-nitrogen with the highest level being 53 ppm (Table 2). All eight wells having excessive nitrate-nitrogen were located in one county. At five well sampling sites, broiler manure had been stacked outside the broiler house within 50 feet of the well head. Any rainfall runoff would have been from the stacked manure toward the well.

Table 2. Summary of Laboratory Analysis for Nitrate-Nitrogen in 105 Well Water Samples

Levels of Nitrate-Nitrogen (ppm)	Numbers of Well Water Samples
0 to .99	50
1 to 4.99	41
5 to 9.99	6
10 to 60	8

The broiler growers were made aware of the water test results. Since 51 percent of the broiler growers used a well as a common water source for the chickens and family use, suggestions were given about ways to improve conditions around wells and methods to purify water.

One year later well water samples were taken from: (1) 29 wells that previously had greater than 10 CFU/ml fecal coliform; (2) seven wells that had tested positive for Salmonella and (3) seven wells that had greater than 10 ppm nitrate nitrogen. Sixteen of 29 wells (55 percent) still had fecal coliform present. Only two of seven wells still had Salmonella present. All of the wells retested for high nitrate-nitrogen showed virtually no change in the level of nitrate-nitrogen for the second test.

During the course of the survey, well location was noted as a potential factor contributing to contamination. Many wells were drilled in areas without serious consideration given to potential contamination sources including: (1) home septic tank and leachfield; (2) nearby cattle, hog and dairy feedlots; (3) litter stacked outside broiler houses; (4) drainage from the chicken house and other building roofs and (5) surface drainage area surrounding the well site. In addition, two wells were located inside the broiler houses with the well head exposed to the chickens and chicken manure. All of the previously mentioned conditions could contribute to contamination which caused high fecal coliform bacteria counts and/or elevated nitrate-nitrogen levels.

Other concerns noted were: (1) top of the well casing below ground level; (2) well casings with no protective seal on top; (3) no protection around the casing to prevent surface water penetration and (4) fertilizer, oil and other chemicals stored near the well head.

SUMMARY

Well location is an important factor to consider when drilling a well. On many farms it appeared that convenience was the primary consideration when drilling a new well. Apparently, very little consideration was given to possible contamination of wells from nearby sources. In the poultry farms survey, none of the growers had tested their well water during the previous five years. After the results of the first water test were given to the growers, over 35 percent of the growers improved conditions around the well.

As a result of the survey, broiler growers were advised to test their water on a regular basis if:

- There are unexplained family illnesses
- There are unexplained changes in poultry or livestock performance
- There are changes in water taste, odor, color or clarity
- There is a spill of chemical or petroleum products near the well
- Broiler litter is stacked or spread on the land near the well.

REFERENCES

1. American Public Health Association. 1975. Standard Methods for the Examination of Water and Wastewater, 14th Edition.

2. Barton, T.L., L. Hileman and T. Nelson. 1989. A Survey of Water Quality on Arkansas Broiler Farms and its Effect on Performance. University of Arkansas, Fayetteville, Arkansas, Fayetteville, AR.

3. Food and Drug Administration. 1992. <u>Salmonella</u>, Bacteriological Analytical Manual. 7th Edition, AOAC International, Arlington, VA.

4. French, D. and N. Dale. 1988. Georgia Poultry Farms Water Quality Surveyed. Poultry Digest, pp 44-49. Jan.

5. Ritter, W.F. 1989. Ground Water Contamination from Poultry Manure. The Clean Water Act and the Poultry Industries Symposium. Sept. 7-8, Washington, D.C.

ALLIGATOR PRODUCTION IN SWINE FARM LAGOONS AS A MEANS OF ECONOMICAL AND
ENVIRONMENTALLY SAFE DISPOSAL OF DEAD PIGS

W.R. Walker, T.J. Lane, E.W. Jennings,
R.O. Myer and J.H. Brendemuhl*

ABSTRACT

Dead animal disposal for large commercial swine farms is becoming more difficult because of increasing regulation. Increased production of alligators in the southeastern United States has increased demand for economical feed sources suitable for alligators. The use of dead pigs as a food source for alligators may provide an environmentally safe solution to both problems. An initial trial was conducted to assess the feasibility of rearing alligators in a swine waste lagoon and fed dead pigs. Seventy-five alligators were reared in the single stage lagoon (100 x 37 m) of a commercial 2000 sow farrow-to-finish swine farm. Dead pigs were the sole source of provided nutrients for these alligators. Growth performance of these alligators were compared to 25 alligators reared on a nearby commercial alligator farm. The control alligators were fed a diet consisting of a mixture of meat and fish by-products. After one year, alligators reared in the swine farm lagoon were 177% heavier (P<.01) and 97% longer (P<.001) than the controls. No detrimental affects were noted for either the hides or meat from the swine farm reared alligators. The results of this initial study suggest that alligator production on swine farms in the southern United States can be an environmentally safe method of dead pig disposal.
KEYWORDS. Swine, Alligators, Feeding, Dead pig disposal.

INTRODUCTION

The swine industry in Florida is extremely diverse ranging from small backyard operations to very large farrow to finish confinement operations. In a swine farrow to finish operation, in which pigs are raised from birth to market size of 110 kg (240 lb), it is not unusual to have dead pigs as a waste product that must be disposed of. Preweaning mortality, from 0 to 28 d of age including stillbirths, is usually from 15 to 25% of the pigs farrowed (born). For nursery age pigs (28 to 60 d) death loss is 2 to 5% while death loss of growing-finishing pigs (60 to 150 d) will average 1% (Pond and Maner, 1984). Death loss in the breeding herd is usually quite low (<1%), however, mature pigs weigh from 200 to 450 kg (450 to 1000 lb). In a large swine operation, even those well managed, proper disposal of waste products such as dead pigs can be quite costly. Problems with dead pig disposal is not unique to Florida, it is a nationwide problem (Bell, 1993).

The normal methods of disposal of dead pigs include burial, open pit burning or incineration. Environmental regulations in many areas now prohibit burial, leaving burning as one of the few available options. Open pit

---

*Walker, Associate Professor (present address: Land-O-Lakes, Inc., Ft. Dodge, IA), and Brendemuhl, Associate Professor, Department of Animal Science; Lane, Associate Professor, College of Veterinary Medicine; University of Florida, Gainesville. Jennings, Agricultural Extension Agent II, Pasco County, Dade City, FL. Myer, Associate Professor, University of Florida, North Florida Research and Education Center, Marianna, FL.

burning is less expensive than incineration but increases the environmental hazard. For this reason, open pit burning may not be an option in the future. Incineration requires regulatory permits and specialized equipment subject to frequent inspection. The cost of permits, equipment and fuel make incineration the least attractive alternative. Rendering or recycling into tankage, which can be utilized as a feed source, is a viable option, however, rendering is not available in many areas.

Alligator farming is currently a growing industry in Florida. Edible waste products from livestock and seafood processing industries provides a major portion of the food supply on many alligator farming operations. Dead pigs could also be used as a food source for alligators, however, due to high transportation and storage costs, the alligator and swine farms would need to be in close proximity for this to be economical. Wild alligators commonly live in swine farm waste lagoons throughout Florida. Therefore, a possible solution to dead pig disposal problems would be to rear alligators in existing swine farm lagoons. This would provide an environmentally safe means of dead pig disposal with the possibility of added income to the swine producer through alligator production. However, many questions must be answered as to the feasibility and overall cost effectiveness of alligator production in swine farm lagoons.

Reported here is the results of an initial study assessing the feasibility of rearing alligators in a commercial swine farm lagoon. The specific objective of this initial study was to determine the effect of swine waste lagoon rearing and dead pig feeding on growth rate and health safety of meat of farm raised alligators.

EXPERIMENTAL

This study was conducted on a large scale commercial swine farm located in Dade City, Florida. This farm is a total confinement farrow-to-finish operation with approximately 2000 brood sows producing in excess of 30,000 pigs per year. The waste system on this farm is a automated manure handling system that transports all manure to a common holding tank. Manure from this tank is pumped to a mechanical separator. The solids portion of the manure is recycled back to the brood sow herd as a feed source (blended with dry feed before feeding). The liquid extract containing about 10% solids is discharged into a single stage lagoon (100 m x by 37 m; 3.7 m deep). Water from the lagoon is used to irrigate nearby pasture and orange grove. The lagoon is constantly replenished with waste water from the swine buildings. The lagoon is bordered on one side by a concrete block retaining wall and was fenced on the other three sides with welded galvanized steel fencing 1.2 m tall. The fenced area included an embankment of 3 to 4 m around three sides of the lagoon.

Treatments in this study consisted of rearing alligators either on a commercial alligator farm (control) or in the above mentioned swine farm lagoon. Alligators on the alligator farm were kept in a pen approximately 7 m^2 with the back two-thirds of the pen recessed and filled with water. This is typical of the rearing environment on most Florida alligator farms.

The feed utilized on the alligator farm consisted of a mixture of meat and fish (mostly edible wastes and by-products), while that for alligators on the swine farm consisted of dead pigs chopped to appropriate feeding size. In this study, 25 alligators were reared on the alligator farm and 75 alligators were reared in the swine farm lagoon.

All alligators used in this study came from two commercial alligator farms on a cooperative agreement between the alligator and swine farm owners. All alligators were individually tagged, weighed and the length measured prior to being placed in the lagoon or in the pens. Alligators were weighed and measured again at the time of marketing. After slaughter, representative samples of alligator meat and organs were collected and sent to the University of Florida Meat Science Laboratory and Veterinary College for testing for possible bacterial and antibiotic contamination, respectively. All dead pigs fed to the alligators were weighed. Representative samples of all feed fed to alligators were subjected to laboratory analysis (AOAC, 1984) for common nutrients.

This study began in August of 1989 and continued through September of 1990. All data were analyzed by appropriate statistical procedures (SAS, 1985). One alligator died during the course of this study. This alligator was reared in the swine farm lagoon and the cause of death is unknown. Four additional alligators somehow managed to escape the confines of the swine farm lagoon.

RESULTS AND DISCUSSION

Average composition of diets fed to the alligators reared on the alligator farm or in swine farm lagoon is given in Table 1. The diets for both treatments were similar in dry matter content. Diets fed for both treatments, as expected, were high in protein content. The swine farm diet was overall higher in energy (caloric density) as indicated by a higher average fat content and lower ash content.

Table 1. Average Chemical Composition of Diets Fed to Alligators at the Alligator and Swine Farms.

Item	Alligator Farm Diet[a]		Swine Farm Diet[b]	
	Wet[c]	Dry	Wet[c]	Dry
Dry Matter, %	35	100	30	100
Crude Protein, %	17.4	49.8	12.9	42.8
Crude Fiber, %	0.1	0.2	0.1	0.2
Fat, %	7.9	22.4	9.5	31.6
Nitrogen Free Extract, %	0.7	2.0	4.0	13.3
Calcium, %	3.0	8.5	1.0	3.3
Phosphorus, %	1.5	4.1	0.6	2.0
Ash, %	8.7	24.9	3.6	12.1

[a]Combination of fish and meat by-products.
[b]Pig carcasses ranging from 1 to 25 kg.
[c]As fed basis.

Growth parameters for alligators are summarized in Table 2. Differences in initial size were accounted for in the statistical analysis. Alligators reared in the swine farm lagoon and fed dead pigs as the sole source of nutrients gained weight (P<.01) and length (P<.001) faster than those reared on the alligator farm. The relatively large increase in average daily weight gain (177%) and average daily length (97%) for alligators reared in the swine farm lagoon was unexpected. However, little is currently known about nutritional and environmental needs for alligator production. This study did not ascertain whether the improved performance was due to nutrition or environment.

Table 2. Effect of Rearing Alligators on Confinement Alligator Farm or in Swine Farm Lagoon on Gain in Weight and Length[a,b].

Item	Farm	
	Reared on Alligator Farm	Reared on Swine Farm
Initial Number	25	75
Final Number	25	70
Initial Weight, kg	9.1	12.9
Final Weight, kg	18.1	44.3
Initial Length, cm	133.8	145.3
Final Length, cm	158.7	199.8
Average Weight Gain[c], kg	11.9	31.3
Average Daily Weight Gain[c], g	29.8	82.4
Average Length Gain[d], cm	29.8	56.8
Average Daily Length Gain[d], mm	0.76	1.50

[a]Least square means.
[b]Initial weight was a significant covariate.
[c]Means differ ($P<.01$).
[d]Means differ ($P<.001$).

Male alligators tended to grow faster (Table 3) in both weight ($P<.1$) and length ($P<.05$) than female alligators regardless of treatment. It is generally recognized that male alligators grow faster than female alligators.

Table 4 shows the influence that environmental temperature can have on alligator feed consumption. As temperatures declined in the winter months, feed intake declined accordingly. This table also provides an indication of total feed intake for alligators reared in the swine farm lagoon. Based on this data, the calculated feed-to-gain ratio for the lagoon reared alligators was about 4.5 kg of dry matter intake per kg of weight gain.

Economic value in alligators is derived through the production of both hides and meat. No detrimental effects were noted for either hides or meat due to swine lagoon rearing. Samples of alligator meat and organs tested negative for both antibiotic residue and bacterial contamination.

For livestock producers in Florida and perhaps other southern states facing environmental decisions regarding the disposal of dead animals, alligator production may be an economical alternative. In addition to swine lagoon rearing, confinement rearing of alligators on the swine farm or nearby to provide a more controlled environment may be a viable option.

Table 3. Effect of Type of Farm Rearing and Sex of Alligator on Gain in Weight and Length[a,b].

Item	Farm			
	Reared on Alligator Farm		Reared on Swine Farm	
	Sex		Sex	
	Male	Female	Male	Female
Initial Number	6	19	26	49
Final Number	6	19	26	44
Initial Weight, kg	9.0	9.1	13.7	12.5
Final Weight, kg	19.1	18.3	48.9	41.6
Initial Length, cm	136.9	132.8	148.9	143.4
Final Length, cm	164.7	156.8	209.0	194.4
Average Weight Gain, kg[c,d]	12.3	11.5	33.8	28.8
Average Daily Weight Gain, g [c,d]	30.8	28.8	88.9	75.9
Average Length Gain, cm[c,d]	31.6	27.9	60.7	52.9
Average Daily Length Gain, mm[c,e]	0.81	0.71	1.60	1.40

[a]Least square means.
[b]Initial weight was a significant covariate.
[c]Farm x sex interaction not significant (P>.1).
[d]Sex effect (P<.1).
[e]Sex effect (P<.05).

Table 4. Influence of Swine Farm Lagoon Temperature on Alligator Food Consumption.

Year	Month	Average Lagoon Temperature, °C	Total Feed (Wet), kg	Total Feed (Dry), kg
1989	September	30	2412	724
1989	October	26	1812	544
1989	November	23	1373	412
1989	December	18	964	289
1990	January	21	1602	480
1990	February	21	1930	579
1990	March	23	1888	566
1990	April	25	2289	687
1990	May	30	2450	735
1990	June	31	3532	1060
1990	July	29	4216	1265
1990	August	30	4153	1246
1990	September	28	4185	1255
Total			32,806	9,842

REFERENCES

1. AOAC. 1984. Official Methods of Analysis (14th Ed.). Association of Official Analytical Chemists, Arlington, VA.

2. Bell, A. 1993. Know your carcass disposal options. Pork 93. (Vance Livestock Publ., Shawnee Mission, KS), December, p. 24-30.

3. Pond, W.G. and J.H. Maner. 1984. Swine Production and Nutrition. AVI Publishing Co., Westport, CT. p. 96-97.

4. SAS. 1985. SAS User's Guide. Statistics. SAS Inst., Inc., Cary, NC.

DEAD BIRD COMPOSTING
Nga. N. Watts[1]

ABSTRACT

The increasing awareness of protecting the environment is forcing the poultry industry to find a better method of disposing of the dead bird carcasses and poultry litter. Composting is one of the preferred methods of utilizing this waste.

At the present time, most poultry growers dispose of dead poultry carcasses by burying them in a pit; and, most store poultry litter in uncovered and areas. There are 129 poultry operations in the Middle Suwannee River Basin Watershed alone. Due to the Karst topography of the river basin, there is a potential for pollution of the ground water from improper disposal of poultry waste. Composting dead birds and storing the litter in a covered area decreases the potential for ground and surface water contamination.

Composting are approved by United States Department of Agriculture for cost share, one-stage and two- stage composting. One-stage composting is usually limited to the smaller operations with limited equipment, such as front-end loader. The one-stage composter generally are hand operated with little addition of extra nitrogen source (litter) for the composting. The two-stage composting is recommended for the large poultry growers that have the proper equipment available and requires more litter for the composting process. In Florida, more of the two-stage composters have been built.

Final distribution of the compost on crops at an agronomic rate will reduce the potential for nutrients to leach or run off the fields. If the litter must be stored, it is recommended that it is stored in a covered structure.

COMPOSTING

Composting can be economical and is an environmentally accepted method of disposing of dead birds. It is the biological decomposition of organic matter which can be accelerated by mixing organic waste with other ingredients for optimum microbial growth. Composting converts dead birds into a stable organic product which could be used as a soil amendment and is usually done under aerobic conditions. This process reduces the odor and fly problem, destroys weed seeds, and pathogens, reduces the bulk, and improve the organic material's handling properties.

Since the Floridan Aquifer is the primary source of drinking water for the Middle Suwannee River Basin Watershed area, it is important to protect it. It has been proven that livestock and poultry operations are the only significant uncontrolled sources of contamination (SCS, 1993). Waste associated with poultry operations include manure and dead poultry. Mortality rate for poultry ranges from 3% to as high 25%. (See Table 1)(SCS, 1992). In the Middle Suwannee River Basin, the majority of the poultry farms are broilers and burial is the disposal method most commonly used. Because of the large numbers of dead birds associated with large poultry operations, the disposal of dead birds is a major concern. Poultry facilities must have adequate means for disposal of dead birds in a sanitary manner.

---

[1]Nga N. Watts is an Agricultural Engineer for the USDA Soil Conservation Service

TABLE 1 Poultry mortality rates

Poultry type	Loss rate (%)	Flock life (days)	Cycles per year	Market weight (lb)
Broiler	4.5-5.5	42-49	5.5-6.0	4.2
Roaster				
females	3	42	4	4.0
males	8	70	4	7.5
Laying hens	14	440	0.9	4.5
Breeding				
hens	10-12	440	0.9	7-8
males	20-25	300	1.1	10-12
Turkey				
females	5-6	95	3	14
males	9	112	3	24
feather prod.	12	126	2.5	30

One method approved by USDA cost share is one-stage composting. This method offers an opportunity for smaller poultry operations that do not have a bucket type loader, but it is more labor intensive. It is recommended that the one-stage composter be limited to operations when the maximum weight of birds is 2.3 kg (5 lbs) or less and one 1.5 m X 2.4m (5' X 8') unit with two 1.5m X 2.2m (5' X 4') cells be provided for each 44,000 bird farm. Other conditions for this type of composting are: bird mortality in the composter be covered with at least one part litter for each part recycled compost and each composter unit be emptied twice yearly.

The more popular method of composting, also approved by USDA cost share, is the two-stage composter. Dead birds and amendments are initially added in the first stage, also called the primary composter. The mixture is moved from the first stage to the second stage when the compost temperature begins to decline below 60°C (140°F). Sizing of the bins for each stage should accommodate the peak disposal period usually close to the birds' market weight. The second stage volume is equal to or greater than the sum of the first stage bins. Equation EQN-1 (SCS, 1992) is used to determine the volume for each stage.

$$Vol = B \times \frac{M}{T} \times W \times \frac{VF}{100} \qquad \text{EQN-1}$$

where:
- Vol = Volume required for each stage (m³(ft³))
- B = Number of animals
- M = Percent normal mortality of animals for the entire life cycle expressed as percent
- T = Number of days for animal to reach market weight (days)
- W = Market weight of animals (kg (lb))
- VF = Volume factor - 0.156 m³ (ft³) composter volume per kg (lb) of weight loss

The number of bins required for each stage can be determined by EQN-2 (SCS, 1992):

$$\# \text{ Bins} = \frac{\text{Total 1st stage volume (m}^3\text{)}}{\text{Volume of single bin (m}^3\text{)}} \qquad \text{EQN-2}$$

The dimensions of the bins are typically 1.5 m (5 feet) high, 1.5 m (5 feet) deep, and 2.4m (8 feet) across the front. However, the width across the front should be determined by the equipment size used by the landowner. The first and secondary bins should be covered and concrete lined. The roof and raised concrete floor will keep rain water from entering and will control percolation due to added water. These steps help achieve the ideal moisture content of the compost. The proper proportion of waste and amendments for composting is commonly called the "recipe." For optimum microbial activity, the carbon-nitrogen ratio, moisture content, and the internal temperature of the compost mixture is critical. The mixture consists of a nitrogen source (litter and carcass), a carbon source (straw, peanut hulls, cotton seed hulls, etc.), water, and oxygen (Table 2) (SCS, 1992).

Table 2 - Material Proportions for Composting

Ingredient	Volumes	Weights	C:N Ratio
Poultry Carcasses	1.0	1.0	5
Manure	1.5	1.2	15
Straw	0.5	0.75	0
Water	0.5	0.75	0

The material is placed in the first stage bins in sequential layers:
1. Thirty cm (1 ft) dry manure on the floor of the bin. This manure is not part of the recipe.
2. Fifteen cm(6 in) layer of loose straw on top of the manure.
3. A layer of carcasses, at least 15 cm (6 inches) between birds and wall.
4. Water if needed.
5. Manure over the carcass according to the recipe.
6. Continue by repeating steps 2-5 for each subsequent batch. The final layer of manure should be about 15 cm (6 in) deep to minimize odors.

The C:N ratio of the original mix in the first stage should be between 25:1 and 40:1 (SCS, 1992). The moisture content should be maintained between 40% and 60%. If the mixture gets too wet, the compost will convert to anaerobic activity which will produce foul odor and will not function properly. The compost should reach the optimum operating temperature between 60°C (140°F) to 71°C(160°F) approximately within 7 days and remain elevated for up to 14 days. When the temperature starts to decline in the first stage, the compost should be mixed and aerated by using a front end loader to move the material into the second stage bins. Again, the temperature should elevate to the optimum operating temperature. The high temperature is necessary to destroy the pathogenic bacteria, viruses, weed seeds, and fly larvae. If the temperature exceeds 74°C (165°F), it is necessary to remove the material from the bin and cooled to keep it from spontaneous combust. If the temperature falls below or does not reach 54° (C130°F), during composting period and odors develop, investigate for moisture content, porosity, and thoroughness of mixing. The temperature should be monitored daily.

After the second stage, the compost can be applied directly onto the land; but it is recommended to store the material for at least 30 days to further dry the compost for ease of spreading. The material should be stored no higher than 7 feet (2.1 m) to reduce the potential for spontaneous combustion. It also should not come in contact with any manure stored in the same facility. The compost should be stored in a covered and lined structure and be applied to the land at recommended agronomic rates to reduce the chance of nutrients leaching or running off.

SUMMARY

Most of the poultry operations in the Middle Suwannee River Basin Watershed are burying the dead birds. By composting on a lined and covered are, one or two stage, and storing the litter in a covered structure, the contamination of the ground and surface water can be reduced. By following a simple "recipe" of hay, carcass, and litter, and reaching the ideal temperature and moisture content, dead birds waste could be transformed into a stabilized soil amendment.

REFERENCES

1. Soil Conservation Service, U.S. Department of Agriculture. *South East Middle Suwannee River Area, Watershed Protection Plan and Environmental Assessment.* April 1993.

2. Soil Conservation Service, U.S. Department of Agriculture. *Agriculture Waste Management Field Handbook.* April 1992.

The United States Department of Agriculture (USDA) prohibits discrimination in its programs on the basis of race, color, national origin, sex, religion, age, disability, political beliefs and marital or familial status. (Not all prohibited bases apply to all programs). Persons with disabilities who require alternative means for communication of program information (Braille, large print, audio tape, etc.) should contact the USDA Office of Communications at (202) 720-5881 (voice) or (202) 720-7808 (TDD). To file a complaint, write the Secretary of Agriculture, U.S. Department of Agriculture, Washington, D.C. 20250, or call (202) 720-7327 (voice) or (202) 720-1127 (TDD). USDA is an equal employment opportunity employer.

PRELIMINARY WATER QUALITY RESULTS

FOR

A ROCK REED FILTER HOME TREATMENT SYSTEM

Gene Dougherty[*]

ABSTRACT

The continuing increase in population in Florida is forcing the development of sites where soil conditions significantly reduce the effectiveness of the typical residential onsite sewage disposal system (septic tank) using a drain field.

As part of an effort to develop alternative onsite treatment systems, the USDA-Soil Conservation Service (SCS) in cooperation with the US-Environmental Protection Agency (EPA), constructed a rock reed filter at the USDA SCS Plant Materials Center in Brooksville, Florida. The purpose of the installation is to evaluate the effectiveness of using the rock reed filter as a method to treat residential waste water at sites where poor site conditions reduce the effectiveness of conventional drain fields. A secondary purpose is to evaluate the esthetics of the treatment system through the use of different varieties of aquatic plants.

The rock reed filter uses aquatic plants growing on a rock substrate to improve the water quality of the effluent by filtering the suspended solids and utilizing the nutrients for plant growth. Plants being evaluated include; Golden canna (<u>Canna</u> <u>flaccida</u>), Duck potato (<u>Sagittaria</u> <u>lancifolia</u>), Soft rush (<u>Juncus</u> <u>effusus</u>), Southern blue flag (<u>Iris</u> <u>virginica</u>), Giant bulrush (<u>Scirpus</u> <u>californicus</u>) and Soft stem bulrush (<u>Scirpus</u> <u>validus</u>).

Initial monitoring of the effluent shows a reduction in suspended solids and reductions in phosphorous, nitrogen and coliform bacteria. It is expected that further testing will indicate that the rock reed filter is a potential alternative to conventional septic tank AND DRAINFIELDsystems at a reasonable cost.

OVERVIEW OF THE ROCK REED FILTER

<u>Design Criteria</u>
The system was designed using the guidelines contained in the Tennessee Valley Authority (TVA) publication TVA/WR/WQ--91/2[1]. The rock reed filter consists of a 6 m x 6 m x 0.5 m (20 ft x 20 ft x 1.5 ft) filter bed lined with a 30 mil Polyvinyl Chloride (PVC) liner with a rock substrate consisting of Florida Department of Transportation (FDOT) No. 6 and No. 3 aggregate. This is in addition to a septic tank, dosing tank and drain field. The system is designed to treat 1135 L (300 gallons) per day and has a 529 square foot surface area. The large size was necessary to handle the additional waste water from a conference center at the site.

---

[*] Gene Dougherty is a Civil Engineer with the U.S. Department of Agriculture, Soil Conservation Service Gainesville, Florida

PRELIMINARY WATER QUALITY RESULTS

Test apparatus and results

SCS and Hernando County, Florida, Department of Health personnel began water quality testing of the influent and effluent in the rock reed filter in September, 1993 using a Hach DR\700 Colorimeter and testing for nutrients and lab testing for coliform bacteria. Samples were taken at three points in the filter with well #1 located at the inlet, well #2 located half way between the inlet and outlet and well #3 located at the outlet as shown in Figure 1.

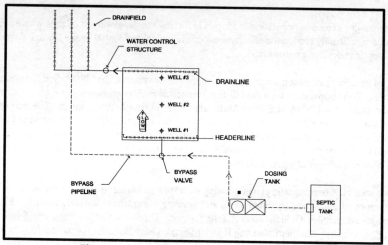

Figure 1 ROCK REED FILTER SCHEMATIC

Fecal coliform

Fecal coliform counts in the influent ranged from 540 to ≥ 2400 MPN / 100 ml. Reductions in fecal coliform ranged from 80% to 97% which is within the expected range and is reflected in figure 2. Variations in the influent values may have been caused by a leaking toilet during October, which increased the volume of water flowing through the system.

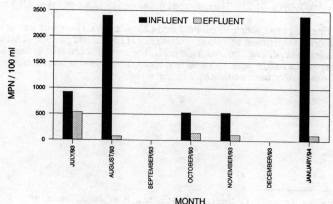

Figure 2: FECAL COLIFORM

Total Coliform

The total coliform count remained relatively constant at ≥ 2400 MPN / 100 ml for the influent. Effluent values ranged from 920 MPN / 100 ml to ≥ 2400 MPN / 100 ml with an average reduction of 33%. The reduced influent value reflected in Figure 3 can probably be attributed to the leaking toilet that affected the fecal coliform values.

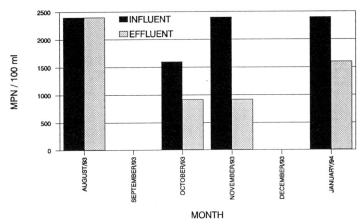

Figure 3: TOTAL COLIFORM

Phosphorous

The total phosphorous testing gave the most consistent results and showed that 75% to 80% of the plant available phosphorous is removed in the first half of the filter bed. Approximately 89% to 95% (see figure 4) of the plant available phosphorous was removed from the effluent prior to discharge to the drain field. These reductions are a result of vigorous growth of several of the plant species growing on the filter and the fact that limestone was used for the substrate.

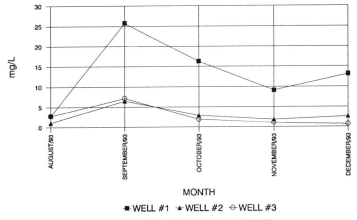

Figure 4: PHOSPHOROUS REACTIVE

Conclusions

Preliminary results fall within the ranges given in studies performed by Green and Upton (1993)[2], Davies, Cottingham and Hart (1993)[3] and Steiner and Combs (1993)[4] which have shown reductions of 16 % - 87 % for dissolved phosphorous and 79% to 99% removal of fecal coliform.

Although there is an insufficient number of samples to reach a final conclusion, it appears that several preliminary conclusions can be reached from the available data. The significant reductions in fecal coliform and phosphorous indicate that the quality of water discharged from an onsite sewage disposal system can be greatly improved and the environmental and health risks of polluting the ground water decreased.

The preliminary water quality results achieved at the Plant Materials Center show the need for further study and evaluation of rock reed filter technology and its applicability for use in high risk areas of Florida.

REFERENCES

1. GENERAL DESIGN, CONSTRUCTION, AND OPERATION GUIDELINES; CONSTRUCTED WETLANDS WASTEWATER TREATMENT SYSTEMS FOR SMALL USERS INCLUDING INDIVIDUAL RESIDENCES, TVA/WR/WQ--91/2, March 1991, Water Resources, River Basin Operations, Resource Development, Tennessee Valley Authority

2. Moshiri, G.A. (ed), 1993, Constructed Wetlands for Water Quality Improvement, CRC Press, Boca Raton, Fl. 517 - 524

3. Ibid, 577 - 584

4. Ibid, 491 - 498

The United States Department of Agriculture (USDA) prohibits discrimination in its programs on the basis of race, color, national origin, sex, religion, age, disability, political beliefs and marital or familial status. (Not all prohibited bases apply to all programs). Persons with disabilities who require alternative means for communication of program information (Braille, large print, audio tape, etc.) should contact the USDA Office of Communications at (202) 720-5881 (voice) or (202) 720-7808 (TDD). To file a complaint, write the Secretary of Agriculture, U.S. Department of Agriculture, Washington, D.C. 20250, or call (202) 720-7327 (voice) or (202) 720-1127 (TDD). USDA is an equal employment opportunity employer.

Tree-Crop Polyculture: Small Farming in the Suburbs

J. A. Rogers*

ABSTRACT

Suburban tree-crop polyculture affords flexible designs for reforesting and bringing into production the already disturbed areas of human residential acreage. Most of these designs fit best in a decentralized, human scale economy that cultivates and is sustained by a diverse yield of food, forage and timber. Most applications would start as part-time hobby farms. Marketing strategies include various forms of subscription gardening or cooperative wholesaling to niche markets.
KEYWORDS: Tree crops, polyculture, permaculture, suburban farming, human scale

GENERAL PHILOSOPHY

This model draws heavily from the inter-disciplinary field called permaculture, which seeks to design sustainable human settlements that can live within their bioregions. The strategies always include living near one's food supply and working toward renewable energy. Cheap fossil fuels and petro-chemical inputs are too uncertain and too wasteful of topsoil to continue to provide the energy and fertility of modern agriculture.

When we propose solutions that are inter-disciplinary, we are acknowledging that sustainable solutions require that many elements work together. It is this interweaving that makes ecosystems resilient. A climax forest or a natural wetland are good models. Each has attracted complex guilds of plants and animals that function at full employment with no waste. Contrast these stable ecosystems with the US. social ecosystem that has recently been set up to expect grain from Kansas, apples from Washington and vegetables from California or Mexico (with the help of an interstate highway network, private autos and foreign oil fields). The latter system is fragile because any broken links could shut it down. It looks like someone is playing Monopoly with the food supply and the economy.

DESIGN GUIDELINES

If you live in a place that was once a forest, like most of eastern North America, then **reforesting** may be easier than **deforesting** to increase cultivated acreage. This gives tree crops a certain priority in species selection. A judicious selection of tree crops could be assembled to enhance the diet rather than diminish it. All the food groups are present, especially if some of the yield is fed to farm animals. A number of tree crops are worthy substitutes for certain grains. Culinary skills that rise to match a region's unique abundance earn a prominent role in the attending cultural shift. Doesn't much of the charm of Europe depend on the human scale of the villages and regional variations in the foods, wines and cheeses?

---

*John Rogers is member of the Brevard Rare Fruit Council. Though an engineer by profession, he is transitioning toward hunter-gatherer.

Tree crops have certain advantages worth noting. First, by their very nature they form the basis for a permanent agriculture. They do not require the yearly plowing and planting like most conventional crops. The soil under them is best left undisturbed except by leaf fall, cover crop and earthworm. Plows, tractors and tractor payments are not required, especially when kept on a small scale and planted with cooperative diversity. Tree crops lend themselves to marginal areas that are too steep or too small for conventional plow agriculture. Through leaf and branch fall, trees create their own fertility. Through transpiration, trees create their own rain and cycle ground water faster than the same acreage of smaller plants or grasses. Insofar as organic litter belongs on the forest floor, densely treed sections of a neighborhood provide places to aesthetically recycle plant residues and other organic wastes like municipal sewage sludge. Such a forested neighborhood increases biodiversity and increases wildlife habitat. (Animals are necessary as the legs and wings of a forest.)

A polyculture gets its resistance to disease and pest attacks through its own diversity and that of the species it hosts. A monoculture is a huge collection of one species vulnerable to being taken by force en mass by the highly resistant pests modern agriculture has been selectively breeding over the last fifty years.

In a polyculture, the harvest is more human scale because it happens as many small harvests spread over the entire year. The yields are usually small enough to be marketed locally whereas in a monoculture the harvest window is very small, and the yield is probably too large to be marketed locally. Polycultures are more small-farmer friendly. As such, they can provide the basis of a local food supply, and part of the local economy.

The convenience of trading within a community offers the grower the ability to choose varieties more on the basis of flavor and disease resistance and less on the basis of storage tolerance and shippability. For this reason alone, local growers can out compete the tough, almost flavorless produce on the supermarket shelf.

Vertical stacking different species in the forest more fully utilizes the acreage cultivated and the plants assembled. Robert Hart (1991) identifies seven layers in his forest garden in England: canopy, under story, shrub layer, herb layer, ground cover, root (and tuber) zone and the vine layer. Models this dense approximate the productivity of a climax forest.

ENVIRONMENTAL BENEFITS

Adopting forest farming techniques in the suburbs has a number of beneficial side effects. Forests temper the climate and soften extremes in weather. They abate noise and wind. They provide an excellent storage place for carbon and create oxygen in the process. Planted in sufficient quantity, agro-forests can help offset the increased carbon dioxide loads from slash-and-burn tropical farming practices, clear cutting, and first world fossil fuel habits. A healthy forest with sufficient topsoil absorbs storm water faster than other upland ecosystems. The same forest also transpires storm water much faster that a retention pond evaporates it, having the advantage of a vast leaf area. And not to be discounted is the simple but universally recognized fact that forested neighborhoods are nicer places to work and live.

CULTURAL BENEFITS

The model I am sketching values the ingenuity of local subcultures and celebrates their diversity and adaptability. It is probably the best antidote to a sprawling, homogenous mass culture that shops in a few huge chain discount department stores and eats out in the same dozen franchise restaurants no matter where you are on the continent. From a human labor point of view, it rebels against the boredom of factory production quotas, the monotony of industrial farming, and the low status of clerical sales jobs at the consumer end of the production machine. If the trademark of uniform industrial production and consumption is that it can be imposed on any bioregion without regard to its special features and climatic subtleties, then the trade mark of this human scale model is that it works best when you take into account these regional differences.

There are psychological benefits when a community begins to take responsibility for its most basic needs. It gives a neighborhood more control of its food supply and a greater appreciation of its topsoil and water table. Storm water becomes a resource, not a nuisance requiring an expensive infrastructure. Grass clippings and autumn leaves become too valuable to throw away, when they can with less effort add to next year's topsoil. Niche markets and cottage industries are created as resource loops are closed. A community is empowered as its self reliance develops.

Presently, the average bite of food travels about 1300 miles before it lands on our plate. So there is a lot we don't know about it. We depend almost entirely on the diligence of the FDA and the EPA to keep honest the profit-driven multi-national corporate farming operation, and their minimum wage workers who may not speak or read English as a first language. Now contrast this with a system where much of your food is grown within a few miles of where you live. It is possible to know almost everything about it. Irresponsible dumping of hazardous waste would not be tolerated by a town that eats from its surroundings. This is built-in earth care with a component of self regulation and supervision.

There are economic advantages to increasing trade within a neighborhood. Much of every dollar spent at a locally owned business stays local. Not so with the large corporate owned stores, where most of every dollar spent goes to out-of-town suppliers, transporters and owners whose commitment is more to the bottom line than to the community. The separation works both ways. It is easier to cheat or steal from an impersonal corporation than from people that you live near and trade with regularly. The peer pressure to maintain good business ethics is high in face to face transactions among principals. Not so in the anonymity of a mass marketplace where it's a coveted skill to be able to take advantage of the system before it takes advantage of you.

A meandering network of locally owned small farms woven in and around cities and towns is perhaps our best shot at re-humanizing and personalizing human settlements. Besides the many cultural benefits and economic advantages from such a system, there are ecological spin-offs from producing food, forage, fiber and timber near to where it is used. Resource loops are closed, vast amounts of energy is saved, and solid waste is reduced as new cottage industries step in to treat wastes as resources. As a community begins to increase trade within its boundaries, the local economy becomes information intensive rather than energy intensive.

How will such a model get started? Probably a thousand different ways in ten thousand different places. Will everyone farm in their yards? No, not everyone enjoys that activity, but they may eventually link up in some way with the local trading that it promotes. On every residential block there may only be a few people who consider themselves gardeners. On the average this will be enough human effort to oversee the farmer-friendly tasks of tree crop polyculture, even if all the non-gardeners offer their yards for a share of the harvests. The ratio of local enterprising gardeners is sufficient even with the added responsibilities of overseeing publicly owned commons and rights-of-way, as these are added.

Are there obstacles to such a plan? Dozens, from blind faith in high-input industrial agriculture to the mind set of suburban lawnmower jockeys who see trees as obstacles to their weekend sport. What will change such ingrained patterns? Probably the terror that comes from acknowledging the guaranteed scarcities of the present system and the adventure that comes from creating and living in the Garden of Eden.

REFERENCES

Hart, Robert A, 1991. Forest Gardening. Green Books, Bideford Devon. 212p.

HYDRAULIC PROPERTIES OF RE-CYCLED SHREDDED RUBBER TIRES AS CONDUCTIVE FILL IN OPEN DITCHES

Stephen J. Langlinais, P.E., P.L.S.

ABSTRACT

The environmental problems related with the disposal of rubber tires by burning, has forged the need for alternative methods of disposal. Shredding of tires has become a common method for handling rubber tires, and the need for finding uses for these shredded tires which are practical and yet environmentally safe has become of grave concern. One of the proposed uses, as part of this study, is to use the shredded tires as a conductive fill in roadside ditches upstream from highway culverts. This method of disposal could serve as a safety cushion for absorbing the impact from vehicular traffic involved in highway culvert accidents. This method of tire disposal could become practical and feasible only if the solution to a traffic safety problem does not cause, as a side affect, a drainage or flooding problem.

The objective of this research project involves the initial investigation of the open channel hydraulic parameters which affect the drainage flow through the shredded tires. Two of the parameters investigated in this project were 1) the determination of the water surface profile with water flowing through an 18 inch (46 cm) depth of shredded tires in a trapezoidal open channel, and 2) the determination of a comparable n value for Manning's equation. The results of the water surface profile data indicated the difference between downstream and upstream heads above the shredded tires ranged from 0.4 ft. (12.2 cm) of head loss with 4 ft. (122 cm) of shredded tires, to 0.9 ft. (27 cm) of head loss with 40 ft. (12.2 m) of shredded tires. The results of Manning's coefficient in the shredded tire section showed that n value for the concrete lined open channel was found to be 0.013, but with the shredded tires, the n values were computed to range between 0.613 with 4 ft. (122 cm) of shredded tires to 0.737 with 40 ft. (12.2 m) of shredded tires.

The Manning's coefficient calculated from this initial study could be used by design engineers in predicting head losses through the shredded tires when used in open channel configurations along road ditches.

KEY WORDS: Manning's n, re-cycled shredded tires, water surface profiles, drainage design

INTRODUCTION & OVERVIEW

The disposal of used tires has become a severe problem in recent years as the traditional method of burning has caused grave environmental concerns with regard to air pollution and contaminants. As a result of this concern, the method of burning as a means of disposal has been banned. Consequently, other alternatives to the tire disposal problem must be sought which are environmentally safe and yet not cause other problems as side effects to the tire disposal method

---

Stephen J. Langlinais, P.E., P.L.S., Associate Professor, Agricultural Engineering, University of Southwestern Louisiana, Lafayette, LA 70504, Phone 318-231-6486

sought. Presently, a $2.00 fee per tire is being assessed for every tire to be disposed. This fee generates several million dollars in the State of Louisiana alone. The Louisiana Transportation Research Center (LTRC) is assigned as part of its agenda, the task of finding and developing safe methodologies for disposing of these tires.

This research project involves the initial investigative studies for one proposed methodology of tire disposal, i.e. the use of shredded tires as a highway traffic safety device to cushion the impact of vehicular accidents involving highway ditches and culverts. Several million dollars in legal defense costs are spent each year in defending lawsuits against the State of Louisiana for highway culvert accidents. If the shredded tires could be used as a safety impact device, this would reduce accident losses and would also provide one additional source for recycling of these millions of old tires, thereby serving a twofold objective.

This project is a study of the hydraulic parameters of these shredded tires in open channels, and their cause and effects on water surface profiles when designing drainage ditches and culverts with shredded tires. The purpose of this research effort was to experimentally determine open channel flow hydraulic parameters for ditches filled with used shredded tire parts. This required placing different lengths of used shredded tire parts in a concrete trapezoidal channel and measuring water surface elevations at sections at the beginning and end of the shredded tires. The constant discharge through the channel was measured with a pre-calibrated rectangular weir.

A literature survey conducted revealed no experimental results reported on hydraulic characteristics of open channels filled with shredded tires or facsimiles thereof.

METHODOLOGY & EQUIPMENT

The Agricultural Engineering Hydraulics Field Laboratory on the University of Southwestern Louisiana campus, as shown in Figure 1, was used to conduct the hydraulic parameter tests on the shredded tires used in this research. The field test was set up to utilize a variable discharge water supply source from a fire hydrant which was fed through a rectangular shaped reservoir measuring device to which a rectangular weir (Israelson and Hansen, 1962) and a Parshall flume were used to measure the flow rates. Downstream from the Parshall flume is a 100 ft. (30.5m) concrete trapezoidal channel having a 6 inch (15 cm) bottom width and 1.5 to 1 side slopes. The schematic drawing and details of the equipment used is shown in Figure 1. The depth of the channel is approximately 18 inches (46 cm) deep. The capacity of the fire hydrant supplying the water ranged from 0 to 0.998 CFS (.028 M^3/sec).

The first test consisted of measuring the flow characteristics of the open channel without any shredded tires in the channel. This later served as a control test for comparison of flow parameters with and without shredded tires. The Manning's N value for the concrete channel without tires was computed using this data. Elevation readings were also taken at all points labeled H1, H2, H3, H4, H5, H6, H7, H8, and H9 as shown in Figure 2.

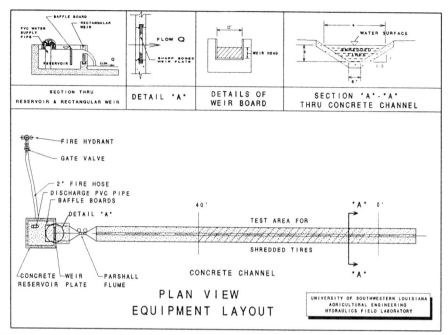

Figure 1. Schematic layout of equipment and test site used to conduct hydraulic tests with shredded tires placed in concrete lined channel.

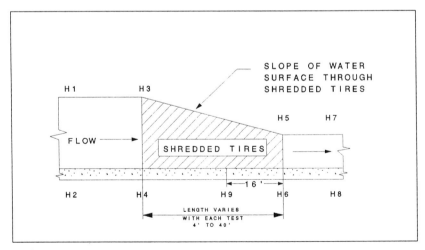

Figure 2. Nomenclature showing location of all data points taken during tests, and used in performing all calculations.

The following lengths of tires were used for each of the runs labeled 5 through 15 as shown in TABLE I.

TABLE I
LENGTH OF TIRES USED FOR EACH RUN

Run 5	No tires
Run 6	4 ft of tires (1.2m)
Run 7	8 ft of tires (2.43m)
Run 8	12 ft of tires (3.65m)
Run 9	16 ft of tires (4.87m)
Run 10	20 ft of tires (6.09m)
Run 11	24 ft of tires (7.32m)
Run 12	28 ft of tires (8.53m)
Run 13	32 ft of tires (9.75m)
Run 14	36 ft of tires (10.97m)
Run 15	40 ft of tires (12.20m)

A specially designed feature was added on the bottom of a Philadelphia rod in order to obtain an accurate reading on the surface of the water with high precision. A sharply pointed nail was attached to the bottom of the Rod. The sharp pointed nail allowed the rod men to carefully observe when the nail just pierced the surface tension of the water surface when the reading was taken with an engineering level. It was at this precise point that the instrument man was instructed to take his elevation reading on the Philadelphia rod.

The Manning coefficient, **n**, (Vennard, 1963) was calculated using <u>average flow depths</u> and the slope of the hydraulic grade line. The Manning coefficient, **n**, (King, 1060) calculated using Manning's equation is given as:

$$n = \frac{(1.49) \times (A_m) \times (R_m)^{2/3} \times S^{1/2}}{Q}$$

where:

Am = mean area based on the average flow depths at the beginning and end of the shredded tire test section for each run

Rm = mean hydraulic radius based on average flow depths for each run

S = slope of the hydraulic grade line in the test section for each run

The slope of the hydraulic grade line in the test section was calculated as the difference in the water surface elevations at the beginning and end of each test section divided by the length of the test section for each run. Manning's equation applies to uniform flow. However, for gradually varied flow in regular channels of short reach, it has been applied with comparable accuracy for analyzing non-uniform flow when average depth values were used.

DISCUSSION OF RESULTS

Water Surface Profiles:

The water surface profiles for the different lengths of shredded tires are shown in Figure 3, and shows the **elevation** changes between the water surfaces measured upstream and downstream of the shredded tires in the concrete channel. The profile line labeled "NO TIRES" is the water surface profile taken before any of the tests were run with the shredded tires in the channel. The average hydraulic gradient slope measured was S = .001 under free flow conditions without any obstructions in the channel. The line labeled "4'TIRES" is the profile line superimposed over the same channel position as the the "NO TIRES" profile. Each of the additional test runs for 8', 12', 16', 20', 24', 28', 32', 36', and 40' tires are subsequently superimposed over the same channel position for each test run. The elevations of the water surface profiles upstream of the shredded tires varied from 97.55 ft. elevation for 4 feet of tires to a maximum of 98.04 ft. elevation for 40 feet of tires. The elevation of the control section with "NO TIRES" for the entire 40 ft. channel was 97.20 feet upstream and 97.16 feet downstream.

The elevation of the section **downstream** of the tires fluctuated very little and varied from 97.14 to 97.16. This indicates that the addition of tires has very little influence on the downstream channel profile. However the section upstream of the tires is influenced considerably by the addition of the shredded tires.

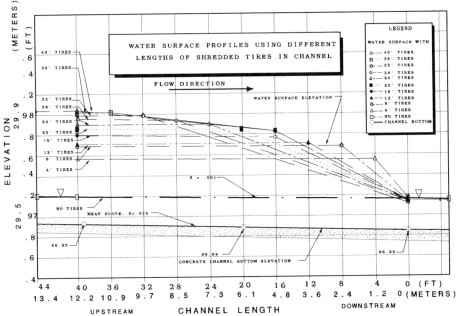

Figure 3. *Results of water surface profiles using different lengths of shredded tires in open channel.*

Head Loss Across Shredded Tires:

The effects of the head loss (H3-H5) versus the length of tires in the channel is illustrated in Figure 4. The head loss for 4 feet of tires was .38 feet for the minimum value and the maximum head loss was .90 feet across 40 feet of tires. This head loss (H3-H5) appears to vary linearly with the length of tires in the test section, and the linear graph was fitted on the curve according to Figure 4. Additional test results could provide additional curve points in order to perform a linear regressional analysis on the data. The R^2 factor could then be determined with a greater degree of significance if more points were available. However the results of Figure 4 indicate the probable manner in which a linear relation could result.

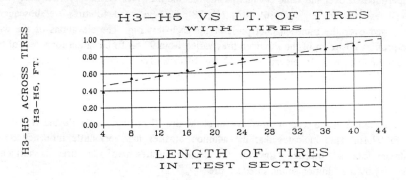

Figure 4. Drop in head across tires versus length of tires in test section.

The drop in head (H1-H3) across the **upstream** section versus length of tires in the test section is shown in Figure 5. The change in head loss upstream of the tires varied from .02 ft maximum to a -.01 ft. and appears to have no effect on the profile upstream of the tires. This seems to indicate that the maximum head loss occurs across the tires, and no apparent additional head loss trend occurs further upstream of the tires as additional tires are added in the channel.

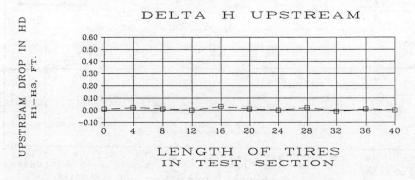

Figure 5. Drop in head across upstream section versus length of tires in test section.

CONCLUSIONS

The following conclusions may be drawn based upon the data collected and the analysis conducted on the field data. The flow regime in the shredded tire test section is non-uniform. The slope of the hydraulic grade line sloped downward in the direction of flow. Either Manning's equation using mean values or the non-uniform equation could be used in analyzing hydraulic parameters in the shredded tires. Manning's coefficient, n, for flow in the shredded tires was about 0.73.

In <u>coastal</u> <u>prairie</u> <u>areas</u> and <u>flatlands</u> of South Louisiana, the water surface profiles having head losses of up to .9 ft. (.27m) in 40 ft. (12.2m) of shredded tires would present a severe drainage problem where the hydraulic gradient slopes are below s= .003. A typical scenario which would present a problem would be along typical State Highways in crowded urban areas where driveways may be located at 200 ft. (61m) intervals for example. Given the scenario of five driveways located 200 ft. (61m) apart, this could cause a typical backwater problem of 5 x .7ft. or 3.5 ft. (1.06m) This kind of condition could result in severe flooding problems for the homes located upstream where the head water would be at its highest peak in the ditch.

In the <u>rolling</u> <u>hills</u> and <u>upland</u> <u>areas</u>, having similar ditch sizes as conducted in this study, where the hydraulic gradient slopws could be as high as s=.01 or 1%, and homes are apaced further apart, i.e. 500 ft. (152m) -600 ft. (183m), the flooding problems may not be as critical as in the coastal and flatland areas. Along Interstate Highway systems where culverts may be placed at 1000 ft. (305m) to 2000 ft. (610m) intervals, the back water profiles would not be as conducive to flooding if shredded tires were placed upstream of the culverts if used only on the culverts at the lower end of the watershed area.

RECOMMENDATIONS

It is recommended that future studies be conducted in the following areas:

a) Experimental modeling studies could be conducted at sites which have steeper slopes than the ones used in these tests, provided that test sites are available under controlled flow conditions.

b) Further hydraulic analysis could be made to develop a Froude number which hopefully could use dimensional analysis in hydraulic modeling. This could then be applied to the results to make projections on several different channel slopes, and varying depths of flow.

c) A detailed hydraulic analysis should be made to develop equivalent culvert sizes and head losses which could carry the same capacity as the channel studied in this project. A computer software package developed by Professor Langlinais entitled DRAINCALC Release 4.0 could be used in calculating the culvert types, diameters and lengths capable of handling the same discharges as used in these field studies.

ACKNOWLEDGMENTS

The authors wish to acknowledge the following for their contributions to this project: 1) Mr. Steve Cumbaa, Research project leader with the Louisiana Transportation Research Center, who acted as a consultant, assisted in the collection of field data, and organizing and setting up the transportation of shredded tires to and from the Agricultural Engineering Hydraulics field Laboratory on the U.S.L. campus, 2) Mr. Daniel J. Broussard, Research Supervisor with the Louisiana Transportation Research Center who also assisted in collecting the data and coordinating the transportation of shredded tires to and from the hydraulics field laboratory, 3) Employees of the Louisiana Department of Highways who assisted in handling the shredded tires during the field data collection process, 4) Members of the Project Review Committee, and 5) the administration and staff of the Louisiana Transportation Research Center.

REFERENCES

1. Chow, V.T., "Integrating the Equation of Gradually Varied Flow", Proce. A.S.C.E., Vol. 81, Nov. 1955.

2. Israelson & Hansen, *Irrigation Principles & Practices*, Third Edition, John Wilen & Sons, Inc. 1962 P 113, Table of Orifice and Weir Formulas P 81

3. King, *Handbook of Hydraulics*, McGraw-Hill Book Company n Values tables for Manning's equation

4. Vennard, J.K., *Fluid Mechanics*, 4th ed., Wiley, New York, 1963, pp. 384-393.

Petroleum Storage Tanks in the Agriculture Industry

Christopher G. Ward, C.P.G.
John R. Harriman, P.E.
Andre R. Fontaine

ABSTRACT

Several million facilities in the United States (US) store petroleum products, primarily motor fuels and heating oils. According to the Environmental Protection Agency (EPA), tens of thousands of these tanks and their piping currently are leaking; and many more are expected to leak in the future. Leaking petroleum products from storage tanks can cause fires or explosions that threaten human safety. In addition, a leaking petroleum storage tank (PST) can contaminate local groundwater. Contaminated potable water would devastate operations at almost any agricultural site. Since more than 50 percent of the U.S. population depends on groundwater as a source of drinking water, the federal government created legislation to protect groundwater resources from leaking PSTs.

Owners and operators of PSTs can minimize the risk of releasing petroleum products to the environment by complying with the federal and state regulations that define safe methods of petroleum storage. Many state regulatory agencies have guidelines for leak detection requirements, spill prevention contingency planning, and inventory control methods. Some states exempt farm-use PSTs from these regulations. What requirements does your state impose on the PSTs at your farm?

This paper defines the current requirements for PSTs used for agricultural operations and identifies pending federal legislation for owners of aboveground PSTs. Alternate methods of PST monitoring and inventory control are presented for farm operators who are exempt from federal and state regulations, fire prevention and safety is addressed.

UNDERGROUND STORAGE TANKS

In 1984, Congress enacted Subtitle I of the Resource Conservation and Recovery Act (RCRA). Subtitle I required EPA to develop regulations to protect groundwater resources from leaking underground storage tanks (USTs). In 1988, EPA issued final federal regulations to control the storage of petroleum products in USTs. In general, the regulations required that all new USTs be equipped with leak detection, spill/overfill protection, and corrosion protection. The regulations also scheduled a "phase-in" for existing USTs, allowing various amounts of time for existing USTs to upgrade into compliance.

The following problems cause most leaking USTs:

> *No Corrosion Protection* — Most of the UST systems already in the ground have tanks and piping made of bare steel. Unprotected steel below ground can be eaten away by corrosion.

> *Spills and Overfills* — Spills and overfills cause many UST releases. Even though many spills and overfills may seen small in quantity (i.e., a few drips each time), these drips become significant amounts over time. Also, an improper delivery truck hose disconnect creates a spill.

Installation Mistakes — Tanks and piping also will leak when not installed correctly. Poorly-selected backfill materials, inadequate tank burial depths, and loose pipe fittings can result in a leaking system.

Piping Failures — EPA studies show that most leaks result from piping failure. Piping is assembled on-site with numerous connections, which usually are near the ground surface. Piping is more susceptible to excessive surface loads, the stresses of underground movement, and corrosion than is a buried tank.

Regulatory Requirements

As of December 22, 1993, all regulated USTs must have:

1. **The tank and piping must be installed according manufacturer's specifications and industry standards,** such as the National Fire Protection Association (NFPA), the Steel Tank Institute (STI), and the Petroleum Equipment Institute (PEI).

2. **The UST must be equipped with spill and overfill prevention devices,** including high-level alarms, automatic shut-off devices, a ball float valve on the vent, and a spill containment basin on the fill.

3. **Both the tank and the piping must be protected against corrosion.** This protection can be done by coating the steel tank and providing either a sacrificial anode system or impressed current system, installing composite materials composed of steel bonded with fiberglass reinforced plastic (FRP), or installing a tank and piping constructed of FRP.

4. **Both the tank and the piping must be equipped with leak detection.** Various methods may be used to comply with the leak detection requirements of the federal regulations, including Automatic Tank Gauging (ATG), vapor monitoring, interstitial monitoring, and groundwater monitoring.

Another provision of the federal regulation for UST systems is financial responsibility. Financial responsibility means that if you own or operate an UST, you must ensure , either through insurance or other means, that there will be money to help pay for the costs of third-party liability and corrective action caused by a leak from your tank. These costs may include recovering leaked petroleum, correcting environmental damage , supplying drinking water, and compensating people for personal injury or property damage.

The financial responsibility requirements of the regulations are broken down as follows:

— At least $1 million of "per occurrence" coverage and at least $1 million of aggregate coverage if you are a petroleum marketer.

— At least $500,000 of "per occurrence" coverage and at least $1 million of aggregate coverage if you are not a petroleum marketer.

A tank owner can demonstrate financial responsibility for an UST system through insurance coverage, a guarantee issued by a parent company or firm, a surety bond, a letter of credit, by setting up a trust fund, or by any combination of these methods. Some states have even established programs to pay for cleanup costs from petroleum leaks. In most states, however, funds will pay only for part of cleanup costs, and rarely will pay for third-party damages caused by a leak.

In addition to the federal regulations governing the storage of petroleum products in USTs, each state has adopted its own set of regulations and standards. Many states adopted the federal regulations for their standards; others have amplified them. For example, the Commonwealth of Virginia adopted the EPA standards, but included all heating oil tanks of 5,000-gallons or greater in its UST regulations. The State of New Jersey regulates all "residential building" heating oil tanks of 2,000 gallons or greater, and all heating oil tanks not used for on-site consumption at a "residential building." Farm owners/operators should check the regulations of the state of residence to determine if an UST system falls under the provisions of state regulations as well as federal.

Leak Detection Techniques

Automatic Tank Gauging (ATG) is a computer-automated system which records the liquid level in the tank on a daily basis, and performs inventory reconciliation based on the amount delivered to the UST and the amount dispensed out. An ATG system can provide a daily printout report summarizing the quantity of fuel and water present in the tank, the temperature of the product, the amount of fuel delivered, and the difference between the computed volume of product and the amount gauged by the system. The last capability of an ATG system can be used as a method of leak detection. In addition, an ATG is often combined with other leak detection methods, such as interstitial monitoring.

Interstitial monitoring is a leak detection method used for double-wall tank and piping systems, where an inner pipe or tank is completely contained by a second pipe or tank. A brine solution is present in the interstitial space between the inner and outer tank (double-wall piping does not have a solution present). A sensor at the top of the tank detects any change in the liquid level of the interstitial fluid, and signals an alarm to the operator. Double wall piping is installed so that all piping slopes back to the tank manway, where a sump is equipped with a liquid refraction sensor. The presence of any liquid on this sensor also would signal an alarm to the operator, at which time the sump would be examined for product.

Vapor monitoring is a method of leak detection that requires more interaction from the operator. A series of wells are installed around the UST system to a depth just above groundwater. On a monthly basis, the operator must use a meter sensitive to petroleum vapors to determine if petroleum compounds are present in the vapors of the wells. If vapors are detected, additional investigation must be completed to determine the source of the vapors. At a greater expense a sensor can be installed in the well to provide continuous monitoring.

Groundwater monitoring is another method of leak detection that requires monthly activity by the operator. A series of wells are installed around the UST system that penetrate down into the groundwater beneath the UST system. On a monthly basis, each well must be checked for the presence of petroleum products. As with the vapor monitoring, if petroleum compounds are detected, the source of the petroleum must be determined (either the tank or the piping, or which tank/piping system among many). This leak detection method may be the most economical for a large tank field (i.e. many tanks in one location), but offers the least protection against identifying a leak before groundwater is impacted.

Exemptions

Exempt from the federal UST regulations are "farm or residential tanks of 1,100 gallons or less capacity used for storing motor fuel for non-commercial purposes" and "tanks used for storing heating oil for consumptive use on the premises where stored." A farm tank is defined as a tank located on a tract of land devoted to the production of crops or raising animals, including fish. To be exempt from UST regulations, a farm tank must be located on the farm property. "Farm" includes fish hatcheries, rangeland, and nurseries with growing operations. "Farm" does not include laboratories where animals are raised, land used to grow timber, pesticide aviation operations, and retail stores or garden centers where the products of nursery farms are marketed, but not produced (40 CFR 280).

ABOVEGROUND STORAGE TANKS

Many regulations have been implemented to control the storage of petroleum in USTs, but what about aboveground storage tanks (ASTs)? Most ASTs are easily inspected to detect leaks and defects, with the exception of the side contacting the ground surface. Some states have pushed ahead of the federal government and enacted regulations for ASTs similar to those for USTs. Many states now have specific design standards and require registration of ASTs as well as leak detection equipment. Additionally, NFPA, STI, and PEI have composed industry standards covering ASTs. However, no uniform (i.e., federal) regulations have been developed for ASTs at this time.

Existing Regulatory Requirements

In 1974, Congress enacted federal regulations for Oil Pollution Prevention (40 CFR 112). These regulation apply to any facility which has an aggregate aboveground storage capacity greater than 1,320 gallons, any single aboveground tank greater than 660 gallons, an aggregate underground storage capacity greater than 42,000 gallons, or any other facility that "could reasonably be expected to discharge oil in harmful quantities into or upon the navigable waters of the United States or adjoining shorelines."

The requirements for facilities included in this regulation are the preparation and implementation of Spill Prevention Control and Countermeasure Plans (SPCC Plan). The SPCC must include, but is not limited to, the following:

- Facility name, address, and telephone number;
- Owner/operator name, address, and telephone number;
- Facility type;
- Operation hours;
- Facility history, including oil usage;
- Past spill events;
- Emergency contacts;
- Tank information;
- Person responsible for training/implementation;
- Use of fuel;
- Inventory and records;
- Fuel supplier;
- Frequency of deliveries and amounts;
- Piping and, pumping sizes and materials;
- Spill containment devices;
- Spill containment and absorbent materials;
- Overfill prevention equipment;
- Leak detection;
- Security and safety features;
- Current SPCC training/procedures;
- Facility drainage;
- Inspections and records;
- Security;

- Personnel training and spill prevention procedures;
- Detailed site plan;
- Contingency plan — stop flow and contain; and
- Emergency telephone numbers.

Although the SPCC Plan is not required to be submitted to EPA for review, the SPCC Plan must be prepared by a Registered Professional Engineer, and kept on file at the facility for all personnel to use when needed. Civil penalties can be imposed by EPA if a release occurs at a facility and the SPCC Plan is determined to have been inadequate.

Proposed Federal Regulatory Requirements

Recently, Congressman James Moran (Virginia) introduced a bill to Congress entitled the "Safe Aboveground Storage Tank Act of 1993" in the House of Representatives. The purpose of the bill is to "establish a comprehensive program for the regulation of aboveground storage tanks to promote environmental and fire protection." Important provisions of the Safe Aboveground Storage Tanks Act include:

- **Notification** — Would require registration to a federal or state agency of existing tanks, those tanks taken out of operation, and newly installed tanks.

- **Release Detection** — Would require a release detection system for facilities at which tanks are located.

- **Preventions of Releases** — Would require certified inspection of tanks, maintaining records of visual inspection, corrosion protection of tank bottoms and piping in contact with the ground, labeling of tanks, and spill and overfill prevention devices.

- **Reporting** — Would require reporting of releases of 1 barrel (42 gallons) of petroleum and those corrective actions taken in response to a release.

- **Performance Standards** — Would establish standards for new and existing tanks, including design, construction, installation, maintenance, inspection, secondary containment, labeling, corrosion protection, integrity, and compatibility standards.

The current text of this bill would exclude motor fuel tanks of 1,100 gallons or less used at a farm or residence for non-commercial purposes, as well as heating oil tanks of 1,100 gallons or less. Many agricultural facilities are equipped with multiple aboveground tanks, each ranging in size from 275 to 550 gallons. Currently, these types of tanks are exempt from federal regulations, and would remain exempt under the Safe Aboveground Storage Tank Act, if enacted.

MANUAL TANK GAUGING

Manual Tank Gauging (MTG) is an economical way for a farm operator to perform leak detection on any underground or aboveground tank. MTG is a release detection procedure in which the level of the contents of a tank are compared before and after a specified period of tank inactivity. This method requires that the tank be inactive during the test period. MTG is an EPA-approved method of leak detection for USTs up to 1,000 gallons in capacity. USTs from 1,001 to 2,000 gallons can combine MTG with annual tank tightness testing.

Manual tank gauging can be performed with the following conditions:

- The tank should be gauged at the beginning and end of the test period.

- No product can be added or removed from the tank during the test period.

- The level of product must be measured at the start of the test period and again at the end of the test period using a gauge stick capable of measuring ⅛inch increments.

- Use an average of two stick readings. API recommends that an average of two readings that are within ¼inch of each other be used.

- The test should be conducted once a week, and averaged monthly.

- The records should be maintained for a minimum of one year.

The following form can be used by a farm operator to perform manual tank gauging.

MANUAL TANK GAUGING RECORD RECONCILIATION FORM

LOCATION: _____   TANK NO.: _____
ADDRESS: _____   CONTENTS: _____
_____   DATE: _____

	STICK READINGS (Inches)					TANK CONVERSION (Gallons)	WEEKLY VARIATION (Gallons) (Test End Less Test Start)
	DATE/TIME	1ST	2ND	SUM	AVERAGE		
WEEK 1 TEST START							
TEST END							
WEEK 2 TEST START							
TEST END							
WEEK 3 TEST START							
TEST END							
WEEK 4 TEST START							
TEST END							

TOTAL MONTHLY VARIATION (Add Weeks 1 Through 4) _____
MONTHLY AVERAGE VARIATION (Total/4) _____

Note:
1. Tank test must be done weekly; no product may be put in/removed during the test period. (See chart below for the required test period.)
2. The stick reading measurements must be an average of two stick readings (see C.2) to be considered valid for the test start and end.
3. See chart below for allowable monthly and weekly variations.

TANK SIZE (Gallons)	ALLOWABLE VARIATION (Gallons)		TEST PERIOD
	WEEKLY VARIATION	MONTHLY VARIATION	
If Manual Tank Gauging is the ONLY leak detection method used:			
550 or Less	10 Gallons	5 Gallons	36 Hours
551–1000[a] (When Largest Tank Is 64" x 73")	9 Gallons	4 Gallons	44 Hours
1000[a] (If Tank is 48" x 128")	12 Gallons	6 Gallons	58 Hours
If Manual Tank Gauging is combined with Tank Tightness Testing:			
551–1000	13 Gallons	7 Gallons	36 Hours
1001–2000	26 Gallons	13 Gallons	36 Hours

Does this week's variation exceed weekly allowable above? _____
Does variation of monthly average exceed monthly allowable shown? _____
(A "Yes" answer to either question should be reported as a suspected leak.)

[a] Refer to EPA reference document in 1.4 entitled *Straight Talk on Tanks*.

FIRE PREVENTION AND SAFETY

Paramount to most farm operations is the safety of the personnel, livestock, and structures. To greatly reduce the threat of fire from a fuel dispensing system, fire preventive measure should be incorporated into the fueling system. The following five documents are used as the foundation of fire hazard prevention in an agricultural environment:

- NFPA 30, "Flammable and Combustible Liquids Code"

- NFPA 30A, "Automotive and Marine Service Station Code"

- NFPA 31, "Standard for the Installation of Oil Burning Equipment"

- NFPA 70, "National Electrical Code"

- NFPA 395, "Standard for the Storage of Flammable and Combustible Liquids on Farms and Isolated Construction Projects"

Many potential hazards can be eliminated by upgrading a fueling system. For example, upgrades might include:

- An explosion-proof emergency stop button to disconnect the electrical power to the fueling system.

- A shear valve located on the dispenser fuel supply line to close automatically? if the dispenser is sheared from its mounting.

- Provide an approved fire extinguisher intended for use with the type of fuel.

- Removal of the locking device that keeps a fueling nozzle open.

- Provide fuel-absorbent material nearby for ease of fuel spill cleanup.

- Upgrade the electrical system so that fuel vapors do not come in contact with electrical switches that create sparks.

For additional improvements to a particular fueling system, state and local regulations, codes, and standards should be consulted, or ask the local fire marshall for an inspection. A qualified engineering consultant with extensive engineering and field practice in this area also can be contacted.

SUMMARY

In summary, federal and state regulations have been enacted to protect the environment and to foster safe practices for the storage of petroleum products. Although exemptions exist for many agricultural facilities, owners and operators of these facilities need to apply proper handling and management procedures to reduce the possibility of a petroleum product release with its accompanying potential for environmental damage.

REFERENCES

Bulk Liquid Stock Control At Retail Outlets. API Recommended Practice 1621. Fifth Edition. May 1993.

Environmental Protection Agency Regulations on Oil Pollution Prevention. 40 CFR 112; 38 FR 34164, December 11, 1973; Amended by 39 FR 31602, August 29, 1974; 41 FR 12657. March 26, 1976.

Underground Storage Tanks; Technical Standards and Corrective Action Requirements. Virginia Regulation 680-13-02. August 1, 1989.

Underground Storage Tanks; Technical Requirements and State Program Approval; Final Rules, 40 CFR Parts 280 and 281. September 23, 1988.

Musts For USTs. United States Environmental Protection Agency, Office of Underground Storage Tanks. Document 530/UST-88/008. July 1990.

The Safe Aboveground Storage Tank Act of 1993. 103rd Congress, 1st Session. H.R. 1360. The Honorable James Moran.

Industrial Waste Streams as Nitrogen Sources for Crop Production

S.P. France*, B.C. Joern and R.F. Turco

ABSTRACT

Nitrogen (N) mineralization kinetics of four industrial waste streams were evaluated under both field and laboratory conditions. Four waste streams were applied at a rate of 224 kg plant available N ha^{-1} in all studies. In the field experiments, corn (*Zea mays* L.) was grown for two consecutive years following waste stream application. Dry matter accumulation, grain yield and crop N uptake averaged 19.7 Mg ha^{-1}, 11.5 Mg ha^{-1} and 213 kg ha^{-1}, respectively, the year of waste stream application. No significant differences in dry matter accumulation, grain yield, .or crop N uptake were observed between N sources the year of application. Plant N availability during the residual year was related to the amount of organic N initially applied. Laboratory data support differences in waste stream N mineralization rates observed in the field. Approximately 37 - 51% of the organic N fraction from the industrial waste streams mineralized in the laboratory. Grain yield, plant N uptake and dry matter accumulation under field conditions and mineralization rates under laboratory conditions indicated that the organic component of these industrial waste streams mineralizes faster than the value of 20% used by Indiana regulatory agencies.

Keywords: Mineralizable nitrogen, Sludge, Waste disposal, Antibiotic processing waste.

RATIONALE

As available landfill space decreases, alternative disposal strategies for organic residual materials must be developed. One of the best alternatives to disposal is to use these materials as fertilizers to recycle nutrients through the soil-plant system.

Several articles have been written about the utilization of sewage sludge in agriculture (Kelling et al., 1977a; Kelling et al., 1977b; Khaleel et al. 1981; Mitchell et al., 1978; Soon et al., 1978). When organic wastes low in metal and toxic organic contaminants are applied to land, application rates are usually based on the nitrogen (N) requirement of the crop to be grown. As with any other N source, applications of organic waste materials in excess of crop N needs increase the potential for nitrate (NO$_3^-$) movement to groundwater.

In order to estimate plant available N (PAN), the N that will be available to a crop during the year of application, it is necessary to determine the mineralization rate of the organic N present in the material. Based on data in the literature from Jenny (1941), Stevenson (1965), and Stanford and Smith (1972), it has been commonly assumed that N mineralization follows approximate first-order kinetics. Laboratory incubation experiments designed to estimate N mineralization rates in the field have been partially successful in predicting N availability to crops from soils amended with organic wastes (O'Keefe, 1983).

S.P France, Graduate Student; B.C. Joern, Assistant Professor; and R.F. Turco, Associate Professor; Agronomy Department, Purdue University, West Lafayette, IN 47907-1150.

The Indiana Department of Environmental Management (IDEM) calculates PAN for sewage sludges and industrial sludges using eq. (1). This equation is based on data from Parker (1981) and Parker and Sommers (1983) who found that approximately 20% of the organic N present in anaerobically digested sewage sludges mineralizes during the first year. Nitrogen mineralization kinetics of many undigested organic waste streams have not been documented. The objectives of this study were to (1) determine the ability of four industrial waste streams to supply N to two successive corn crops at field scale and (2) measure the mineralization rates of the four waste materials in a controlled laboratory environment.

$$PAN = [(20\% \text{ organic N}) + (NH_4\text{-}N + NO_3\text{-}N)] \quad (1)$$

MATERIALS AND METHODS

Field experiment

Four antibiotic processing wastes and an inorganic fertilizer (anhydrous ammonia) were applied to two soils (Toronto silt loam, Fine-silty, mixed, mesic, Udollic Ochraqualf; and Elston loam, Coarse-loamy, mixed, mesic, Typic Argiudoll) in north central Indiana. A randomized complete block design with four replications of each treatment was used. The materials were subsurface injected at a rate of 224 kg PAN ha^{-1} either in the fall or spring and corn was grown for two consecutive growing seasons. Selected analyses of the antibiotic processing wastes are presented in Table 1.

Table 1. Selected analyses of the four industrial wastes*.

Assay	Material 1	Material 2	Material 3	Material 4
		mg kg^{-1}		
Solids	80000	12000	12000	3000
TKN	4091	6452	6005	5455
NH$_4$-N	624	1230	1550	2654
NO$_3$-N	3	12	12	15
P	4332	1033	3481	2410
K	209	1053	620	624
Cu	2	1	2	2
Zn	4	6	6	1
Ni	<1	1	<1	<1
Cd	<1	<1	<1	<1

*Analyses based on an "as received" basis.

Figure 1 shows the total N application rate needed to achieve 224 kg PAN ha^{-1} using eq. (1). Materials composed primarily of organic N required the highest total N application rate.

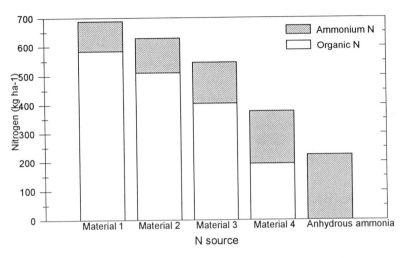

Figure 1. Total N needed to achieve 224 kg PAN ha^{-1}.

Plant tissue N was monitored in the above ground portion of randomly selected plants within the harvest rows of each plot throughout the growing season. Fifteen whole plants were taken at six leaf stage (V-6), 10 ear leaf samples were taken at silking (R-1), and 10 whole plants were taken at physiological maturity (R-6). Samples taken at physiological maturity were used to estimate both dry matter accumulation and total N uptake. The four center rows of each plot were harvested with a combine to obtain grain yield and grain N uptake. All tissue samples were dried at 60°C and ground to < 1mm. Total N in all tissues was determined using the micro-Kjeldahl technique outlined by Nelson and Sommers (1973).

Laboratory experiment

The surface 25cm of the two soils used in the field experiment was collected, sieved through a 2mm screen and stored moist at 4°C until used. A method similar to that of Stanford and Smith (1972) as modified by Smith et al. (1980) was used to determine the N mineralization rates for the treated soils. Each treatment combination was run in duplicate. The four industrial waste materials, a fresh dairy manure and an inorganic N source ($(NH_4)_2SO_4$) were applied at a rate of 224 kg PAN ha^{-1} using eq. (1).

Mineral N initially present was removed by leaching with 100ml 5mM $CaCl_2$ solution in 3-50ml increments. Excess water was removed under vacuum (30cm Hg). The columns were then covered with parafilm and a 3mm diameter hole was placed in the parafilm for aeration. The columns were then weighed and incubated at 30°C. Moisture content was adjusted 4 days after leaching and columns were leached weekly for 10 weeks. Leachate was assayed for inorganic N (Keeney, 1982).

A nonlinear least squares method developed by Smith et al. (1980) was used to estimate the first order rate constant (k) and potentially mineralizable N (N_o) for each material after subtracting background N mineralized by the control using eq. (2)

$$N_m - C_m = N_o [1-\exp^{(k \cdot t)}] \qquad (2)$$

where N_m is the cumulative N mineralized by the treated soil column and C_m is the cumulative N mineralized by the control column at a specific time t.

RESULTS AND DISCUSSION

Field experiment

Since trends between soil types were similar and time of application (fall or spring) had minimal effect, the data presented is one soil type (Elston loam) with treatments averaged over time. No significant differences in plant tissue N concentration were observed between N sources the year of application, but all N sources were significantly greater than the control (Table 2). Tissue N concentration averaged 38.6, 25.6, 7.9, and 13.7 g N kg^{-1} for V-6, R-1, R-6 and grain samples, respectively. No significant differences in dry matter accumulation, grain yield or crop N uptake were observed the year of application but all treatments were significantly greater than the control (Table 3). Dry matter accumulation, grain yield, and crop N uptake averaged 19.7, 11.5 Mg ha^{-1}, and 213 kg ha^{-1}, respectively for the N sources.

Table 2. Plant N tissue concentration the year of application*.

N source	Whole plant V-6	Ear leaf R-1	Stover R-6	Grain harvest
	---------------- g N kg^{-1} ----------------			
Material 1	39.1 a	26.0 a	8.4 a	13.8 a
Material 2	38.4 a	25.8 a	8.0 a	14.1 a
Material 3	39.3 a	25.1 a	7.9 a	13.7 a
Material 4	39.0 a	25.5 a	7.7 a	13.7 a
NH_3	37.2 a	23.9 a	7.4 a	13.2 a
Control	33.1 b	20.2 b	5.5 b	10.5 b

*Means followed by the same letter are not significantly different using SNK @ alpha=.05

Table 3. Selected yield components the year of application*.

N source	Dry matter accumulation	Grain yield	Crop N uptake
	-------- Mg ha^{-1} ----------		- kg ha^{-1} -
Material 1	19.6 a	11.3 a	218 a
Material 2	19.7 a	11.6 a	218 a
Material 3	19.4 a	11.6 a	211 a
Material 4	19.5 a	11.3 a	208 a
NH_3	20.2 a	11.6 a	208 a
Control	15.2 b	6.9 b	115 b

*Means followed by the same letter are not significantly different using SNK @ alpha=.05

In the year following application, an additional inorganic N control (anhydrous ammonia applied at 224 kg N ha^{-1}) was added as a reference for optimum crop growth. Plant tissue N concentration in Materials 1 and 2 tended to be slightly higher than the other organic N sources early in the growing season during the second year, but these differences were not statistically significant (Table 4). Yield components for Materials 1 and 2 were generally higher than for Materials 3 and 4 the second year (Table 5). No significant differences in dry matter accumulation were observed between Material 2 and the 224 kg N ha^{-1} N control the second year. Grain yield for Materials 1 and 2 were significantly higher than all other N sources except the 224 kg N ha^{-1} N control the second year. Based on crop N uptake data, Materials 1 and 2 contribute approximately 35 and 45 kg N ha^{-1} the second year. To avoid excessive N loading and potential ground water contamination by NO$_3^-$-N, proper N credit from the residual N pool for Materials 1 and 2 should be managed the second year.

Table 4. Plant tissue N concentration the second year*.

N source	Whole plant V-6	Ear leaf R-1	Stover R-6	Grain harvest
	---------- g N kg^{-1} ----------			
Material 1	31.6 b	19.2 b	5.8 ab	13.1 a
Material 2	31.1 bc	20.7 b	6.0 ab	12.6 a
Material 3	29.9 bcd	17.6 b	5.2 b	13.1 a
Material 4	29.5 cd	17.3 b	5.3 b	13.7 a
NH$_3$	29.0 d	16.0 c	5.5 b	12.7 a
Control	29.4 bcd	17.9 b	5.0 b	11.7 a
N control	36.8 a	27.8 a	7.0 a	13.1 a

*Means followed by the same letter are not significantly different using SNK @ alpha=.05

Table 5. Selected yield components the second year*.

N source	Dry matter accumulated	Grain yield	Crop N uptake
	---------- Mg ha^{-1} ----------		- kg ha^{-1} -
Material 1	20.7 b	5.8 b	112 bc
Material 2	23.5 a	6.1 b	122 b
Material 3	19.7 bc	4.8 c	95 cd
Material 4	17.6 c	4.2 cd	92 d
NH$_3$	16.9 c	3.5 e	76 d
Control	18.8 bc	3.7 de	72 d
N control	24.2 a	9.6 a	173 a

*Means followed by the same letter are not significantly different using SNK @ alpha=.05

Laboratory experiment

The composition of the industrial waste materials is listed in Table 1. The amount of total N needed to achieve an application rate of 224 kg PAN ha^{-1} is illustrated in Fig. 1. The amount of N mineralized over time is related to the quantity of N initially applied for the 4 materials (Fig. 2). Materials 1, 2 and 3 had the highest application rates of total N and released the greatest quantity of N during the incubation period. Net immobilization was observed in the fresh dairy manure for the first 5 weeks of the incubation period (Fig. 2). Since the N was immobilized during this period, the fresh dairy manure did not approximate first-order kinetics.

Using eq. (3), the fraction of organic N mineralized after the ten week incubation period (N_r) was higher for the four industrial waste materials than the fresh dairy manure (Fig. 3). The N_r for the four materials ranged from 37-51% compared to 6% for the fresh dairy manure.

$$N_r = [(N_T - (N_m - C_m))/N_T] \times 100 \qquad (3)$$

where N_r is the percent organic N mineralized, N_T is the total organic N initially applied, N_m is the cumulative N mineralized by the treated soil column, and C_m is the cumulative N mineralized by the control column at a specific time t.

The initial leaching process may also remove a significant proportion (>25) of the organic N from sludge amended soils (Parker, 1981). Similar levels of leachable organic N were obtained by Smith et al. (1980) for non-sludge amended soils. Since this leachable organic N probably consists primarily of low molecular weight compounds that would be readily degraded by microorganisms, the incubation and leaching procedure developed by Stanford and Smith (1972) may underestimate N_o in soil if soluble organic N present in the leachate is not analyzed.

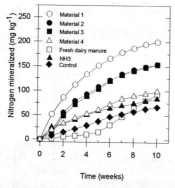 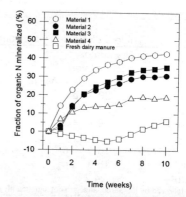

Figure 2. Cumulative N mineralized   Figure 3. Fraction of organic N recovered

The k values for the materials shows that at 30°C, the mineralizable fraction is released at a rate between .20-.37 weeks^{-1} based on the quantity of mineralizable N remaining after each succeeding week of incubation (N_o-N_t) (Table 6). The time required to recover one-half of the potentially mineralizable N, $t_{1/2}$, at 30°C for each waste stream using eq. (4) is also presented in Table 6.

$$t_{1/2} = 0.693/k \qquad (4)$$

where k is the mineralization rate constant from eq. (2).

Table 6. Potentially mineralizable nitrogen (N_o), mineralization rate constants (k), percent organic N mineralized (N_r) and time required to recover one-half of the potentially mineralizable N ($t_{1/2}$) for the four materials.

Source	N_o	k	N_r	$t_{1/2}$
	mg N kg^{-1} soil	$weeks^{-1}$	%	$weeks^{-1}$
Material 1	136	.37	51.0	1.9
Material 2	96	.26	38.2	2.7
Material 3	103	.20	47.9	3.5
Material 4	32	.36	36.8	1.9

Based on the N_r values generated under controlled laboratory conditions in Table 6, all four Materials mineralize much faster than the value of 20% used by IDEM in eq. (1). By substituting these calculated mineralization rates into eq. (1), approximately 400, 315, 335, and 255 kg N ha^{-1} would be made available to a crop during the first growing season for Materials 1, 2, 3, and 4, respectively. Optimum yields will be obtained the year of application as was observed in the field experiment (Tables 2 and 3), however the residual N remaining in the soil profile after harvest could potentially leach to ground water prior to crop establishment the following year. The mineralization rate constant used in eq. (1) should be adjusted for the material applied to more accurately predict PAN.

References

1. Jenny, H. 1941. Factors of Soil Formation. p. 249-251. McGraw Hill, New York.

2. Keeney, D.R., and D.W. Nelson. 1982. Nitrogen-inorganic forms. p 643-693. In L.A. Page, R.H. Miller, and D.R. Keeney (eds.). Methods of soil analysis. Part 2. American Society of Agronomy. Madison, Wisconsin.

3. Kelling, K.A., A.E. Peterson, L.M. Walsh, J.A. Ryan, and D.R. Keeney. 1977a. A field study of the agriculture use of sewage sludges: I. Effect on crop yield and uptake of N and P. J. Environ. Qual. 6:339-345.

4. Kelling, K.A., L.M. Walsh, D.R. Keeney, J.A. Ryan, and A.E. Peterson. 1977b. A field study of the agriculture use of sewage sludges: II. Effect on soil N and P. J. Environ. Qual. 6:345-352.

5. Khaleel, R., K.R. Reddy, and M.R. Overcash. 1981. Changes in soil physical properties due to organic waste applications: A review. J. Environ. Qual. 10:133-141.

6. King, L.D. 1979. Waste wood fiber as a soil amendment. J. Environ. Qual. 8:91-95.

7. Mitchell, M.J., R. Hartenstein, B.L. Swigt, E.F. Neuhauser, B.J. Abrams, R.M. Mulligan, B.A. Brown, D. Craig, and D. Kaplas. 1978. Effects of different sewage sludges on some chemical and biological characteristics of soil. J. Environ. Qual. 7:551-559.

8. Nelson, D.W., and L.E. Sommers. 1973. Determination of total nitrogen in plant material. Agron. J. 65:109-112.

9. O'Keefe, B.E. 1983. Evaluation of nitrogen availability indexes on a compost amended soil. M.S. thesis, Univ. of Maryland.

10. Parker, C.F. 1981. Carbon and nitrogen mineralization in soils treated with sewage sludge. M.S. Purdue Univ.

11. Parker, C.F., and L.E. Sommers. 1983. Mineralization of nitrogen in sewage sludges. J. Environ. Qual. 12:150-156.

12. Smith, J.L., R.R.Schnabel, B.L. McNeal, and G.S. Campbell. 1980. Potential errors in the first-order model for estimating soil nitrogen mineralization potentials. Soil Sci. Am. J. 44:996-1006.

13. Soon, K.Y., T.E. Bates, E.G. Beauchamp, and J.R.Moyer.1978. Land application of chemically treated sewage sludge: I. Effect on crop yield and nitrogen availability. J. Environ. Qual. 7:264-280.

14. Stanford, G., and S.J. Smith. 1972. Nitrogen mineralization potentials of soils. Soil Sci. Soc. Am. Proc. 36:465-472.

15. Stevenson, F. J. 1965. Origin and distribution of nitrogen in soil. In Soil nitrogen Agronomy 10:1-42.

ENVIRONMENTALLY SOUND AGRICULTURE THROUGH REUSE AND RECLAMATION OF MUNICIPAL WASTEWATER

Andrew Roberts and Walter Vidak
Pascua Florida Corporation - Agronomic Wastewater Reuse Division
Tallahassee, Florida

ABSTRACT

Few municipalities have met the national goal of elimination of wastewater discharge into surficial and navigable waters mandated in the Federal Water Pollution Control Act Amendments of 1972. In addition, national goals of sustainable agriculture and conservation require our focused efforts to reclaim and reuse all forms of nutrient resources. Treated municipal wastewaters represent such a nutrient resource that can be reclaimed and used through agriculture yet in such a manner as to protect the environment. Urban life and agriculture can interrelate in a mutually compatible and advantageous situation through agriculture's management, reuse and removal of nutrients in treated effluent.

Tallahassee, Florida achieved this mandate by the 1985 deadline by developing one of the country's largest and longest operating successful spray irrigation programs using treated municipal effluent for agriculture production. Pascua Florida Corporation has managed this operation since 1981 and has reported the complexities and history in reports to American Society of Agricultural Engineers (1988), USDA Soil Conservation Service, the first Environmentally Sound Agriculture Conference (Vidak and Roberts, 1991), Water Environment Federation (Vidak and Roberts, 1992) and other gatherings of national and international interested parties.

The extensive historical experience from the 1720 acre agricultural operation managing an average 12 to 18 million gallons daily effluent application provides the successful example that allows calculation of needed similar operations for other municipalities. Integration of urban goals of waste removal, environmental protection and aesthetic improvement or preservation have been achieved along with agricultural goals of efficient management, nutrient application and effective production.

Sustainable interrelationships of sprayfield operation are complex and the specifics of routine production, crop rotation, unique pest and disease control, micronutrient adjustments, alternative crop evaluations and projected nutrient removals are unique and constantly evolving. At the first conference, initial reports of 12 month triple cropping of corn, soybeans, rye/rye grass for hay production and grazing were made. Constant environmental monitoring as well as desired reduction in fertilizer inputs have led to alternative crop evaluations of canola and increased animal feeding.

Although the partnership operation between Pascua Florida Corporation and the City of Tallahassee along with the Institute of Food and Agricultural Sciences and other scientists continue to research means to obtain efficient agricultural production, sometimes proper management of large wastewater quantities may result in unprofitable agricultural decisions to achieve environmental protection. The continuing record of operation illustrates that urban/agricultural wastewater reuse is an achievable goal for all municipalities.

INTRODUCTION

Florida must treat and properly dispose of over 1 billion gallons of domestic wastewater each day. Currently, 47 percent is disposed of through onsite treatment with 53 percent through central treatment systems. Leon County utilizes approximately 34.4 million gallons of groundwater per day treating an average 15.9 million gallons per day through wastewater treatment facilities and then applying this to 1720 acres through agricultural production.

In 1991, we provided an initial report of the Southeast Sprayfield with approximately 1750 acres under cultivation in 13 center pivots irrigation systems. The maximum design capacity of the current system is 24.2 million gallons per day at a maximum operating pressure of 60 pounds per square inch. The pivots with accompanying end guns, have a design capacity ranging from 1090 to 1740 gallons per minute with 9 pivots rated at 1450 gallons per minute. Actual operations deliver 12- 17 million gallons per day. The effluent is pumped eight miles from the Thomas P. Smith Wastewater Treatment Facility located near the municipal airport. Lined holding ponds at the sprayfield can contain 100 million gallons of wastewater to provide for storage for cold weather shutdown, maintenance of system, harvest of crops and planting.

AGRONOMIC WASTEWATER REUSE GOALS

The management of the sprayfield system continues to be one of balance and difficulty. Our primary goals of wastewater reuse are nutrient and water management, prevention of environmental contamination, and production of crops that result in maximum nutrient removal from the effluent and environment.

ENVIRONMENTAL PROTECTION AND MANAGEMENT

Prevention of environmental contamination has been the first goal of both Pascua and the City and the requirements of the State and federal government are closely followed. The sprayfield does provide a beautiful environment for our enjoyment as well as a habitat for many endangered species. Geologically, the sprayfield lies within the sensitive Woodville Karst Plain. The soils are chiefly Lakeland Sand which is very highly permeable. Soil characteristics have significantly changed over our 12 year management period presenting us with both environmental and management challenges. Soil pH is now approximately 7 down to 15 ft (Allhands, 1989). Chloride levels have dramatically increased from an original background level of 2.8 - 3.1 mg/L in 1981 to routine findings of 44 mg/L in 1992. Micronutrient trends being followed by Allhands and Overman show increasing magnesium, potassium and sodium levels. Management of this changing environment to prevent contamination yet allow crop production has been difficult. As yet we have not detected phytotoxicity with the chloride levels encountered. In fact, chlorides at a continuing ratio of 3.5 times the nitrogen in the effluent enables us to track nitrogen sources.

Fertilizers, herbicides, insecticides and limited fungicides must be applied to allow adequate crop production. These agrichemicals are applied through the irrigation pivots in a chemigation system with appropriate backflow prohibition as required by Florida regulations. Chemigation is usually conducted at night because of reduced winds and reduced volatilization of chemicals due to lower night time temperatures. Principals of Integrated Pest Management are utilized in pest management to protect.

No-till production is utilized to prevent wind erosion of the land and damage to tender young plants, as well as to prevent leachability of effluent and chemicals used in the highly permeable soils. The Southeast Farm has for many years been the largest no-till operation in the state of Florida. No-till does present problems of immobile nutrient stratification, particularly with phosphates after three years. Layered soils sampling indicates 85% of the phosphate in the top 3 inches and also stratification of organic matter therefore tillage must be utilized on a rotated basis to prevent problems.

With soils with high leachability, for prevention, the sprayfield environment is constantly monitored by onsite visits by state and local officials as well as groundwater monitoring by 58 strategically located monitoring wells. Pruitt (1988) has published data on the historical monitoring with 20 wells tapping the Surficial aquifer and 38 wells placed to monitor the deeper Floridian aquifer from which the majority of groundwater use in Florida is withdrawn.

The Environmental Protection Agency has established a maximum concentration for nitrate in drinking water at 10 parts per million (mg/L). Microorganisms convert the effluent nitrogen and the nitrogen applied as ammonia, fertilizers or animal wastes to nitrates and nitrites. Decisions on nitrogen applications to crops and management of effluent must prevent groundwater contamination above this level with either nitrates or any of the other agrichemicals such as fungicides and herbicides that are used. Monitoring data show no indication of environmental contamination with pesticides. Most monitoring wells show compliance with nitrate levels; however, limited problems with nitrate in specific monitoring wells have led to change in management decisions in crop production. Also, investigation of specific wells revealed that the excessive nitrate levels may have been due to faulty well construction, because another well drilled close by revealed acceptable nitrogen levels of 1.82, 5.06 and 5.68 mg/l nitrate.

Many factors affect the environmental nitrogen concentration found in these monitoring wells. Factors affecting include:

- irrigation rates
- nitrogen concentration applied
- nitrogen distribution
- depth to groundwater
- well depth

* subsurface geology
* ground movement of water

Berndt (1990) has shown that the nitrates from applied fertilizer may be distinguished from other nitrogen sources due to a different isotopic profile. Monitoring wells are sampled at frequent intervals to prevent any off-site migration of nitrates or other chemicals.

Some would propose that no management be conducted with land application of effluent, and that somehow this would be the better alternative environmentally. A portion of the Tallahassee site is deliberately being left unmanaged and has already demonstrated a weed choked, unbalanced system that may quickly be unable to handle the nitrogen and phosphorus applied through the 125 inches of effluent applied each year. Continued observations will document the effect of no management.

NUTRIENT AND WATER MANAGEMENT

The goal of adequate nutrient and water management is integrally tied with environmental protection and crop production cannot be achieved without this management. Many years ago certain authorities felt inaccurately that providing a farmer with 125 inches of precipitation a year with varying low nutrient levels was a gift of significant value. Sprayfield management has shown that the opposite is correct and that management of this vast water and nutrient resource is the most difficult task at hand.

An annual rainfall of 63.29 to 104.18 inches combined with effluent rates of 1.94 - 2.39 inches per week requires us to manage crop production at total hydraulic loadings of 170 inches water per year. When we realize that tropical rain forests are categorized at 140 inches precipitation per year, you can recognize the difficulties involved. Although 3.16 inches/week is the permitted loading capacity, we do not anticipate additional loading at this time. An sprayfield expansion proposal has undergone an Environmental Impact Statement scrutiny but as yet we do not know when Tallahassee will go forward with this proposal.

Nitrogen and phosphorus loading for this acreage is also difficult to completely assess. The chemical characteristics of the effluent supplied has remained fairly constant over the past ten year period until recent treatment plant modifications increased treatment and reduced nitrogen levels to now 11-12 mg/L. Research and documentation of crop yields continue in attempts to determine how to effectively manage a continuing dilution of applied nutrients. Allhands (1989) indicated that a total of 340 lbs nitrogen per acre was required for adequate commercial production. Decrease in effluent nitrogen content will then elevate the amount of fertilizer that must be applied.

The sprayfield soils have insufficient nutrients for crop production without application either by effluent or commercial fertilizer application. The amount required will vary from crop to crop and as alternative crops are investigated research must be conducted to determine management strategies.

Increased cattle grazing in 1991, 1992 and 1993 is being closely monitored to determine nitrogen loading from animal wastes, necessity for varying nutrient management and additional needs to prevent environmental contamination. Preventive decisions on nitrogen loading are difficult because changing crop management practices or changes in effluent do not impact groundwater concentrations for 1 to 2 years.

Nitrogen loading from animal wastes from cattle grazing compared to nitrogen removal in live cattle shipped is very difficult to assess. According to ASAE projections a 1400 lb dairy cow will excrete 230 lbs of nitrogen in wastes per year. Utilizing the size of cows and the feed consumed during 1991, we might calculate that 16,485 pounds of nitrogen was loaded onto the land from animal wastes. A total of 263,000 pounds was gained by the cows and additional 140,000 pounds removed for 400 calves born averaging 350 lbs each. That year we feed 600 calves averaging 450 to a weight of 650, and 650 cows averaging 800 to a weight of 1000 for a period of 210 days. The cows were allowed to graze on an unlimited basis and the lighter calves were fed an additional 1% of their body weight in corn each day. The corn consumed was a net removal since it was produced on the sprayfield and the rye/rye grass grazing was also a net removal.

FDACS reports up to 150 lbs/day of clover consumed by one cow, yet only 70 lbs/day of Pangola grass. Our experience has been between 35 and 70 lbs/ grass consumption per day. Additionally some researchers estimate as much as 50% of nitrogen lost to volatilization from animal wastes. As you can see from all factors to be considered, there is still no consensus at the current time as to the ultimate nitrogen removal from such a system of cattle grazing. Additional experience in 1992 with contract feeding of over 2100 cows with and an additional 400 calves produced, as well as 1993 with over 2000 head currently being fed will give us hopefully the necessary numbers to provide proper statistical evaluation.

CROP PRODUCTION THROUGH WASTEWATER REUSE

The second major goal of the sprayfield farm is to produce crops that result in the maximum nutrient removal from the land in conjunction with environmental protection. Crop selection through the years has been based on regulatory permitted crops, research conducted by Allhands and Overman as well as individual research by Pascua and trial and error.

Crops selected must be able to withstand application of large quantities of water, ability to grow in the environment available at the spray field and resistance to plant pests and diseases which present unique difficulties to sprayfield management. In addition, crops must be selected which will utilize the land area for 12 months of the year.

Based on these criteria, we reported in the first Environmentally Sound Agriculture Conference in 1991 our experiences in the triple cropping or production of three consecutive and overlapped growing and harvesting periods for chiefly corn, soybeans, and rye/rye grass combination and other grasses. In the first conference, we reported in the detail the unique problems associated with no-till, pests, weeds, and other unique production problems. Each crop has specific nutritional needs and each crop will remove a certain amount of nitrogen and other nutrients at harvest.

The efforts of the past three years have concentrated on more thorough evaluation of alternative crops such as canola as well as a greater utilization of contract grazing of cattle. Other crops utilized besides corn, soybeans, and rye/rye grass include Bahia, alyssa bermuda, clover and grazing sorghum. Limited trials with alfalfa and perennial peanuts have not been as successful.

Canola is a specific type of rapeseed that is becoming increasingly popular as a source of vegetable oil with low molecular weight unsaturated fatty acid composition. Canola can be produced at a rate of 2000 lbs per acre which equates to 40 bushels per acre. The canola requires a 200 lb nitrogen/acre uptake but harvesting removes only 90 lbs/acre at a cost of $105/acre. Canola, a member of the mustard family, presents difficulties in harvesting and storage also due to the minute size of the seeds.

Published values and experience exist for nitrogen uptake and removal based on average yield for corn, soybeans, and other crops. Continued efforts with IFAS have reduced nitrogen inputs into corn with a continuing harvesting level of approximately 158 bushels per acre. Leaf tissue and soil samples were taken weekly for selected pivots of corn and soybeans in order to determine nutrient needs but little statistical correlation for a total season could be determined. Rapid nutrient movement through sandy soils does not permit optimum crop nutrient utilization under the high water effluent and precipitation rates encountered. We are now utilizing a Best Management Practice of withholding water at certain stages in development to encourage a more extensive root system for nutrient uptake.

The situation is complicated by high levels of calcium in city water wells pumped from limestone formations. Calcium ions attach more readily than ammonium and potassium ions, further depleting the soil's capability to hold nitrogen and potassium in the root zone for plant absorption. This calcium phenomena causes constant micronutrient deficiencies particularly with manganese as earlier reported (Vidak and Roberts, 1991).

Extensive and unique pest and disease problems continue to effect crop production in the humid environment of the sprayfield. Options for disease control through Integrated Pest Management are limited and our need for additional research is pressing. For continued effective land application of effluent whether at the Tallahassee location or any other geographical location, we need additional research in the area of:

* effects of dilution of nutrients and how they are utilized as well as effects on specific crops

* most desirable effluent quantity and quality for effective management of large quantities of municipal effluents

* identification of best application methods and timing for most efficient nutrient removal and utilization

* changing microbial activity in sprayfield operations

* determination of effluent effects on plant pests and diseases including development of management strategies

* evaluation of possible organic incorporation

* investigation of long term effects of animal production on the sprayfield environment with animal waste loading and removal

* incorporation of all previous investigations into a data base for easier use in environmental and management strategies

* evaluation of environmental fate without agricultural production

SUMMARY

We believe that the primary goal of wastewater land application sites is nutrient and water management not profitable agriculture production. Reclamation of our water resources and reuse of the nutrients in treated wastewaters is essential to prevent discharge to and contamination of surface and groundwaters.

Our twelve years of experience, while showing us the complexity needed to manage municipal effluent application, has also shown us that this can be achieved. If everyone recognizes that land application of effluent must be managed to prevent environmental contamination, then we must also accept that some of the agricultural production decisions we must make will be unprofitable to achieve environmental protection.

Reuse and reclamation of treated municipal wastewater through sustainable agricultural production can and must be an achievable goal for all municipalities.

REFERENCES

1. Allhands, M., 1989, Effects of Municipal Effluent Irrigation on Agricultural Production and Environmental Quality.

2. Berndt, M. P., 1990, Sources and Distribution of Nitrate in Ground Water at a Farmed Field Irrigated with Sewage Treatment Plant Effluent, Tallahassee, Florida. U. S. Geological Survey, Water Resources Investigation Report 90-4006.

3. Environmental Protection Agency, 1990 Draft Environmental Impact

Statement Supplement, Tallahassee, Leon County Waste Management. Environmental Protection Agency, Atlanta, Georgia.

4. Overman, Allen R., 1979, Effluent Irrigation of Coastal Bermuda Grass, Journal of the Environmental Engineering Division, American Society of Civil Engineers.

5. Overman, Allen R. 1981, Irrigation of Corn with Municipal Effluent, Transactions of the American Society of Agricultural Engineers.

6. Pettygrove, E. S., Asano, T., 1988, Irrigation with Reclaimed Municipal Wastewater, Lewis Publishers Inc., Chelsea, Michigan.

7. Pruitt, Janet B., Elder, John F., Johnson, I., 1988, Effects of Treated Municipal Effluent Irrigation on Groundwater Beneath Sprayfields, Tallahassee, Florida. U. S. Geological Survey, Water-Resources Investigation Report 88-4092.

8. Vidak, W. and A. Roberts, 1991, Development of Maximum Agricultural Production Utilizing Land Application of Municipal Wastewater Effluent and Efficient Crop Choice and Rotation, Proceedings for the Environmentally Sound Agriculture Conference, University of Florida Press, Gainesville, FL.

9. Vidak, W. and A. Roberts, 1991, Operation - Tallahassee System, Water Reuse Seminar, Proceedings American Society of Agricultural Engineers, April, 1991, Jacksonville, FL, University of Florida Press, Gainesville, FL.

10. State Water Use Plan, FL Dept. of Environmental Regulation, Tallahassee, Florida.

11. Camp Dresser and McKee, 1992, Optimization of the Existing Agricultural Spray Field Systems, City of Tallahassee, Florida.

12. Florida Department of Agriculture and Consumer Services, 1976, Beef Cattle in Florida.

Diagnostic Evaluation of Wastewater Utilization in

Agriculture, Morelos State, Mexico

Rodriguez Zavaleta, Carlos[1], Oyer, Linda[2], and Cisneros, Xochitl[1]

ABSTRACT

The application of wastewater for irrigating agricultural crops is an agronomic practice which exists in Mexico since the beginning of the century. Actually, the use of this type of water in agricultural activities is very extensive, approximately 370,000 has. being irrigated with wastewater in areas close to large cities. The State of Morelos occupies second place on a national level with 42,797 ha., after the State of Hidalgo where the largest irrigation district in the world is located. The general objective consisted of conducting a diagnostic evaluation of the use of wastewater in irrigation with emphasis in one irrigation district and determining the most important agroecosystems. Factors selected were: frost-free period, soil type, water quality, and irrigation area. Eleven agroecosystems were identified and, based on their characteristics, technical guides for improving the productivity of these agricultural areas were developed. The socioeconomic analysis indicated that the producers have the technical capacity and the desire to apply simple, cost-effective tecnologies for improving the productivity and sustainability of the defined agroecosystems. Soil and crop management as well as wastewater treatment were considered.

Keywords: Wastewater utilization, agroecosystems, untreated wastewater.

[1] Mexican Water Technology Institute (IMTA), Paseo Cuauhnahuac No. 8532, Col. Progreso, Jiutepec, Morelos, Mexico.
[2] USDA Soil Conservation Service, Jiutepec, Morelos, Mexico; USDA-SCS, P.O. Box 3087, Laredo, TX 78044.

INTRODUCTION

Due to the demographic explosion in the central region of Mexico and the increasing demand for white water by the large population nuclei, the lack of white water for irrigation has sharpened. This has increased markedly the use of wastewater for agricultural purposes.

In Mexico, the utilization of wastewater for irrigation of crops has existed since the beginning of this century. Actually, Mexico occupies second place in the world after the Republic of China in the use of wastewater. Approximately, 370,000 has. in Mexico are irrigated with wastewater, the State of Hidalgo contributing 22% of this surface area, followed by the State of Morelos with 42,797 ha.

Agricultural systems are comprised of structure and function. The operation and functioning of the agricultural system depends principally on the structural elements that define it. An important structural element is the ecological medium, which upon manipulation by humans, generates the agroecosystems. The knowledge and understanding of the agroecosystem must be based on a clear socioeconomic scheme, and the precise determination of the opportunities and limitations of the ecological medium. In this way, it will be possible first to analyze and then through the scientific method to achieve an optimal utilization of these agricultural systems.

This study was conducted to determine the existing agroecosystems in the State of Morelos (with initial emphasis on one irrigation unit) with the use of wastewater to optimize them in terms of soil, water, and plant resources and estimate the potential environmental and socioeconomic impact caused by their utilization.

LITERATURE REVIEW

Calderon (1991) indicated that only 8% of the municipal discharge that is generated in Mexico receives an adequate treatment in a total of 361 treatment plants. Approximately, 15% of industrial discharge is treated and, in general, the treatment for wastewater is only 10%. It is estimated that to treat all of the wastewater of domestic origen (without including the cities of Mexico and Guadalajara) is an additional 1.5 billion pesos. To achieve 65% treatment of wastewater in 1994, an annual investment of 0.8 billion pesos is required, signifying tripling the actual investment. In this way, the treatment of residual discharge would be covered in a period of eight years. In the State of Morelos, there are seven water treatment plants: six municipal with an installed capacity of 43 lps and one for industrial water with capacity of 200 lps (Arango cited by Torres, 1992).

The federal government of Mexico has initiated the "Clean Water Program" in which the production of crops, principally vegetable crops, in direct contact with wastewater, such as vegetable root crops (onion, beet, carrot) is prohibited.

Irrigation of agricultural crops with wastewater can have varying impacts on some physical, chemical and biological soil properties. In general, physical properties, such as pore space, stability, aggregate size, and moisture retention capacity, are benefited with the use of wastewater. Some chemical properties are improved (organic matter, C:N ratio, enzymatic activity, nutrient content) while others deteriorate (major concentration of salts and heavy metals).

Torres (1992) evaluated the impact of heavy metals in wastewater originating from a large industrial center (CIVAC) in the State of Morelos on soil and crops and determined that the levels of Fe exceed the permissible limits for water, soil, and crops studied. Lead and Zn exceeded Cd and Cr but were less than those of Fe.

Among the potential impacts of utilization of wastewater on the plant resource are increase in yields, increase in disease and insect incidence, nutrient imbalance, as well as decrease in crop quality. Day and Tucker (1960) compared the effect of irrigation with wastewater with and without fertilizer and the control with white water on the yield of the following grasses: oats, barley, and wheat. The best results were obtained upon utilizing wastewater without fertilizer. Similarly, Clapp (1984) and Bole et al. (1985) mentioned that forages have a good response to wastewater. Wastewater mixed with secondary effluents can be utilized to grow crops like alfalfa, cotton, rice, forages, and sorghum. Rodriguez (1987) in research conducted in the State of Hidalgo, Mexico found no response to fertilization for corn, alfalfa, oats, wheat, barley, tomato, and bean when irrigated with wastewater.

MATERIALS AND METHODS

For the determination of agroecosystems, maps and existing information on climatological, edaphic, and agronomic aspects were obtained and analyzed, including temperature, precipitation, evaporation, frost-free period, soil depth, soil texture, water quality monitoring data, e.g. electrical conductivity, chlorides, sodium adsorption ratio, total dissolved solids, fecal coliforms, Vibrio cholerae, surface areas of crops and yields, and hydrologic infrastructure.

For the agroeconomic study, ecological, technological, and socioeconomic data for the Irrigation Unit 1, District 16, Morelos were obtained, analyzed, and interpreted. Forty-four in-depth interviews, of which 28 were recorded on video, were conducted with representative producers by irrigation water source and by crop. Direct observation in the field was also utilized to collect data on predominant soils, technologies, and production systems.

With the ecological and socioeconomic aspects and the form of resource use (available technology), human-environmental interactions were characterized, which served to establish the agroecosystems. Technical guides were elaborated for each agroecosystem identified.

RESULTS AND DISCUSSION

CLIMATE.

Data obtained on climatological aspects indicate that the dominant climate in the study area is, according to Koeppan, modified by Garcia, is Aw, warm with rain in summer, with annual average temperatures greater than 22^0 C. The precipitation ranges from 900 to 1200 mm annually. In the entire State of Morelos, the rains fall between May and September in the summer, the rainfall exceeding the infiltration capacity of the soil, causing runoff in the watersheds.

The altitude is less than 1300 m above sea level. From an agricultural standpoint, periods of frost do not exist in the study area, for which reason a great variety of agricultural crops can be grown in the winter with irrigation. An extremely hot and dry period presents itself in the middle of the summer, which can be corrected with supplementary irrigations.

SOILS.

The predominant soils in the area covered by Irrigation Unit 1 are presented in Table 1.

Table 1. Predominant soil types according to FAO classification in irrigation unit 1, Morelos state.

Soil Unit	Symbol	Surface Area[a]	%
1. Pelic Vertisol	Vp	14,936	35.6
2. Calcaric Feozem	Hc	10,191	24.3
3. Luvic Feozem	Hl	1,097	2.6
4. Haplic Feozem	Hn	269	0.6
5. Rendzina	E	3,790	9.0
6. Lithosol	I	3,036	7.3
7. Calcic Cambisol	Bk	2,542	6.1
8. Haplic Castañozem	Kn	1,780	4.3
9. Calcic Castañozem	Kk	1,170	2.8

[a] Surface area represents the total irrigated and non-irrigated area.

WATER QUALITY.

With the exception of springs, all the sources of irrigation water supply utilized in the Irrigation Unit 1 are contaminated by municipal, domestic and industrial wastes. The municipalities of Cuernavaca, Emiliano Zapata, Jiutepec, Temixco, Tetecala, Xochitepec, and Puente de Ixtla contaminate the water supplies with urban-industrial wastes. The Apatlaco, Tembembe, and Chalma Rivers are contaminated by these towns, resulting in the restriction of the production of vegetable crops with these waters. According to the information obtained in the State Agency of the National Water Commission (CNA), in the Irrigation District I and field interviews with producers, three zones of urban-industrial contamination were estimated:

  a. Zone of High Contamination. Apatlaco River, Las Fuentes Spring, Gachupina and Tlahuapan.
  b. Zone of Intermediate Contamination. Chapultepec and Palo Escrito Springs, Chalma River.
  c. Zone of Low Contamination. San Ramon, El Limon, Saborit I and II, El Dorado, and Santa Rosa Springs, Tembembe River. In these springs, water quality monitoring for chemical characteristics indicated these are apt chemically for irrigation.

DEFINITION OF AGROECOSYSTEMS.

One of the objectives of this work was to determine the existing agroecosystems in this irrigation area. The factors selected were:

  a. Frost-free period.
  b. Soil.
  c. Water quality.
  d. Area of influence of the Irrigation Unit.

Due to the nonexistence of a frost-free period, the agroecosystems were defined based on the soil type, water quality and the area of influence of the Irrigation Unit. In all, 11 agroecosystems were defined:

  1. High Contamination Vertisol.
  2. Intermediate Contamination Vertisol.
  3. Low Contamination Vertisol.
  4. Highly Contaminated Feozem.
  5. Intermediate Contamination Feozem.
  6. Low Contamination Feozem.
  7. Intermediate Contamination Rendzina.
  8. Intermediate Contamination Lithosol.
  9. Intermediate Contamination Cambisol.
  10. Intermediate Contamination Castañozem.
  11. Low Contamination Castañozem.

AGROECONOMIC STUDY.

In the Irrigation Unit 1 of the Irrigation District 16, Morelos State, predominant land tenure is of the "ejido" system (70%) and the remaining 30% corresponds to small-farmers. Principal crops in order of importance are, by planted surface area, are: corn, sugar cane, forage sorghum, rice and others (vegetables and flowers).

Of the total of ejidos (51) that are within the Irrigation Unit, it is estimated that 56% utilize improved seed, 84% agrochemicals in the control of weeds, insects, and diseases and 86% apply fertilizers; only 43% receive official technical assistance and 50% have access to agricultural credit.

Of the total population that inhabits the zone of influence of the Irrigation Unit, 73% is considered as urban and the rest as rural, the population concentration being located in the municipality of Cuernavaca and surrounding municipalities. Average family size is 5, similar to the state average.

The economically active population is calculated at 188,937 persons. Conditions of life are in the low and medium strata. Only 60% of the dwellings are constructed with material like brick and flagstone. The rest are constructed with materials of lesser quality. Eight-six percent of the total dwellings have potable water, 88% with electric energy and only 50% with drainage.

The sources of irrigation water supply of the Irrigation Unit 1 are contaminated in distinct levels; the problem has sharpened in the past thirty years, with industrialization and the anarchical growth of human dwellings, which in addition to a lack of planning of public services caused thousands of drain discharges to run directly into the rivers and gulleys, causing damage principally to the agricultural activity.

The traditional cropping pattern which included basic grains and diverse vegetable crops has been affected drastically by the contamination of the water supply; gradually, the growing of vegetable crops has been restricted, cases of cholera have increased in the Irrigation Unit, and yields have decreased to the point of almost disappearance.

Collateral effects of the contamination of the irrigation water in the study zone have included: reduction in income in 90% of the ejidos that are dedicated to agricultural activity; drastic restriction in agricultural credit opportunities; high grade of crop failure; renting and sale of agricultural lands.

From the factorial analysis of production conducted on the principal crops of the Irrigation Unit 1, there appears to be an effect of the use of wastewater on production, but it is not direct. The behavior of the irrigated, winter-grown crops (corn, tomato, squash) analyzed show a decrease in yield over the past 8 years, with the exception of common bean. The decrease in yields, according to field observation and interviews with producers, is due to a generalized increase in the frequency and intensity of insects and diseases, for which it is necessary to review the crop management and to continue monitoring water quality of the irrigation water to establish the direct causes of the problem.

Of the summer crops (rice, corn, and beans) analyzed, only rice shows an increase in its yield index, for which it is believed to be a crop with greater resistance to the effects of the wastewater; nevertheless, its harvested area has decreased gradually, affected among other things because of its susceptibility to insects and diseases, its high cost of production, and the decrease in availability of credit. It is necessary to investigate in greater depth the problems affecting agricultural production and establish the direct and indirect causes of the problem.

The tendency of the production systems predominant in the past years has been toward less diversity and monoculture, as an indirect effect of contamination and the establishment of ecological technical standards that prohibit the growing of vegetable crops irrigated with wastewater. For this reason, it is recommended to research alternative cropping systems, both on experiment stations and on producers' fields with the objective of making more efficient and sustainable wastewater use and minimizing the negative impacts of wastewater in the soil, water, and plant resource base.

It is necessary to conduct these studies to characterize with greater exactitude the effects of the wastewater in the frequency and intensity of insect and disease incidence and to determine in more detail the effect of wastewater on the fertility and contamination of the soil. It is also essential to conduct research/demonstrations on producers' fields on cropping systems and management in order to modify the technological packets that are actually applied on lands irrigated with wastewater.

In 28 videotaped interviews conducted with producers, the consensus with respect to the use of wastewater was the following: the producers of the Irrigation Unit 1 consider the participation of the Government essential in solving the problematics of the water in their different supply sources since the producers are actually in a state of crisis due to the global decapitalization of the sector and due to the low yields of the crops. The contamination of the irrigation water has had a negative impact on the economy of the agricultural sector of the state, specifically on that of the sector located in the irrigation area of the Unit 1. The negative impact is manifested in the restriction in the traditional cropping pattern without offering alternatives and in the decrease of the productivity shortening the harvest period. In addition to these problems, there exist low prices and unfavorable marketing channels, especially for rice, sugar cane, and vegetable crops. It is indispensable to determine with greater accuracy and distinguish the effect of wastewater use on these factors; therefore, it is recommended to continue the study of impact of wastewater in agriculture and monitor the productive yields of 10 plots from the different contamination zones (High, Intermediate, and Low Zone). In the collection of data, certain soil, water and crop factors will be included.

TECHNICAL GUIDES.

Some recommended alternative technical practices and systems for each agroecosystem in order to improve their productivity and sustainability follow:

1. High Contamination Vertisol.
    a. Utilize industrial, ornamentals, and forest crop.
    b. Add organic matter to improve the soil structure.
    c. Establish drainage systems.
    d. Design irrigation system.
    e. Water treatment if desired to take advantage of the high agricultural potential of these soils.
    f. Establish conservation tillage.

2. Intermediate Contamination Vertisol.
    a. Add organic matter to improve the soil structure.
    b. Establish drainage systems.
    c. Design irrigation system.
    d. Establish crops: rice, sugar cane, corn, and forage sorghum.
    e. Establish conservation tillage.
    f. Rotation of crops.

3. Low Contamination Vertisol.
    a. Practically all crops including fruits.
    b. Add organic matter to improve the soil structure.
    c. Land leveling and irrigation system design.
    d. Establish drainage systems.
    e. Establish conservation tillage.
    f. Rotation of crops.

4. High Contamination Feozem.
   a. Industrial, ornamental, and forest crops.
   b. On sloped land, establish contour farming.
   c. Land leveling and irrigation system design.
   d. For highly productive soils with non-contaminated water, it is convenient to invest in water treatment.

5. Intermediate Contamination Feozem.
   a. Practically all crops and vegetable crops that are not restricted by the Ecological Technical Standards.
   b. On sloped land, establish contour farming or contour stripcropping.
   c. If the layer of clay is not very deep, it is the Luvic Feozem, utilize subsoiler.
   d. Land leveling and irrigation system design.
   e. Establish rotation of crops.

6. Low Contamination Feozem.
   a. All crops: basic, industrial, forage, ornamentals, fruit and forest.
   b. On steeply sloped land, established contour farming or contour stripcropping.
   c. Since the soils are of high agricultural potential, invest in land leveling and irrigation system design.
   d. Tube existing springs and utilize pressurized irrigation systems.
   e. Establish crop rotation.

7. Intermediate Contamination Rendzina.
   a. Utilization of vegetable crops that are not restricted by the Ecological Technical Standards.
   b. Use contour stripcropping in soils with moderate to strong slopes.
   c. Apply subsoiler if the calcareous layer is superficial.
   d. Add organic matter to the soil.
   e. Establish conservation tillage.

8. Intermediate Contamination Lithosol.
   a. Use for forage and ornamental crops (shallow soil).
   b. Add organic matter.
   c. Subsoil to increase the moisture retention capacity.
   d. Contour farming and stripcropping.
   e. Establish conservation tillage.

9. Intermediate Contamination Cambisol.
   a. All types of crops: agricultural, aquacultural and forestal with the exception of those restricted by the Ecological Technical Standards.
   b. Tube the spring water and use better irrigation systems to make the water use more efficient.
   c. Conservation tillage.
   d. Crop rotation.

10. Intermediate Contamination Castañozem.
    a. Use for forage crops and shallow-rooted vegetable crops not restricted by the Ecological Technical Standards.
    b. Add organic matter.
    c. Subsoil to improve the moisture retention capacity.

11. Low Contamination Castañozem.
    a. Agroecosystem without crop restrictions.
    b. Tube existing springs and improve the efficiency of water use with pressurized irrigation systems.

CONCLUSIONS AND RECOMMENDATIONS

The major conclusions from this study included:

1. Eleven agroecosystems of the Irrigation Unit were identified based on climate, soil and degree of contamination.

2. For each one of the agroecosystems obtained, recommendations for more appropriate technical alternatives to improve their productivity were made.

3. The agroeconomic study indicated that there exists a negative impact of the wastewater on the traditional cropping pattern. Rice is the only crop that the producers recommend as resistant to the wastewater.

General recommendations included:

1. Conduct on-station and on-farm research/demonstrations on alternative cropping systems to make more efficient the use of wastewater and minimize the negative effect of this in soil, water and crop resources.

2. Conduct training courses for producers of the Irrigation Unit of study on the application of the alternative practices and systems proposed for each agroecosystem.

REFERENCES

1. Bole, J.B., Carson, J.A., and Gould, W.D. 1985. Yields of forage irrigated with wastewater and fate of added nitrogen-15 labeled fertilizer nitrogen. Agron. J. 77(5):715-719.

2. Calderon, J.L. 1991. Programa Agua Limpia y el Manejo del Agua Residual. Memoria del Taller Agua y Desarrollo Agrícola Sostenido. pp. 43-62.

3. Clapp, C.E. 1984. Effects of municipal wastewater and cutting management on root growth of perennial forage grasses. Agron. J. 76 (4): 642-647.

4. Day, A.D. and T.C. Tucker. 1960. High production of small grains utilizing city sewage effluent. Agron. J. 52(4):238-239.

5. Rodriguez, Z.C. 1987. Investigación agrícola con aguas residuales. IN Memorias del 20 Congreso Nacional de la Ciencia del Suelo. Zacatecas, Zac., p. 64.

6. Torres, M.J.C. 1992. Evaluación del impacto de los metales pesados del agua residual proveniente de la Ciudad Industrial del Valle de Cuernavaca (CIVAC) en el recurso suelo y en cultivos en los Municipios de Jiutepec, Emiliano Zapata and Xochitepec, Edo. de Morelos. Thesis. Universidad Autonoma del Estado de Morelos.

NEWSPRINT AND NITROGEN SOURCE INTERACTION ON CORN GROWTH AND GRAIN YIELD[1]

Ningping Lu, J.H. Edwards, R.H. Walker and J.S. Bannon[2]

ABSTRACT

Communities are faced with the problem of municipal solid waste (MSW) disposal with declining landfill space. Newsprint and yard waste comprise 58% of the MSW stream; they are targeted to help achieve recycling goals. A land-application study was initiated in spring, 1993, to evaluate the effects of ground noncomposted newsprint and different nitrogen (N) sources on corn (*Zea Mays* L) growth and grain yield. The soil was a Wickham fine sandy loam (Typic Hapludults). Ground newsprint was applied at the rate of 2.44 kg carbon m^{-2} of soil and uniformly incorporated to a depth of 10 to 15 cm. Ammonium nitrate, urea, anhydrous ammonia and poultry litter were the N sources used to adjust the newsprint C:N ratio to 30:1. When ground newsprint was applied to soil in combination with inorganic N sources, corn seedlings were stunted during the first 4 to 6 weeks after seedling emergence when compared to poultry litter as N source. When the corn was sidedressed with urea or ammonium nitrate, the apparent phytotoxicity of the newsprint was reduced. When poultry litter was the N source, ground newsprint increased the corn grain yield by 37% when compared to the other N sources. When anhydrous ammonia was the N source, ground newsprint decreased the grain yield by 78% when compared to no newsprint application. Aluminum concentration of 40-d old corn plant tissue was 2.5 times higher than the normal range (1079 mg kg^{-1}) when anhydrous ammonia was the N source, Al was in the normal range (<400 mg kg^{-1}) when poultry litter was the N source.

INTRODUCTION

A critical problem facing cities and large industries today is the safe disposal of large amounts of municipal solid waste (MSW) with limited landfill space. Waste paper comprises about 41% the MSW stream (USEPA, 1989). Alabama has passed laws requiring a 25% reduction in the volume of material in landfills, and other states have mandated reductions as high as 60%, by 1995. There likely will be increased emphasis on recycling paper products; however, today recycling of old newsprint accounts for only 17% of the waste paper fraction (USEPA, 1989).

The averaged cost of disposal of MSW in landfills in the U.S. ranged for $46 to $107 ton^{-1} in 1986, $59 to $194 ton^{-1} in 1991, and is predicted to increase to $128 to $219 ton^{-1} by 1996 (Greshman, 1992; Glenn, 1992). This increasing cost of disposal in landfills, and environmental problems associated with alternative disposal methods, are the primary reasons for renewed interest in application of MSW to agricultural lands. Application of noncomposted newsprint to agricultural cropland may be one potential disposal option.

---

[1]Contribution of Department of Agronomy and Soils, 202 Funchess Hall, Auburn University, AL 36849-5412; USDA/ARS, National Soil Dynamics Lab, Auburn, AL, 36849-3469; and the Alabama Agric. Exp. Stn. Journal Series No. 3-943700.

[2]Ningping Lu, Postdoctoral Research Associate, Department of Agronomy and Soils; J.H. Edwards, Soil Scientist, USDA/ARS, National Soil Dynamics Lab, Auburn, AL; R.H. Walker, Professor, Department of Agronomy and Soils; J. S. Bannon, Director E.V. Smith Research Center, Auburn University, AL.

Most MSW, such as newsprint, wood products, and yard waste, contain a substantial amount of organic carbon. The average organic matter content of MSW composts was in the range of 50 to 60% (He et al., 1992). Judicious application of MSW to soil may improve soil physical properties, i.e., organic matter content, bulk density, soil aeration, porosity, and water infiltration, and in turn help to control soil erosion and enhance soil productivity. Edwards et al. (1993b) reported that soil organic matter content was increased from 11.9 g kg^{-1} to 23.8 g kg^{-1} in a 50:40:10 backfill mixture of soil, newsprint, and poultry litter that was placed in a vertical trench within seven months after application. Soil organic matter content was increased from 9.5 g kg^{-1} to 18 g kg^{-1} after two annual spring applications of newsprint, yard waste, or gin trash; soil organic matter content was increased to 14.9 g kg^{-1} with application of woodchips (Edwards et al., 1994). However, presence of trace metals in newsprint and other MSW may be of concern as a result of repeated application of newsprint to the same site. High concentrations of trace metals can impair crop growth an d excessive accumulation of trace metals in soils and other media may eventually contaminate both human and animal food chains (Mays et al., 1973; Petruzzelli et al., 1989).

In Alabama, 53% of over 875 million broiler chickens (*Gallus gallus*) grown annually is concentrated in just 8 counties in northern Alabama (Hinton, 1991). About 1 million tons of a total 2 million tons poultry litter are concentrated in this area. Increased environmental awareness has focused more attention on the utilization of poultry litter as a source of nutrients for crops. However, poultry litter proved to be a less efficient source of N for corn than commercial NH_4NO_3-N (Wood et al., 1991). But when poultry litter was used to adjust the C:N ratio of soil-applied newsprint to 30:1 in another of our experiments, it proved to be a better N source than ammonium nitrate (Edwards et al., 1993a).

The loading rates for applying animal manure or sewage sludge to agricultural land are calculated based on total nitrogen content of the material (O'Keefe et al., 1986), or on plant-available nitrogen (PAN) content of the waste (King, 1984). These loading rates are based on indices that estimate soil N needed to maximize crop production and minimize the potential for environmental pollution from different forms of N. The loading rates for applying newsprint to soil should be determined, they are important because of the high C:N ratio of newsprint (about 100:1) and its ability to immobilize soil N. Incorporation of C and N sources into the soil biomass and ultimately into the soil organic matter fraction is the desired pathway for C and N. Thus, it is possible that C:N ratio of newsprint can be balanced by using poultry litter as N source when newsprint is applied to agricultural land. The objectives of our research were to determine the effects of application of noncomposted ground newsprint and several N sources on corn growth and grain yield.

MATERIALS AND METHODS

A field study with corn was initiated in spring, 1993, at the E.V. Smith Research Center, Tallassee, AL. The soil is a Wickham fine sandy loam complex (Typic Hapludults) alluvial soil. The productivity of this soil is poor, with a low water-holding capacity (24%) and low organic carbon content of 4.7 g kg^{-1} (Table 1). The experimental area was surface limed to pH 6.1 (based on soil test).

The experimental design was a randomized complete block with three replications. Whole plots were four N sources (poultry litter, ammonium nitrate, urea, and anhydrous ammonia) and ground newsprint. The ground newsprint had not undergone any microbial degradation. Six-row plots 15.2 m long, with row spacing of 0.75 m were used in the experiment. Corn (Dekalb-689) was planted at 24,640 seeds ha^{-1} on March 24, 1993. Unless otherwise noted, all statistical tests were reported at a 0.05 level of probability.

Table 1. Selected chemical and physical properties* of surface soil of the experimental area.

Water-holding capacity	CEC	Total N	Total C	Double-acid extractable			
				Ca	Mg	K	P
%	Cmol kg^{-1}	mg kg^{-1}	g kg^{-1}	-------- mg kg^{-1} --------			
24	6.3	417	4.5	558.5	131.0	68.3	53.2

* Based on moisture-free weight.

Ground newsprint was obtained from a commercial insulation company and contained no fire retardant. Newsprint was ground using a hammer mill equipped with a series of three screens; the smallest was approximately 0.63 cm in size. Poultry litter, a mixture of excreta and wood chips used as a bedding material, was obtained from a local poultry house; the average moisture content of the litter was 14.1 percent.

Elemental concentrations of the ground newsprint and poultry litter were determined by dry-ashing method (Hue and Evans, 1986). The concentrations of Ca, Mg, K, P, Mn, Fe, Cu, Fe, Zn, Pb, AL, Na, Cd, and Cr were measured by Inductively Coupled Argon Plasma (ICAP) spectrophotometry. The chemical composition of ground newsprint and poultry litter used in this experiment is given in Table 2.

Table 2. Chemical composition of ground newsprint and poultry litter.

Elements	Ground newsprint	Poultry litter
	g kg^{-1}	
C	61.5	39.9
N	1.2	27.9
S	0.8	--
Ca	0.8	27.1
Mg	0.1	5.7
K	0.1	28.5
P	0.08	20.2
Al	4.5	2.6
Na	0.9	7.0
	mg kg^{-1}	
Cu	22.5	550.5
Mn	31.2	632.5
Zn	51.2	533.4
Ni	0.7	7.6
Ba	17.2	31.6
Cr	1.5	8.5
Cd	0.3	2.4
Pb	8.4	14.6

The ground newsprint was applied at the rate of 2.44 kg C m^{-2} of soil and uniformly incorporated into a depth of 10 to 15 cm. Three commercial inorganic N sources (ammonium nitrate, urea and anhydrous ammonia) and poultry litter were used to adjust the C:N ratio to 30:1, requiring 336 kg N ha^{-1}. Nutrients needed by the corn plants (N, P, K, etc.) for optimum growth were supplied in addition to the N needed to adjust the C:N ratio of ground newsprint. Standard production practices were used. Whole plant samples were taken from each plot at 40-d and 115-d after corn seedling emergence.

Plant samples were taken from 1 m row of each plot for dry matter determination. Plant tissue was oven-dried at 60°C, weighed, and ground to pass a 40-mesh screen. Corn grain was harvested from 3 rows of each plot. Mineral elements in plants tissue were obtained by dry-ashing method (Hue and Evans, 1986); concentrations of elements in plant tissue were determined by ICAP spectrophotometry.

RESULTS AND DISCUSSION

Corn Growth and Grain Yield

With all inorganic N sources, application of ground newsprint stunted corn seedling growth during the first 4 to 6 weeks after seedling emergence when compared to poultry litter as N source (Fig. 1). When ground newsprint was applied in combination with anhydrous ammonia, corn plants were severely stunted, corn maturity was delayed about 14 to 17 days. The dry matter weight of 40-d old plants decreased 75% and the grain yields decreased 72% when compared to no newsprint application. When the corn was sidedressed with urea or ammonium nitrate as N sources, the apparent phytotoxicity of the newsprint was reduced; however, corn seedling growth was stunted and plant maturity was delayed about 7 to 10 days when compared to no newsprint application (Table 3).

Table 3. Corn plant growth period, dry matter weight, and grain yield as effected by application of ground newsprint and nitrogen source.

Treatment	N source	Growth period[a]	Dry matter weight[b]		Grain yield
			40 days	115 days	
		Days	kg ha^{-1}		
Newsprint	Ammonium nitrate	160	22	13,160	5,380
	Urea	160	22	16,351	6,545
	Anhydrous ammonia	170	6	13,379	2,237
	Poultry L	153	208	28,369	9,062
	No N	153	6	3,102	1,562
Mean			53	14,872	4,945
No newsprint	Ammonium nitrate	153	29	16,054	4,989
	Urea	153	35	13,110	6,378
	Anhydrous ammonia	153	24	11,347	7,940
	Poultry L	153	214	29,556	6,627
Mean			76	17,517	6,483
LSD$_{05}$[d]			22	ns	ns
LSD$_{05}$[e]			31	5,462	2,760

[a] Growth period represented the days between planting and harvesting.
[b] Calculation of dry matter weight was based on plant tissue oven-dried at 60°C.
[c] Corn grain yield was based on 13% moisture content.
[d] LSD$_{05}$ for newsprint means.
[e] LSD$_{05}$ for N source x newsprint.

Application of ground newsprint had no effect on corn grain yield when ammonium nitrate and urea were used as N source. When ground newsprint was applied to soil with poultry litter as N source, the highest dry matter weights of 40-d and 115-d old plants were obtained when compared to newsprint treatments containing the three inorganic N sources. Application of newsprint with poultry litter as N source did not significantly affect the dry

Figure 1. Corn Seedling Growth at 14 Days After Surface Applying Ground Newsprint With Different N Sources to Adjust C:N Ratio to 30:1. A) No N; B) Anhydrous Ammonia; C) Urea; D) Ammonium Nitrate (on the left) Versus Poultry Litter (On the right).

matter weight of corn plants, but did significantly increase the grain yield by 37% when compared to poultry litter alone (Table 3).

In a field experiment conducted on a sand loam soil with an application rate of 300 kg N ha^{-1}, Wood et al. (1991) found that corn grain yield was much higher from NH_4NO_3-N (5229 kg ha^{-1} of corn grain) than from poultry litter (4956 kg ha^{-1}). In field experiments with corn grown on loamy sandy soils at 168 kg ha^{-1} of potentially available nitrogen (PAN), Sims (1987) reported that corn grain yield was 3200 kg ha^{-1} from NH_4NO_3-N and 2700 kg ha^{-1} from poultry manure. However, in this experiment, when poultry litter was used to adjust the C:N ratio of newsprint to 30:1, poultry litter proved to be a better N source than NH_4NO_3-N; corn grain yield was 5380 kg ha^{-1} when NH_4NO_3-N was the N source and 9062 kg ha^{-1} when poultry litter was the N source (Table 3).

Elemental Composition of Plants

With all three inorganic N sources, manifestation of corn seedling nutrient disorder was observed with application of ground newsprint. When newsprint was applied in combination with anhydrous ammonia, N, Mg, and P foliar deficiency symptoms were observed during the first 6 weeks after corn seedling emergence. Plant tissue analysis of 40-d old whole corn plants showed that N concentration was less than the 3.5 to 5.0% normal level (Plank, 1989). When ground newsprint was applied in combination with all inorganic N sources, plant P concentration was below the 0.3 to 0.5% normal level, and K concentration was below the 2.5 to 4.0% normal level for corn whole plant (Plank, 1989). However, Al concentration of 40-d plant tissue was 2.5 times higher than the normal range when anhydrous ammonia (1079 ppm) was the N source (Plank, 1989), 2 times higher than the normal range when urea (899 ppm) and ammonium nitrate (773 ppm) were the N sources, and in the normal range when poultry litter (< 400 ppm) was the N source. Corn roots exhibited signs of aluminum toxicity, i.e. severe stunting and thickening at the root apex, with few lateral roots. Iron concentration in 40-d old corn plant tissue was 2.8 times higher than the normal range when anhydrous ammonia (718 ppm) was the N source, 2.5 times higher when urea (657 ppm) and ammonium nitrate (631 ppm) were the N sources, and in the normal range when poultry litter (< 250 ppm) was the N source (Fig. 2).

Induced Nutrient Disorder

The manifestation of plant nutrient disorder is an indication of an improper nutrient balance in the soil solution, limited soil nutrient concentrations, or toxicities from excess soil nutrient levels. Any of these conditions can affect nutrient absorption by the plant. Application of newsprint to soil will increase organic matter and elemental concentrations such as Ca, Al, etc. The interaction between newsprint and different N sources may influence the soil pH and the solubility of some elements in soil shortly after newsprint application. When anhydrous ammonia is the N source, the soil pH may increase and create alkaline conditions in the injected area for a short time (a few hours or 1 to 2 days) after application due to NH_3. The soil pH will decrease due to conversion of NH_3 to NH_4^+, and NH_4^+ will undergo nitrification. The changes in soil pH would influence the solubility of many elements such as Ca, Mg, P, and Al added through ground newsprint. Total soil solution Al, as measured by ICAP, was partitioned into monomeric Al (Al^{3+} + hydroxy-Al species) and complexed Al (Al-organic acid complexes) (Hue et al., 1986). The intensity of the phytotoxicity to plant seedlings has been shown to be highly correlated with monomeric Al activity in soil solution (Adams and Lund, 1966) because of inhibited cell division at the root apex. Since the uptake of a given nutrient depends partly on the volume of roots present, any condition that affects root growth will alter nutrient uptake and will be manifested by the corn plants as nutrient deficiency or nutrient toxicity.

CONCLUSION

When noncomposted ground newsprint was land-applied in combination with inorganic N sources to adjust the C:N ratio to 30:1, corn seedlings were stunted during the first 4 to 6 weeks after seedling emergence when compared to poultry litter as N source. When the corn was sidedressed with urea or ammonium nitrate as N sources, the apparent phytotoxicity of the newsprint was reduced. However, ground newsprint increased the corn grain yield by

37% with poultry litter as N source, whereas newsprint decreased the grain yield by 78% with anhydrous ammonia as N source. Aluminum concentration of 40-d old plant tissue was 2.5 times higher than the normal range (1079 mg kg^{-1}) with anhydrous ammonia as N source; however, Al concentration was in the normal range (<400 mg kg^{-1}) when poultry litter was the N source. These results suggest that ground newsprint applied to agricultural land in combination with poultry litter as N source to balance the C:N ratio has enhanced corn growth and corn grain yield.

Figure 2. Corn seedling growth at 40 days after surface-applying ground newsprint with different N sources to adjust C:N ratio to 30:1. A) no N; B) anhydrous ammonia; C) urea; D) ammonium nitrate (on the left), versus poultry litter (on the right).

REFERENCES

Adams, F., and Z.F. Lund. 1966. Effect of chemical activity of soil solution aluminum on cotton root penetration of acid subsoils. Soil Sci. 101:193-198.

Edwards, J.H., E.C. Burt., R.L. Raper, and D.T. Hill. 1993a. Recycling newsprint on agricultural land with the aid of poultry litter. Compost Science & Utilization 1(2):79-92.

Edwards, J.H., R.H. Walker, and J.S. Bannon. 1994. Effects of repeated application of noncomposted organic wastes on cotton yield. Proc. 1994 Beltwide Cotton Conf. (In press).

Edwards, J.H., R.H. Walker, C.C. Mitchell, and J.S. Bannon. 1993b. Effects of soil-applied noncomposted organic wastes on upland cotton. Proc. 1993 Beltwide Cotton Conf. 3:1354-1356.

Glenn, J. 1992. The state of garbage in America: 1992 nationwide survey. Biocycle:Journal of Waste Recycling 33(4):46-55.

Greshman, H. 1992. Municipal waste costs going up. Biocycle:Journal of Waste Recycling 33(3):13.

He, X., S.J. Traina, and T.J. Logan. 1992. Chemical properties of municipal solid waste composts. J. Environ. Qual. 21:318-329.

Hinton, S.A. 1991. Poultry review. In: Alabama agricultural statistics. Alabama Agricultural Statistic Service. Bulletin 34.

Hue, N.V. and C.E. Evans. 1986. Procedures used for soil and plant analysis by the Auburn University Soil Testing Laboratory, p 22-23.

Hue, N.V., G.R. Craddock, and F. Adams. 1986. Effect of organic acids on aluminum toxicity in subsoils. Soil Sci. Soc. Am. J. 50:28-34.

King, L.D. 1984. Availability of nitrogen in municipal, industrial and animal wastes. J. Environ. Qual. 13:609-612.

Mays, D. A., G.L. Terman, and J.C. Duggan. 1973. Municipal composts: Effects on crop yields and soil properties. J. Environ. Qual. 2:89-92.

O'Keefe, B.E., J. Axley, and J.J. Meisinger. 1986. Evaluation of nitrogen availability indexes for a sludge comport amended soil. J. Environ. Qual. 15:121-128.

Petruzzelli, G., L. Lubrano, and G. Guidi. 1989. Uptake and chemical extractability of heavy metals from a four year compost-treated soil. Plant Soil 116:23-27.

Plank, C.O. 1989. Plant analysis handbook for Georgia. Cooperative Extension Service, The University of Georgia College of Agriculture. Athens, GA.

Sims, J.T. 1987. Agronomic evaluation of poultry manure as a source for conventional and no-tillage corn. Agron. J. 79:563-570.

U.S. Environmental Protection Agency. The solid waste dilemma: an agenda for action. Washington, DC: USEPA, 1989.

Wood, C.W.., C.D. Cotton, and J.H. Edwards. 1991. Broiler litter as a nitrogen source for strip and conventional tillage corn. In: A.B. Bottcher, K.L. Campbell, and W.D. Graham (eds.) Proc. of the Conf. on Envor. Sound Agric. pp. 568-575. April 16-18, 1991. Orlando, FL.

PRELIMINARY EFFECTIVENESS OF CONSTRUCTED WETLANDS FOR DAIRY WASTE TREATMENT

Charles M. Cooper, Samuel Testa, III, and Scott S. Knight[1,2]

ABSTRACT

During the past two decades the beneficial role of wetland ecosystems for improving water quality has been thoroughly documented. Recent research interests have begun to focus on applied uses of natural and constructed wetlands in the area of waste processing. Processing and disposing of concentrated on-farm animal waste, a major water quality concern, is a primary focus of the Soil Conservation Service and regulatory agencies. A constructed wetland for treatment of dairy wastewater was built in DeSoto County, MS during 1990 by the Soil Conservation Service and the Agricultural Research Service. Three parallel wetland cells, planted with giant bulrush (*Scirpus validus*), were monitored for eighteen parameters. Measures of physical and chemical water quality, BOD, and coliform bacteria were recorded, and average seasonal nutrient-trapping efficiencies were calculated. Total phosphorus reduction ranged from 42% to 87%, and ammonia was reduced 76% to 96% by treatment. Nitrate trapping varied from -278% to 40%; negative efficiency was due to conversion of ammonia to nitrate. Biological oxygen demand decreased more than 70%. Accumulation of senescent plant biomass in the treatment cells after two growing seasons caused reduced growth of bulrushes in some areas, suggesting a biomass removal strategy be implemented to maintain the culture. Constructed wetlands have considerable potential as cost-effective on-farm waste management systems. Further research is needed to evaluate their long-term effectiveness and range of applicability. **KEYWORDS**: Water quality, Nutrients, Coliforms, Bulrushes.

INTRODUCTION

During the past two decades the beneficial role of aquatic plants for improving water quality has been thoroughly documented (Boyd, 1970; Sheffield, 1967). The production-trapping system of wetlands can remove nutrients, organic chemicals, heavy metals, and sediments from inflowing waters. Environmental engineers have recommended the re-establishment of wetlands where water quality has deteriorated since wetland removal (Kloetzli, 1981; Jones and Lee, 1980). Seidel (1976), Wolverton and McDonald (1981) and Rogers et al. (1991) documented the efficiency of aquatic plants in removing organic chemicals from water. Simpson et al. (1983), Peverly (1985) and Lan et al. (1992) demonstrated the effective role that wetlands play in trapping heavy metals. Research on natural wetlands indicates that physical, chemical, and biological processes that occur in wetlands are similar to those occurring in mechanical sewage treatment plants. Thus, many recent nutrient uptake and

---

Supervisory Ecologist, Biologist, and Ecologist of the Water Quality and Ecological Processes Research Unit, USDA-Agricultural Research Service, National Sedimentation Laboratory, Oxford, MS.

Names of commercial products are included for the benefit of the reader and do not imply endorsement or preferential treatment by the author or USDA. "All programs and services of the U. S. Department of Agriculture are offered on a nondiscriminatory basis without regard to race, color, national origin, religion, sex, age, marital status, or handicap."

cycling studies conducted on wetlands have been concerned with their potential use for natural sewage treatment (Simpson et al., 1983; Dolan et al., 1981) or as water purifiers (Nichols, 1983).

Research interests have begun to focus on applied uses of natural and constructed wetlands (Reed, 1991). Small municipalities are finding constructed wetlands an alternative to conventional waste treatment plants (Gearheart et al., 1989). Processing and disposing of concentrated on-farm animal waste, a major source of water quality deterioration, is a concern of the Soil Conservation Service and regulatory agencies. Thus, several projects for evaluating the ability of constructed wetlands to process animal waste have been initiated across the United States (Holmes et al., 1992; Ulmer et al., 1992; Payne et al., 1992). The Mississippi Soil Conservation Service (SCS) and the Agricultural Research Service (ARS) National Sedimentation Laboratory in Oxford, Mississippi are cooperating on an on-farm dairy waste treatment project which uses a constructed wetland for processing. The purpose of this paper is to present findings from the first two years of operation.

STUDY AREA AND METHODS

A constructed wetland for treatment of dairy farm wastewater was implemented in DeSoto County, Mississippi during 1990 on a 100 cow dairy operation. The Soil Conservation Service designed and constructed a 40 m x 52 m lagoon and three 6 m x 24 m wetland cells in April, 1990 (Fig. 1). All cells were immediately planted in bulrush (*Scirpus validus*) at 0.3 m intervals with rhizome cuttings purchased from a wildlife supply company. Subsequent rains, supplemented with water pumped from the lagoon and well water, maintained standing water in the cells for the remainder of the year. The anaerobic lagoon received inputs from milking equipment and tank cleanings, milking barn washings, loafing area runoff, and rainfall (Table 1). Water level in the lagoon increased slowly because of high evaporation rates and lateral seepage through levees until the basin sealed. An insufficient amount of water accumulated in the lagoon to allow a gravity fed water supply to the cells during 1990.

TABLE 1. SCS CALCULATED DATA FOR LOADING ESTIMATES

Rainfall Runoff Area		
Roof and Concrete	351 sq. m	(3784 sq. ft.)
Anaerobic Lagoon	2132 sq. m	(22950 sq. ft.)
Wetland Cells	557 sq. m	(6000 sq. ft.)
Dairy Waste Production		
(Based upon 100 cow dairy herd)	10 cu. m/day	(365 cu. ft/day)

Plant growth in the cells was rapid. By September the cells were covered by a uniformly dense monoculture with the majority of culms supporting flowering/seeding heads. Natural senescence occurred in November and December. Re-emergence of bulrushes from rhizomes occurred in February, 1991, through the litter created by the previous year's growth. Duckweed (*Spirodela polyrhiza*) spread to cover nearly all available water surface by May, 1991. In April, gravity flow from the lagoon to the wetland cells began functioning. Discharges to each cell were calibrated to yield 3.0 L/min.

Rapid water level decline in the anaerobic lagoon during summer, 1991, prompted a reduction of cell inflow rates to 0.5 L/min, but, settling and clogging of pipes and valves accelerated at the lower rate. Standpipes were fitted with threaded end caps with orifices. End caps with different sized orifices could be used to achieve desired flow rates. Original valves were opened fully to prevent occlusion. Using this method, a cell inflow rate of 1.0 L/min was implemented, and the frequency of remedial action was greatly decreased. Also during summer, 1991, another cell, Cell 4, was constructed in series with Cell 1 to allow greater loading capacity and assessment of further treatment (Fig. 1). A constant head tank was placed on the lagoon levee and plumbed in to the outflow pipe to the cells. An electric pump controlled by a timer maintained water in the tank, allowing the cells to receive consistent pressure and, consequently, consistent inflow.

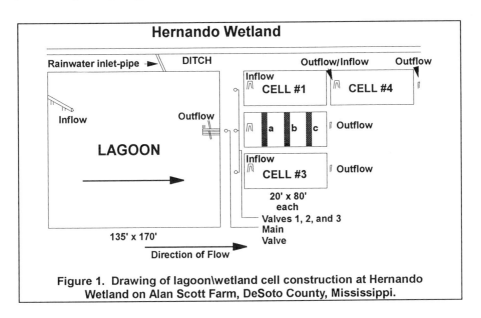

Figure 1. Drawing of lagoon\wetland cell construction at Hernando Wetland on Alan Scott Farm, DeSoto County, Mississippi.

Eighteen parameters were monitored at biweekly intervals from May, 1991 to April, 1993. Physical and chemical variables were measured at cell inflow and outflow. In addition, 3 walkways were constructed equal distances apart in Cell 2 so that in-cell measurements could be taken. Lagoon samples were taken from the end of the outflow control platform at a depth of 0.3 m. Water quality parameters were measured according to APHA (1989) guidelines (for further details, see Cooper et al., 1993).

RESULTS AND DISCUSSION

Physical parameters varied with season and as the system matured. An overall reduction in temperature was seen as effluent moved through the constructed wetland cells. Water temperatures, on average, were 13.9% lower at outflow stations than upon entering the cells because of physical processes such as shading by the wetland plants and the shallower mass of water within the wetland cells. This was most evident during winter when outflow water temperatures were nearly 25% lower than inflow temperatures. Inflow extremes ranged from a maximum of 30.3° to a minimum of 7.2°. Outflows ranged from 27.3° to a minimum of 1.5°C.

Conductivity exhibited an overall reduction of 27.5%. Highest reduction of conductivity occurred during the cooler months. Conductivity values were distinctly higher in 1992 than during 1991. Maximum conductivity measured from inflow was 773 µmhos/cm, and the minimum was 28 µmhos/cm. Outflows ranged from 785 to 103 µmhos/cm.

Dissolved oxygen (DO) concentrations decreased markedly with water's passage through the cells (34.7% reduction). Expected increases were seen only during the initial three months of operation. Reduced oxygen levels were attributable to biochemical demand, shading from the dense bulrush stand, and duckweed which quickly colonized even small open areas of the water surface. Although greater reductions of dissolved oxygen occurred during 1992, overall DO increased relative to 1991 concentrations. Recorded measurements for treatment cells ranged from 7.8 to 0.03 mg/L for inflows, while outflows ranged from 5.8 to 0.03 mg/L.

A small decrease in pH was observed for water flowing through the wetland cells (average 10.5% decrease). Highest reductions in pH have been observed during the winter months. Inflow pH ranges were from 8.5 to 5.7; outflow values ranged from 7.4 to 5.7.

For the two year period, 32.5% of inflowing total solids (TS) were removed by the wetland cells. Total solids maintained higher levels in 1992 than 1991, but trapping efficiencies (TE) were considerably higher during 1992 despite the elevated loading. This increase in TE probably resulted from the large amount of senescent plant material added to the water column during the winter months. TE was highest during summer 1992, when it reached nearly 43%. The sharp decline in TS removal efficiency in the fall of 1992 was from export of suspended solids from cell 2. Cell inflow concentrations for TS ranged from 749 to 202 mg/L while the outflow station measurements varied from 605 to 161 mg/L. Mean dissolved solids (DS) removal by the cells averaged 22.4% over the study period. TE following initiation of wetland operation was less than 10% during the summer of 1991, but climbed to greater than 25% for most of the remainder of the study period (slightly less in summer 1992). Dissolved solids fluctuated seasonally, falling during the winter months, then returning to growing season highs. Dissolved solids at inflow stations were measured between 588 and 102 mg/L, and outflow stations had a maximum of 527 and a minimum of 107 mg/L. Mean suspended solids TE was 58.0%. TE was low (ca. 22%) during the first season of operation. It then increased to >60% in the fall and remained relatively high until declining in the fall of 1992. Interpretations of the sharp decrease in efficiency during the fall of 1992 must be in context of net export from Cell 2. Suspended solids concentrations peaked during warmer months. Inflow values ranged from 466 to 0.0 mg/L. Outflows varied from 255 to 0.0 mg/L.

Filterable ortho-phosphorus (FOP) concentrations were reduced an average of 56.2%. TE rose from near 70% to >85% during the first six months of operation, but declined afterward, averaging about 42% for the most recent 12 months (Fig. 2). FOP concentrations entering the treatment cells rose gradually from inception of the project, with peaks during the winter and spring during plant senescence and decay. Maximum inflow concentration was 15.6 mg/L and minimum inflow measured was 0.9 mg/L. Outflow values ranged from a maximum of 14.8 to a minimum of 0.1 mg/L.

Total phosphorus removal averaged 61.3%. Reduction efficiencies during the first three months of operation were slightly over 50%, but increased to over 80% in the following winter and spring season before declining to <50% in the summer and fall of 1992 (Fig. 3). Inflow concentrations varied from 69.0 and 2.8 mg/L while outflow ranged from 21.8 to 0.2 mg/L.

Ammonia reduction by the wetland system was consistently high, with an overall average for the three treatment cells of 87.2%. TE averaged about 90% for 1991. Removal was elevated to >95% in the spring of 1992, followed by a decline to 80% for the remainder of the study period (Fig. 4). Actual concentrations showed several elevated peaks during summer and fall, 1991, which were not observed during 1992. Inflow concentrations ranged from 30.8 to 0.2 mg/L ammonia while the outflows varied from 10.8 to <0.01 mg/L.

Nitrate concentrations decreased an average of 27.7% in cells 1 and 2 but showed an increase of 211% in cell 3. Inflow concentrations exhibited occasional spikes but generally were lowest during summer. An expected seasonal release of nitrate from senescent and decaying plants occurred most noticeably in cell 3 during the fall and winter, 1991, and spring, 1992, resulting in large negative TE during winter and spring months (Fig. 5). Nitrate concentrations also rose within cell 2 during the same periods, but decreased to below inflow levels before discharge from the cell. Maximum inflow of nitrate was also in cell 3, measured at 0.3 mg/L. Undetectable levels of nitrate occurred at least once at all inflows. Maximum outflow concentration of nitrate, also from Cell 3, was 3.3 mg/L, and the minimum value at each outflow station was <0.01 mg/L.

Chlorophyll values for treated water declined an average of 76.2% from cell inflow to outflow. Eutrophic algal growth in the primary settling lagoon yielded high concentrations of chlorophyll entering the cells. Shading by bulrushes and duckweed caused rapid reductions in primary productivity. Concentrations of chlorophyll were reduced <30% in the cells during the first season of operation, but subsequent seasons had reductions of near 90%, except the fall of 1992 which had less than 60% reduction in chlorophyll within the cells, coincident with the relatively early senescence of bulrushes that year. Maximum recorded chlorophyll concentration for inflow was 1505 mg/L, and the minimum inflow value was 13.4 mg/L. Outflow concentrations ranged from 759 to 1.2 mg/L.

Carbonaceous biochemical oxygen demand ($CBOD_5$) 5-day tests indicated that there was an overall reduction in $CBOD_5$ of 75.6% during this study period. After 42.3% reduction for the initial three months of operation, $CBOD_5$ was reduced nearly 80% for the remainder of the analysis period. Maximum $CBOD_5$ measured for inflow was 184.4 mg/L; minimum was 9.7 mg/L. Outflow demand concentrations ranged between 48 and 0.3 mg/L.

Coliform bacteria reduction averaged 85.9%. Reduction for the first season of operation was >70%, followed by reduction of >99%. Reduction was >90% during summer of 1992, but declined in the fall 1992 to only 37.1%. Maximum seasonal inflow occurred during winter, while maximum outflows occurred during the fall. Inflow values recorded were between 101,000 and 0 colony forming units (CFU's). Outflow densities ranged from 8500 to 0 CFU's.

Use of catwalks in Cell 2 (stations 2a, 2b, and 2c respectively) allowed assessment of in-cell processing. Several parameters showed sharp declines between inflow and 2a. Temperature decreased after inflow, then remained fairly constant through the cell. Conductivity and pH values exhibited distinct declines from inflow to 2a, followed by gradual declines to outflow. Dissolved oxygen dropped sharply, then increased through the cell. Mean suspended solids were higher at station 2a than at inflow. Phosphorus showed a gradual decline through the cell. Ammonia decreased nearly linearly to mid-cell (2b), then reduction slowed to outflow. Nitrate concentrations increased from inflow to station 2a, then gradually decreased until outflow. Fecal coliform bacteria exhibited a decline from inflow to station 2a and reached outflow levels by station 2b. $CBOD_5$ and chlorophyll exhibited similar longitudinal patterns in the cell, decreasing distinctly by station 2a, declining further by station 2b, with slight increases at 2c before reaching lowest levels at outflow.

After construction of Cell 4 in 1991, further wastewater treatment was possible beyond the lagoon and primary wetland cells. Dissolved oxygen increased 92% at the fourth cell outlet (versus 42% average decrease at outflow from the primary treatment cells). Nitrate outflow concentrations were 18% lower than inflow values, even with increased removal of ammonia (10% added removal in Cell 4). Suspended sediments, $CBOD_5$ and coliform concentrations remained nearly the same with the additional treatment. Removal of total phosphorus and chlorophyll increased by approximately 1/3 with the additional treatment, and total solids and filterable ortho-phosphorus concentrations declined by an added 50%. Conductivity decreased an additional 3/4 over that seen in the primary treatment cells. Dissolved solids removal was doubled.

Greatest reductions in the three original treatment cells were seen for ammonia (96%), coliform bacteria (86%), chlorophyll concentrations (76%), and, to a lesser degree, for biochemical and chemical oxygen demand (76 and 64% respectively), suspended solids (58%), and total and filterable ortho-phosphorus (61 and 56% respectively). In view of the relatively small size of the treatment cells involved (6 x 24 m), the measured reduction in pollutants indicates a successful point source management plan.

As researchers continue to search for methods of treating agricultural and other contaminants using wetlands, more efficient and ecologically sound designs and maintenance programs will certainly be forthcoming. Long term effects on surface and ground water, adjacent land, wildlife, and humans have yet to be explored. Our system has proven to be effective for the treatment of concentrated on-farm animal waste, and following subsequent evaluation and modification, may provide further useful information for a variety of applications.

ACKNOWLEDGMENTS

This paper was prepared as a part of the water quality research of the Agricultural Research Service at the National Sedimentation Laboratory, Oxford, Mississippi. Partial funding was received from USDA Soil Conservation Service, Jackson, MS. SCS technical assistance was received from Ross Ulmer, Lon Strong, and Jimmy Wilson. SCS New Albany Area and DeSoto County District staff also helped with construction aspects. The farm cooperator, Alan Scott, was helpful whenever needed. The authors wish to thank these people and the following ARS personnel: Pat McCoy, Terry Welch, and Betty Hall.

LITERATURE CITED

1. Amer. Public Health Assoc. 1989. Standard methods for the examination of water and wastewater. APHA, Washington, D. C.

2. Boyd, C. E. 1970. Vascular aquatic plants for mineral nutrient removal from polluted waters. Econ. Bot. 24:95-103.

3. Cooper, C. M., S. Testa III, and S. S. Knight. 1993. Evaluation of ARS and SCS constructed wetland/animal waste treatment project at Hernando, Miss. in cooperation with the Miss. Soil Conservation Service 1991-1992. National Sedimentation Laboratory Research Rpt. No. 2., Oxford, Mississippi, 55 p.

4. Dolan, T. J., S. E. Bayley, J. Zolteck, Jr., and A. J. Hermann. 1981. Phosphorus dynamics of a Florida freshwater marsh receiving treated wastewater. J. Appl. Ecol. 18:205-219.

5. Gearheart, R. A., F. Klopp, and G. Allen. 1989. Constructed free surface wetlands to treat and receive wastewater: Pilot project to full scale. pp. 121-138. In: D. A. Hammer, (ed.), Constructed wetlands for wastewater treatment. Lewis Publishers, Chelsa, Michigan. 831 p.

6. Holmes, B. J., L. R. Massie, G. D. Bubenzer, and G. Hines. 1992. Design and construction of a wetland to treat milkhouse wastewater. 1992 Intl. Winter Meeting. Am. Soc. Agric. Engrs., 2950 Niles Rd., St. Joseph, MI.

7. Jones, R. A. and G. F. Lee. 1980. An approach for the evaluation of efficiency of wetlands-based phosphorus control programs for eutrophication related water quality improvement in downstream water bodies. Water Air Soil Pollut. 14:359-378.

8. Kloetzli, F. 1981. Some aspects of conservation in over-cultivated areas of the Swiss Midlands. Intl. J. Ecol. Environ. Sci. 7:15-20.

9. Lan, C., C. Guizhu, L. Li, and M. H. Wong. 1992. Use of cattails in treating wastewater from a Pb/Zn mine. Environ. Management. 16(1):75-80.

10. Nichols, D. S. 1983. Capacity of natural wetlands to remove nutrients from waste water. J. Water Poll. Control Fed. 55(5):495-505.

11. Payne, V. W. E., T. A. McCaskey, and J. T. Eason. 1992. Constructed wetland for treating swine lagoon effluent. Presented at the 1992 Intl. Winter Meeting. Am. Soc. of Agric. Engrs., 2950 Niles Rd., St. Joseph, MI.

12. Peverly, J. H. 1985. Element accumulation and release by macrophytes in a wetland stream. J. Environ. Qual. 14(1):137-143.

13. Reed, S. C. 1991. Constructed wetlands for wastewater treatment. Biocycle 32(1):44-49.

14. Rogers, K. H., P. F. Breen, and A. J. Chick. 1991. Nitrogen removal in experimental wetland treatment systems: evidence for the role of aquatic plants. J. Water Poll. Control Fed. 63:934-994.

15. Seidel, K. 1976. Macrophytes and water purification. p. 109-121. In: J. Tourbier and R. W. Pierson, Jr. (ed.) Biological control of water pollution. University of Pennsylvania Press, Philadelphia, PA.

16. Sheffield, C. W. 1967. Water hyacinth for nutrient removal. Hyacinth Control J. 6:27-30.

17. Simpson, R. L., R. E. Good, R. Walker, and B. R. Frasco. 1983. The role of Delaware River USA fresh water tidal wetlands in the retention of nutrients and heavy metals. J. Environ. Qual. 12(1):41-48.

18. Ulmer, R., L. Strong, T. Cathcart, J. Pote, and S. Davis. 1992. Constructed wetland site design and installation. Presented at the 1992 Intl. Winter Meeting. Am. Soc. Agri. Engrs., 2950 Niles Rd., St. Joseph, MI.

19. Wolverton, B. C. and R. C. McDonald. 1981. Natural processes for treatment of organic chemical waste. The Environ. Prof. 3:99-104.

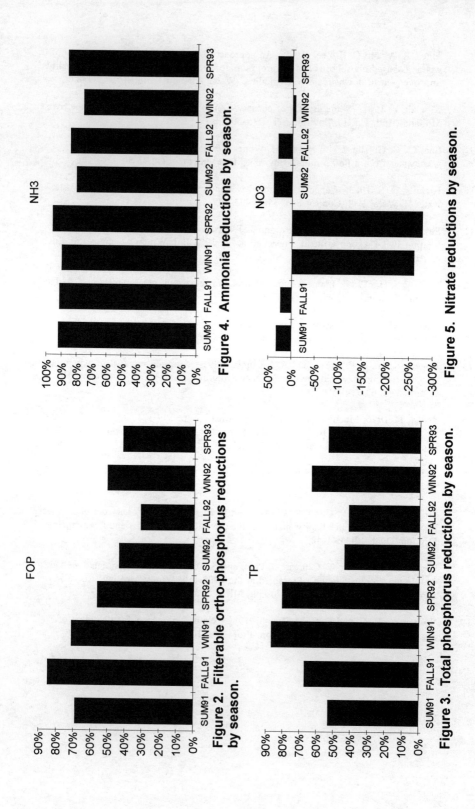

Figure 2. Filterable ortho-phosphorus reductions by season.

Figure 3. Total phosphorus reductions by season.

Figure 4. Ammonia reductions by season.

Figure 5. Nitrate reductions by season.

RESTORATION OF WETLANDS THROUGH THE WETLAND RESERVE PROGRAM

A MISSISSIPPI PERSPECTIVE

Ramon L. Callahan, Assistant State Conservationist (Technology)
L. Pete Heard, State Conservationist
Soil Conservation Service, Jackson, Mississippi

ABSTRACT

The Wetland Reserve Program (WRP) was established by congressional action to restore converted agricultural wetlands. The first sign-up for WRP was in 1992. This program initially involved nine states and funding for 50,000 acres. Accepted participants sign perpetual easements to the USDA, and agree to apply and maintain essential planned restoration practices. In Mississippi, these practices primarily include tree planting and structural practices to restore hydrology. Through good interagency cooperation and landowner involvement, Mississippi had a large response to the program and eventually the highest number of selections -- 47 farms totaling 14,885 acres. Eighty percent of the converted wetlands to be restored will be planted to various hardwood species. Compatible uses of the easement area are determined and listed in the easement. Suggestions for future program direction are discussed.

INTRODUCTION

Mississippi was one of nine states chosen to participate in the 1992 pilot Wetland Reserve Program (WRP). This paper will discuss an overview of the implementation process of the program in Mississippi and technical issues such as reforestation practices and hydrology restoration techniques utilized in WRP. Even though eligible landowners statewide were allowed to submit intentions to participate in the program, over ninety percent of intentions, and all but two of the 47 bids selected came from counties in the Mississippi River Alluvial Flood Plain. Therefore, my presentation will focus primarily on WRP activities in this region of the state, commonly known as the Mississippi Delta.

SETTING

The Delta once comprised a vast and unique wetland ecosystem of bottomland hardwood forest, natural lakes, deep fertile alluvial soils, and experienced prolonged periods of flooding. These characteristics, along with Mississippi's long growing season produced a habitat mecca for fish and wildlife species, including wintering feeding grounds for the largest population of migratory waterfowl in North America.

HISTORY

At the time of our nation's founding the U.S. Fish and Wildlife Service estimates there were 9.8 million acres of wetlands in Mississippi; today there are less than 4 million acres (National Research Council, 1992). In the Delta region of Mississippi, it is estimated there was once over three million acres of bottomland hardwood wetlands; today, there are only an estimated 1,000,000 acres remaining. The wetland hydrology of these stands has often been significantly altered. This conversion of Delta bottomland hardwood forests took place at an accelerated rate during that period between 1947 and 1977. During this thirty year period, approximately one million acres were converted, primarily for agricultural production. Availability of capital, mechanized land clearing equipment, increased demand for soybeans and cotton, and the installation of elaborate flood control systems including channels, levees, pumping stations, and flood storage reservoirs all contributed to this dramatic rate of conversion. It is estimated that 1.5 million acres of converted wetlands are eligible for the WRP program in the Mississippi Delta alone.

BASICS OF THE WETLAND RESERVE PROGRAM

The WRP was authorized and funded by Congress for fiscal year 1992 to obtain perpetual leases on converted wetlands that would be restored to a functioning wetland ecosystem. A program cap of 50,000 acres was established for the nine pilot states. Eligible wetlands were prior converted cropland (PC) and farmed wetlands (FW). Prior converted cropland is the term used to describe wetlands converted and farmed prior to 1985. They have been altered to the point that they do not exhibit significant wetland functions. Farmed wetlands are those that also were converted prior to 1985 but still flood for 14 consecutive days during the growing season on a 50% occurrence. They retain significant wetland functions, even though farmed during the summer months. A small amount of existing natural wetlands, nonwetlands and buffer areas were sometimes accepted to create a contiguous, manageable easement area. The prospective participant signed up for the program and following eligibility determinations made by the Soil Conservation Service and U.S. Fish and Wildlife Service technical staff, a restoration plan was prepared with the landowner. The purpose of the plan was to identify the hydrology and vegetative restoration practices to be accomplished. In the Delta, at least 80% of the restored area had to be planted to various bottomland hardwood species within three years of bid acceptance. The tax liability and maintenance of essential elements of the plan remain the responsibility of the landowner. The filed easement lists the compatible uses of the easement area that the landowner and planners have agreed upon.

PROGRAM PARTICIPATION

Mississippi landowners submitted intentions to participate in WRP on 375 farms totaling 117,470 acres. This was second highest among the nine states. Two hundred forty-five of these landowners submitted bids on 64,957 acres; again, second highest among the nine pilot states. Mississippi had the highest number of accepted bids with 47 farms totaling 14,885 acres, almost 30% of the total acres allowed for the program.

BID SELECTION

Easement payment bids submitted by and accepted of Mississippi landowners were among the lowest of the nine pilot states, second only to Louisiana. The average easement payment was $561.00 per acre. The cost of the easements in other states were dramatically higher. Upon completion of the essential practices in the Wetland Plan of Operations the landowner receives full payment. There are provisions for partial payments to the landowner during the implementation period.

ENVIRONMENTAL BENEFIT INDICES

The environmental benefit index (EBI) is an evaluation system developed to rank the bids so the best offerings for the tax dollars spent could be identified. Such factors as size, type of wetlands, value as riparian corridors, proximity to other protected state or federal areas, hydrology restoration planned etc., were rated by the planning team and submitted with the bid. In Mississippi, the EBI's were very high. The reasons for this were: (1) 64% of accepted bids were on prior converted cropland (PC's) and will have the hydrology restored on more than 50% of the area beyond that level necessary to meet Food Security Act (FSA) wetland hydrology criteria; (2) several accepted tracts are adjacent to National Wildlife Refuges or State Wildlife Management Areas; and (3) most farms accepted are in areas that will provide quality waterfowl habitat for the North American Waterfowl Management Plan.

RESTORATION PRACTICES

On 92% of the approved WRP acreage the planned restoration practice is reforestation to adapted hardwood species similar to those present on the site prior to conversion. This will be accomplished by direct seeding with acorns, and planting seedlings.

Direct Seeding: Direct seeding of acorns has been shown to be an effective cost-efficient reforestation method on Delta soils. This practice is being used extensively on National Wildlife Refuges in the Mississippi Delta. This is done by using a modified soybean planter planting 1200 acorns per acre at a spacing of 3 1/2' x 10'. Germination rates of 25-33% are typical which will provide an initial stand of 300-360 seedlings per acre. This equates to using 3-4 lbs. per acre of small acorns such as water, willow, and cherrybark oak, and 10-12 lbs. per acre of larger sized acorns such as overcup, nuttall, Shumard, and swamp chestnut oak.

Direct seeding is primarily used on sites which are inundated for long periods during the winter months (January - April). This prolonged period of inundation does not allow for planting seedlings during the normal tree planting season.

Planting Seedlings: Hardwood seedlings must be planted from December through March to avoid high mortality rates. Reforestation with seedlings is planned on most prior converted fields where prolonged inundation is not common. Hardwood seedlings will be planted on 10' x 12' row spacings providing 365 seedlings per acre. All reforestation sites will be planted to adapted hardwood species and usually will include at least three species of oaks. These will be randomly mixed to provide for future diversity.

HYDROLOGY RESTORATION

Planned practices for restoration of hydrology in WRP primarily include surface ditch plugs and construction of levees. Flashboard risers will be installed on many drainage outlets for the management of moist soil plants and to provide waterfowl management capability.

COST-SHARING

Planned essential practices are cost-shared to help defray the landowners cost and to provide technical assistance and quality control by the assisting agency -- usually the Soil Conservation Service. The cost-share rate in Mississippi for all essential practices is 75%.

OPERATION AND MAINTENANCE

Trees planted as essential practices will be reasonably protected from nuisance wildlife such as beaver and nutria.

Structural measures will be operated and maintained in a manner which will fulfill their intended function(s) of hydrology restoration and/or waterfowl habitat enhancement.

No haying or grazing is permitted without prior written authorization from the technical agencies.

Any chemicals used on the easement area will be applied according to state and federal laws, and if authorized by the technical agencies. Any applications will be done in accordance with all label requirements and restrictions.

All essential practices, both structural and nonstructural, will be maintained throughout the easement period. However, if a practice fails or becomes nonfunctional, the participating technical agencies may deem the discontinuance or modification of the practice if it is determined the practice is no longer necessary to the continued function of the restored wetland.

COMPATIBLE USES

The following have been determined by the participating technical agencies to be compatible with the purpose and function of the WRP easement. Only those activities listed as compatible uses are allowed. Agreed to compatible uses are listed in the filed easement document.

1. <u>Timber Harvest</u> - in accordance with a timber management plan developed and/or approved by the participating technical agencies.

2. <u>Greentree Reservoirs</u> - these are temporarily flooded timber stands and may be constructed in the future for waterfowl use as long as the construction and maintenance is not detrimental to the values and functions of the restored wetlands. Must be constructed in accordance with all state and federal permits. The size and location will be approved by SCS and USFWS.

3. <u>Hunting, Fishing, and Trapping</u> - is allowed in accordance with state and federal regulations.

4. <u>Crawfishing and Catfishing</u> - incidental harvesting for personal consumption is allowed.

5. <u>Environmental Education Activities</u> - these are allowable.

6. <u>Other Compatible Uses</u> - are considered for approval by the participating technical agencies and documented in the easement if approved.

INTERAGENCY COOPERATION

Throughout all phases of the WRP, interagency cooperation was extremely successful. The three lead agencies that assisted in providing technical information and guidance to the program were the Soil Conservation Service, Agricultural Stabilization and Conservation Service, and the U.S. Fish and Wildlife Service. Other agencies involved were the U.S. Army Corps of Engineers, Mississippi Department of Wildlife, Fisheries, and Parks, Mississippi Forestry Commission, and Mississippi State University's School of Forestry and Wildlife Resources.

WORKLOAD

For the pilot program, SCS in Mississippi has spent over eleven thousand staff hours providing direct technical assistance to the WRP. Since the implementation phase of the program is still ahead, it is anticipated that this time will double over the next three year period. Participants have three years to fully implement essential practices.

CONCLUSIONS

The Wetland Reserve Program offers a carrot approach to enlisting the support of landowners in restoring wetlands converted to agriculture. The program received wide acceptance among Mississippi landowners. Considering that the easements are perpetual and the landowner continues to pay the taxes and is responsible for maintenance of the restoration practices, this program is very cost effective. The pilot program worked satisfactorily; however, as expected, some changes are expected in the 1994 Program.

RECOMMENDATIONS

Recent meetings with participating agencies, environmental concern groups, and farm organizations were held to develop recommendations to improve the effectiveness of future WRP programs. Some agreed to recommendations expressed at these meetings include: (1) The need to establish a preselected easement payment rate. Only 23 percent of submitted bids were selected. This equates to several thousand staff hours that were spent working with landowners developing plans and evaluations on farms that would have been eliminated from the consideration process if easement payment rates had been established before signup. (2) The need for an established time table for accepted landowners to file easements. (3) Future WRP funding should be distributed to individual states (e.g. Waterbank Program), and states allowed to select lands based on wetland values and environmental factors/benefits which are unique to that state.

At the time of writing this paper, it is anticipated that there will be another Wetland Reserve Program in 1994. The signup will probably occur in February. There will be some changes in the program, but these changes have not yet been announced.

ACKNOWLEDGEMENTS

The following individuals provided assistance with this paper and their contributions are gratefully acknowledged:

Floyd A. Wood, Biologist, SCS, Jackson, MS
Alan Holditch, Forester, SCS, Jackson, MS
Ken E. Murphy, Assistant State Soil Scientist, SCS, Jackson, MS
Reginald (Kim) Harris, Assistant State Conservation Engineer, SCS, Jackson, MS
Robert W. Stobaugh, Public Affairs Specialist, SCS, Jackson, MS
Laura T. Anderson, Public Affairs Assistant, SCS, Jackson, MS
Ernest E. Dorrill, Programs Analyst, SCS, Jackson, MS
Bob Strader, Wildlife Biologist, USFWS, Jackson, MS
Thomas Owens, Conservation Programs Specialist, ASCS, Jackson, MS

LITERATURE CITED

1. National Research Council, 1992. Restoration of Aquatic Ecosystems, National Academy Press, Washington, DC. 276 pp.

HABITAT PROTECTION AND LAND DEVELOPMENT
EXPLORING A SYNERGETIC APPROACH

L.M. Zippay and J.W.H. Cates

ABSTRACT

Biodiversity is rapidly declining in Florida as well as in the United States. Habitat fragmentation and isolation are recognized by many biologists as the primary causes of this biodiversity decay. Land use pressures from development and agriculture further this degradation and contribute to a landscape mosaic in which the few remaining natural habitat features are stream systems and their riparian ecosystems. With population growth increasing worldwide, development and agricultural pressures are certain to escalate -- creating the need for a synergetic approach to habitat protection and crop production.

This paper presents a case study of how the southern phosphate district of Florida can produce a future land use plan which incorporates a balance of intensive and non-intensive land uses, maintenance/protection of regional water resources, and replacement/protection of critical, native plant and animal habitats. District lands that are subject to mining disturbance are separated into two categories: (1) those included in the Integrated Habitat Network (IHN), and (2) those included in the Coordinated Development Area (CDA). This paper illustrates how, through the IHN/CDA approach, development and agricultural productivity can coexist with habitat protection in a manner that is mutually beneficial.

The conclusion is that the IHN/CDA (ecosystem management) approach to development and habitat protection would be feasible and beneficial in the southern phosphate district of Florida as well as in other areas of Florida.

KEYWORDS. Biodiversity, Habitat Fragmentation, Ecosystem Management, Integrated Habitat Network, Coordinated Development Area,

OVERVIEW OF THE CONFLICT BETWEEN
LAND DEVELOPMENT AND HABITAT PROTECTION

Effects of Land Development on Wildlife

Human settlements in America began as isolated entities within a large network of undisturbed natural habitats. Today, however, the opposite is true. The population growth of the nation dictates the development of increasing amounts of natural areas, and as a result, vast expanses of forest and wetland systems are being replaced with urban areas, pastureland, and agricultural land. As currently designed, these land uses do not provide suitable habitat for most wildlife species and, therefore, result in habitat loss and the consequent decay of biodiversity. Careful holistic planning and management of remaining natural areas and future development is necessary if existing wildlife species and ecological processes are to persist.

Biological diversity (biodiversity) is defined as follows: "Biological diversity refers to the variety and variability among living organisms and the ecological complexes in which they occur. Diversity can be defined as the number of different items and their relative frequency. Thus, the term encompasses different species, ecosystems, genes, and their relative abundance" (U.S. Congress, Office of Technology Assessment, 1978), (Hudson, 1991). Biodiversity is important to the health of the environment as each species contributes to the overall balance of an ecosystem. Biodiversity degradation is a formidable concern that researchers in the U.S. attribute to habitat fragmentation and destruction.

Habitat Fragmentation can be described as the segmentation of natural habitats into small, isolated subsections of the original habitat. Fragmentation limits not only the available nutritional and sheltering resources for a species, but also the exchange of genetic material which results in inbreeding depression (the accumulation of deleterious alleles in succeeding populations of a species, resulting in reduced adaptability to environmental changes and an increased likelihood of vulnerability to disease, etc.) (Simberloff, 1988). Habitat Fragmentation is a serious problem that, realistically, will persist with development -- indefinitely. Therefore, the problem must be addressed in terms of how to offset the effects of fragmentation rather than how to cease development.

Florida as a Case Example

The State of Florida is a prime example of the conflict between land development and wildlife habitat. Florida is a large agricultural producer which has a high rate of population growth, and the second highest number of endangered and threatened species in the United States (Cox, 1992). Included on this list are Florida's two largest mammals, the Florida panther (endangered) and the Florida black bear (threatened), both of which are dependent on a diversity of habitats for their survival. In the past 50 years, 8 million acres of habitat has been converted to urban and agricultural uses (Kautz, 1992) -- 4.5 million acres of which constitutes forest land (Kautz, 1992). Much of the land that is earmarked for conversion to citrus is currently rangeland or pastureland interspersed with forested uplands and wetlands. Citrus production is economically beneficial to the State, and as outlined above, engulfs critical habitat for the Florida panther and black bear. Therein lies the problem and possible solution -- cooperation and coordination of private and public initiatives.

The Florida panther and the Florida black bear have both been sighted within citrus groves -- their occurrence hypothesized to be attributable to the proximity of forested habitat (IFAS, 1993). Rather than threaten the private property rights of the farmer, the ecosystem management approach would encourage the construction of wildlife corridors through the grove (via existing drainways) which would not only reduce the negative effects of fragmented habitats (by providing access to additional prey and mates in adjacent habitat areas), but also filter agricultural runoff to protect water quality.

The designing of citrus groves for wildlife usage will need to incorporate adjacent land uses as the spacial arrangement and ratio of different land uses within and surrounding the grove will be of critical importance. Factors identified as possibly affecting the species diversity within groves include; grove bed size, interspersion of other critical habitats, and the proximity of the grove to large natural areas (IFAS, 1993).

COOPERATIVE METHODS OF LAND DEVELOPMENT

Ecosystem Management

Citrus production in southwest Florida is growing immensely, with 80,000 ha projected to be in production by the year 2000 (Currently it is 60,000 ha) (IFAS, 1993). Because much of the proposed and existing citrus development occurs in an area bordered by the Everglades National Park and the Big Cypress National Preserve (and occupied by many state and federally listed plants and animals), the development lends itself easily to the adoption of an ecosystem management approach.

Public lands were once thought of as a reasonable form of long-term protection for wildlife -- evidence now shows that no national park in North America is maintaining its original populations of species (Sullivan, 1992). The condemnation of critical private land for the purpose of acquiring habitat is neither economically feasible (from an acquisition point of view) nor wise (from an agricultural point of view). Rather, the habitat protection approach needs to flow from a statewide comprehensive environmental directive which considers all environmental aspects and impacts, and which is implemented through private property/non-profit/governmental coordination. This directive is embodied in the initiative now known as the "ecosystem-based management approach".

The Ecosystem Management Approach, as proposed by the Department of Environmental Protection Merger Plan Summary, is a multi-faceted, flexible program which attempts to coordinate production and environmental parameters on a landscape scale. "Ecosystem Management is holistic -- involving the explicit management of the environment for a broad array of uses, products, and services with a goal of sustaining the long-term health and productivity of the resources while protecting public values and private property rights." (Soc. Am. Foresters, 1993). To be successful, this management approach necessitates an unprecedented level of cooperation and coordination among owners and managers of public and private land. Individual management decisions need to be analyzed in the context of adjacent, proximal properties within the landscape -- not just single parcels.

Man's productive exploitation of any single resource is dependent upon the simultaneous use of other resources for stasis conversion, transportation, and assimilation. Crops cannot grow without nutritional soil, clean air and water. Mineral extraction cannot proceed without water for transport and processing. Fertilizer for the crops cannot adequately exist without mineral extraction and is not useful without adequate water for solubilization. Cities must have land for buildings and roads - and water for drinking, bathing, cooking, and the transportation of waste. The analogies are endless. Until recently we have overlooked the finite nature of the world's resources and assumed that there would always be enough - because there appeared to be so much. Now we better understand the interwoven nature of resource use and conversion. Years ago a farmer needn't worry if the run-off from his field was over enriched or turbid because the adjacent wetland cleaned the water before it reached the lake and the run-off eventually returned as clean rain. Today however, the wetland is drained and converted to a jiffy-store parking lot; and the lake is surrounded by a subdivision whose septic tanks drain into the lake. "Downstream-users" must constantly concern themselves with the slop-over effects of "upstream use". Land development, agriculture, all forms of resource exploitation and conversion must realize that there is no longer a resource margin which allows for slop-over. The equation must become evermore precise and continually calculate the economic and environmental cost/benefit ratios of resource use for everyone in the cycle.

The IHN/CDA Model

One example of the Ecosystem Management approach, as it applies to general agriculture and the environment, can be found in the landscape planning effort currently underway in the phosphate mining district of west central Florida. This phosphate district (Bone Valley area) covers approximately one-half of the entire watershed and headwaters portions of each of five rivers within the region (Peace, Alafia, Manatee, Little Manatee, and Myakka). The area of economic mineability or "mineable limit" covers approximately 1,265,000 acres (1,977 sq. miles) in five counties. The landscape is highly altered - having been ditched, drained and converted to agriculture over a period of decades. The remaining, highly fragmented, natural features are generally only streams, associated riparian systems and areas of rangeland. Florida's reclamation program, like most environmental regulatory programs, is almost twenty years old and was structured on the premise that maintaining the status quo with respect to the environment is a sound goal. The past reclamation program therefore set criteria requiring the reclamation of wetlands and other habitat features within the same general area as the pre-existing habitats. The landscape which has emerged is one which replicates the premining fragmented landscape. The post-reclamation landscape is one of scattered and often incompatible landforms and land uses. This somewhat chaotic "shotgun pattern" of landforms and land uses renders effective, comprehensive planning virtually impossible. The interaction of the incompatible land uses within this mosaic inevitably produce an increased negative impact on non-intensive, environmental uses by exacerbating the effects of intensive uses.

In July of 1992, utilizing years of experience, the Florida Department of Environmental Protection (then Natural Resources), Bureau of Mine Reclamation published a document entitled <u>A Regional Conceptual Reclamation Plan for the Southern Phosphate District of Florida</u>. The document outlines a broad, landscape-scale scenario for the comprehensive planning of the phosphate district and subsequent incorporation into a west central Florida regional planning effort. <u>Guidelines for the Reclamation, Management and Disposition of Lands within the Southern Phosphate District of Florida</u> (currently in draft form) is the next phase of this planning effort and documents general criteria intended for implementation of the district-wide reclamation plan concept.

The objective is to produce a district-wide conceptual reclamation plan which is also a comprehensive, district-wide landscape plan incorporating maintenance/protection of regional water resources, a balance of intensive and non-intensive land uses, and replacement/protection of critical, native plant and animal habitats. The goals of the district-wide conceptual reclamation plan are to demonstrate the efficacy of the above statement by means of replacing drainage and hydrologic function, organizing land uses to maximize their longevity and water-cleansing capacities, and to provide quality, interactive wildlife habitat. The objective of the guidance document is to facilitate implementation of the district plan by providing guidelines for reclamation, management and disposition of mined lands. As the district-wide plan is presented in the guidance document, the district lands subject to mining disturbance are separated into two categories: (1) those to be included in the Integrated Habitat Network (IHN), and (2) those to be included in the Coordinated Development Area (CDA). Reclamation will be guided by state-of-the-art knowledge in earthmoving, post-mining soils, hydrology, revegetation, sustainable agriculture and silviculture, citriculture, animal science, general ecology and wildlife management. Data gathering, in every applicable field, will be constantly ongoing and will be used to "fine-tune" reclamation methodology.

The Integrated Habitat Network (IHN) is a series of connected analogs of naturally occurring communities. The integrated connections are planned to optimize interspersion and juxtaposition of habitats while maintaining individual community identity and, concurrently, inter-habitat connectivity. The IHN proposes to use unmined and reclaimed tributaries and rivers as the nucleus of the network system. The protection afforded these landscape features provides the foundational basis for long-term maintenance, protection and habitat viability. The main goals of the IHN are: (1) the replacement and long-term protection of analogs of natural communities which have been disturbed or destroyed by mining activities, (2) the maximization of habitat quality within the individual community/habitat analogs, (3) the maximization of inter-habitat connectivity on a regional basis and connection of the regional network with inter-regional community/habitat conglomerations, and (4) the integration of habitat networks into a comprehensive regional landscape plan which optimize the positive effects of postmining landforms/land uses on water quality and quantity.

The IHN incorporates a "wildlife corridor" function, may incorporate a recreational "greenway" or passive "greenspace" function and yet is broader in scope. Because of the ecological need for fully functional communities, analogs are designed on a "whole habitat" basis rather than being designed around the specific needs of one or two species. Creation of functional soil, hydrology and floral precursors serve as the basis for food-web development. All components are selected to maximize expansion of this base, precursor food-web. Community analogs are designed to produce a foundation level of functionality in their initial stages and to produce a maximum initial "diversity potential," based upon the premise that, with proper management, the initial input will yield, over time, maximum "ultimate diversity."

The Coordinated Development Area (CDA) consists of those lands outside of the IHN whose reclamation is planned for intensive or semi-intensive land uses. General land uses may include, but are not limited to, intensive agriculture (i.e., citrus, row crops), semi-intensive agriculture (i.e., pasture, silviculture), and intensive and semi-intensive development (i.e. power plants, housing).

The post-mining development potential of the west central Florida mining district is significant. Within the last decade formerly mined areas of Polk and Hillsborough counties have been converted to shopping centers, light industrial facilities, office complexes, and housing developments. Currently, one power plant is under construction and two larger plants are planned. Additionally, in the near future, several major transportation corridors will bisect the district. Agricultural production and marketing research has progressed to the point that field scale production trials are beginning on several hundred reclaimed acres. Increasing statewide population pressures and the resultant need for clean water will place pressure on historically important winter farming areas. As muck-farm lands in the Everglades, Upper St.Johns River, Upper Ocklawaha River, Lake Apopka and similar areas are removed from production or as additional water quality and quantity restrictions are placed on these areas, the potential and need for viable, alternative sites increases. The likelihood of agricultural and other development within the phosphate district provides an opportunity for innovative planning never before realized in the state of Florida. The location and disposition of the mining district,in a statewide context, is <u>strategic</u> for completion of an IHN system and commercial and agricultural development.

As part of a larger comprehensive plan, the CDA also shares the need (with the IHN) for a rational basis of institution and guidelines for implementation. In order for the land uses within the IHN and the CDA to be coordinated and compatible, a common goal for both areas must be realized. The successful long-term viability of both areas is dependent upon an adequate supply of clean water. Therefore the maintenance, enhancement and protection of water quality and quantity within the district/region becomes the common, end goal. All planned uses of both areas must embrace this goal.

Operating phosphate mines account for exiting water quantity and quality factors by means of in-mine retention and by use of NPDES (National Pollution Discharge Elimination System) permitted discharge points. Theoretically, when a mine is reclaimed, "point source" pollution changes to "non-point source" pollution as in any other agricultural or developmental area. As non-point source pollution occurs in four main forms - turbidity, nutrient overloads, chemical contaminants and biological contaminants, the most effective controls rely on prevention and management methodologies. The best sources of applicable methodologies are the Sustainable Agriculture, Coastal Zone Management (CZM) and Best Management Practices (BMP) guidelines. New and more pertinent hydrology modeling, mutually designed and applied by industry and DEP, will form the basis of construction, function, and management analysis. Under the consolidated mine permitting scenario proposed to the legislature, factors relative to the management and storage of surface waters (MSSW permits), dredge & fill, and reclamation will be regulated by the Department of Environmental Protection.

Therefore, the major factors affecting postmining/postreclamation water quantity and quality can be reduced to three general categories to be addressed in the general and specific guidelines for the IHN and the CDA. Water quantity is controlled by both (1) watershed replacement and restructuring and by (2) stormwater retention/control requirements. Water quality is controlled by (3) the replaced landforms, landuses and the various methodologies utilized to prevent and manage non-point source pollution.

Summary

"The regulatory paradigm, under which many agencies operate, is outdated. Environmental permitting/regulation is reactionary and not based on a holistic visionary goal. The purpose of the IHN/CDA concept is to create a visionary goal; one to which the industry and state can realistically react and respond" (Craft, 1993). Likewise, comprehensive and integrated planning for all parameters affecting the environment has been overlooked or stymied because it appears too long-term and complicated. The status quo slaps a band-aid on a problem area without looking at all of the ramifications and cause/effect parameters. The ecosystem-based management approach, through such concepts as the IHN/CDA, promises to provide a reachable, equitable goal by accounting and assimilating all of the cause/effect parameters into an incorporated, integrated plan of development and protection. With the goal and plan in place, the environmental machine will have a chance of returning to equilibrium.

REFERENCES

1. Cates, J. July 1992. <u>A Regional Conceptual Reclamation Plan for the Southern Phosphate District of Florida</u>. Tallahassee, Florida: Florida Dept. of Natural Resources.

2. Cates, J. and Zippay, L. 1993. <u>Ongoing Projects and Programs which are Interrelated with the Implementation of the Integrated Habitat Network/ Coordinated Development Area.</u> Tallahassee, Florida: Florida Department of Natural Resources.

3. Cates, J. and Zippay, L. 1993. <u>Guidelines for the Reclamation, Management and Disposition of Lands within the Southern Phosphate District of Florida.</u> (in draft) Tallahassee, Florida: Florida Dept. of Natural Resources.

4. Craft, J. and Cates, J. 1993. "A Regulatory Approach to the Protection of Riparian Systems and the Restoration of Adjacent Habitats" (in press). Proc. Conf. Riparian Ecosystems in the Humid U.S.

5. Cox, Jim. 1992. <u>Wildlife Corridor Design and Placement: Where Will Corridors Be Most Effective?</u> Tallahassee, Fl: Florida Game and Fresh Water Fish Commission

6. Hudson, Wendy E. (ed.). 1991. <u>Landscape Linkages and Biodiversity</u>. Washington, D.C.: Island Press.

7. Institute of Food and Agricultural Sciences (IFAS) and the Cooperative Fish and Wildlife Research Unit, University of Florida. 1993. <u>An Evaluation of the Regional Effects of New Citrus Development on the Ecological Integrity of Wildlife Resources in Southwest Florida.</u> West Palm Beach, Fl.: South Florida Water Management District

8. Kautz, Randy S. 1992. "Satellite Imagery and GIS Help Protect Wildlife Habitat in Florida." <u>Geo Information Systems</u>: 37-42.

9. Simberloff, Daniel. 1988. "The Contribution of Population and Community Biology to Conservation Science." <u>Annual Review of Ecology Systems</u> 19: 473-511.

10. Sullivan, Rick. 1992. <u>Tying The Landscape Together: The Need for Wildlife Movement Corridors.</u> Gainesville, Fl: Cooperative Extension Service, University of Florida, Institute of Food and Agricultural Sciences.

THE IMPACT OF THE CITRUS CONVERSION PROCESS
ON GROUND AND SURFACE WATER QUALITY: A CASE STUDY

Ashok N. Shahane, Ph.D, P.E., P.H., CGWP*

ABSTRACT

The freezes experienced during the past several years have damaged citrus groves in central and northern counties of Florida. As a result, a trend has been observed to develop land with flatwood soils for citrus in the southern parts of Florida. Land has been planted for citrus in several counties including DeSoto, Collier and Hillsborough. Citrus growers believe that flatwood soil and modern microjet type irrigation system can yield higher citrus production per year in these counties. While agricultural production is increased, the impact of the citrus conversion process on the water quality of ground and surface water needs to be assessed. With cooperation from a citrus grower, monitoring of pesticides and nutrients was undertaken routinely during the various phases of the citrus conversion process in DeSoto County.

The objectives of this paper are to present base line and routine sampling results for pesticides and nutrients. The site characterization, monitoring program and the location of sampling points are discussed. The monitoring results to date indicate that the citrus conversion process during the first three phases of the project appears to have no adverse impact on the quality of shallow and deep ground water of the grove. Although bromacil, aldicarb, simazine and diuron are sporadically observed in the surface water, the surface water leaving the grove site was found to contain no pesticides. This indicated low potential for off-site adverse water quality impact. Explanations include dilution, degradation and best management practices. These results are helpful in assessing short and long term impact.

KEYWORDS: Pesticide monitoring, surface and ground water, citrus management and water quality impact.

INTRODUCTION

It is estimated that citrus production in the Florida counties north of I-4 (an interstate highway which divides the state approximately in half) has declined by about 79 percent as a result of number of recent freezes in central and northern Florida. At the same time, the citrus production in the counties south of I-4 increased by about 31 percent(Florida Agricultural Statistics Service, 1994). A trend has been observed to develop land with flatwood soils for citrus in the southern parts of the state. Specifically, land has been planted for citrus in several counties including DeSoto, Collier and Hillsborough. It is obvious that citrus is moving south as new citrus groves are planted rapidly in southern counties. This rapid citrus conversion is based on the belief of the citrus growers that flatwood soil and modern microjet type irrigation system can yield higher citrus production per year.

---

* Dr. Ashok N. Shahane, Hydrologist, Scientific Evaluation Section, Bureau of Pesticides, Division of Agricultural and Environmental Services, Florida Department of Agricultural & Consumer Services, 3125 Conner Blvd, MD-1, Tallahassee, Florida 32399-1650.

While the higher citrus production is useful in minimizing the adverse economic impact of freezes on Florida citrus industry, the impact of citrus conversion process on the water quality of ground and surface water needs to be assessed. With cooperation from a citrus grower, the routine monitoring of pesticides and nutrients was undertaken during various phases of the citrus conversion process on a site in DeSoto County. The main objectives of this paper are to present a base line and routine sampling results of the past several years for pesticides and nutrients at this citrus grove in DeSoto county. The description and the characteristics of the site are also briefly presented.

SITE DESCRIPTION

The site, a square mile section (640 acres), is located near Arcadia in Desoto County. The project site is zoned for agricultural uses. The site plan is shown in Figure 1 with the information on the type of citrus that is planted in the different parts of the site.

The project site is being developed in four phases. In each phase, bedded citrus is planted on a quarter section of the site which is called a unit. The first phase started in 1987. The citrus plantings, 200 trees per acre, were completed in December, 1987 in Unit B (see Figure 1). The second phase was completed in June 1988 when Unit A was developed. The third phase was completed in 1990 after developing unit C. The fourth phase is scheduled to be developed in either 1994 or in 1995 to complete Unit D. These four phases will complete the citrus conversion process at the site.

As shown in Figure 1, the site is bordered by four canals. The water detention areas and the associated drainage system have been permitted by the Water Management District. To the north side of the site, there are mature citrus groves maintained by another owner. Similarly, there are sod production, cattle grazing and citrus operations under different ownership on the east side of the site. The land on the west and south side of the site is part of a ranch owned by the citrus grower and is used for cattle grazing (Shahane, April,1989).

There is a house with a drinking well 1/4 mile from the northern boundary of the site. The manager's office and barn is about 2 1/2 miles south of the site.

SITE CHARACTERISTICS

The site lies entirely within the Desoto plain. The elevations on the site range from 68 feet above mean sea level to 62 feet above mean sea level. Elevations decrease as one goes from north to south and from west to east (Campell,1985).

Holocene, Pleistocene, and Pliocene surficial deposits consist of sand, terrace sand, phosporite and an undifferentiated deposits. The sequence of surficial deposits is generally less than 100 feet thick in Desoto County (Campbell,1985,SWFWMD,1988, SCS,1988). The Hawthorn formation of Miocene age underlies the surficial deposits and it's thickness ranges from about 160 to 370 feet in Desoto County. The Hawthorn consists principally of beds of sandy, phosphatic limestone, dolomite, and sandy chalky to granular phosphatic marl, clay and crystalline limestone at the lower level. The Tampa Limestone of Miocene age is a granular phosphatic limestone with varying amounts of interbedded sand and clay and it's thickness ranges from about 50 to 100 feet (SCS,1988). The Suwannee Limestone of oligocene age is a granular fossiliferous limestone with beds of crystalline dolomite and is encountered at between 350 or 700 feet below Nautical Geodic Vertical Distance(NGVD). The thickness of the Suwannee limestone ranges from approximately 100 feet to over 250 feet within the County. The Ocala Limestone of Eocene age is composed of chalky, fossiliferous limestone and granular limestone with beds of crystalline dolomite. The thickness of the Ocala ranges from about 260 to 400 feet in Desoto County (SWFWMD,1988, SCS,1988). The Avon Park formation of middle Eocene age consists of limestone interbedded with hard brown highly fractured dolomite. The thickness of the Avon Park formation is from 200 to 470 feet (Wilson 1977)

The permitted irrigation wells at the site penetrate through these geologic formations.

The climate of the site is humid sub-tropical, characterized by high mean annual rainfall and temperature. The mean annual air temperature is about 73°F. Temperatures from a front rarely remain below freezing during the day and generally last only 2 - 3 days. The average annual rainfall is 52.3 inches, about 60% of the annual rainfall occurs in June through September. The dry season is from October to May during which the irrigation of row crops and citrus occurs in this area.

There are five major soil types that are found on the site. These soil types include: Delray mucky fine sand depressional, Eau Gallie fine sand, Farmton fine sand, Malabar fine sand, Malabar fine sand high, Malabar fine sand depressional. The description of these soil types and their physical-chemical and other characteristics are provided in Reference 3(SCS,1988).

Ground water is obtained from the surficial aquifer system, the intermediate aquifer system and from the Floridan Aquifer. The aquifers are separated by confining layers which restrict vertical water movement between the aquifer systems (Campbell,1985, SWFWMD,1988). The surficial aquifer system underlies essentially all of Desoto County and is utilized primarily for domestic, lawn irrigation and stock water supplies (SWFWMD, 1988). Wells developed in the Floridan Aquifer yield large quantities of water, often in excess of 1,000 gallons per minute. The primary use of water from the Floridan is large scale irrigation (SWFWMD,1988).

Potential maps of the Floridan aquifer clearly indicates that movement of ground water in the Floridan aquifer is from Northeast to Southwest in the vicinity of the Site. Due to the interactions between canal water levels and the shallow water table aquifer, ground water direction in the surficial aquifer cannot be generalized since it can vary locally on the Site.

The site drains via sheetflow, sloughs and canals southwesterly toward Hawthorne Creek (a tributary of Joshua Creek). The existing canal and berm along the eastern boundary of the site prevent off-site flow from entering the property and convey water southward which appears to drain ultimately into the Prairie Creek watershed. The principal components of the proposed surface water management system are bed swales, lateral ditches, and retention reservoirs with lift pumps as shown in Figure 1. The reservoirs have been designed to provide for peak flow attenuation of storm runoff, water quality treatment, and wetland preservation/mitigation. An elaborate engineering system is designed for handling the agricultural drainage of the site. Irrigation wells, microjet irrigation system, bedded citrus, tile drainage, flashboards, existing and future detention ponds with lift pumps are some of the components of the engineering system that is approved for the project. The approved sizing of canals, citrus beds, pumps, detention areas were designed by an engineering consultant. Some of that information is included in the reference No. 7.

The agricultural runoff, stormwater drainage and the vertical percolation of rain and irrigation water are carried by the tile and stormwater drains. These underground pipes, staggered at regular intervals, drain into the canals which move the water to the east and west to the main canals by pumping it through the water detention areas. The main canals carry the drainage to the off-site beyond the property of the citrus grower.

The existing and proposed systems are designed to facilitate the beneficial interactions between the surface water and surficial aquifer through the control of water levels if necessary. The tile drainage can lower the water table by draining the subsurface flow into canals which drain into the creeks in the southside of the site and the creeks eventually drain into the Peace River System. By increasing water levels in the canals through the use of flashboards, the water table can be elevated so that soil moisture in the root zone can be maintained if necessary.

The general directions of the agricultural drainage and flow of the water in the system are indicated by arrows in Figure 1.

SAMPLING PROGRAM FOR THE SITE

1. <u>Sampling Program:</u>

The number of sampling points for monitoring pesticides, nutrients and copper on the site was dependent on the phase of the Citrus Development project. Currently, there are 14 sampling points on the site as shown in Figure 2. These sites are labeled as ANS 1 through ANS 14. These sampling sites are selected in a manner to generate the water quality information at various points in the system. Furthermore, these sampling sites are intended for monitoring surface water, surficial aquifer and the deep aquifer to the extent possible. Special consideration was given to the limited laboratory and manpower resources in selecting optimum number of sampling points by maximizing the benefits and minimizing the total cost.

The first three sampling sites (ANS 1, 2 and 3) were selected to compare the quality of surface water as it passes from the northern part to the central part of the site. Four tile drain sampling points (ANS 4,5,6 and 7) were selected to sample the percolated water beyond the root zone before it mixed with surface water. Sampling locations at 8 and 9 represent two shallow drinking water wells at the northern and southern sides of the site tapping the surficial aquifer. The sampling point 10 is at the large scale irrigation well which is in the middle of the grove. Sampling this well provides information on the quality of the deep Floridan Aquifer. Sampling stations ANS 11, 12, 14 and 31 are in two primary canals that flow north and south and carry the drainage off-site. The monitoring at these points will reveal the effects of dilution, degradation process at the site and the water quality improvements (if any) by the detention units as the drainage water moves through the system. The specific details about the locations of sampling points are given in the reference No. 6 (Shahane, January,1989).

2. <u>Monitoring Program:</u>

Baseline sampling was conducted to determine the background levels of pesticides and nutrients in the flowing waters of the site. The base line sampling was conducted in January 1989 when Units A and B were initial stages of operations. Considering the components of the system at that time, it was decided to collect base line samples at locations 1, 2, and 3 for surface water; at locations 4, 5, 6, and 7 for tile drainage, at two drinking water wells at locations 8 and 9 for shallow aquifer and at a location 10 of a large irrigation well location in Floridan Aquifer. Based on the surrounding land uses and the agricultural uses of the site, the baseline samples were analyzed for aldicarb, atrazine, bromacil, fenamiphos, metalaxyl, simazine, fluridone, chlorpyrifos, nitrate, phosphates and copper (Shahane, April,1989).

Routine sampling was conducted to monitor the system at different times of the year. The frequency of the sampling was variable as the priorities and work load of the Laboratories changed over the period of time. During the last five years (from 1989 to 1993) the routine sampling was conducted thirteen times. Majority of the field sampling was planned during rainfall conditions that were conducive to the dripping of tile drains. This condition was essential for collecting representative samples from tile drains. The information on rainfall conditions and on the pesticide applications obtained from the citrus grower was used in planning the routine sampling schedule and in selecting the analytes for laboratory analyses. Rainfall and irrigation data were collected from the grower and used to calculate water balances using the GLEAMS model(Knisel and Leonard, 1989).

The removal of nitrates in the eastern and western retention ponds at the site was monitored. This was done by comparing nitrate concentrations at the influent and effluent points.

The procedures for collecting samples from canals, drinking water wells, irrigation well and tile drains are in accordance with the approved Quality Assurance and Quality Control (QA/QC) plan of the Department for sample collection in the field.

3. Sample Analyses:

Depending upon the pesticides applied during various times of the year, the routine samples were analyzed for aldicarb, atrazine, bromacil, fenamiphos, metalaxyl, simazine, fluridone, chlorpyrifos, ethion, diuron and glyphosate during some times of the last five years. Analysis for pesticides was done by the pesticide laboratories of the Florida Department of Agriculture and Consumer Services. Analysis for nitrate concentrations was also done in the laboratory. Field samples were also tested in the field for nitrate using a HACH test kit model No.41100-12 for estimating nitrate removal efficiencies. The non-availability of equipment associated with the laboratory methodology was a factor for excluding some pesticides for analysis.

RESULTS AND DISCUSSIONS

The analytical results for base line and routine sampling are compiled in table 1. Although the results of the table are self-explanatory, some points need to be noted.

Samples in which pesticides were not found at the limit of detection are expressed by zeros in this table, so that the data can be used in the computational software with graphics capabilities.

As mentioned earlier, citrus conversion occurred in phases. As a result, more stations were added as more units were planted for the site. For example, the sampling locations 11, 12, 31 and 14 were added to the sampling program in 1989, 91 and 93. The base line data set of Table 1 did not include these sampling locations.

By considering the characteristics of the site as described in the earlier part of the paper, the annual water balance calculations for an average year of 1988 estimated an annual rainfall of 54 inches with annual runoff of about an inch, 14 inches of percolation, 39 inches of evapotranspiration and about 5 inches of irrigation on the site.

The base line sampling conducted in January 1989,(the beginning of the several years of the subsequent monitoring efforts), detected no pesticide concentrations in the surface water and the ground water of the site. These base line results suggest that no significant pesticide contributions from the surrounding agricultural activities (external sources) have occurred. No corrections are required to the calculation of the subsequent monitoring results because base line water quality indicated a relatively clean environment at the site.

Sampling point 11 represents surface water leaving the site and only one detection of pesticide occurred (5.10 ppb of bromacil). This indicates very low potential for impact on off-site water quality.

Water samples from drinking water wells at locations 8 and 9 and an irrigation well at location 10 were found to contain no detectable levels of pesticides for the last five years with exception of September 8, 1993 when the drinking water well at location 8 and the irrigation well were found to contain 0.32 ppb of ethion. The surface water samples collected at locations 4,5, and 14 were found to contain 0.60, 0.60 and 0.50 ppb of ethion respectively. It is noted that these observed concentrations of ethion are below the health advisory levels as determined recently by the Department of Health and Rehabilitative Services(HRS). Some trace concentrations of nitrate were sporadically found in these drinking water wells and an irrigation well.

The highest concentrations of bromacil, simazine and aldicarb (78.00 ppb, 3.31 ppb and 6.6 ppb respectively) were observed in surface water locations No. 2 and 6. Diuron was detected once in surface water samples at locations 1, 2,3 and 6 with diuron concentrations of 2.55 ppb, 0.86 ppb, 0.67 ppb and 0.42 ppb respectively.

No uniform trend was observed for the removal efficiencies of nitrate for the western and eastern detention ponds because the scatter of efficiency numbers, based on the nitrate data collected in the field, is large and sometimes erratic. Among the ten calculated efficiency numbers, seven numbers were positive indicating 100, 0, 57, 89, 30, 18, 36 percentages of removal efficiencies for nitrate. The remaining three efficiencies were negative indicating perhaps the luxury uptake and release of the nutrients. This point will be further evaluated in future as the data base is expanded. It is hoped that after completion of the four phases of the project in 1994 or 95, the water quality improvements units (detention ponds) can help reduce the water quality impact of agricultural drainage.

The cooperator in this project is always willing to do his part in minimizing the adverse impact of pesticide use on the environment and public health. The grower works with the Institute of Food and Agricultural Sciences (IFAS) research staff to find effective ways to implement best management practices in applying pesticides. The pesticides are applied at a rate consistent with the label. Several years, the pesticides rates were reduced as the pest threats were reduced. The best management practices, prudent use of pesticides, degradation processes and the dilution provided in the water transport system are some of the factors at the site that helped to keep the water quality impact to the minimum level during the first three phases of the project.

CONCLUSION

The routine sampling of the surface water, shallow and deep drinking and irrigation wells at fourteen locations of this site for about five years (January 1989 to September 1993)generated water quality information including pesticides, nutrients and copper. This information indicated that, during the first three phases of the citrus conversion process, bromacil, aldicarb, simazine and ethion are sporadically observed in the surface water. Since the surface water leaving the grove site was found to contain only one pesticide at low concentration, it appears that the citrus conversion process had no significant off-site adverse water quality impact.

ACKNOWLEDGEMENTS

The following individuals, who provided constant support and cooperation to this long term project, are acknowledged.

Mr. Richard J. Budell, Mr. Steven E. Dwinell, Mr. George Fong, Dr. Marion H. Fuller, Mr. Nacho Garza, Mr. Marshall Gentry, Mr. Roberto Guinnerez, Dr. Roger C. Inman, Mr. George

Owen, Dr. William E. Pace, Mr. Mike Page, Mr. Brady Pfeil, Mrs. Patricia Pfeil, Dr. Martha Roberts, and Mr. Steven J. Rutz.

Many others who directly or indirectly assisted in completing this phase of the project are also gratefully acknowledged.

REFERENCES

1. Campbell, K. M. 1985. Geology of Desoto County, Florida" OFR-11, Florida Geological Survey, Tallahassee, Florida.

2. Ground Water Resource Availability Inventory: Desoto County, Florida. April 1988. Prepared by Resource Management and planning Departments of the Southwest Florida Water Management District, Brooksville, Florida.

3. Interim Soil Survey Maps and Interpretations, Desoto County, Florida. 1988. Unpublished report prepared by the Soil Conservation Service.

4. Knisel W.G. and Leonard R.A. 1989. GLEAMS User's Manual, Southeast Watershed Research Laboratory, Tifton, Georgia.

5. Shahane A.N. April,1989. Protocol for a study of the impact of the citrus conversation process on ground and surface waters in DeSoto county. a report of the Bureau of Pesticides of the Florida Department of Agriculture and Consumer Services, Tallahassee, Florida.

6. Shahane A.N. January,1989. A proposed plan of sampling program for a site in DeSoto county. a memorandum report of the Bureau of Pesticides of the Florida Department of agriculture and Consumer Services, Tallahassee, Florida.

7. Surface Water Management Design for the Site. 1987. prepared by Agricultural Management Services Company.

8. Wilson, W. E. 1977. Ground Water Resources of Desoto and Hardee Counties, Florida. A report prepared by United States Geological Survey in cooperation with Southwest Florida Water Management District and Bureau of Geology, Tallahassee, Florida.

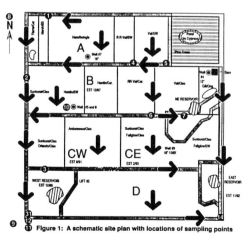

Figure 1: A schematic site plan with locations of sampling points

TABLE 1: Summary of observed pesticide concentrations at a site in DeSoto county.

Location No. 1

Time	Bromacil (ppb)	Nitrates (ppm)	Copper (ppm)	PO$_4$ (ppm)	Simazine (ppb)	Aldicarb (ppb)
1/20/89	0.00	0.00	0.000	0.040	0.00	0.00
7/5/89	0.00	1.00	0.022	0.120	0.00	0.00
9/19/89	0.00	0.00	--	--	0.00	0.00
12/28/89	2.80	0.00	0.051	0.100	0.00	0.00
2/22/90	0.00	0.00	0.000	0.120	2.80	0.00
9/27/90	0.00	0.00	0.031	0.090	0.00	0.00
1/24/91	3.66	0.00	--	--	0.00	0.00
4/25/91	3.72	0.00	0.008	0.000	0.00	0.00
8/8/91	5.62	0.15	0.008	0.000	0.00	--
12/19/91	0.00	0.00	0.004	0.000	0.00	--
4/2/92	0.00	0.00	0.009	0.000	0.00	--
9/18/92	4.30	0.0002	0.007	0.000	0.00	0.00
9/8/93	--	0.01	0.003	0.000	0.00	0.00

Location No. 2

Time	Bromacil (ppb)	Nitrates (ppm)	Copper (ppm)	PO$_4$ (ppm)	Simazine (ppb)	Aldicarb (ppb)
1/20/89	0.00	0.00	0.000	0.050	0.00	0.00
7/5/89	2.30	8.40	0.015	0.130	0.00	41.30
9/19/89	13.40	1.00	--	--	0.00	3.90
12/28/89	11.20	3.80	0.037	0.170	0.00	0.00
2/22/90	4.50	0.50	0.033	0.150	0.00	0.00
9/27/90	0.00	0.50	0.014	0.045	0.00	0.00
1/24/91	8.14	3.40	--	--	0.00	0.00
4/25/91	9.39	0.30	0.007	0.000	3.31	0.00
8/08/91	14.62	0.16	0.005	0.000	0.00	--
12/19/91	0.00	0.00	0.010	0.000	0.00	--
4/02/92	16.80	3.00	0.014	0.000	0.00	5.80
9/18/92	10.10	1.30	0.007	0.20	--	6.60
9/8/93	--	4.23	0.004	0.000	0.000	0.000

Location No. 3

Time	Bromacil (ppb)	Nitrates (ppm)	Copper (ppm)	PO$_4$ (ppm)	Simazine (ppb)	Aldicarb (ppb)
1/20/89	0.00	0.80	0.000	0.050	0.00	0.00
7/5/89	16.70	0.00	0.020	0.120	0.00	9.10
9/19/89	10.20	0.50	--	--	0.00	0.00
12/28/89	11.90	3.80	0.027	0.170	0.00	0.00
2/22/90	3.30	0.00	0.045	0.140	0.00	0.00
9/27/90	2.26	0.00	0.002	0.055	0.00	0.00
1/24/91	8.99	3.90	--	--	0.00	0.00
4/25/91	10.16	0.30	0.009	0.000	2.03	0.00
8/8/91	3.39	0.16	0.004	0.000	0.00	--
12/19/91	0.00	0.70	0.011	0.000	0.00	--
4/2/92	11.20	4.70	0.016	0.000	0.00	4.00
9/18/92	7.00	0.90	0.007	0.000	--	3.70
9/8/93	--	3.40	0.027	0.000	0.00	0.00

TABLE 1 (Contd)

Location No. 4

Time	Bromacil (ppb)	Nitrates (ppm)	Copper (ppm)	PO$_4$ (ppm)	Simazine (ppb)	Aldicarb (ppb)
1/20/89	0.00	0.00	0.000	0.30	0.00	0.00
7/5/89	32.50	13.00	0.032	0.75	0.00	0.00
9/19/89	10.40	0.00	--	--	0.00	0.00
12/28/89	15.80	0.00	0.032	0.10	0.00	0.00
2/22/90	0.00	0.00	0.006	0.15	0.00	0.00
9/27/90	21.57	0.00	0.040	0.04	0.00	0.00
1/24/91	12.79	0.00	--	--	0.00	0.00
4/25/91	16.98	0.00	0.012	0.00	0.00	0.00
8/8/91	32.92	0.16	0.005	0.00	0.00	--
12/19/91	9.20	0.00	0.000	0.00	0.00	--
4/2/92	0.00	1.40	0.013	0.00	0.00	0.00
9/18/92	0.00	0.20	0.005	0.20	0.00	0.00
9/8/93	--	2.20	0.000	0.07	0.00	0.00

Location No. 5

Time	Bromacil (ppb)	Nitrates (ppm)	Copper (ppm)	PO$_4$ (ppm)	Simazine (ppb)	Aldicarb (ppb)
1/20/89	0.00	0.00	0.000	0.050	0.00	0.00
7/5/89	24.00	18.40	0.014	0.350	0.00	0.00
9/19/89	21.40	0.00	--	--	0.00	0.00
12/28/89	14.30	2.30	0.023	0.165	0.00	0.00
2/22/90	8.50	0.20	0.010	0.160	3.3	0.00
9/27/90	16.55	0.00	0.011	0.055	0.00	0.00
1/24/91	15.99	0.00	--	--	0.00	0.00
4/25/91	22.74	0.00	0.007	0.000	0.00	0.00
8/8/91	46.15	0.15	0.002	0.000	0.00	--
12/19/91	13.40	0.00	0.000	0.000	0.00	--
4/2/92	0.00	9.50	0.007	0.000	0.00	0.00
9/18/92	19.90	2.40	0.004	0.000	0.00	0.00
9/8/93	—	4.50	0.011	0.030	0.00	0.00

Location No. 6

Time	Bromacil (ppb)	Nitrates (ppm)	Copper (ppm)	PO$_4$ (ppm)	Simazine (ppb)	Aldicarb (ppb)
1/20/89	0.00	0.00	0.000	0.450	0.00	0.00
7/5/89	78.00	13.60	0.014	0.280	0.00	0.00
9/19/89	55.20	3.00	--	--	0.00	0.00
12/28/89	21.90	4.30	0.020	0.300	0.00	0.00
2/22/90	26.30	0.00	0.003	0.230	0.00	0.00
9/27/90	50.17	11.00	0.005	0.095	0.00	0.00
1/24/91	37.56	10.60	--	--	0.00	0.00
4/25/91	35.59	6.30	0.008	0.000	0.00	0.00
8/8/91	49.06	3.51	0.003	1.500	0.00	--
12/19/91	0.00	0.00	0.014	0.000	0.00	--
4/2/92	4.70	8.10	0.007	0.000	0.00	0.00
9/18/92	18.00	4.00	0.012	0.500	0.00	0.00
9/8/93	—	9.60	0.007	0.080	0.00	0.00

TABLE 1(contd)

Location No. 7

Time	Bromacil (ppb)	Nitrates (ppm)	Copper (ppm)	PO$_4$ (ppm)	Simazine (ppb)	Aldicarb (ppb)
1/20/89	0.00	0.00	0.000	0.050	0.00	0.00
7/5/89	26.40	15.60	0.023	0.180	0.00	0.00
9/19/89	28.40	2.60	--	--	0.00	0.00
12/28/89	16.10	7.20	0.010	0.225	0.00	0.00
2/22/90	11.00	0.00	0.000	0.150	1.00	0.00
9/27/90	13.08	10.00	0.032	0.055	0.00	0.00
1/24/91	17.35	10.80	--	--	0.00	0.00
4/25/91	17.32	6.60	0.013	0.000	0.00	0.00
8/8/91	22.50	4.10	0.005	2.100	0.00	--
12/19/91	0.00	0.00	0.002	0.000	0.00	--
4/2/92	21.20	8.30	0.009	0.000	0.00	0.00
9/18/92	15.60	3.60	0.006	0.800	--	0.00
9/8/93	--	2.40	0.004	0.000	0.00	0.00

Location No. 11

Time	Bromacil (ppb)	Nitrates (ppm)	CU (ppm)	PO$_4$ (ppm)	Simazine (ppb)	Aldicarb (ppb)
1/20/89	--	--	--	--	--	--
7/5/89	--	--	--	--	--	--
9/19/89	--	--	--	--	--	--
12/28/89	5.10	1.40	0.009	0.155	0.00	0.00
2/22/90	0.00	0.00	0.005	0.250	0.00	0.00
9/27/90	0.00	0.00	0.000	0.090	0.00	0.00
1/24/91	0.00	0.00	--	--	0.00	0.00
4/25/91	0.00	0.00	0.000	0.000	0.00	0.00
8/8/91	0.00	0.00	0.003	0.000	0.00	--
12/19/91	0.00	0.00	0.005	0.000	0.00	--
4/2/92	0.00	0.00	0.005	0.000	0.00	0.00
9/18/92	0.00	0.20	0.012	0.300	0.00	0.00
9/8/93	--	0.02	0.006	0.020	0.00	0.00

Location No. 12

Time	Bromacil (ppb)	Nitrates (ppm)	Copper (ppm)	PO$_4$ (ppm)	Simazine (ppb)	Aldicarb (ppb)
1/20/89	--	--	--	--	--	--
7/5/89	--	--	--	--	--	--
9/19/89	--	--	--	--	--	--
12/28/89	8.60	2.40	0.018	0.170	0.00	0.00
2/22/90	0.00	0.30	0.000	0.190	0.00	0.00
9/27/90	0.00	0.00	0.004	1.950	0.00	0.00
1/24/91	2.39	0.00	--	--	0.00	0.00
4/25/91	2.03	0.40	0.004	0.000	0.00	0.00
8/8/91	0.00	0.15	0.002	0.000	0.00	--
12/19/91	0.00	0.00	0.000	0.080	0.00	--
4/2/92	0.00	0.10	0.005	0.000	0.00	0.00
9/18/92	2.90	0.20	0.014	0.200	0.00	0.00
9/8/92	--	0.02	0.003	0.030	0.00	0.00

Location No. 31

Time	Bromacil (ppb)	Nitrates (ppm)	Copper (ppm)	PO$_4$ (ppm)	Simazine (ppb)	Aldicarb (ppb)
1/24/91	3.69	0.00	--	--	0.00	0.00
4/25/91	5.25	0.50	0.005	0.000	4.40	0.00
8/8/91	3.53	0.16	0.003	0.000	0.00	--
2/19/91	0.00	0.30	0.002	0.050	0.00	--
4/2/92	10.00	0.50	0.000	0.000	0.00	0.00
9/18/92	4.30	0.40	0.004	0.200	0.00	0.00
9/8/93	--	1.02	0.002	0.000	0.00	0.00

SPRAY DEPOSITION AND DRIFT IN CITRUS APPLICATIONS

Masoud Salyani*

ABSTRACT

Pesticide application is a very inefficient process and only a small portion of the applied agrochemicals may reach the target and contribute toward biological effect. The rest of the materials may be lost by misapplication, volatilization, runoff from leaf surface, and drift. The magnitudes of these losses depend on the design and operational parameters of spray equipment, physical properties of pesticides, characteristics of the plant and pest targets, and weather conditions. Air-carrier ground sprayers are the primary type of spray equipment used in Florida citrus and treat about 95% of the acreage. The remainder is sprayed partially or fully with aerial applicators. The objective of the paper is to characterize spray deposition and drift in citrus applications and identify practices that may maximize on-target spray deposition and minimize environmental contamination.

In general, ground sprayers provide variable amounts of deposits within the tree canopy. The highest and least variable deposits were found at canopy locations near the sprayer. Low-volume (674 L/ha) applications resulted in less runoff from leaf surface and more deposition than high-volume (5083 L/ha) dilute applications. Both ground and aerial applications generated measurable drift deposits as far as 195 m from the edge of the grove. More than 70% of the drift originated from sprays applied to the last 2 rows of the trees downwind of the applications. Minimum fallout and airborne drift deposits were generated from high-volume applications.

It appears that spray applications to the last 1 or 2 downwind rows have the greatest contribution to drift deposits. Therefore, treating the last few downwind rows of a grove with less drift-prone equipment under favorable weather conditions may be a feasible practice to minimize drift.

KEYWORDS. Spray, Deposition, Drift, Citrus, Ground/aerial applications.

INTRODUCTION

Pesticide application technology reports indicate that a large portion of the applied pesticides does not deposit on the intended target and may be lost by misapplication, volatilization, runoff from leaf surface and drift (von Rümker and Kelso, 1975; Herrington et al., 1981). These losses not only result in substantial increase in the cost of pest control but also pollute air, soil, and water resources. The magnitudes of the on-target deposition and off-target losses depend on the design and operational parameters of the spray equipment, physical properties of pesticides, characteristics of the plant and pest targets, and weather conditions during pesticide applications (Steiner, 1977; Göhlich et al., 1979; Morgan, 1983). In general, trees have larger and more complex canopy structure than field crops and create a serious challenge for pesticide application. Especially citrus trees, with evergreen, large, and densely foliated canopies (that often touch the ground), represent a very difficult target for spray deposition.

---

*Associate Professor of Agricultural Engineering, University of Florida, IFAS, Citrus Research and Education Center, Lake Alfred, FL 33850, USA.

Air-carrier sprayers are the primary type of spray equipment used in Florida citrus and treat about 95% of the acreage (Whitney et al., 1978). The remainder is sprayed fully or partially with aerial equipment. Ground sprayers are normally equipped with axial-, centrifugal-, or cross-flow fans to generate air volume to propel the spray droplets toward the target. The volume, velocity, energy, and turbulence of the air can affect the droplet transport and deposition (Randall, 1971; Hale, 1978). However, the air jet and droplet trajectories can be influenced by prevalent wind direction. Both horizontal and vertical components of wind velocity are important in spray cloud movement (Threadgill and Smith, 1975), but drift deposits increase under stable (inversion) conditions (Yates et al., 1966).

Droplet size spectrum is an important factor in spray application and can affect spray deposition (Salyani, 1988) and drift (Akesson and Yates, 1988). The spectrum may be varied by changes in the pressure of the hydraulic nozzles, orientation of air-shear nozzles in air stream, or rotational speed of rotary atomizers (Akesson and Yates, 1988). It is also a function of the spray mixture surface tension, viscosity, and density (Bouse et al., 1990). Spray additives can alter the mixture properties and improve spray deposition (Ware et al., 1970) or reduce drift (Salyani and Cromwell, 1993). Electrostatic charging of the droplets may improve spray deposition under certain conditions (Law and Lane, 1981), but it was not found effective in citrus application (Whitney et al., 1986). Air temperature and relative humidity have direct effects on evaporation; thus, the size of water-based droplets can decrease significantly under hot and dry conditions (Brann, 1964).

This paper describes 2 field experiments that were designed to characterize on-target spray deposition and drift from citrus applications. The objective was to identify practices that may maximize on-target spray deposition and minimize environmental contamination.

MATERIALS AND METHODS

On-target Deposition

An engine-driven airblast sprayer (FMC Model 1087), equipped with an axial-flow fan, air tower, and ceramic disc-core nozzles, was used to apply spray mixtures at volume rates of 470-9400 L/ha. The mixtures, containing cupric hydroxide (50% metallic copper) as deposition tracer, were sprayed on plots of Valencia orange trees, at a ground speed of 2.4 km/h, in 4 replications. The tracer concentration in the mixtures varied so that a constant rate of 6.7 kg/ha was applied at all volume rates. Leaf samples were collected from one quadrant of 2 center trees of each plot at 3 heights (1.2, 2.4, 3.6 m), 3 azimuths (90, 45, and 0° to the trunk line, TL), and center of the trees. The samples were washed with 0.05N nitric acid solution and the amount of copper deposit was determined by colorimetry (Salyani and Whitney, 1988). Copper deposit was calculated in $\mu g/cm^2$ of the sample area and variability of deposition among samples was expressed as the coefficient of variation (CV). The data were analyzed as a split-plot design, with spray volume as main plot and sample locations as subplot effects. Schematic view of the sprayer, nozzle arrangement, and detail of the sampling technique are given in Salyani and McCoy (1990).

Spray Drift

Two aerial and 2 ground sprayers were used to spray the 4 downwind tree rows of an orange grove in 3 replications. The spray equipment consisted of a fixed-wing aircraft

(FW), a helicopter (HE), and a PTO-powered airblast ground sprayer which was used as a high-volume (GH) and low-volume (GL) sprayer. Spray solutions, containing a water-soluble fluorescent tracer dye (BASO Red NB 546), were applied with the 4 equipment at 125, 159, 5083, and 674 L/h, respectively. The aerial sprayers had a swath width of 15.2 m and sprayed 2 rows in one pass. The FW and HE made 4 and 2 passes over each pair of rows to complete a replication. The ground sprayers (GH and GL) started with one-sided spraying the outer side of the first row (1W) then sprayed middles of rows 1 and 2 (1E+2W), 2 and 3 (2E+3W), 3 and 4 (3E+4W), and finally one-sided spraying of other side of row 4 (4E) for each replication. The total amount of the discharged dye was nominally constant (1000 g/ha) for all applications.

Spray drift was sampled on 3 lines (perpendicular to the tree row lines) at 5 distances downwind of the applications. Mylar plastic targets and high-volume air samplers (fitted with cellulose filter paper) were used to measure fallout and airborne drift deposits, respectively. The samples were analyzed by fluorometry. Weather parameters including temperature, relative humidity, wind velocity, wind direct, and solar radiation were recorded during the applications. The data were corrected for differences in wind direction and solar radiation. More detailed information about the spray equipment, application procedures, sampling techniques and fluorometry, data analysis, and weather information are given in Salyani and Cromwell (1992).

RESULTS AND DISCUSSION

Statistical significance of the effects, where indicated, refers to F values at 5% level.

<u>On-target Deposition</u>

Spray volume and sample location had significant effects on deposition. Overall, the amount of deposit increased as spray volume decreased and mean deposit at the lowest volume was about 2.3 times that of the highest volume (fig. 1). For all volumes, the higher levels had more deposition than the lower levels (due to the nozzle arrangement) and the interaction of the height with volume was significant (fig. 1). At the higher levels, there were substantial reductions in deposition as spray volume increased. But, the trend was not as steep at the 1.2 m level.

Radial location of the samples showed a significant effect on deposition and the interaction with height was also significant (fig. 2). Maximum deposits were observed at the 90° azimuth (TL 90°) which was nearest location to the sprayer. The center of the tree received the least amount of deposition. At all levels, spray droplet clouds were filtered by the canopy foliage and very little amounts could reach the tree center. The interaction of radial location and spray volume was significant (fig. 3). While mean deposit at all 3 azimuths decreased with increasing spray volume, no such trend was observed at the tree center.

The increase in deposition at lower volumes could partially be attributed to decrease in the amount of runoff from leaf surface. Comparatively large spray droplets, produced by large nozzles of high-volume applications, can either shatter or coalesce and run off the leaf surface. They mostly fall on the ground and result in lower deposition. On the other hand, smaller and more concentrated droplets of low-volume applications do not coalesce readily and distribute more uniformly on the targets. In addition, they are entrained by the sprayer air and transported to the hard-to-reach canopy locations and increase the deposition (figs. 1 and 3).

In general, lower volumes produced more variable deposition (higher CVs); however, there was not a consistent trend in all sample locations. Comparatively high variability of the 9400 L/ha deposits is not in agreement with previous results (Salyani et al., 1988; Salyani and McCoy, 1989). The discrepancy could be attributed to the difference in sample locations and resulting runoff effects.

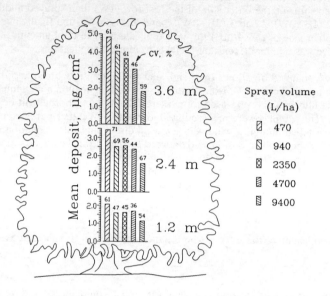

Figure 1. Spray Deposition at the 3 Heights for Different Spray Volumes.

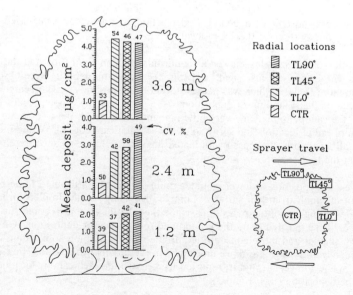

Figure 2. Mean Deposition at the 3 Azimuths and Center of the Tree.

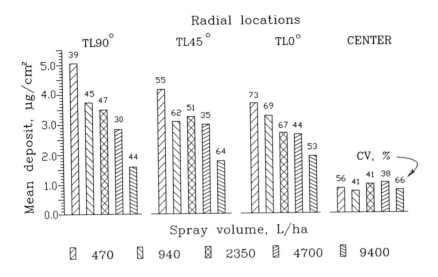

Figure 3. Deposition at the 3 Azimuths and Tree Center for Different Spray Volumes.

Spray Drift

Apart from meteorological effects (Yates et al., 1966; Salyani and Cromwell, 1992), both spray equipment and target distance had significant effects on drift deposits. The amount of drift deposits decreased as downwind target distance increased (fig. 4). Both fallout (mylar) and airborne (filter) deposits at the first 3 downwind locations were significantly different; but, deposits at the 2 farthest locations were not different. Beyond the 15.2 m sample station, there was no significant difference in the mean fallout deposits of the 4 equipment (fig. 4a); however, the effect of equipment on airborne deposits were significant (fig. 4b). The low- and high-volume ground applications (GL and GH) produced the highest and lowest airborne deposits, respectively, in all sample locations. The mean airborne deposits of the aerial sprayers were significantly less/more than that of the GL/GH, respectively. Mean fallout and airborne deposits of the 4 equipment are shown in fig. 5. Fixed-wing sprayer produced significantly higher mean fallout deposits, but there was no difference among helicopter, low-, and high-volume ground sprayers.

A great portion of the airborne drift deposits originated from spraying the tree rows closest to the outside of the grove, downwind of the applications. Airborne drift deposits, resulting from aerial spraying of rows 1+2 and 3+4 are shown in fig. 6. At the 15.2 m sample location, 84.5% of the FW deposits and 79.4% of the HE deposits originated from rows 1+2 and the balance from rows 3+4. However, smaller droplets could move to farther distances and at the 48.8 m sample station, drift contributions of rows 1+2 and 3+4 became 60.1 and 58.4%, for the FW and HE, respectively.

Drift contributions from spraying each tree row with the ground equipment is shown in fig. 7. At the 15.2 m sample location, 72.3% of the GH deposits (fig. 7a) and 79.8% of the GL deposits (fig. 7b) originated from rows 1+2. However, at the 48.8 m sample station, contributions of rows 1+2 became 63.2 and 70.0%, respectively. Drift deposits from spraying middles of the 4 rows (i.e. 1E+2W, 2E+3W, 3E+4W) show that, as the

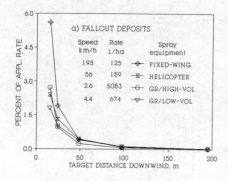
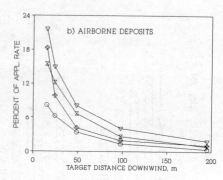

Figure 4. Fallout and Airborne Drift Deposits from Ground and Aerial Applications.

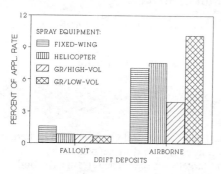

Figure 5. Mean Fallout and Airborne Deposits from 4 Equipment.

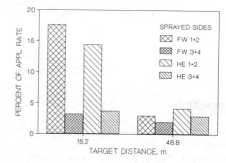

Figure 6. Airborne Drift Deposits from Spraying Pairs of Tree Rows with the Fixed-wing (FW) and Helicopter (HE) Sprayers.

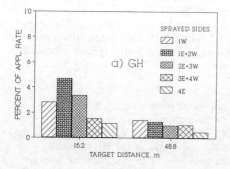
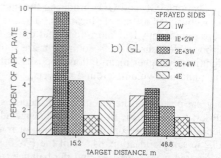

Figure 7. Airborne Drift Deposits from Spraying Each Tree Row with the High-volume (GH) and Low-volume (GL) Ground Sprayers.

sprayer moved deeper inside the grove, most of the displaced swath was captured by the downwind rows and the amount of drift out of the grove substantially decreased.

The data suggest that the last one or 2 downwind tree rows have greatest contribution to swath displacement and drift deposits. The high-volume ground sprayer produced minimal drift and appeared to be the least drift-prone among the 4 spray equipment. Since wind direction is normally variable and drift deposits can be scattered in many directions (Salyani and Cromwell, 1993), spraying the outer 2 rows of a grove with less drift-prone equipment under favorable weather conditions may be a feasible practice to minimize drift.

CONCLUSIONS

1. Spray volume has a significant effect on deposition, but the effect may be different at different canopy locations.

2. Mean deposition increases as volume decreases, but, within the canopy, low-volume depositions appear to be more variable.

3. Spray volume of ground sprayer will have a significant effect on drift. Low-volume applications appear to be more susceptible to drift than high-volume applications.

4. There is not substantial difference between aerial and ground sprayers in producing drift and both can generate comparable airborne and fallout drift deposits under drift-prone conditions.

5. Spray applications to the outer 2 tree rows have the greatest contribution to the overall drift.

Acknowledgements

Florida Agricultural Experiment Station Journal Ser. # N-00879. The author thanks Mr. Roy Sweeb for his technical assistance in the preparation of the paper.

REFERENCES

1. Akesson, N.B. and W.E. Yates. 1988. Spray atomization parameters for optimizing pest control efficacy. Proc. Int. Conf. Liq. Atom. Spray Sys. 4:471-481.

2. Bouse, L.F., I.W. Kirk and L.E. Bode. 1990. Effect of spray mixture on droplet size. Trans. ASAE 33(3):783-788.

3. Brann, J.L., Jr. 1964. Factors affecting the thoroughness of spray application. Proc. N.Y. State. Hort. Soc., pp 186-195.

4. Göhlich, H., M. Hosseinipour and R.V. Oheimb. 1979. Effect of climatic and application factors on drift. Nachrichtenbl. Deut. Pflanzenschutzd 31(1):1-9.

5. Hale, O.D. 1978. Performance of air jets in relation to orchard sprayers. J. Agric. Engng. Res. 23:1-16.

6. Herrington, P.J., H.R. Mapother and A. Stringer. 1981. Spray retention and distribution on apple trees. Pestic. Sci. 12:515-520.

7. Law, S.E. and M.D. Lane. 1981. Electrostatic deposition of pesticide spray onto foliar targets of varying morphology. Trans. ASAE 24(6):1441-1445, 1448.

8. Morgan, N.G. 1983. Tree crop spraying - Worldwide Proc. Conf. on Tree Crop Spraying with Ground Equipment, Lake Alfred, Florida, pp. 1-44.

9. Randall, J.M. 1971. The relationships between air volume and pressure on spray distribution in fruit trees. J. Agric. Engng. Res. 16(1):1-31.

10. Salyani, M. 1988. Droplet size effect on spray deposition efficiency of citrus leaves. Trans. ASAE 31(6):1680-1684.

11. Salyani, M. and R.P. Cromwell. 1992. Spray drift from ground and aerial applications. Trans. ASAE 35(4):1113-1120.

12. Salyani, M. and R.P. Cromwell. 1993. Adjuvants to reduce drift from handgun spray applications. ASTM STP 1146: Pest. Form. and Appl. Sys. 12:363-376.

13. Salyani, M. and C.W. McCoy. 1989. Deposition of different spray volumes on citrus trees. Proc. Fla. State Hort. Soc. 102:32-36.

14. Salyani, M. and C.W. McCoy. 1990. Spray deposition in citrus trees for different spray volumes. Int. Conf. on Agric. Engng. (AgEng 90), Berlin, Germany, 12 p.

15. Salyani, M., C.W. McCoy and S.L. Hedden. 1988. Spray volume effects on deposition and citrus rust mite control. ASTM STP 980:Pest. Form. and Appl. Sys. 8:254-263.

16. Salyani, M. and J.D. Whitney. 1988. Evaluation of methodologies for flied studies of spray deposition. Trans. ASAE 31(2):390-395.

17. Steiner, P.W. 1977. Factors affecting the efficient use of orchard airblast sprayers. Trans. Ill. State. Hort. Soc. 110:57-64.

18. Threadgill, E.D. and D.B. Smith. 1975. Effects of physical and meteorological parameters on the drift of controlled-size droplets. Trans. ASAE 18(1):51-56.

19. von Rümker, R. and G.L. Kelso. 1975. A study of the efficiency of the use of pesticides in agriculture. U.S. EPA Rep. No. EPA-540/9-75-025, pp. 1-115.

20. Ware, G.W., W.P. Cahill, P.D. Gerhardt and J. Witt. 1970. Pesticide drift IV. On-target deposits form aerial application of insecticides. J. Econ. Ent. 63(6):1982-1983.

21. Whitney, J.D., R.F. Brooks and R.C. Bullock. 1978. Pesticide application methods for citrus in Florida. Proc. Int. Soc. Citriculture pp. 163-167.

22. Whitney, J.D., D.B. Churchill, S.L. Hedden and R.P. Cromwell. 1986. Performance characteristics of pto airblast sprayers for citrus. Proc. Fla. State. Hort. Soc. 99:59-65.

23. Yates, W.E., N.B. Akesson and H.H. Coutts. 1966. Evaluation of drift residues from aerial applications. Trans. ASAE 9(3):389-393, 397.

PERENNIAL PEANUT IN CITRUS GROVES - AN ENVIRONMENTALLY SUSTAINABLE AGRICULTURAL SYSTEM

J. J. Mullahey, R. E. Rouse, and E.C. French[*]

ABSTRACT

Perennial peanut (*Arachis glabrata* Benth.) is a tropical legume that has environmental and ecological benefits that could benefit Florida's citrus industry. A two-year study using perennial peanut was initiated in a southwest Florida citrus grove to evaluate the effect of herbicide (fusilade), fertilizer (K-Mag, K-Mag + 34-0-0), and herbicide + fertilizer treatments on hastening establishment of a perennial peanut as a ground cover. Fusilade treatment had the highest ground cover (93%) followed by K-Mag alone (87%), fusilade + K-Mag (81%), check (76%), and fusilade + K-Mag + 34-0-0 (72%). Suppression (mowing, herbicide) of grassy weeds greatly enhanced perennial peanut ground cover and annual fertilization helped maintain plant vigor.

There are many potential environmental and ecological benefits from using perennial peanut in citrus production. The potential environmental and ecological benefits from perennial peanut in citrus are discussed. Research is needed to further document these benefits and determine cultural and management practices for perennial peanut in citrus groves.

Keywords: *Arachis glabrata*, citrus production, citrus management, sustainable agriculture.

OVERVIEW OF PERENNIAL PEANUT

Citrus producers in the flatwoods of south Florida utilize grass covers in citrus-row middles to stabilize the soil and to provide access for vehicles and farm labor. Maintenance of grass row-middles require frequent mechanical mowing (3-6 times/year), or possibly chemical mowing, to suppress the vegetation. Use of grasses in citrus row middles is not an environmentally sustainable agricultural system.

Perennial peanut (*Arachis glabrata* Benth.) offers several environmental and ecological advantages over the standard bahiagrass and bermudagrass presently used in citrus groves. The rhizoma (perennial) peanut is a tropical perennial legume introduced to Florida from Brazil in 1936 (French and Prine, 1991). This rhizomatous legume has primarily been utilized as a forage crop (Saldivar et al., 1992) though it can be used as an ornamental, conservation cover, and living mulch. 'Florigraze' rhizoma peanut, released in 1979, is commercially grown on 8,000 acres of pasture land in Florida (French et al., 1993). Perennial peanut requires low levels of fertilizers (potassium, sulfur, magnesium) generally applied to agricultural crops, and does not require any applied nitrogen (thus no nitrogen leaching) or pesticides for control of diseases or insects (French and Prine, 1991). Nitrogen produced by this leguminous plant

---

[*]Asst. Prof., Wildlife and Range Sci. Dept., and Assoc. Prof., Horticulture Dept., Univ. Florida, IFAS, Southwest Florida Res.& Educ. Cent., P.O. Drawer 5127, Immokalee, FL 33934; Prof., Agronomy Dept., Univ. Florida, IFAS, Gainesville, FL 32611

could reduce fertilizer needs for the citrus trees, resulting in energy savings and less pollution from fuels that would be used in the manufacture of fertilizer. Perennial peanut is quite drought tolerant and can be produced without irrigation (French et al., 1993).

An annual energy savings of 52,000 gallons of diesel fuel is realized for every 1,000 acres of Coastal bermuda (*Cynodon dactylon*) replaced with perennial peanut (French and Prine, 1991, 1992). With no requirement for pesticides or applied nitrogen, perennial peanut is an environmentally sound, ecologically important component for sustainable agricultural systems in citrus. Establishment methods and cultural practices for perennial peanut in south Florida citrus groves are being developed. Further evaluation of the ecological and environmental benefits from perennial peanuts is needed.

MATERIALS AND METHODS

To develop guidelines for the establishment of perennial peanut in citrus row middles, a two-year study (1992-93) was conducted to evaluate the effect of herbicide and fertilizer inputs on ground cover spread in a citrus grove. Conceptual discussion was initiated on potential environmental and ecological benefits from perennial peanut in citrus.

Treatments in the establishment study were: 1) fusilade; 2) K-Mag; 3) fusilade + K-Mag; 4) fusilade + K-Mag with 34-0-0; and 5) check (no herbicide or fertilizer). The fertilizer treatment in the first year was 448 kg ha^{-1} of 0-10-20 with 34 kg ha^{-1} of minor elements applied in July. In year two, a split application was applied with 224 kg ha^{-1} of K-Mag (22% potassium, 11% magnesium, 22% sulfur) in February and 23 kg ha^{-1} of K-Mag in July for added sulfur. Treatment 4 included 34 kg N ha^{-1} applied in July of 1992 and February of 1993. Current University of Florida, IFAS fertilizer guidelines recommend K-Mag for perennial peanut. Fusilade treatments were applied (6.9 l ha^{-1} application^{-1}) twice in 1992 and 1993 (July, August) to control grassy weeds such as common bermudagrass.

In the commercial grove where this test was done, perennial peanut was planted in 1991 at 1409.6 ℓ (10^{-3}m^3) which is one-half the recommended planting rate. This study was initiated in the second growing season and continued through the third growing season. Density of perennial peanut (i.e., percentage ground cover) was measured using a 1.0 m^2 frame prior to beginning the study (1992) and at 6 month intervals through 1993. The 1.0 m^2 frame was divided into 100 squares (10 cm x 10 cm) and squares from which perennial peanut shoots originated were recorded as filled. Percentage ground cover was based on the number of squares containing peanut shoots out of 100 total. For this paper, only the initial and final percentage ground cover measurements are discussed.

RESULTS AND DISCUSSION

Establishment Study

Measurements in 1992, prior to applying herbicide or fertilizer treatments, showed that all plots had similar peanut ground cover (mean ground cover of 42%) (Figure 1). After two years the fusilade treatment had the highest ground cover density (93%) and the fusilade + K-Mag + 34-0-0 the lowest density (72%). Peanut ground cover in plots receiving K-Mag alone (87%), fusilade + K-Mag (81%), and the untreated check (76%) were similar but not significantly lower than the fusilade treatment (93%). Applying N appeared to stimulate growth of common bermudagrass (the competition) thus reducing the perennial peanut ground cover.

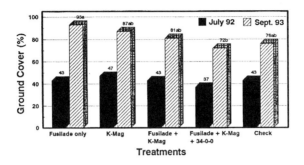

Figure 1. Effect of Herbicide And Fertilizer on Establishment of Perennial Peanut in Citrus. No significant difference among treatments in 1992. Bars having the same letter are not significantly different (p>0.05) as determined by Fisher's F-protected L.S.D.

The use of fusilade to remove competition from bermudagrass enhanced the spread of perennial peanut. In addition, perennial peanut responded favorably to fertilizer (K-Mag). It is unexplained why the response to fusilade + fertilizer in this study was not higher (81% ground cover). In order for perennial peanut to spread during establishment, the plant must be healthy (fertilized) and weed competition should be eliminated or reduced. Mowing to suppress weeds, rather than using herbicides, may be an option. Growers should adjust the mower height to cut the weeds, not the peanut. It is suggested that three to five mowings per season during establishment may be needed to produce a solid stand of peanuts after 2 or 3 years.

Environmental Benefits

There are several potential environmental benefits from incorporating perennial peanuts into a citrus operation (Figure 2). One major environmental benefit of perennial peanut is that it does not require the application of nitrogen in the fertilizer. This reduces inputs to the environment thus eliminating concerns for nitrogen leaching into surface and ground water. There are potential environmental benefits in energy savings (no applied nitrogen) and less pollution from fuels that would be used in the manufacture of fertilizer. The extensive rhizome and root system of perennial peanut could allow this crop to serve as an inceptor of pesticides allowing for their degradation before entering the soil and leaching. This could also be a factor with respect to filtering nutrients (e.g., nitrogen, phosphorus) thus improving ground water quality.

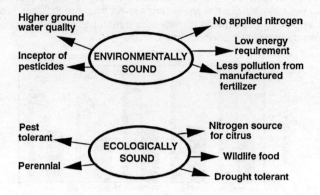

Figure 2. Potential Environmental And Ecological Benefits of Perennial Peanut For a Sustainable Agricultural System.

Ecological Benefits

Perennial peanut is a legume which obtains nitrogen from atmospheric source. Estimated nitrogen production from perennial peanut is approximately 235 kg ha^{-1}, or 78 kg ha^{-1} when planted with citrus where one-third of land surface is planted with peanut. Assuming all of the nitrogen was available, a grower could lower the amount of nitrogen fertilizer applied to citrus by 78 kg ha^{-1} due to the release of nitrogen from perennial peanut. Periodic mowing would be necessary to benefit from the nitrogen in the perennial peanut. Perennial peanut has no known economically damaging disease, insect or nematode pest in the U.S.A. (French et al, 1993). It is the only long-lived (perennial), high quality forage available to Florida producers today. Once established, it survives Florida's drought periods and prospers under high rainfall and temperatures on well drained soils. Perennial peanut persists with minimal inputs which makes it an excellent candidate for sustainable agricultural systems. Wildlife, such as deer, have been observed feeding on perennial peanut in young citrus groves resulting in minimal damage to the citrus trees from the deer. The high forage quality should improve the diet quality of many wildlife species.

Research on the environmental and ecological benefits of perennial peanut are needed to substantiate some of the theories and observations presented.

REFERENCES

1. French, E.C. and G.M. Prine. 1991. Perennial peanut: An alternative forage of growing importance. Fla. Coop. Ext. Ser. Agron. Facts SS-AGR-39.

2. French, E.C., G.M. Prine, W.R. Ocumpaugh, and R.W. Rice. 1993. Regional experience with forage Arachis in the United States. In: P.C. Kerridge and B. Hardy (eds.) Biology and Agronomy of Forage Arachis. CIAT, Cali, Columbia. Chapter 15 pp. 167-184.

3. Saldivar, A.J., W.R. Ocumpaugh, R.R. Gildersleeve, and G.M. Prine. 1992. Growth analysis of 'Florigraze" Rhizoma peanut: Shoot and rhizome dry matter production. Agron. J. 84:444-449.

Biological Control of Silverleaf Whitefly:
An Evolving Sustainable Technology

Philip A. Stansly[1], David J. Schuster[2] and Heather J. McAuslane[3]

University of Florida/IFAS

[1]S.W.Florida Research and Education Center, Immokalee

[2]Gulf Coast Research and Education Center, Bradenton

[3]Department of Entomology and Nematology, Gainesville

Bemisia tabaci ("sweetpotato" or silverleaf whitefly"), has become the key pest of cotton, vegetables and ornamentals in much of the tropics and subtropics of the world in recent years. Damage is caused by plant debilitation through sap removal, sooty mold and sticky lint (cotton) caused by honeydew secretions, plant disorders induced by salivary toxins, and the transmission of plant viruses. The recent status as a key pest is due, in part, to the appearance of a new biotype or species known as the silverleaf whitefly for its ability to induce a non-infectious silvering disorder on many squash and pumpkin cultivars through nymphal feeding. First detected in Florida in 1986 on poinsettia, the silverleaf whitefly quickly displaced its predecessor throughout the state, probably by virtue of a broad host range enabling it to infest many new crops including tomato, curcurbits and crucifers.

Heavy infestations on tomato in 1987 and 1988 were implicated in a new disorder, irregular ripening, which caught many growers unaware until after shipment of produce to market. By 1989 a new disease vectored by the whitefly, tomato mottle geminivirus (TMoV) had swept through most production areas causing widespread crop damage. Losses and control costs in Florida tomato for whitefly and virus combined were estimated at $141 million for the 1990-91 season.

Peanut, a crop that had no history of whitefly pests, was first observed infested by B. tabaci whitefly in Florida in 1988. Anecdotal reports suggested that peanut yields were reduced in some fields from an average of 3,500 lb/acre to 1,000 lb/acre due to whitefly feeding pressure. However, the true cost of the whitefly to peanut growers must be measured in terms of greatly increased usage of broad-spectrum insecticides, with frequencies of once a week not uncommon.

The crisis has abated in tomato over the last few years, due in part to intensive chemical use. Twice-a-week applications with broad-spectrum insecticides are not uncommon. In spite of the barrage of chemicals, biological control by natural enemies of the whitefly is a major factor in the abatement, particularly in the southwest and west central areas of the state. This is because natural biological control can operate unhindered in weeds during fallow periods. This activity can provide an effective reduction of whitefly populations because growers have in large part responded to the recommendation to remove crop residues immediately after harvest and to maintain fields free of volunteer plants that could serve as sources of virus inoculum during the crop-free period. The fallow practice forces the whitefly to depend largely on unsprayed weeds for survival during the summer months. Consequently, pest populations are subject to heavy mortality by a wide array of insect predators and parasites. At least 15 species of wasp parasites attacking B. tabaci were recently identified in Florida and the Caribbean (G. Evans, personal communication). At least 11 species of predator were identified in a survey of predator species attacking B. tabaci on unsprayed tomato in Florida, including ladybeetles, true bugs, lacewings and spiders. Over 100 of these predators have been counted from a single unsprayed tomato plant.

The potential of these natural enemies to control whitefly populations was illustrated in a field study carried out in west-central Florida in 1992. Combined rates of predation and parasitism rose from near zero to between 50% and 100% within one month on the 3 weed species monitored (Fig. 1). The study illustrated two important points: the importance of predation and parasitism in reducing whitefly populations during the fallow period, and the potential for biological control in the absence of broad-spectrum insecticides. The

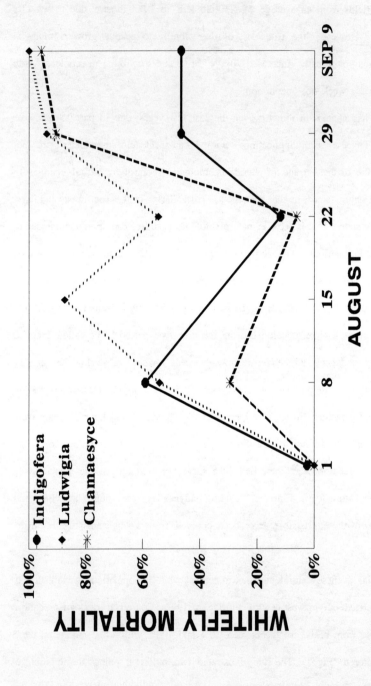

D. Schuster, GCREC Bradenton.

Figure 1. Combined levels of predation and parasitism of B. tabaci nymphs on 3 weed species in west-central Florida.

realization of this potential in the tomato crop itself is a special challenge, given the high value of the crop and the threat of TMoV, even with low whitefly populations.

Florida peanuts do not yet suffer from whitefly-vectored viral diseases so natural biological control is more easily realized. In field experiments in 1992 and 1993 we recorded considerable parasitism of B. tabaci whitefly in peanut fields that received only Bacillus thuringiensis instead of broad-spectrum insecticides for control of pest Lepidoptera. Whitefly populations never reached damaging levels, most likely due to parasitism by three aphelinid wasp species, Encarsia nigricephala, E. pergandiella, and E. transvena. Parasitism increased from 0% to 90% and then to 100% within one to two months on these unsprayed plots (Fig. 2).

In vegetables, work on an organic farm, "Pine Island Organics" showed the activity of whitefly natural enemies in an intensively cropped system where synthetic pesticides were not used. The farm consisted of 45 acres divided into blocks of 1 to 2 acres separated by sugarcane windbreaks and planted in rotated sequences to tomato (about 50%), pepper, eggplant, cucumber and squash. We mapped the block into 100x20 foot cells and sampled each cell for whitefly and other pests. Our objective was to document the influence of crop associations on whiteflies and their natural enemies. Although damaging levels of some pests such as spider mites, pepper weevils, and potato aphids were been observed, we have not yet seen damaging levels of whitefly or TMoV. Whiteflies appeared to migrate into the crops in the fall and reach their highest levels early in the season. However we soon observed increasing activity of parasitic wasps such as Encarsia pergandiella, and a corresponding decline in whitefly populations ensued (Fig. 3). Although circumstantial, the negative correlation between whitefly populations and their hymenopteran parasitoids is evidence of effective biological control.

Efforts have also been made to enhance the levels of biological control. We applied the insect pathogenic fungus, Paecilomyces fumosoroseus, to individual plants in a flowering tomato crop. The fungus was recovered up to seven weeks later and peaked at an infection

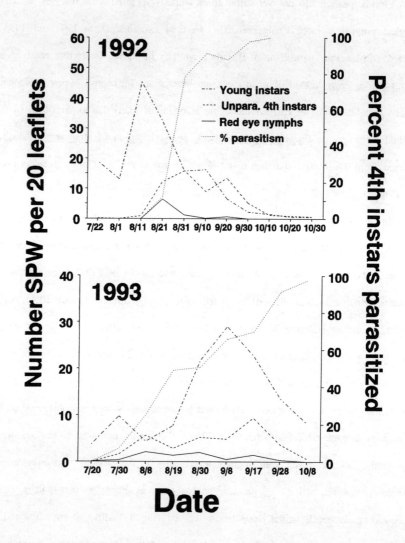

Figure 2. Density of large whitefly nymphs and pupae and parasitism rates on unsprayed peanuts in Gainesville Florida.

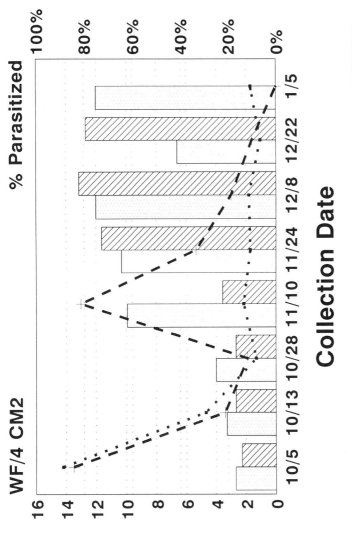 

Figure 3. Density of large whitefly nymphs and pupae and parasitism rates on tomato and eggplant on an organic farm in southwest Florida.

rate of 32% of whitefly nymphs three weeks after the application. Also, a combination of whey, yeast and sucrose was applied weekly to tomato plants in the field to attract predator insects. The nutritional sprays resulted in increased oviposition by lacewings compared to unsprayed tomato plants.

Another experimental method of enhancing natural enemies could be through the use of refuge crops. Results of a survey of crop hosts in southwest Florida showed that parasitization of okra averaged 56% over a 4 week period, twice or more that observed on cotton (31%), eggplant (25%), collards (25) or cowpea (12%). Thus some plant species may be more conductive to parasitization than others. If the refuge crop was also extremely attractive to whitefly it could function as a trap crop as well. Greenhouse choice tests have shown that zucchini squash is almost twice as attractive to whiteflies as tomato. Another advantage of the squash is that they are not susceptible to TMoV and therefore could act as a sink for viruliferous whiteflies. Levels of whitefly and TMoV in 3 rows of tomato adjacent to 6 rows of squash were approximately half that seen in tomatoes grown next to tomatoes in a field trial carried out in west-central Florida. Experiments are now underway to determine if squash can also act as a refuge and source of natural enemies that could move into nearby tomato.

Even though we are confident that natural enemies can control silverleaf whitefly, the realization of their potential in heavily sprayed commercial farms is a difficult task. One approach we have taken is the use of so-called "biorational" insecticides. According to our definition, an insecticide could be considered biorational if it was selectively toxic to pest species but relatively innocuous to beneficial organisms, in particular natural enemies of the pest. We have been testing materials considered biorational such as mineral oils, soaps and detergents for activity against whitefly and their natural enemies in the laboratory and field. Mineral oil has proven to be an effective contact toxicant to all stages of silverleaf whitefly, and to have repellent activity against adults. Detergents and soaps are most effective against nymphs as opposed to adults and especially eggs. Activity against adults is limited to before sprays have dried. We have also tested these materials against whitefly predators known as lacewings. Mineral oil is highly toxic to lacewing eggs, moderately toxic to lacewing larvae and non toxic to lacewing adults. Insecticidal soap is not toxic to lacewing eggs nor adults. Therefore, these materials may still be termed "biorational" according to our yet considerable laboratory and field testing that lies ahead.

In summary, biological control of silverleaf whitefly is already a reality in unsprayed weeds, in at least one crop not affected by whitefly vectored virus, peanut, and on tomato under organic vegetable production practices. The extension of these results to conventional vegetable farms is a challenge which we feel can eventually be met with success.

USE OF FIRE AS A MANAGEMENT TOOL IN ALFALFA PRODUCTION ECOSYSTEMS

W. C. Stringer and D. R. Alverson
Agronomy and Entomology Departments, respectively
Clemson University

ABSTRACT

Alfalfa weevil, a key pest of alfalfa throughout its range, is currently controlled to sub-economic population densities in many areas only by the use of chemical pesticides in conjunction with biological and climatic regulating agents. The necessity for alternative measures in alfalfa culture led to the evaluation of mowing and burning standing winter residues which house weevil eggs and developing populations. Mowing alone provided measurable control, but was not as effective as current methods. Burning of winter-mowed alfalfa was statistically as efficacious for alfalfa weevil control as a single application of the systemic pesticide, carbofuran, applied at the recommended population density threshold. Even in pesticide application regimes, burning enhanced subsequent weevil suppression. The canopy of the crop was modified by fire, reducing the number of stems, but allowing growth patterns consisting primarily of long stems. Thus, overall yields remained unaffected if not improved.

RATIONALE

Alfalfa weevil (AW), *Hypera postica* Gyllenhal, is a difficult pest in alfalfa, *Medicago sativa* L., ecosystems in the southern states. Though it causes serious crop losses over much of its North American range, it is particularly difficult to manage in an area where pest activity and development occur over a long and sporadic season. Spring defoliation by the AW reduces the first and subsequent yields and quality of the forage (Berberet 1981, Berberet and McNew 1986, Wilson and Quisenberry 1986, Alverson et al. 1991). Adult females are in alfalfa fields from late autumn through spring, laying eggs whenever temperatures permit. In southern states, larvae (feeding stage) first appear on the tips of healthy green shoots in February or earlier. These winter shoots are important contributors to the maintainance of stored energy levels. Damage to these shoots can accentuate dependence on autumn-stored energy reserves. Larvae developing from late winter and spring oviposition can impose severe defoliation on the rapidly developing spring shoots. This two-phase pest outbreak, if left uncontrolled, can seriously damage alfalfa productivity and stand persistence; thus efficacious control of this pest is vital to alfalfa culture in southern states.

Control options for AW are limited. Several insecticides are very effective, but are potentially toxic to the environment, livestock consumers, and the applicator. At least six species of parasitic insects attack AW in various portions of its range (Puttler et al. 1961, Brunson and Coles 1968, Dysart and Coles 1971, Drea et al. 1972, Dysart 1972, Weaver 1976). In our area, an introduced ichneumon wasp, *Bathyplectes curculionis* (Thomson), is the most important established biological control agent (Morill 1979). Others, such as *B. annurus* (Thomson), acting together in some regions, may regulate population desities to sub-economic levels. The pathogenic fungus, *Erynia phytonomi*, occurs in South Carolina and adjoining states (G. R. Carner and D. R. Alverson, unpubl.; Gardner 1982), but it has not been shown to have effective economic impact. On the whole, biological control agents have not consistently provided economic levels of AW control in parts of the Southern USA.

As we decrease dependence on pesticides, the need arises to assess other biological or cultural control practices. Winter grazing by livestock has been shown to significantly reduce damage by AW. However, grazing is not always practical in alfalfa production systems.

In an earlier study, we characterized the oviposition rates and development of AW in relation to crop development throughout the winter months in SC. The results of this characterization showed that oviposition occurred continuously over the season, resulting in a high population of mature AW eggs by mid-February. We thus hypothesized that effective control could be effected by culturally removing or destroying the egg-containing stems in late winter. We evaluated the potential impact of late-winter burning and mowing on subsequent AW populations and the development and yield of the spring alfalfa crop.

MATERIALS AND METHODS

A randomized complete block experimental design was established to evaluate winter mowing and burning against and in combination with early spring insecticide treatments for control of AW. A 4-yr-old stand of 'Cimmaron' alfalfa was divided into four blocks, each subdivided into six treatments. The last havest was taken in late September prior to cool-season treatments. Plots receiving either burning or mowing treatments were mowed with a sickle-bar mower in late February. Mowing was prerequisite to burning so that the residue would help sustain a fire. Applications of the insecticide, carbofuran (Furadan 4F), were made on March 14 at the rate of 1.12 kg active ingredient/ha to one of each pair of mowed, burned, and unmowed plots in each replicate. Burning was accomplished by igniting the plot edges with a drip torch, allowing the fires to propagate into the plots. Re-lighting was required where mulch was of insufficient quantity to sustain the burn.

AW densities were determined by beating five randomly selected shoot terminals over the side of a pail and counting the number of larvae dislodged. Counts were made once weekly until first harvest. A flail harvester was operated through a 3X17-ft swath near the center of each plot at harvest, and the yield was collected and weighed. A subsample of several kg was taken from the harvest in each plot. This was weighed and dried at $65^{\circ}C$ in a forced air dryer for 48 h. The dry-weight sample was reweighed for calculation of plot yields.

RESULTS AND DISCUSSION

AW larvae began eclosure from eggs in early February and reached economically treatable levels (1^+larvae/stem) by mid-March. Table 1 shows the effects of winter mowing and burning on AW larval density in each of the subsequent chemical treatment regimes. On all sampling dates, there was no significant difference ($p=0.05$) between pesticide-treated plots (normal practice) and the non-treated (chemically) plots where winter residues were mowed and burned. In other words, burning was statistically as effective as spraying for AW control. Mowing tended to reduce AW density also, though not consistently significant, particularly in unsprayed plots. In addition mowing and mowing/burning enhanced the control of AW in pesticide-treated plots.

Table 1. Alfalfa weevil larval density as affected by winter mowing, spraying and burning.

Treatment	Mar 21	Mar 28	Apr 4	Apr 11	Apr 18	Apr 25
	------------------------AW larvae per 5 stems------------------------					
Unsprayed						
Check	10.0 a*	23.0 a	18.0 a	10.8 a	3.3 a	0.3 b
Mowed	6.8 b	16.0 b	12.0 b	8.8 a	4.8 a	1.0 ab
Mow/burn	1.0 c	2.3 c	4.5 c	2.5 b	5.3 a	2.3 ab
Sprayed						
Check	0.0 c	0.3 c	5.5 c	6.3 ab	5.0 a	2.8 ab
Mowed	0.0 c	0.0 c	2.3 c	8.8 a	2.5 a	3.5 a
Mow/burn	0.0 c	0.8 c	0.0 c	2.5 b	1.8 b	1.8 ab

* Means within any column (dates) followed by the same letter are not significantly different at P= 0.05 level.

Treatments that resulted in more effective AW control also resulted in a modified plant canopy. As the degree of control increased, there were fewer stems per plant, but individual stems were longer (Fig. 1). This may have been a result of more AW feeding, stimulating the plants to produce or retain more stems. Burning probably destroyed some shoots in late February, but the plants were able to compensate due to decreased intra-plant shoot competition.

      Highest yields resulted from the mow/burn/spray management regime, as expected, since AW control was highest here and shoot lengths were uniformly long (Table 2). Based on AW control alone, the expected yields from the untreated plots should be lowest of all treatments. However, the high percentage of short stems and unremoved winter residue could account for the yields recorded. No data on forage quality were taken. Yields from other mowed/sprayed plots can be attributed to both AW control and removal of early spring shoot competition or winter residues.

In summary, the non-pesticidal cultural treatments evaluated here offer efficacy against AW and contribute to yields comparable to pesticide-treated fields. Advantages include increased environmental, consumer, and operator safety through reduced pesticide use. The principal disadvantages would be the difficulties of obtaining a uniform burn and the potential for smoke production during burning. An alternative method of burning is the use of tractor-mounted propane burners (Ruen 1991). Smoke and burning have always been a part of natural ecosystems, and burning is currently used as a conventional low-input practice in many agricultural, range and forest systems. What is needed is an environmental and economic cost-benefit comparison between pesticide-based management and cultural practices which may include fire.

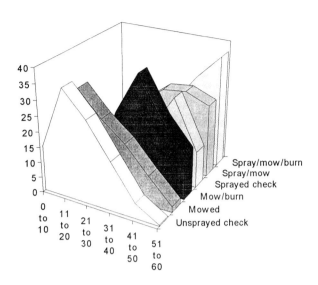

Figure 1. Shoot length distribution of alfalfa canopies as affected by cultural practices. (Y axis shows percentage of total shoots, and X axis shows length classes, in cm.)

Table 2. Effect of winter mowing, spraying and burning on first cutting yields of alfalfa.

Mow	Mow, burn	Spray	Mow, spray	Mow, burn, spray	Check
----------------------------kilograms per hectare----------------------------					
2589 b*	2871 ab	2755 b	2666 b	3740 a	3203 ab

LITERATURE CITED

Alverson, D. R., W. C. Stringer and A. J. Vybiral. 1991. Effect of fall cutting and alfalfa weevil on alfalfa yields and forage quality. J. Agric. Entomol. 8: 137-146.

Berberet, R. C. and R. W. McNew. 1986. Reduction in yield and quality of leaf and stem components of alfalfa forage due to damage by larvae of *Hypera postica* (Coleoptera: Curculionidae). J. Econ. Entomol. 79: 212-218.

Berberet, R. C., R. D. Morrison, and K. M. Senst. 1981. Impact of the alfalfa weevil, *Hypera postica* (Gyllenhal)(Coleoptera: Curculionidae), on forage production in non irrigated alfalfa in the southern plains. J. Kansas Entomol. Soc. 54: 312-318.

Brunson, M. H. and L. W. Coles. 1968. The introduction, release, and recovery of parasites of the alfalfa weevil in eastern United States. USDA (ARS) Production Report 101.

Drea, J. J., R. J. Dysart, L. W. Coles, and C. C. Loan. 1972. *Microctonus stelleri* (Hymenoptera: Braconidae, Euphorinae), a new parasite of the alfalfa weevil introduced into the Unitedweevil. Environ. Entomol. 1: 386-388.

Dysart, R. J. and L. W. Coles. 1971. *Bathyplectes stenostigma*, a parasite of the alfalfa weevil in Europe. Ann. Entomol. Soc. Amer. 64: 1361-1367.

Gardner, W. A. 1982. Occurrence of *Erynia* sp. in *Hypera postica* in central Georgia. J. Invert. Pathol. 40: 146-147.

Puttler, B., D. W. Jones, and L. W. Coles. 1961. Introduction, colonization, and establishment of *Bathyplectes curculionis*, a parasite of the alfalfa weevil in the eastern United States. J. Econ. Entomol. 54: 878-890.

Ruen, J. 1991. Burn 'em, beat 'em, or feed 'em. Hay and Forage Grower. Feb.

Weaver, J. E. 1976. Parasites of the alfalfa weevil in West Virginia. WV University Agric. Exp. Sta. Current Report 67.

Wilson, H. K. and S. S. Quisenberry. 1986. Impact of feeding by alfalfa weevil larvae (Coleoptera: Curculionidae) and pea aphid (Homoptera: Aphididae) on yield and quality of first and second cutting of alfalfa. J. Econ. Entomol. 79: 785-789.

SOIL AND WATER MANAGEMENT FOR SUSTAINABILITY

CHARLES A. BOWER

LOCATION: IDAHO   LATAH

SECTION 16, TOWNSHIP 39   RANGE 2, WEST BOISE MERIDIAN

DURATION: MAY 1989 TO OCTOBER 1989

Objective: Reduce chemical use to conserve moisture, improve yield and profitability; commonly known as Low Input Sustainable Agriculture (LISA) on farm testing under the guidance of Washington State University, University of Idaho, and other agricultural agencies.

History: 20 years of rotation of wheat, barley, oats, peas, clover, and rape seed; commercial fertilizer and herbicides in continuous use.

Comparison: Solid seeded dry edible beans versus row seeded beans.

Winter wheat stubble was plowed using the "reverse" method (starting in the center and working to the outside) in Fall 1988; in the spring of 1989, the field was divided into four plots. One six acre plot was solid seeded to pinto beans; one six acre plot was solid seeded to white navy beans; one three acre plot was seeded to pinto beans with 42 inch spacing; one three acre plot was seeded to white navy beans with 42 inch spacing. Pre-emergence herbicide was used on the solid seedings with herbicide banding and seeding on the rowed beans in a single operation. The solid seedings were sprayed with a mustard control herbicide with little effect on the rape.

Visual moisture checks during the growing season showed more moisture in rowed beans than in solid seedings at any given time. The solid seedings on wet ground showed visible mosaic damage while the rowed seedings were disease free. The row seeded beans were cultivated twice to control weeds and rape.

At harvest time, it was necessary to swath the solid seeding because of uneven ripening. The rowed beans were pulled at the same time.

Yield:			
	Solid seeded pinto	450 lb/acre	seed rate 90 lb/acre
	Solid seeded navy	325 lb/acre	seed rate 60 lb/acre
	Row seeded pinto	900 lb/acre	seed rate 65 lb/acre
	Row seeded navy	645 lb/acre	seed rate 40 lb/acre

Conclusion: Row seeded beans yielded nearly twice that of solid seeded beans with less seed and two thirds less herbicide. Seed bed preparation and harvesting costs of the comparisons were equal. The extra two cultivations offset the cost of herbicide application on the solid seedings. Savings on herbicide cost was $15 per acre for pre-emergence. Full cost of the mustard herbicide was $18 per acre. A total savings of $33 per acre resulted.

Herbicides are labeled for maximum crop tolerance, often far more than necessary to kill the weeds. Example: pig weeds and wild oats in peas and potatoes. Potatoes are labeled at 3 1/4 pints per acre. Peas are labeled at 3/4 pint per acre. The potato farmer pays for 2 1/4 pints he doesn't need, and the environment also suffers.

Chemical firms should be required to label the minimum amount actually needed to kill a particular weed.

RESERVOIR TILLAGE

Duration: October 1991 to October 1993

This term refers to the results of a subsoil-riper with alternating chisel shanks (ridge till). The fracturing of the soil to a depth of 16 inches forms water storage areas under the surface. For maximum benefit, this is done on a contour.

In Fall 1991, a 50 foot strip was subsoiled across a 20 percent slope immediately after seeding winter wheat on rape stubble. This steep slope showed no erosion on the subsoiled area. A wet area was nearly double in yield with a measurable increase in yield on the steep ground comparable with the check side.

In Fall 1991, 25 acres of fall wheat stubble was subsoil/ridge tilled for planting to rowed dry edible beans. In 1992, four acres adjacent was moldboard plowed for check.

The subsoil/ridge till was ready a full week ahead of the moldboard plowed area.

At time of pulling, the subsoil/ridge till beans were pale yellow, a sign of adequate moisture. The plowed check plot showed marked premature ripening. The yield was not checked.

The fields were seeded to winter wheat at harvest. In 1993, the subsoil/ridge till yielded three bushels per acre more than the checks.

This ongoing farm testing shows great potential for the control of erosion, especially on low residue crops, for reducing chemical use, and for improving yield and profitability, especially in soils 18-24 inches deep with annual precipitation of 24 inches.

Subsoil/ridge till provides the deep tillage necessary for the soils in this area while maintaining adequate surface residue.

Research scientists may be taking a wrong approach to soil-borne disease control by advocating delayed seeding, which is an invitation to erosion. The aeration of deep tillage may destroy or reduce soil-borne diseases by exposing them to oxygen and excess moisture from the surface as evidenced by reduced smut on the subsoiled/ridge till fields. This could allow for early fall seeding, a proven erosion control. This will be the next addition to current farm testing for Solutions to Environment and Economic Problems (STEEP) in cooperation with the Washington State University, University of Idaho, and other agricultural related agencies.

Previous experience with early seeding in the 50's and 60's indicates it is possible and economically feasible to reduce chemical usage and to provide a measure of safety to the environment and living creatures, including man.

BROMIDE AND FD&C BLUE #1 DYE MOVEMENT THROUGH UNDISTURBED AND PACKED SOIL COLUMNS

K. K. Hatfield, G. S. Warner*, K. Guillard

ABSTRACT

Bromide and FD&C Blue #1 Dye were applied to undisturbed and packed columns of a Woodbridge soil. Bromide, a conservative tracer, has been shown to mimic nitrate. FD&C Blue #1 Dye has been shown to be a moderately adsorbed solute that can mimic moderately adsorbed pesticides such as atrazine. Water and solute application occurred at four hydraulic heads of -10 cm, -5 cm, +1cm and +5 cm. Variables considered included volumetric flow rate (q), peak relative bromide and FD&C Blue Dye #1 concentrations (C/Co), and time to peak relative bromide and FD&C Blue Dye #1 concentrations. Significant differences were found with respect to column type (undisturbed vs packed) for all variables. In addition, linear hydraulic head effects were found with respect to q (ml/min) and the time to peak relative bromide concentration (C/Co). Significant column x head interaction was found for peak relative bromide concentration. Further analysis of breakthrough curves resulting from this experiment is planned as well as soil column dissection and visual inspection.
KEYWORDS. Undisturbed and packed soil columns, bromide, FD&C Blue #1 Dye, breakthrough curves

INTRODUCTION

In 1990, the United States Environmental Protection Agency completed a National Pesticide Survey. The main objectives of this study were to associate pesticide application patterns and ground water vulnerability to pesticide contamination in drinking water wells nationwide and to determine the extent of contamination of drinking water wells nationwide (EPA, 1990). The contamination that the EPA found provides ample evidence to support the fact that pesticides are reaching national drinking water wells and that the mechanisms by which these pesticides are reaching drinking water wells deserve further study.

The mechanisms by which water and solutes move through soil are not well understood. In the past, it was assumed that water and solutes moved through the soil matrix as a uniform wetting front according to Darcy's Law. Current models describing solute transport are usually based on the assumption of uniform flow through a homogeneous medium. If, however, the medium is not homogeneous, as is the case in most soils, flow cannot adequately be described by these models and the differences between flow through a homogeneous medium and a heterogeneous medium must be assessed.

Understanding the effects of soil structure in this process would be helpful in determining best management practices as well as providing information to create more accurate models of solute movement in agricultural soils. One tool for understanding these effects is the analysis of breakthrough curves of solutes through soil columns. Laboratory analyses of soil columns offers the advantage of assessing the importance of different variables on water and solute movement in soils under controlled conditions. For example, in a field situation, the amount and intensity of rainfall can change solute movement significantly. In the laboratory setting, the application of water and solutes to soil columns can be strictly controlled. In addition, McMahon and Thomas (1974) have shown that the use of undisturbed soil columns is a better approximation of field movement of water and solutes than packed columns.

The objective of this study was to compare hydraulic and transport characteristics of undisturbed vs packed soil columns. Two tracers, bromide and FD&C Blue #1 Dye were used to measure the -

---

K. K. Hatfield, G. S. Warner, Dept. of Natural Resources Management and Engineering, Univ. of Connecticut. K. Guillard, Dept. of Plant Science, Univ. of Connecticut. Department of Natural Resources Management and Engineering, U-87, Univ. of Connecticut, 1376 Storrs Road, Storrs, CT 06269-4087. Scientific contribution no. 1519 of the Storrs (Conn.) Agric. Exp. Stn., Storrs, CT 06269. Funding for this research was from the Special Grants Water Quality Research Program of the CSRS, USDA. *Corresponding author.

transport parameters. Bromide has been shown to behave similarly to nitrate (Smith and Davis, 1974). FD&C Blue #1 Dye has been shown to mimic moderately adsorbed pesticides (Andreini and Steenhuis, 1988). Specific transport parameters measured were volumetric flow rate (q), peak relative concentrations and times to peak relative concentrations.

MATERIALS AND METHODS

Soil Description

Four undisturbed soil columns were taken from an experimental agricultural field in Storrs, Connecticut. The soil at the site was of the Woodbridge series (Table 1).

Table 1. Morphological characteristics of the soil pedon, a coarse, loamy, mixed, mesic, Aquic Dystrochrept.

Horizon	Depth	Color a (moist)	Mottles b	Texture c	Structure d	Consistence e (moist)	Boundary f	Notes g
Ap	0-18cm	10YR 3/3		fsl	1fsbk parting to 1fg	vfr	aw	12% cf (8% gravel, 4% cobbles)
Bw1	18-43cm	10YR 4/6		fsl	1msbk	vfr	cs	<10% cf, common krotavina (1 cm diameter), common fine pores
Bw2	43-61cm	10YR 5/4	cfd 10YR 5/2 10YR 6/1 10YR 6/2 cfd 7.5YR 4/6 fcd 7.5YR 5/8 cmd 10YR 5/2 ffd 10YR 6/1	fsl	1csbk	vfr	gw	<10% cf
BC	61-79 cm	10YR 5/3	ffp 7.5YR 4/4 cc 7.5YR 5/4 fff 10YR 5/2	fsl	1csbk	fr	gw	<10% cf
Cd	79-111 cm	10YR 6/2	cc 7.5YR R4/4	fsl	1mpl	vfi		<10% cf, common saprolytic pebbles

a Munsell color notation
b c=common and coarse, f=fine, few and faint, d=distinct, m=many
c fsl=fine sandy loam
d 1=weak, f=fine, m=medium, c=coarse, sbk=subangular blocky, pl=platey, g=granular
e v=very, fr=friable, fi=firm
f a=abrupt, c=clear, s=smooth, g=gradual, w=wavy
g cf=coarse fragments

Although this soil's texture, a fine sandy loam, is usually easy to work with, the presence of many large coarse fragments (gravel and cobbles) made excavation extremely difficult at times. Also, there was an extremely hard compact glacial till layer at approximately 60-cm depth.

Soil Column Preparation

Large soil pedestals approximately 1m by 1m by 1.5 m deep were created by excavating with a tractor mounted mechanical excavator. The pedestals were then manually carved down to a diameter of 30 cm using care to avoid cave-ins due to numerous stones in the pedestals. The columns ranged in length from 50 to 100 cm. A length of corrugated polyethylene drain tile was slipped over the pedestals and a minimally expanding foam sealant (commonly used for home insulation)was sprayed in between the soil and the drain tile to secure the soil and to prevent flow down the sides of the columns during the experiment. A steel frame was attached to the columns for transport to the laboratory. Once in the laboratory, a layer of glass spheres (0.25 mm in diameter) was added to the top and bottom of the soil columns to provide even contact surfaces. A PVC grid system was installed at the bottom of the columns which divided the outflow from the column into 32 separate cells 5 cm by 5 cm square. Perimeter cells were approximately half this size once the grid was cut into a circle to fit the column base. The columns were placed on a truss approximately 90 cm above the laboratory floor to allow easy access to samples.Four packed soil columns were also made from the same soil as the undisturbed columns. Soil was excavated from the site in 15 cm depths and brought to the laboratory and packed into a 30 cm by 30 cm by 100 cm wooden box. The bulk densities of the soil horizons were approximated while packing the box to conform to published bulk densities in the soil survey of the area. Once the soil was packed to the desired bulk density, the wooden box was removed and the standing soil pedestal was carved down to the 30 cm in diameter and prepared in the same manner as the undisturbed soil columns.

Water and Solute Application

Two chemical tracers, bromide and FD&C Blue #1 Dye, were applied to the columns at four different hydraulic heads of -10 cm, -5 cm, +1 cm and +5 cm. Bromide, a conservative tracer, was chosen for its non-adsorptive properties and ease of measurement. FD&C Blue #1 Dye was chosen for its ability to mimic moderately adsorbed pesticides (Andreini and Steenhuis, 1988). Deionized water and tracers were applied through a disc permeameter (29-cm diameter) manufactured from Plexiglas. A 40 μm nylon mesh with an approximate air entry value of 30 cm H_2O was attached to the bottom of the disc permeameter to maintain small negative heads. The permeameter was placed inside the top of the corrugated polyethylene tile of each soil column and pressed firmly against the layer of glass spheres. Two Marriotte columns, the first of which contained deionized water and the second of which contained 0.1% solutions of both KBr- and FD&C Blue #1 Dye, were connected to the disc permeameter. A builder's level was used to determine the exact height of the soil surface in the soil columns in relation to the bottom of the air inlet tubes in the Marriotte columns. The Marriotte column system was placed on a large scale which was monitored by a datalogger so that the exact inflow to each soil column was known. The hydraulic head of deionized water and tracers entering the soil columns was then controlled by raising and lowering the Marriotte column system in relation to each soil column.

Deionized water was applied at the desired hydraulic head until steady state was reached. At that time, the valve from the deionized water Marriotte column was closed while the valve from the Marriotte column containing bromide and FD&C Blue #1 Dye was simultaneously opened. An approximate 500 ml slug of the 0.1% KBr- and FD&C Blue #1 Dye solutions was added to the column. The valve from the KBr- and FD&C Blue #1 Dye Mariotte column was then closed while the valve from the Marriotte column containing deionized water was simultaneously opened. Deionized water was applied to the column until the concentrations of bromide and FD&C Blue #1 Dye had peaked and had receded to a nearly constant level. The hydraulic head was then changed and the experiment repeated. The experimental runs began with -10 cm hydraulic head and progressed up to + 5 cm hydraulic head.

Sample Collection and Analyses

Tubing was attached to the grid system at the base of the columns and run to 32 individual collection jugs on the floor. A small amount of filter material was placed in the tubing as it left the grid system to prevent glass beads and soil from blocking flow down the tubing. Once setup was complete, the four runs on each column began.

Samples were collected from the columns as frequently as the sample analyses would allow. Once samples were removed from beneath the column, each sample jug was weighed to get an approximate volume. Next the samples were analyzed for dye content using a spectrophotometer. Finally, the samples were analyzed for Bromide content using an ion specific probe.

Once all samples were collected for each column, the data were composited to determine the bulk properties of each soil column. Individual cell samples were collected to satisfy a separate objective of this experiment which is still underway.

Statistical Analyses

The experiment was arranged as a 2 by 4 factorial and set out in a completely randomized design with four replicates. Factors consisted of two soil column types (undisturbed and packed) and four hydraulic heads of -10 cm, -5 cm, +1 cm and +5 cm. Data were log transformed then analyzed using AOV. The head effect was further analyzed using single-degree-of-freedom orthogonal polynomials. The effects were then separated into linear, quadratic and cubic effects.

RESULTS

Significant differences between column type (undisturbed and packed) were observed for all variables (Table 2). In addition, linear hydraulic head effects were found with respect to q (ml/min) and the time to peak relative bromide concentration (C/Co). A significant column x head interaction was found for peak relative bromide concentration.

Table 2. ANOVA table for response head effects.

Source of Variance a	d.f.	Mean Square	F	P
		q (ml/min)		
Column	1	1.380	17.36	0.0003
Head	3	0.404	5.08	0.0073
Column x Head	3	0.003	0.04	NS
Head Linear	1	1.186	14.91	0.0007
Head Quad	1	0.006	0.07	NS
Head Cubic	1	0.021	0.26	NS
Column x Head L	1	0.001	0.01	NS
Column x Head Q	1	0.003	0.04	NS
Column x Head C	1	0.005	0.06	NS
		peak relative bromide concentration (C/Co)		
Column	1	2.101	9.00	0.0062
Head	3	0.069	1.30	NS
Column x Head	3	0.349	1.49	NS
Head Linear	1	0.012	0.05	NS
Head Quad	1	0.075	0.32	NS
Head Cubic	1	0.122	0.52	NS
Column x Head L	1	0.969	4.15	0.0528
Column x Head Q	1	0.018	0.08	NS
Column x Head C	1	0.059	0.25	NS
		peak relative FD&C Blue #1 Dye concentration (C/Co)		
Column	1	2.250	7.05	0.0139
Head	3	0.516	0.54	NS
Column x Head	3	0.313	0.33	NS
Head Linear	1	0.335	1.05	NS
Head Quad	1	0.101	0.32	NS
Head Cubic	1	0.080	0.25	NS
Column x Head L	1	0.179	0.56	NS
Column x Head Q	1	0.005	0.02	NS
Column x Head C	1	0.129	0.40	NS
		time to peak (min) relative bromide concentration (C/Co)		
Column	1	1.629	11.39	0.0025
Head	3	1.341	3.13	0.0445
Column x Head	3	0.224	0.52	NS
Head Linear	1	0.907	6.35	0.0188
Head Quad	1	0.217	1.51	NS
Head Cubic	1	0.217	1.52	NS
Column x Head L	1	0.079	0.56	NS
Column x Head Q	1	0.050	0.35	NS
Column x Head C	1	0.094	0.66	NS
		time to peak (min) relative FD&C Blue #1 Dye concentration (C/Co)		
Column	1	2.736	10.21	0.0039
Head	3	0.748	0.93	NS
Column x Head	3	0.473	0.59	NS
Head Linear	1	0.222	0.83	NS
Head Quad	1	0.044	0.16	NS
Head Cubic	1	0.482	1.80	NS
Column x Head L	1	0.086	0.32	NS
Column x Head Q	1	0.00002	0.00	NS
Column x Head C	1	0.387	1.45	NS

The responses of q (ml/min) to hydraulic head change for both undisturbed and packed soil columns are shown in Fig. 1. The responses of peak relative bromide and FD&C Blue #1 Dye concentrations (C/Co) to hydraulic head change are shown in Fig. 2. The responses of time to peak relative bromide and FD&C Blue #1 Dye concentrations (C/Co) to hydraulic head changes for undisturbed and packed soil columns are shown in Fig. 3.

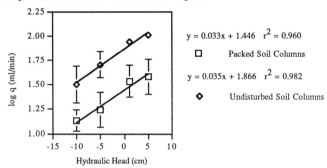

Fig. 1. Mean q (ml/min) values at four hydraulic head levels for undisturbed and packed soil columns.

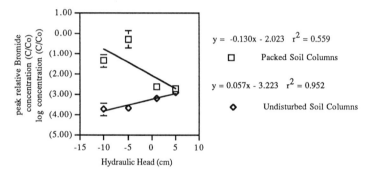

Fig. 2. Mean peak relative bromide concentrations (C/Co) at four hydraulic head levels for undisturbed and packed soil columns.

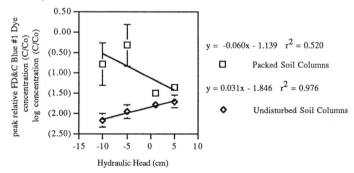

Fig. 3. Mean peak relative FD&C Blue #1 Dye concentrations (C/Co) at four hydraulic head levels for undisturbed and disturbed soil columns.

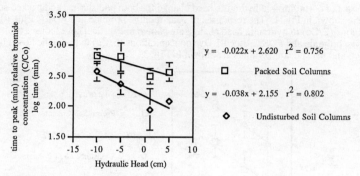

Fig. 4. Mean time to peak for relative FD&C Blue #1 Dye concentrations (C/Co) at four hydraulic head levels for undisturbed and packed soil columns.

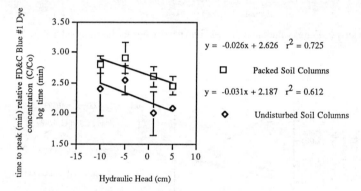

Fig. 5. Mean time to peak for relative FD&C Blue #1 Dye concentrations (C/Co) at four hydraulic head levels for undisturbed and packed soil columns.

Values of q (ml/min) increased with hydraulic head (Fig. 1). Peak relative bromide concentrations also increased with head (Fig. 2) as did peak relative FD&C Blue #1 Dye concentrations (Fig. 3) for the undisturbed columns. For the packed columns, peak relative bromide and FD&C Blue #1 Dye concentrations (Figs. 2,3) decreased as hydraulic head increased. Time to peak relative bromide and FD&C Blue #1 Dye concentrations (Figs. 4,5) decreased as hydraulic head increased.

DISCUSSION

There are significant differences in q (ml/min), peak relative bromide and FD&C Blue #1 Dye concentrations (C/Co), and times to peak bromide and FD&C Blue #1 Dye relative concentrations (C/Co) for the undisturbed and packed soil columns. These differences are probably due to the preservation of soil structure in the undisturbed soil columns. Structured soil may permit greater dispersion of solutes than homogeneous soil (Jury, et al., 1991). This would explain the higher overall peak relative concentrations for the packed soil columns. Aggregate structure, soil cracks, and biopores may all provide for preferential flow in the undisturbed soil columns and, consequently, higher q (ml/min) values (Beven and Germann, 1982). Kluitenberg and Horton (1990) showed that the method of solute application in soils with a significant degree of preferential flow has a large bearing on the transport of the solute through soil. Future dissection and visual inspection of the soil columns will allow for the evaluation of each soil column's structure.

The results of this experiment show increasing q (ml/min) values within both the undisturbed and packed soil columns for the changing hydraulic heads as would be expected. The overall q (ml/min) values of the undisturbed column are higher than the packed columns as well, indicating the effect of preferential flow. The different hydraulic head applied in this experiment are a simulation of a dry field condition, or low rainfall intensity, going to a ponded field condition, or high rainfall intensity.

The decrease in peak relative bromide and FD&C Blue #1 Dye concentration with hydraulic head for the packed soil columns requires further explaination and analyses. One possibility is that the diffusion-dispersion coefficient increases with increasing hydraulic head (Jury, et al., 1991). However, this result can also be explained by changes in the percent mobile water content in the Mobile-Immobile Water Model (van Genuchten and Weirenga, 1976). It is also possible that as the experiment continued through the changing hydraulic heads, the contact between the packed soil in the column and the foam sealant was broken creating a path for flow directly down the sides of the column and causing the discrepancy. These possibilities will be evaluated by the application of a staining dye to the soil columns followed by dissection.

In contrast, the peak relative concentrations of bromide and FD&C Blue #1 Dye increased with increasing hydraulic head for the undisturbed soil columns. One explanation for this observed phenomenum is that larger macropores have flow at higher heads and that the increased flow rates are large enough to negate changes in dispersion within the soil columns. Further analytical analysis of individual cell and composite breakthrough curves may provide insight regarding these flow phenomena and variations in tracer concentrations.

This experiment is part of a larger project, funded by the USDA, begun in the summer of 1991 that is attempting to characterize differences in atrazine movement to ground water due to different pesticide application methods for corn. The overall project includes pan lysimeter studies and stem flow studies as well as the soil column work. This project will allow the findings of the laboratory soil column experiments to be compared to actual field data.

Future analysis of the soil column data will include breakthrough curve analysis, soil column dissection and inspection, cell to cell comparisons within each soil column and geostatistical analysis of the flow through each soil column.

In order to meet the goals of sustainable agriculture, rapid problem identification is crucial. The ability to analyze a field situation in a reliable and economical manner without excessive disturbance is a worthy goal. The use of laboratory soil columns as opposed to extensive field studies is one way to reach that goal.

REFERENCES

1. Andreini, M. S. and T. S. Steenhuis. 1988. Preferential flow under conservation and conventional tillage. Paper No. 88-2633. Amer. Soc. Agric. Engineers. 21 p.

2. Beven, K. and P. Germann. 1982. Macropores and water flow in soils. Water Resources Res. 18: 1311-1325.

3. Jury, W. A., W. R. Gardner and W. H. Gardner. 1991. Soil physics. John Wiley & Sons, Inc., New York. 328 p.

4. Kluitenberg, G. J. and R. Horton. 1990. Effect of solute application method on preferential transport of solutes in soil. Geoderma. 46: 283-297.

5. McMahon, M. A. and G. W. Thomas. 1974. Chloride and tritiated water flow in disturbed and undisturbed soil cores. Soil Sci. Soc. Amer. Proc. 38: 727-732.

6. Smith, S. J. and R. J. Davis. 1974. Relative movement of bromide and nitrate through soils. J. Environ. Qual. 3: 152-155.

7. United States Environmental Protection Agency. 1990. National pesticide survey. Summary results of EPA's national survey of pesticides in drinking water wells. U. S. EPA, Office of Water, Office of Pesticides and Toxic Substances, Fall 1990. 16 p.

8. van Genuchten, M. Th. and P. J. Weirenga. 1976. Mass transfer studies is sorbing porous media. 1. Analytical solutions. Soil Sci. Soc. Am. J. 40: 473-480.

USING GLEAMS TO SELECT ENVIRONMENTAL WINDOWS FOR HERBICIDE APPLICATION IN FORESTS[1]

M. C. Smith[2], J. L. Michael[3], W. G. Knisel[4], and D. G. Neary[5]

ABSTRACT

Observed herbicide runoff and groundwater data from a pine-release herbicide application study near Gainesville, Florida were used to validate the GLEAMS model hydrology and pesticide component for forest application. The study revealed that model simulations agreed relatively well with the field data for the one-year study. Following validation, a modified version of GLEAMS was applied using a 50-year climatic record to determine the periods (windows) for least water quality degradation within the Forest Service's recommended application window for best vegetation control. The pesticide component of GLEAMS was modified to simulate up to 245 pesticides simultaneously. Four herbicides commonly used in the region to control competing vegetation were represented in the model study. Within the application windows for each herbicide, the best application dates, or "environmental" windows were determined to minimize environmental effects for each location. Results of the simulation study are tabulated in the paper for use in the forest industry.

INTRODUCTION

The forest industry in the southeastern United States has successfully used herbicides during the last 10 years to control competing grass and herbaceous vegetation in site preparation for pine (Pinus sp.) plantings and in pine release (Michael et al., 1990). Vegetation control alone and in combination with fertilization has resulted in significant increased pine growth (Neary et al., 1990). Runoff studies have been conducted at a number of locations to measure losses of herbicides to streamflow following site treatments (Michael and Neary, 1993). Field studies of herbicide fate cannot be replicated on the same site in successive years. Efficacy studies have been made to determine the best time period for herbicide application for vegetation control. Results of these studies have been used to estimate the "best" interval within the longer time interval (Miller and Bishop, 1989). The one-time herbicide application on a specific field site does not allow evaluation of climatic and environmental consequences of variable application dates.

---

[1]This study is a part of U.S. Department of Agriculture, Forest Service, Southern Forest Experiment Station, Research Project FS-SO-4105-1.24 (Problem 1).
[2]Assistant Professor, Biological and Agricultural Engineering Department, University of Georgia, Athens 30602. [3]Research Ecologist, U. S. Department of Agriculture, Forest Service, Southern Forest Experiment Station, Auburn University, Auburn, Alabama 36849. [4]Senior Visiting Research Scientist, Biological and Agricultural Engineering Department, University of Georgia, Coastal Plain Experiment Station, Tifton, Georgia 31793. [5]Project Leader, U. S. Department of Agriculture, Forest Service, Rocky Mountain Forest Experiment Station, Northern Arizona University, Flagstaff, Arizona 86001.

A mathematical model called GLEAMS (Groundwater Loading Effects of Agricultural Management Systems) was developed by Leonard et al. (1987) to assess the complex interactions of soil-climate-management for field-size areas on a long-term basis. Although GLEAMS was developed primarily for crop and pasture lands, Nutter et al. (1994) added an option to consider application on forest sites as well. GLEAMS model applications have been made to assess the long-term environmental impact of insecticide use in Southeastern forests (Nutter et al., 1993).

GLEAMS has been validated for agricultural crops (Leonard et al., 1987), and for forested areas (Nutter et al., 1993). A study is currently underway to evaluate forest streamside management zones at the locations included in this paper. Although the results have not been published, the model simulations made thus far compare favorably with observed measurements of runoff and pesticide losses.

Leonard et al. (1992) made 50-year GLEAMS simulations to examine the probabilities of year-to-year pesticide losses for a 20-day planting window for corn (Zea maize, L.). These were compared with 50-year means and standard deviations to consider potential for extreme or "worst case" situations.

The purpose of this paper is to demonstrate the use of the GLEAMS model to determine the best herbicide application periods to minimize potential environmental impacts. A location was selected in the Atlantic Coastal Flatwoods of peninsular Florida where a forest herbicide study provided data for model comparison (Smith et al., 1993). GLEAMS model simulation results are compared with observed data, and a nearby 50-year climatic record was used to determine the best "environmental" window within the "application" window for management recommendations.

METHODS OF ANALYSES

The GLEAMS model was developed to assess edge-of-field and bottom-of-root-zone loadings of water, sediment, and chemicals for comparing alternate management strategies using long-term simulation results. GLEAMS is a continuous simulation model with a daily time step, and consists of hydrology, erosion, pesticide, and plant nutrient components. The hydrology component uses daily climatic data and simulates the water balance components including surface runoff and percolation below the root zone. The erosion component computes soil detachment and sediment transport to the edge of the field. The pesticide and plant nutrient components compute pesticide, nitrogen, and phosphorous transformations, and calculates their transport in the solution and adsorbed phases. Up to 10 pesticides can be represented in a single simulation. Comparisons of long-term simulation results enable the user to make sound management decisions based upon relative loadings. Alternatives that can be evaluated include selection of herbicides and the method and dates of application. GLEAMS model version 2.10 was modified to consider up to 245 pesticides simultaneously in a single computer run. This modification made it possible to consider 1 pesticide applied on as many as 245 days by naming the pesticide with successive numbers and using the same pesticide characteristics for all applications. For example, Roundup was applied on day 1 of the application window as Roundup 1, Roundup 2 was applied on day 2 of the window, and so on to Roundup 245, each with the same characteristics. It is recognized that herbicide half-life may change due to climatic differences within the application window, but the same values were used throughout the window. Losses for each herbicide were kept separate in the simulation and reported separately. Model output includes annual losses and the final total losses in runoff, adsorbed onto sediment, and in percolation.

Herbicide applications are not made each year, but climate is different every year. The model was applied for 50 consecutive years of observed climate, but the same cover (canopy) was assumed for each year. In essence, this gives one treatment and 50 replications in time. The final results represent a significant sample of year-to-year variations in herbicide losses due to changes in climate.

The USDA-Forest Service conducted herbicide efficacy and fate studies in the southeastern United States for site preparation for pine planting and for pine release from competing vegetation. Four herbicides are commonly used for weed and brush control in the region. Pesticide characteristics, soil, and climatic region are factors in determining which herbicide may give the most effective control yet pose the least potential environmental degradation. All herbicides are not applied at each 1-year study site. Characteristics of the four herbicides, their application (efficacy) window, and recommended application rates are given in Table 1. Table 1 also includes the characteristics of the herbicide Garlon (TRICLOPYR) used at the selected study site in Alachua County, Florida.

Four-hectare plots at the study site northeast of Gainesville, Florida were surrounded by drainage ditches approximately 2 m deep and 3 m wide. A flume equipped with a continuous water-level recorder was installed at the outlets of the drainage ditch for discharge measurement. Samples of the discharge were taken during and between storm events for analysis of Garlon (Bush et al., 1990). Shallow groundwater observation wells were installed within the plots to monitor depths to water table and sampling for herbicide determination. The soil on the plots is Pomona fine sand (Ultic Haplaquods, sandy, siliceous, hyperthermic, uncoated), with a surface slope of 0.5%.

Table 1. Herbicide characteristics, and application windows and recommended rates for GLEAMS model simulation.

Herbicide Trade Name COMMON NAME	Water Solubility mg/l	K_{oc} l/g	Half-life Soil days	Half-life Foliage days	Wash-off Fraction	Application Window	Rate[a] kg/ha
Arsenal IMAZAPYR	11,000	100	65	30	0.90	5/1-10/31	2.24
Oust SULFOMETURON METHYL	70	78	20	10	0.65	2/1-5/31	0.42
Roundup GLYPHOSATE AMINE	900,000	24,000	47	3	0.60	8/1-10/31	5.60
Velpar-granules HEXAZINONE	33,000	54	77	[b]	[b]	2/1-4/30	1.68
Garlon TRICLOPYR ESTER	23	780	46	7	0.90	4/20-10/10	1.81

[a] Application rate of active ingredient for site preparation. [b] Not applied on foliage.

RESULTS AND DISCUSSION

Available data were used to develop parameter files for the GLEAMS model simulation for the Garlon study. Soils data were taken from published data (Carlisle et al., 1988) since local data were not available. Herbicide characteristics shown in Table 1 were supplied by the manufacturer. Rainfall was measured at the site for the 4-year study period, 1986-89. Monthly temperature and radiation data were obtained from climatological data at Gainesville.

Runoff (ditch flow) was observed from a 42 mm rainfall event 38 days after Garlon application on October 24, 1986. Runoff samples had Garlon concentrations of 1-2 ppb, with the maximum occurring on the second day (Bush et al., 1988). The small volume of observed runoff along with the Garlon concentration data indicate that the observed flow could have resulted from rainfall in the ditch during the high water table condition and from lateral subsurface flow above the spodic layer. Subsurface flow would be delayed (possibly second day) compared with direct surface runoff. Likewise, subsurface flow containing Garlon from near the channel on the day of the storm could be diluted by the rainfall on the channel compared with subsurface flow on successive days.

The GLEAMS model did not simulate surface runoff in 1986, and Garlon was not simulated to percolate below the 1 m effective root zone. Groundwater samples in the plot did not show Garlon concentrations above the detection limit of 0.7 ppb. Runoff was simulated with GLEAMS in 1988 when about 10 cm was reported (Riekerk, 1989). Again, the reported "runoff" volume could have included both rainfall in the drainage channel and lateral subsurface flow from the plot.

GLEAMS is not intended to be an absolute predictor of water, sediment, and chemical losses. However, the comparison made in the present study indicates the model gives "ballpark" results using published pedon data rather than site-specific soils data. This indicates the model is a useful tool for relative comparisons such as herbicide losses during application windows.

Fifty-year (1925-74) simulations were made for the site for each herbicide listed in Table 1. Since GLEAMS does not consider pesticide toxicity and the health advisory levels do not apply at field's edge or bottom of root zone, only herbicide losses can be examined in this study. Losses with runoff, sediment, and percolation are expressed as percentage of application rate, and are therefore unitized.

A 3-D graph was plotted for each herbicide to show year-by-year losses as a function of application date. Rainfall distribution within the year was reflected in the graphs. Only a simple example with a 2-D graph is shown here to illustrate the procedure. The simulated 50-yr average losses of Arsenal and Velpar granules are shown in Figure 1 by day within the application window. The application window for Velpar granules is 89 days, February 1 to April 30, and for Arsenal is 184 days, May 1 to October 31. Even though the beginning dates are different for the two herbicides, both are shown in the same figure for demonstration purposes. The 50-year simulation resulted in a total of 18 cm runoff for the entire period, or an average of less than 4 mm/year compared with about 0.7 mm for the one year of the Garlon study. Due to the soil-climate-pesticide interactions negligible runoff losses of Arsenal and Velpar were predicted. The losses shown in Figure 1 are essentially all percolation losses below the root zone. Only traces of runoff losses, about 0.01%, were simulated for Roundup and Oust.

In Fig. 1, low, essentially uniform, losses of Arsenal are simulated over the entire 184-day application window. Therefore, there is not a "best" environmental window, that is, there is no time in the 184 days in which simulated losses are significantly lower than any other time. The simulated Velpar losses are lowest at about 70 days into the application window. The best environmental window could be taken as the approximate 2-week period April 5-17 based upon the 50-year simulation results. The recommended application windows and windows for best control (Miller and Bishop, 1989), and the best environmental windows are summarized for the four herbicides in Table 2.

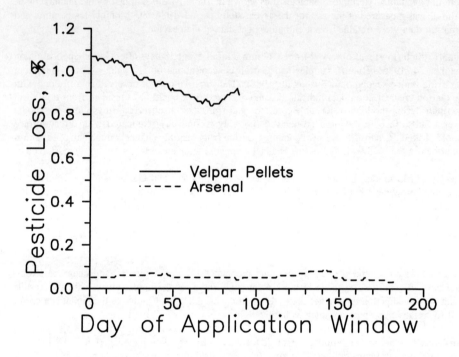

Figure 1. Pesticide loss as percent of application for 50-year GLEAMS model simulation for Pomona fine sand by day of application window: Velpar pellets--89 days beginning February 1; Arsenal--184 days beginning May 1.

Table 2. Herbicide application windows based upon 50-year average runoff, sediment, and percolation losses compared with "best" window for vegetation control.

Window	Herbicide			
	Arsenal	Oust	Roundup	Velpar granules
Application	5/01 - 10/31	2/01 - 5/31	8/01 - 10/31	2/01 - 4/30
Best control	7/01 - 9/30	3/05 - 4/10	8/01 - 10/20	3/05-4/25
Environmental (Alachua Co., FL)	9/24 - 10/31	2/01 - 5/31	8/01 - 10/31	4/05 - 4/17

Simulated year-to-year differences in Velpar loss are shown in Fig. 2. The first day of the application window for Velpar granules, February 1 (Table 1), was selected to demonstrate the variation. The 50-year mean loss for applications on February 1 of each year, 1.07% (Fig. 1), is plotted in Fig. 2. The total loss each year for the February 1 application is shown in Fig. 2. Losses range from a zero low to a maximum of 6.7% in the first 4 years of the 50-year period. Doubtless the high loss in year 4 resulted from significant rainfall on or shortly after the February 1 application date. It was stated above that herbicide applications are made in only one year for site preparation, and therefore field studies are conducted only for that one year. It can be seen from Figure 2 that misleading conclusions might be drawn from field data if the study was conducted in the first year (1925) compared with a study conducted in the fourth year (1928). Another series of differing years occurs from the 40th to the 44th years of simulation. This vividly portrays the significance of long-term simulations with a model such as GLEAMS.

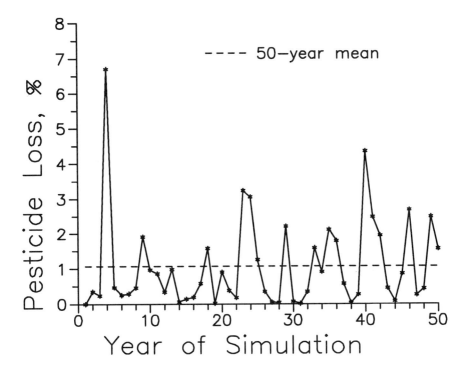

Figure 2. GLEAMS model simulated Velpar loss as percent of February 1 application for each of the 50 years beginning 1925.

SUMMARY

Model simulations in this study show how forest herbicide management alternatives can be assessed with the GLEAMS model. Alternate herbicide selection and recommended application dates were analyzed for different climatic and soil regions. The study indicates that blanket geographical recommendations should be avoided without similar long-term model analyses. Interactions of soils, slope, climate, and pesticide characteristics affect the environmental window.

This presentation represents only one soil-climatic region with the soil being in the extreme hydrologic soil group A. The same results would not be expected for other soils in other climatic regions. The model applications do show that GLEAMS can be used as a tool to examine the consequences of forest management alternatives.

REFERENCES

1. Bush, P. B., D. G. Neary, and J. W. Taylor. 1988. Effect of Triclopyr amine and ester formulations on groundwater and surface runoff water quality in the Coastal Plain. Proceedings of the Southern Weed Science Society 41st Annual Meeting, January 18-20, 1988, Tulsa, Oklahoma. pp. 226-232.

2. Carlisle, V. W., F. Sodek III, M. E. Collins, L. C. Hammond, and W. G. Harris. 1988. Characterization data for selected Florida soils. University of Florida-Institute of Food and Agricultural Sciences, Soil Science Department, Soil Characterization Laboratory.

3. Leonard, R.A., W.G. Knisel, and D.A. Still. 1987. GLEAMS: Groundwater Loading Effects of Agricultural Management Systems. *Transactions Am. Soc. Agri. Engrs.*, v. 30, p. 1403-1418.

4. Leonard, R.A., C.C. Truman, W.G. Knisel, and F.M. Davis. 1992. Pesticide runoff simulations: Long-term annual means vs. event extremes? *Weed Technology*, v. 6, p. 725-730.

5. Michael, J.L., and D.G. Neary. 1993. Herbicide dissipation studies in southern forest ecosystems. *Environ. Toxicol. Chem.* 12:405-410.

6. Michael, J.L., D.G. Neary, D.H. Gjerstad, and P. D'Anieri. 1990. Use, fate, and risk assessment of forestry herbicides in the southern United States. Proceedings of XIXth IUFRO World Congress, Montreal, Canada. Canadian International Union of Forestry Research Organizations, 2:300-311.

7. Miller, J.H., and L.M. Bishop. 1989. Optimum timing for ground-applied forestry herbicides in the South. U.S. Dept. of Agriculture, Forest Service, Cooperative Forestry, Management Bulletin R8-MB 28.

8. Neary, D.G., D.L. Rockwood, N.D. Comerford, B.F. Swindel, and T.E. Cooksey. 1990. Importance of weed control, fertilization, irrigation, and genetics in slash and loblolly pine early growth on poorly drained Spodosols. *For. Ecol. Manage.* 30:271-281.

9. Nutter, W.L., W.G. Knisel, P.B. Bush, and J.W. Taylor. 1993. Use of GLEAMS to predict insecticide losses from pine seed orchards. *Environ. Tox. and Chem.* 12:441-452

10. Nutter, W.L., F.M. Davis, R.A. Leonard, and W.G. Knisel. 1994. Simulating forest hydrologic response with GLEAMS. *Transactions Am. Soc. Agri. Engrs.*, (In press).

11. Riekerk, H. 1989. Forest fertilizer and runoff-water quality. Soil and Crop Science Society of Florida, Proceedings, Volume 48:99-102.

12. Smith, M. C., W. G. Knisel, J. L. Michael, and D. G. Neary. 1993. Simulating effects of forest management practices on pesticide losses with GLEAMS. Proceedings of the International Runoff and Sediment Yield Modeling Symposium, Warsaw, Poland, September 1993. pp. 157-162.

SUSTAINABLE VEGETABLE CULTURE ON ERODIBLE LANDS

M. E. Byers, G. F. Antonious, D. Hilborne and K. Bishop*

ABSTRACT

In Kentucky, the need for sustainable, environmentally sound agricultural systems is acute. Many of Kentucky's farmers (80%) are considered to be "limited resource", and farm small acreages that may be highly erodible. These small farms generally produce traditional crops grown using traditional technology. Once in a while, it is good to examine traditions and see if they fulfill our expaectations for the future. Today it is expected that agricultural systems must produce a profitable product, conserve soil and water, limit off-target movement of ag-chemicals and be logistically feasible. When researchers are concerned with land sustainability and economic profitability they must integrate a variety of information. Soil loss, water conservation, pesticide dissipation, crop science and environmental impact must be incorporated into a single program yielding an appreciation of the "Big Picture". The need for systems research has been established (Francis 1990), and work is indeed being published (Tan et al. 1993) which integrates many aspects of production. This study was designed to incorporate an interdisciplinary approach to the assessment of a vegetable cropping system's influence on herbicide and insecticide dissipation, runoff water loss, sediment yield and crop yield. Provided here is an overview of the sustainable vegetable systems research project, details may be obtained elsewhere.

STATEMENT OF PROBLEM

Soil erosion is implicit in conventional agriculture and is a threat to agricultural sustainability (Laflen et al. 1990). Therefore, the use of inovative soil conserving practices in agriculture as well as horticulture are essential for sustained productivity. Inovative practices may include implementing common technologies such as using plastic mulches, or using uncommon technologies such as intercropped vegetative filter strips to conserve soil, and maintain productivity. The use of plant cover strips intercropped under highly erodible conditions has been successfully accomplished in India and the Dominican Republic, where elephant grass (<u>Pennesetum purpureum</u>) appeared to styme erosion, although no sediment data were quantified (Thomas 1988). These practices may also influence the movement of agricultural chemicals. Benefits of living mulches have been documented (Cook 1982). The influence of fescue strips intercropped with vegetables on soil loss and herbicide movement needs to be quantified. The benefits of plastic mulches on yield (Ricotta and Masiunas 1991, Wien and Minotti 1987, Carolus and Downs 1958) and its effect on soil microclimate (Clarkson 1960, Ashworth and Harrison 1983), have been well documented, but its' effect on soil conservation is not well known.

The use of pesticides in sustainable agriculture is debatable. However, pesticides including herbicides may reduce the need for tillage. Clomazone, a selective herbicide, is labled for weed control in pepper and pumpkin and may be suitable for inclusion into sustainable vegetable cropping systems. Clomazone fate has been studied (Loux et al. 1989a and 1989b), but dissipation through runoff under these cropping conditions is not known.

---

*M.E. Byers, Principal Investigator; G.F. Antonious, Co-Investigator; D. Hilborn and K. Bishop, Research Assistants; Water Quality Program, Kentucky State University, Community Research Service, Frankfort, KY, 40601.

MATERIALS AND METHODS

Field Study

Field plots were located in Franklin County, Kentucky, in the Kentucky River Watershed. Field plot soils (fine silty Lowell Series, Typic Hapludalfs, fine, mixed, mesic with 5% organic matter in the control zone) uniform throughout the plot area. Field plots (22.0 x 3.7-m, 0.008-ha, n = 18) on a 10% grade were planted with green bell pepper transplants (Lady Belle) and seeded with pumpkins (Big Moon) on the contour. In 1992, soil management practices included planting with black plastic (BP) mulch, fescue strips intercropped (FS) between rows (fescue, Kentucky 31), and no mulch (NM) bare soil. In 1993, treatments included fescue intercropped at a reduced rate (every other crop row). Plots were separated using 15.2-cm plastic edging buried to a depth of 10.2-cm. The experimental design included a 2 x 3 factorial with fixed factors including the two crops and three soil management practices. The influence of the factors and combinations on soil loss (kg/ha), water loss (l/ha), and herbicide movement (mg/ha and mg/l) during natural rainfall runoff events was determined by assigning the factor combinations to the plots in a randomized complete block design and conducting an analysis of variance. Natural rainfall quantity and intensity were treated as covariables. Factor combinations were replicated three times.

Runoff water and sediment were quantified using tipping buckets and Omega water meters (Omega Engineering, Stamford CT). Meters allowed for quantifying flow even during low flow events. Samples of water and sediment were collected during first flush (composite from first 30-L of runoff) using borosilicate glass containers. Water samples were transported to the laboratory immediately following collection. Rainfall quantity (cm) and intensity (cm/h) were determined using a Texas tipping bucket rain gauge interfaced with a datalogger (Campbell Scientific CR21X, Logan, UT).

Crop yields were determined through hand-harvesting and grading. Harvests were conducted according to vegetable crop recommendations.

Laboratory

Sediment losses were determined by collecting total filterable solids in a 946 ml sample of runoff water, air drying, and massing the sediment. Sediment (g) per 946 ml was converted to kg/ha based on total runoff water lost per runoff event, per each 0.008-ha plot.

Clomazone in runoff water samples was extracted using a liquid/liquid extraction process using chromatography grade n-hexane. Samples were then dried using anhydrous sodium sulfate, and concentrated using rotary/vacuum and nitrogen stream evaporation. Samples were reconstituted in n-hexane to a volume of 1-ml and stored at -15°C until analysis. Clomazone was extracted from runoff sediment samples by refluxing for 1-hour with 0.25N HCl, neutralizing the aqueous mixture, and performing the liquid/liquid extraction as previously described. Clomazone extraction efficiency from water and sediment was 90% and 87%, respectively. Field spikes conducted in replicate (n=3) indicated no degradation or loss of clomazone in field collected water samples due to transport and storage.

Analysis of concentrated samples for clomazone was accomplished using gas-liquid chromatography (GLC) and GLC coupled mass spectrometry (GC-MS). Samples were injected into GLC (HP Model 5890, Hewlett Packard Co., Palo Alto, CA) in 1-ul quantity. Injector, oven and detector temperatures were 250, 180 and 250°C, respectively. Column used was an HP-1, 100% methyl silicone, 15-m, 0.53-mm i.d. megabore. Carrier gas was helium, and flow was set at 17-ml/min. Detector used was a nitrogen-phosphorus detector (NPD). Mass spectral analysis included injecting 1-ul of sample into an HP 5890 GLC and detecting clomazone in a HP 5971A Mass Selective Detector (MSD) in EI mode. Inlet and detector temperatures were set at 250°C. Oven temperatures were 75 (1-min) to 235°C (hold 5-min), at 20°C per minute. GC-MS column was an HP-1, 15-m, 0.23-mm i.d. GC-MS was used to confirm presence of clomazone. All analyses were conducted using external standards (0.25, 0.5, 0.75, 1.0 ng/ul), generating a standard curve using Lotus 123 (Lotus Development Corporation, Cambridge, Mass.), conducting linear regression analysis of raw data regressing over known concentration, and comparing unknown peak average areas (n=2 areas from two consecutive injections) to the corresponding point on the standard curve.

DISCUSSION

This experiment has indeed caused an interdisciplinary effort to undertaken. Agronomists, chemists, engineers, and KSU students have determined the influence of the described soil treatments on yield, soil loss, water loss, pesticide movement. The use of sustainable techniques such as the intercropped fescue will retard soil loss, water loss, and reduce the concentration of herbicide in runoff water relative to conventional vegetable cropping systems (details found elsewhere). Vegetable yields were variable, but encouraging. Therefore, such techniques may prevent potential impacts of sediment and ag-chemicals introduced to surface water through runoff in conventional systems, and provide growers with an acceptable economic return.

REFERENCES

1. Ashworth, S and H Harrison. 1983. Evaluation of mulches for use in the home garden. Hortscience 18(2):180-182.

2. Carolus, R.L. and J.D. Downs. 1958. Studies on muskmelon and tomato responses to polyethylene mulches. Mich. Ag. Exp. Stat. Quart. Bul. 40:770-785.

3. Clarkson, V A. 1960. Effect of black plastic mulch on soil and microclimate, temperature, and nitrate level. Agron. J. 52(6): 307-309.

4. Cook, T. 1982. The potential of turfgrass as living mulches in cropping systems. Proceedings of Crop Production Using Cover Crops and Sods as Living Mulches Workshop. Oregon State University, Corvallis, Oregon.

5. Francis, C.A. 1990. Future dimentions of sustainable agriculture. in: Sustainable Agriculture in Temperate Zones. C.A. Francis, C.B. Flora, and L.D. King Eds. John Wiley and Sons, NY, NY. pp. 439-464.

6. Laflen, J.M., R. Lal, and S.A. El-Swaify. 1990. Soil erosion and a sustainable agriculture. In: Sustainable Agricultural Systems. Eds; C.A. Edwards, R. Lal, P. Madden, R.H. Miller, and G. House. Soil and Water Cons. Soc. Ankeny, Iowa. pp. 569-581.

7. Loux, M.M., R.A. Liebl, and F.W. Slife. 1989a. Avaiability and persistance of imazaquin, imazethapyr, and clomazone in soil. Weed Science. 37:259-267.

8. Loux, M.M., R.A. Liebl, and F.W. Slife. 1989b. Adsorption of clomazone on soils, sediments, and clays. Weed Science. 37:440-444.

9. Ricotta, J A and J B Masiunas. 1991. The effects of black plastic mulch and weed control strategies on herb yield. HortScience 26(5):539-541.

10. Tan, C.S., C.F. Drury, J.D. Gaynor, and T.W. Welacky. 1993. Integrated soil, crop, and water management system to abate herbicide and nitrate contamination of the Great Lakes. Wat. Sci. Tech. 28(3-5): 497-507.

11. Thomas, G.W. 1988. Elephant grass for soil erosion control and livestock feed. In: Conservation Farming on Steep Lands W.C. Moldenhauer and N.W. Hudson Eds. Soil and Water Cons. Soc. Ankeny, Iowa. pp. 188-193.

12. Wein, H C and P L Minotti. 1987. Growth, yield, and nutrient uptake of transplanted fresh-market tomatoes as affected by plastic mulch and initial nitrogen rate. J. Amer. Hort. Sci. 112(5):759-763.

NEMATODE MANAGEMENT IN SUSTAINABLE AGRICULTURE

R. McSorley*

ABSTRACT

Plant-parasitic nematodes are serious pests of many agricultural crops. Although nematicides have been widely used for managing nematodes, many alternative practices exist, including use of resistant plant cultivars, crop rotation, cover crops, organic amendments, and other cultural practices. Effects of alternative practices for management of the root-knot nematode (*Meloidogyne incognita*) were examined in two series of experiments in north Florida. Maximum yield of cowpea (*Vigna unguiculata*) in sites infested with *M. incognita* was achieved by reducing nematode populations by growing a poor host (rye, *Secale cereale*) as a winter cover crop, and using a cowpea cultivar with some nematode resistance. In a second series of experiments, densities of *M. incognita* in tropical corn (*Zea mays*) were not affected by changing tillage practices from conventional to no-till, but were lowered following rotation with a poor host (sorghum, *Sorghum bicolor*) compared to a good host (soybean, *Glycine max*). Use of individual or integrated cultural practices can become increasingly effective in nematode management, as more data on specific methods, nematodes, and crops are obtained.

MANAGEMENT OF PLANT-PARASITIC NEMATODES

Plant-parasitic nematodes are an important limitation to crop production in many areas of the United States and worldwide (Sasser and Freckman, 1987). In the past, nematode problems in many agricultural crops have been managed by application of chemical nematicides, including broad-spectrum soil fumigants (Hague and Gowen, 1987). With increased regulation and recognition of the environmental hazards associated with use of these materials (Thomason and Caswell, 1987), emphasis is shifting toward alternatives which are more sustainable and less disruptive to the environment. Numerous alternative methods can be effective in nematode management, including use of resistant cultivars, crop rotation, cover crops, organic amendments, soil solarization, and numerous other cultural practices (Johnson, 1982; McSorley and Gallaher, 1991; Overman, 1985; Rodriguez-Kabana, 1986; Trivedi and Barker, 1986). However, much additional research information is needed before most of these practices can be widely used. For example, although nematode-resistant cultivars are available for many crops (Fassuliotis, 1982; Sasser and Kirby, 1978), there are other crops for which resistance is rare (Fassuliotis, 1982), and many crops and cultivars which simply have never been evaluated for their response to nematodes. With most alternative methods of nematode management, a more detailed knowledge of nematode biology and ecology is required than was necessary for management with fumigants and other nematicides. In addition, since some alternative methods, such as soil solarization, may not be as effective as soil fumigation (Overman, 1985), it may be necessary to integrate several of these cultural practices to achieve maximum efficacy.

It is critical to recognize that most agricultural soils contain not one, but several species of plant-parasitic nematodes which may behave quite differently (McSorley and Dickson 1989a; Thomason and Caswell, 1987). Although broad-spectrum soil fumigants may have been

---

*R. McSorley, Professor, Department of Entomology and Nematology, Institute of Food and Agricultural Sciences, University of Florida, Gainesville, FL 32611-0620.

effective against several different nematode species at once, many alternative methods may be effective against only one or a limited number of species. For example, a winter cover crop of hairy vetch (*Vicia villosa*) increased densities of the root-knot nematode (*M. incognita*) but decreased densities of the sting nematode (*Belonolaimus longicaudatus*), while a winter cover crop of rye had the opposite effect on both species (McSorley and Dickson, 1989b). Since it is unlikely that one method will be effective against all nematode species, we have chosen to emphasize methods effective against root-knot nematodes (*Meloidogyne* spp.), which are the key nematode pests in many agricultural systems (McSorley and Gallaher, 1991; Thomason and Caswell, 1987), and are damaging to many crops grown in the southeastern United States (Fassuliotis, 1982; Johnson, 1982).

NEMATODE MANAGEMENT IN NORTH FLORIDA: EXAMPLES

Numerous studies of cultural practices for management of plant-parasitic nematodes have been conducted in north Florida (McSorley and Gallaher, 1991), and two examples are presented (Gallaher and McSorley, 1993; McSorley and Gallaher, 1992; 1993). Both experiments were conducted on sandy soils (90-94% sand, 3-4.5% silt, 2.5-6% clay) in Alachua Co., Florida. Each experiment involved the integrated use of two management methods against the root-knot nematode (*M. incognita*) and ring nematodes (*Criconemella* spp., a mixture of *C. ornata* and *C. sphaerocephala*). *Meloidogyne incognita* is a serious pest of many agricultural crops (Fassuliotis, 1982; Johnson, 1982), whereas many agronomic crops are not affected by *Criconemella* spp. (Barker et al., 1982; McSorley et al., 1987).

Effect of Winter Cover Crop and Cowpea Cultivar

In spring 1991, seven sites in close proximity were sampled for plant-parasitic nematodes (McSorley and Gallaher, 1992). These sites differed in the cover crops planted during the winter of 1990-91 (Table 1). Population densities of *M. incognita* were very low following fallow or grain crops such as rye or wheat, but high fallowing legumes, especially crimson clover or hairy vetch (Table 1). Abundance of *Criconemella* spp. followed a different pattern, with maximum numbers following wheat or surviving fallow, and lowest numbers following lupine. Since *M. incognita* is a more serious plant parasite than *Criconemella* spp., the results and rotation choices minimizing buildup of *M. incognita* are more important for nematode management.

Two of these sites, within 100 m of each other, were selected for experiments conducted during the summer of 1991. The site previously planted to rye was selected as a site with low *M. incognita* density and the site following clover was selected as a high-density site. At each site, seven cowpea cultivars were planted on 24 May in small plots (4 rows, 3.0 m long and 76 cm apart) in a randomized complete block design with four replications (Gallaher and McSorley, 1993). Cowpea can be severely damaged by *Meloidogyne* spp., but resistance to one or more species and races of root-knot nematodes is available in some cultivars (Kirkpatrick and Morelock, 1987; Sasser and Kirby, 1979). Cowpea pod yield at 50% moisture was determined on 23 August, at which time a soil sample was collected from each plot for extraction (Jenkins, 1964) of plant-parasitic nematodes.

Of the cowpea cultivars planted, five were yield-determinant but two were indeterminant (Gallaher and McSorley, 1993). Yields and nematode densities are shown for the five determinant cultivars (Table 2). No differences ≤ 0.05) in population densities of *Criconemella* spp. were observed among the five cultivars, but population densities of *M. incognita* at harvest had built up to very high levels (>600 nematodes/100 cm^3 soil) on three of the cowpea cultivars, even in the site previously planted to rye, in which initial root-knot nematode density was low (6 nematodes/100 cm^3 soil). Population increase of root-knot nematodes was prevented by 'Mississippi Silver' cowpea in the site following rye. In the site following clover, the final population of root-knot nematodes following this cultivar (120 nematodes/100 cm^3 soil) was less than the initial population (353 nematodes/100 cm^3 soil). 'Mississippi Silver' was the highest-yielding cowpea cultivar at both sites, but yield differences among cultivars were greatest at the site following clover, where root-knot nematode pressure was greatest. At this site, 'Whippoorwill' cowpea yielded only 32% of its yield level at the site following rye, whereas the more nematode-resistant cultivar, 'Mississippi Silver,' yielded 86% of its yield level at the site following rye. To achieve maximum cowpea yields, it was necessary to use a suitable winter cover crop (rye) to lower root-knot nematode densities, and to plant a cultivar with some root-knot nematode resistance.

Effect of Crop Rotation and Tillage

A site which had been maintained for many years with conventional-tillage or no-tillage and soybean or sorghum in the summer and a winter cover crop of oats (*Avena sativa*) was used to determine the effect of crop rotation and tillage on nematode densities (McSorley and Gallaher, 1993). The experimental design was a factorial, with two tillage regimes (conventional vs. no-till) x two summer rotation crops ('Centennial' soybean or 'DeKalb BR 64' sorghum) planted during 1989. 'Pioneer Brand X304C' tropical corn was planted in May 1990 in small plots (4 rows, 10 m long), with each treatment replicated four times. Corn was planted in May and harvested in September each year from 1990-1992, and in each winter, a cover crop of oats was maintained. Nematode samples were collected at planting and harvest of each corn crop.

Root-knot and ring nematodes occurred in the corn crop every year, but were not significantly affected by tillage practices (Table 3). Rotation crop had a strong effect on nematode populations, with root-knot nematodes building up by the end of the corn crop every year in plots which had been planted to soybean (a good host) in 1989, while only very low numbers were detected in plots which had sorghum (a poor host) in 1989. Beneficial rotation effects were observed even in 1992, three years after the two different rotation crops were planted. In 1990, ring nematodes built up more in plots previously planted with sorghum than with soybean, but these differences became less evident over time (Table 3). Since ring nematodes do not seem to damage corn (Barker et al., 1982), crop rotation to avoid them is unnecessary, but the benefits of sorghum over soybean as a rotation crop for lowering densities of root-knot nematodes are clear, and consistent with other studies (Gallaher et al., 1991). Although changes in tillage practices have affected nematode populations in some instances, often effects have been minor or inconsistent (McSorley and Gallaher, 1993; Minton, 1986), especially in comparison with a suitable crop rotation (McSorley and Gallaher, 1993).

CONCLUSIONS

Decisions to change tillage practices should be based on agronomic benefits and not on nematode management, because tillage practices are not as effective as crop rotation in reducing nematode populations. The most frequently used and effective nonchemical methods for nematode management continue to be the use of resistant varieties, crop rotation, and cover crops. Successful use of these methods depends on the nematode present and the plant cultivar used. As additional information is obtained on the population dynamics of the various nematode species on a range of hosts, these methods will be applied more readily in nematode management in the future.

Table 1. Nematode Population Densities following Winter Cover Crops at Seven Sites in North Florida.

Winter Cover Crop	Nematodes per 100 cm³ Soil	
	Ring Nematode[a]	Root-Knot Nematode[a]
Wheat, *Triticum aestivum*	181[b]	3
Fallow (clean)	92	5
Rye, *Secale cereale*	31	6
Lupine, *Lupinus angustifolius*	3	20
Lupine, *L. angustifolius*	14	71
Clover, *Trifolium incarnatum*	30	353
Vetch, *Vicia villosa*	54	462

[a]Ring nematode = *Criconemella* spp.; Root-knot nematode = *Meloidogyne incognita*.
[b]Data are averages of four samples per site.

Table 2. Nematode Population Densities and Yields of Cowpea Cultivars at Sites Previously Planted to Rye and Clover.

Cowpea Cultivar	Nematodes per 100 cm³ Soil at Harvest				Yield (kg/ha)		%[c]
	Ring Nematode[a]		Root-Knot Nematode[a]				
	Rye	Clover	Rye	Clover	Rye	Clover	
Whippoorwill	56 a[b]	16 a	876 a	1718 a	5240 ab	1652 c	32
Pinkeye Purplehull	225 a	28 a	852 a	978 a	4409 b	3002 b	68
Texas Purplehull	64 a	35 a	2623 a	697 ab	1936 c	995 c	51
Purple Knuckle	240 a	18 b	81 b	253 bc	4931 b	3887 b	79
Mississippi Silver	92 a	18 c	6 c	120 c	6566 a	5616 a	86

[a]Ring nematode = *Criconemella* spp.; Root-knot nematode = *Meloidogyne incognita*.
[b]Data are means of four replications. Means in columns followed by the same letter are not different ($P \leq 0.05$), by Duncan's multiple-range test.
[c]Cowpea yield at site previously planted to clover as a percent of yield at site previously planted to rye.

Table 3. Effect of Tillage and Rotation Crop on Nematode Population Densities on Tropical Corn during 1990-1992 Seasons.

Tillage	Rotation Crop (1989)	Nematodes per 100 cm^3 Soil on Corn Crop					
		1990		1991		1992	
		May	Sept.	May	Sept.	May	Sept.
Root-Knot Nematode, *Meloidogyne incognita*							
No-till	Soybean	1	3	10	26	1	112
No-till	Sorghum	0	0	0	2	0	0
Conventional	Soybean	0	10	1	104	3	91
Conventional	Sorghum	0	0	0	1	0	3
ANOVA Effects:[a]							
Tillage		-	ns	ns	ns	ns	ns
Rotation Crop		-	*	ns	*	ns	*
Tillage x Rotation		-	ns	ns	ns	ns	ns
Ring Nematodes, *Criconemella* spp.							
No-till	Soybean	40	138	361	374	347	552
No-till	Sorghum	98	202	160	234	460	643
Conventional	Soybean	20	246	149	291	448	1178
Conventional	Sorghum	138	314	391	647	965	999
ANOVA Effects:[a]							
Tillage		ns	ns	ns	ns	ns	ns
Rotation Crop		*	ns	ns	ns	ns	ns
Tillage x Rotation		*	ns	ns	*	ns	ns

[a]Data are means of four replications. * indicates analysis of variance (ANOVA) effect significant at $P \leq 0.05$; dashes "-" = data not analyzed.

REFERENCES

1. Barker, K. R., D. P. Schmitt, and V. P. Campos. 1982. Response of peanut, corn, tobacco, and soybean to *Criconemella ornata*. J. Nematol. 14:576-581.

2. Fassuliotis, G. 1982. Plant resistance to root-knot nematodes. Pp. 33-49. In: Nematology in the Southern Region of the United States, R. D. Riggs (ed.), Southern Coop. Ser. Bull. 276, Arkansas Agric. Exp. Sta., Fayetteville. 206 p.

3. Gallaher, R. N., and R. McSorley. 1993. Population densities of *Meloidogyne incognita* and other nematodes following seven cultivars of cowpea. Nematropica 23:21-26.

4. Gallaher, R. N., R. McSorley, and D. W. Dickson. 1991. Nematode densities associated with corn and sorghum cropping systems in Florida. Suppl. J. Nematol. 23:668-672.

5. Hague, N. G. M., and S. R. Gowen. 1987. Chemical control of nematodes. Pp. 131-178. In: Principles and Practice of Nematode Control in Crops, R. H. Brown and B. R. Kerry (eds.), Academic Press, Orlando. 447 p.

6. Jenkins, W. R. 1964. A rapid centrifugal-flotation technique for separating nematodes from soil. Plant Dis. Rptr. 48:692.

7. Johnson, A. W. 1982. Managing nematode populations in crop production. Pp. 193-203. In: Nematology in the Southern Region of the United States, R. D. Riggs (ed.), Southern Coop. Serv. Bull. 276, Arkansas Agric. Exp. Sta., Fayetteville. 206 p.

8. Kirkpatrick, T. L., and T. E. Morelock. 1987. Response of cowpea breeding lines and cultivars to *Meloidogyne incognita* and *M. arenaria*. Suppl. J. Nematol. 19:46-49.

9. McSorley, R., and D. W. Dickson. 1989a. Effects and dynamics of a nematode community on soybean. J. Nematol. 21:490-499.

10. McSorley, R., and D. W. Dickson. 1989b. Nematode population density increase on cover crops of rye and vetch. Nematropica 19:39-51.

11. McSorley, R., and R. N. Gallaher. 1991. Cropping systems for management of plant-parasitic nematodes. Pp. 38-45. In: Proceedings of the Conference on Environmentally Sound Agriculture, A. B. Bottcher (ed.), Florida Coop. Ext. Serv., Gainesville. 694 p.

12. McSorley, R., and R. N. Gallaher. 1992. Comparison of nematode population densities on six summer crops at seven sites in north Florida. Suppl. J. Nematol. 24:699-706.

13. McSorley, R., and R. N. Gallaher. 1993. Effect of crop rotation and tillage on nematode densities in tropical corn. Suppl. J. Nematol. 25:814-819.

14. McSorley, R., J. L. Parrado, R. V. Tyson, and J. S. Reynolds. 1987. Effect of *Criconemella onoensis* on potato. J. Nematol. 19:228-232.

15. Minton, M. A. 1986. Impact of conservation tillage on nematode populations. J. Nematol. 18:135-140.

16. Overman, A. J. 1985. Off-season land management, soil solarization and fumigation for tomato. Soil Crop Sci. Soc. Fla. Proc. 44:35-39.

17. Rodriguez-Kabana, R. 1986. Organic and inorganic nitrogen amendments to soil and nematode suppressants. J. Nematol. 18:129-135.

18. Sasser, J. N., and D. W. Freckman. 1987. A world perspective on nematology: the the role of the society. Pp. 7-14. In: Vistas on Nematology, J. A. Veech and D. W. Dickson (eds.), Society of Nematologists, Hyattsville, MD. 509 p.

19. Sasser, J. N., and M. F. Kirby. 1979. Crop cultivars resistant to root-knot nematodes, *Meloidogyne* species. Dept. Plant Pathology, N. C. State Univ. and U. S. A. I. D., Raleigh. 24 p.

20. Thomason, I. J., and E. P. Caswell. 1987. Principles of nematode control. Pp. 87-130. In: Principles and Practice of Nematode Control in Crops, R. H. Brown and B. R. Kerry (eds.), Academic Press, Orlando. 447 p.

21. Trivedi, P. C., and K. R. Barker. 1986. Management of nematodes by cultural practices. Nematropica 16:213-236.

SOIL CONTAMINATION CAUSED BY DISCHARGE OF A MINE WATER

Wojcik and M.Wojcik [*]

ABSTRACT

A mine water from a Zinc and Lead ore mines has been discharged on 70 hectares wetland for long time. For last 15 years this discharge has been about 2 cubic meters per second. Extensive sampling and physical, chemical and roentgenographic analyses showed very high concentration of the heavy metals on the wetland. Maximum concentration of Zinc and Lead was 2.5 % and 1.34 %, respectively. This was particularly observed in the upper section of the wetland. Also very high concentration of heavy metals have been recorded in the leaves, stems and roots of the plants. Highest concentration of the Lead monitored in the roots of the Bulrush and Sedge was 800 ppm and 1050 ppm, respectively.

KEY WORDS: Wetland, heavy metals, soil pollution, wastewater

INTRODUCTION

This paper presents some of the results of the investigations of the interactions of the wastewater containing heavy metals with the Biala River wetland in Southern Poland. Industrial westewater containing mainly mine water from Zinc and Lead are mines is discharged into the Biala River Wetland through the Dabrowka Channel (Fig.1). The proportion of the particular waters are as follow:
mine water:about 100 cubic meters pow minute (i.e. about 85 to 90% of total)
waters from the flotation tailing ponds: about 3.5 c.m.p. m.(i.e. about 3% of total)
municipal wastewater after secondary treatment:about 10 m.p.m.(about 7% of total)
The wastewater contains high concentration of Zinc (about 2.0 ppm) and Lead (about 0.7 ppm.), which are present mainly in the forms of suspended solids and colloids.

Several objectives of the research of were specified. One of the main objectives was to characterize the current environmental quality of the wetland. For this purpose investigations of soil, vegetation and water were carried out.

The Biala River Wetland is about 3.5 km long and its width ranges from 70 m to 350 m. The longitudinal slope of the wetland varies from 0.1 % to 0.6 % . The overall surface of the wetland is about 70 ha.
For better understanding of current situation on the wetland it is important to look at the historical data. The story of the effect of human activities on this area is a long one. The Ponikowska Adit, built in 16th century to discharge a mine water from the ore deposits in the Olkusz region, was carrying off the mine waters to the Biala River at the Laski Village. Increased flow was even as much as four times greater than the natural flow of the river. A distinct increase in the amount of water carried off took place in the 1960's and 1970's in connection with the operation of the new mines 'Olkusz' and 'Pomorzany'. Also municipal wastewater from the Olkusz City was added at that time. At present, all the above waters and sewage are carried away together into the Biala River through the Dabrowka Channel. Due to extensive drainage of Pomorzany mine, the sources of the Biala River and its confluents disappeared in 1975. Thus the water running through the wetland is chiefly that disposed wastewater.

---

[*]University of Mining and Metallurgy, al. Mickiewicza 30, 30-059 Kraków, Poland

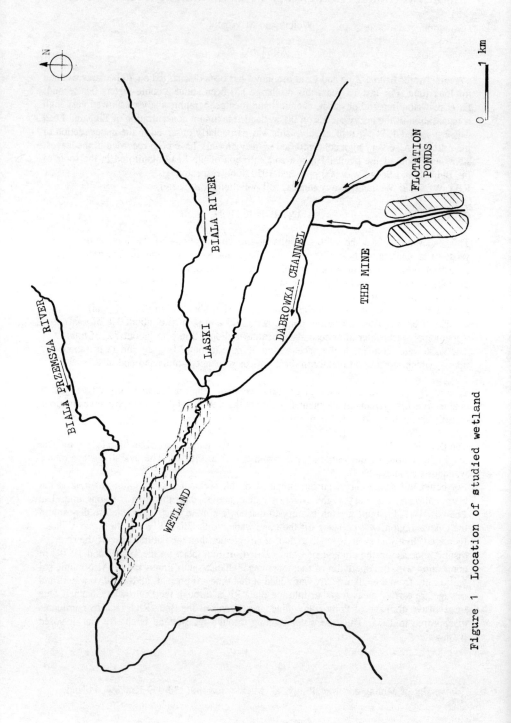

Figure 1 Location of studied wetland

SOIL AND ACCUMULATED SEDIMENTS ON THE WETLAND

The sampling stations for soil analyses were are shown in Fig. 2.
The following characteristic horizons levels were observed in the excavations :
sod horizon, dark, with great amounts of organic fossil remains, most often of the composition of clay and heavy clay;
pseudogley horizon of uniform blue color, occasionally (especially in shallower layers) yellowish spots, usually very moist or even saturated;
black horizon, of various thickness, containing great amount of organic matter decomposed to various degrees, however, most often the plant remains could be easily recognized;
horizon with yellowish or blue sand, occasionally with a composition close to that of dust clays, usually strongly gleized.

Following parameters have been analyzed: organic matter, specific weight, bulk density, porosity, pH, friction contents including sieve curve, Zn, Pb, Cd, Ca, Mn, Mg, Fe, and S. Most characteristic results of the analyses of the physical and chemical properties of the soil samples are shown in Table 1 Figure 3. The shallower layers of soil are characterized by high specific density (up to 3.09 g/cm3 for samples containing organic matter). The soil had high content of Ca and Mg, which attained as much as 28.7% and 8.5%, respectively. The concentration of Zinc, Lead and Cadmium was high up to depth of 1 meter (Fig.3). This was particularly observed in the up-stream section of the wetland. Maximum concentration of Zinc and Lead was 4.4% and 1.34%, respectively.
An interesting regularity has also been observed. At several places the concentration of metals in the first layer of soil was smaller than a little lower, down from the depth of about 10-15 cm (Fig.3). This connected with a break on overloading of the wetland with the mine water in the year 1980.
The grain composition of the soil samples is relatively little diversified. These are heavy clays, or medium bordering on heavy, with a high content of dust fraction, and, according to classification adopted in geology, dusts or dusty clays.
The x-ray examination of the soil samples were also conducted. The samples from downstream part of wetland contain less carbonates and sulfides but more quartz and feldspar.
Also less dolomite was observed in deeper layers of excavations. In upper stream part there is a domination of dolomite and calcite. There was not so many minerals which contain Zinc or Lead (galenite). Probably it is due to fact that x-ray examination is not able to detect amorphous forms of the given elements.
All the results indicate that this soil is not the result of natural soil formation. It is rather deposited sediments from the discharged wastewater. The thickness of accumulated sediments ranges from 0 to 100 cm. Their accumulation is mostly evident along the section from the outlet of the Dabrowka Channel to about 1.5-2.0 km don the valley. There was very little deposit in the water - logged lower parts of the valley, and particularly in exposures close to the valley sides.

VEGETATION

Seventeen plant communities were distinguished on the wetland. For most of them, illustrative phytosociological pictures were made (occasionally several pictures of one community to observe and demonstrate its variation). Most widely spread plant communities were the marsh communities -all types of marshes as well as extensive patches with domination of the grass Desampsia (Desampsia caespitosa) and reeds (Scirpo - Phagmitetum). The flora of the valley of the Biala River shows some peculiar features. One of them is the absence of certain species, which are common elsewhere , e.g. willows , and absence of same species typical for meadows , such as Bellis perennis. It seems this may be associated with the shortage of nutrients on the wetland.

Table 1. Results of analysis of the physical and chemical characteristics of the soil from the valley the Biala River (see Fig.2 for sampling location)

Sample number	Depth [cm]	Fraction contents in [%] Diameter in [mm]						Weight Density [g/cm^3]	Bulk Density [g/cm^3]	pH H2O	pH KCl	Organic Matter [%]
		1-0.1	0.1-0.05	0.05-0.02	0.02-0.006	0.006-0.002	<0.002					
31	0-30	48	20	18	8	6	0	2.11	0.52	5.9	5.8	
31	30-70	75	7	8	2	5	3	2.26	0.87	6	5.7	
31	70-80	51	10	12	10	5	12	2.62	1.53	6.4	6.3	
32	0-30	14	6	6	37	25	12	2.87	1.21	7.5	7.5	4.21
32	30-50	66	14	10	6	2	2	2.32	0.79	7.2	7.2	19.34
32	50-80	75	7	5	5	3	5	2.62	1.65	6	5.6	2.31
33	0-30	20	6	22	42	9	1	2.97	1.28	7.7	7.7	2.62
33	30-60	13	4	7	36	26	14	2.93	1.28	7.4	7.4	2.36
33	60-90	15	7	24	31	15	8	2.79	1.18	7.4	7.2	5.73
34	0-10	0	14	33	39	10	4	2.93	1.36	7.5	7.5	4.33
34	10-20	4	10	36	34	10	6	3.02	1.54	7.8	7.8	0.76
34	20-60	3	7	11	42	23	14	2.82	1.21	7.6	7.5	2.96
34	60-100	13	12	26	28	17	4	2.75	1.13	7.5	7.3	7.87
35	0-25	59	14	19	4	2	2	2.28	0.91	6.4	5.9	23.86
35	25-75	14	22	40	10	5	9	2.64	1.46	6.8	6.5	1.38
35	75-150	88	6	3	0	1	2	2.65	1.76	6.9	6.9	0.2
36	0-15	13	7	8	38	25	9	2.85	1.2	7.5	7.5	2.78
36	15-55	11	2	8	34	31	14	2.89	1.28	7.6	7.6	0.69
36	55-80	28	10	16	16	18	12	2.54	0.84	7.4	7.2	11.79
36	80-130	33	14	30	11	0	12	2.37	0.99	6.5	6.2	20.48
37	0-40	51	31	12	3	2	1	3.09	1.75	7.5	7.3	1.91
37	40-70	29	4	19	25	15	8	2.61	0.33	6.8	6.8	7
37	75-100	90	3	2	0	1	4	2.61	1.44	7	6.6	2.06
40	0-30	14	33	30	17	5	1	3.02	1.72	7.5	7.5	1.46
40	30-70	10	6	19	42	15	8	3.01	1.43	7.5	7.4	0.57
40	70-100	40	18	26	10	0	6	1.94		6.7	6.5	46.61
41	0-30	0	15	47	30	5	3	2.98	1.62	7.8	7.8	0.62
41	30-60	8	5	21	40	17	9	3.06	1.45	7.7	7.7	0.43
41	60-120	38	20	24	10	0	8	2.17	0.68	6.5	6.2	30.78
42	0-15	22	28	27	16	6	1	3.01	1.63	7.5	7.5	2.65
42	15-70	36	16	20	14	12	2	2.54	0.98	7	6.8	13.66
42	7-150	50	45	2	2	1	0	2.63	1.44	7.5	6.2	1.5
43	0-10	15	18	38	14	6	9	2.57	1.32	5.6	5.1	4.69
43	10-50	11	13	47	14	4	11	2.63	1.47	6.1	5.8	1.12
43	50-70	9	12	44	19	7	9	2.67	1.54	5.1	4.1	0.59
43	70-110	8	4	25	28	7	28	2.67	0.57	5	3.8	1.25
45	0-15	77	8	5	2	5	3	2.48	1.17	7	6.7	10.5
45	15-25	95	2	2	1	0	0	2.64	1.71	7.6	6.7	0.5
45	25-50	97	1	1	0	1	0	2.65	1.74	7.5	6.8	0.2
46	0-5	88	3	4	4	1	0	2.12	0.72	6.7	6.2	33.86
46	5-15	33	8	3	25	20	11	2.9	1.14	7.2	7	4.07
46	15-60	88	3	4	5	0	0	2.43	1.25	7.4	7.1	10.93
47	0-25	72	9	8	6	4	1	2.29	0.75	6	5.6	2.27
47	25-50	73	8	8	5	3	3	2.59	1.32	5.9	5.9	4.43
47	50-80	83	4	4	1	4	4	2.64	1.59	6.8	6.1	1.19
48	0-10	74	11	9	4	0	2	2.41	1.13	6	5.4	15.9
48	10-60	93	4	1	2	0	0	2.59	1.54	6.3	6.3	1.09
48	60-80	60	93	2	1	1	0	3	2.64	1.61	6.2	60
49	0-15	81	11	3	2	1	2	2.57	1.3	5.6	5	4.1
49	15-30	83	9	3	1	2	2	2.65	1.41	5.5	4.8	1.96
49	30-50	30	95	4	1	0	0	0	2.64	1.62	6.4	30

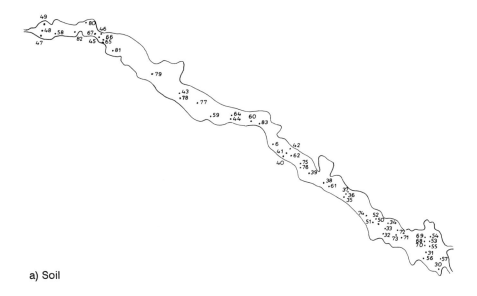

a) Soil

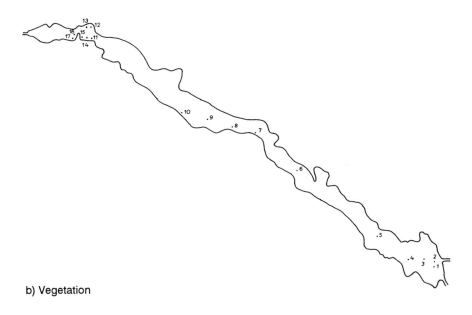

b) Vegetation

Figure 2. Locatio of the sampling station for soil and vegetation analyses

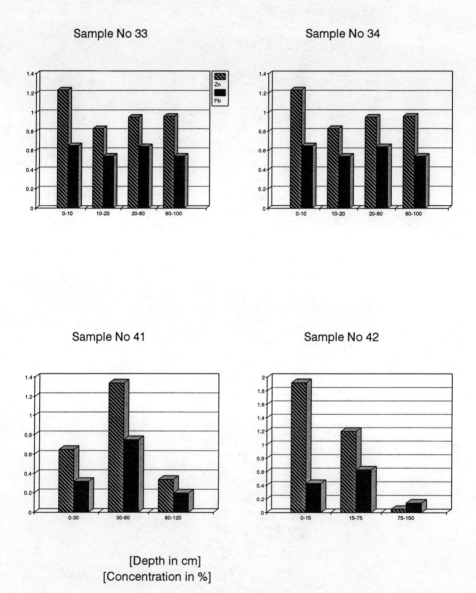

[Depth in cm]
[Concentration in %]

Figure 3a. Changes of Zn and Pb concentration in soil on various depths.

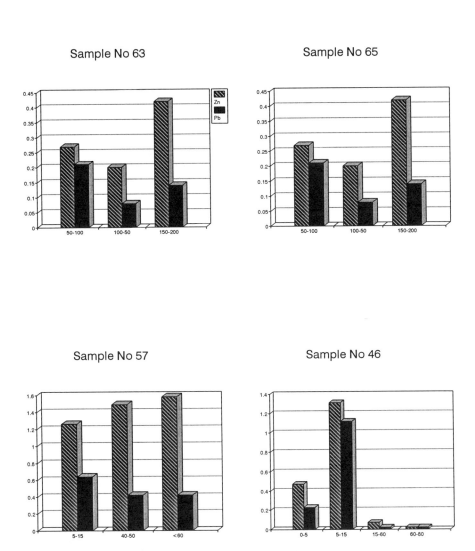

Figure 3b. Changes of Zn and Pb concentration in soil on various depths.

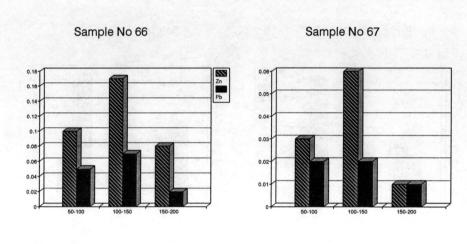

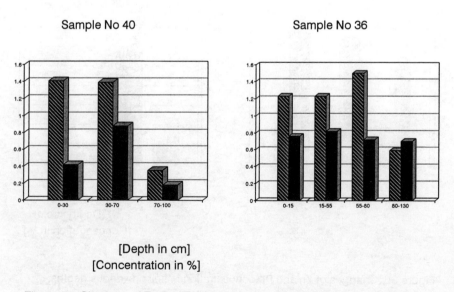

[Depth in cm]
[Concentration in %]

Figure 3c. Changes of Zn and Pb concentration in soil on various depths.

The investigations of heavy metals concentration in the plants growing on the wetland were carried out too. Totally 85 plant samples were taken from 17 sites along the wetland (Fig. 2). The concentration of metals found in plants varied from 25 ppm to 6500 ppm for Zinc, from 5 ppm to 1050 ppm for Lead, and from 0.5 ppm to 48 ppm for Cadmium. Average concentration of heavy metals in plant leaves was for Zn from 29.3 ppm in Typha latifolia up to 270 ppm in Deschampsia caespitosa, for Pb from 16.3 ppm in Typha latifolia up to 91.6 ppm in Mentha aquatica, and for Cd from 1.3 ppm in Phragmites communis and Typha latifolis up to 3 ppm in Scirpus lacustris. In plant stems the average concentration of heavy metals was for Zn from 106 ppm in Mentha aquatica up to 126 ppm in Phragmites communis, for Pb from 20.4 ppm in Phragmites communis up to 33.8 ppm in Mentha aquatica. Concentration of Cd was found on average about 1.6 ppm in Phragmites communis and similar value 1.8 ppm in Mentha aquatica.

The higher concentration of metals was determined in most cases in the underground parts of plants rather than in leaves or stems (Fig.4 and Fig.5). Average concentration of metals in rhizomes were 2 to 3 times greater than in leaves and stems in Phragmites communis for both Zinc and Cadmium, and 3 to 7 times greater for Lead. In Typha latifolia the average concentration was about 5 to 8 times greater in rhizomes for both Zinc and Lead, however for Cadmium values were about the same in above and underground parts. The difference in metals concentration between underground parts versus aboveground parts was even more significant in Carex sp., where it was 10 to 20 times greater for each metal.

The highest concentration 6500 ppm and 1050 ppm of Zinc and Lead respectively, was found in Carex sp. (underground parts), and the highest concentration 48 ppm of Cadmium was found in Potamogeton natans (whole plant).

Interesting is to compare concentration of heavy metals in a soil, in underground, and in above-ground parts of the plants. For this purpose the soil from a root zone of several particular plants were excavated. Such paths of heavy metals transportation showed that the Water Mint had smallest difference between concentration of metals in soil and roots. Also for Cattail this difference was not large. For Sedge it was contrary. there was only two times difference between concentration of Zinc and Lead in the soil and plant roots.

CONCLUSIONS

The soil of the Bials River Wetland is strongly polluted up to the depth of 1 m, mainly by heavy metals. This pollution was caused by sediments of suspended solids discharged in high concentration with industrial wastewater. In the same time high level of Calcium and organic matter, support good growth of the vegetation on the wetland. It means that this pollution does not inhibit a growth of vegetation.

The higher concentration of heavy metals was determined in most cases in the underground parts of plants rather than in leaves or stems. This difference was most significant in Carex sp.. Also highest concentration of Zinc and Lead were observed in these plant- 6500 ppm and 1050 ppm respectively.

ACKNOWLEDGEMENT

This paper has adapted some data from the Sendzimir Family Project in wich Prof. H.T.Odum of the University of Floryda was a pricipal investigator.

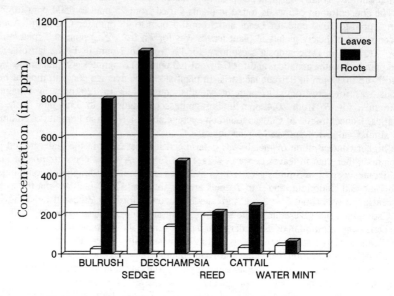

Figure 4. Concentration of lead in selected plants

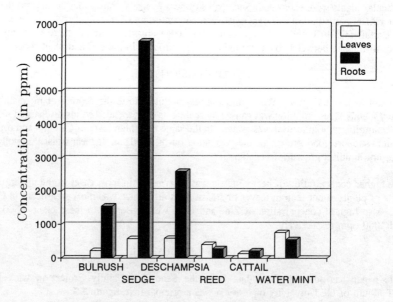

Figure 5. Concentration of zinc in selected plants

POTENTIAL IMPACT OF PROPER SOIL WATER MANAGEMENT ON ENVIRONMENTALLY SOUND AND ANTITOXIC FOOD PRODUCTION

BRANE MATIČIČ [1]

ABSTRACT

Influence of irrigation and fertilization with nitrogen on the yield and on the nitrate and nitrite content in crops and vegetables was observed in an experiment at Agrohydrological station of the Biotechnical Faculty, Ljubljana in the period 1985 - 1993. Experimental plants (cabbage, garden beet, celery, chicory, lettuce, sugar beet, potatoes) were growing in 72 weighing lysimeters, one crop per year. Plants were treated with four nitrogen doses and six water application doses in three replications. Influence of irrigation on yield was significant and higher than the influence of fertilization with nitrogen. Water-fertilizer production functions show a significant impact of irrigation on quantity and quality of the crops. It was established that the influence of irrigation and fertilization with nitrogen on the nitrate and nitrite content in crops and vegetable samples was significant. The NO_2 and NO_3 content in plants increases significantly if plants are under water stress i.e. in conditions with minimal and maximal water content in the soil. Having in mind the necessity of producing healthy food with no toxic amounts of chemical substances, the proper water management in the soil (maintenance of water content in the soil within optimal range) is recommended. Therefore drainage and irrigation should be considered as potential water management measures for environmentally sound and antitoxic food production.

Keywords: irrigation, feltilization, yield, food quality

1. INTRODUCTION

The objectives of the study were to find the optimum soil moisture management for maximum production and monitoring the influence of different nitrogen doses and different water application levels on the nitrate and nitrite content in the plants.

Using high amounts of nitrogen in artificial fertilizers leads to an increase of nitrate in plants,(Eysinga et al., 1985) where the part of the nitrate, not absorbed in plants is washed away by rain and pollutes the ground water (Spalding et al., 1982). Eating vegetable and drinking water with high amount of nitrate and nitrite is toxic especially for children and babies, because it may cause methaemoglobinaemia.

Acceptable daily intake (ADI) for nitrites is 0-0.2 mg/kg of body weight, and for nitrates: 0-5 mg/kg of body weight. For children younger than six months it is not allowed eating food which contains nitrites. /Standards are appointed by common Committee FAO/SZO for nutrition additives (JECFA)/.

The quantity of nitrate present in vegetables depends on the relative speed of two physiological processes: absorption of nitrates by the roots and reduction of nitrate to effect the synthesis of protein. If the speed of absorption of nitrate is greater then the rate of protein synthesis, nitrate automatically accumulates in the plants. If, on the other hand, nitrate is transformed into protein in step with absorption, the nitrate will only be present in small amounts. Disproportion between

---

[1]University of Ljubljana, Biotechnical Faculty, Department of Agriculture, Division for Agricultural Water Management and Engineering
Jamnikarjeva 101, 61000 Ljubljana, Slovenia

nitrogen absorption and protein synthesis may be caused by two reasons which sometimes may be synergetic:
- a) excessive quantities of nitrate in the soil and
- b) a slowing down of protein synthesis.

It is known that the nitrate level in plants depends on the type of soil (Magette et al., 1990), climate conditions (water), conditions of cultivation (Boddy and Baker,1990), age of plants, type of vegetable and also cultivar and variety. Excess nitrate nitrogen in the soil results above all from chemical nitrogenous fertilizers particularly in the form of nitrates, which is easily available for vegetables (Aubert, 1983).

Several factors affect the intensity of protein synthesis; the most important are the following: light, the state of foliage, lack of minerals - particularly of trace elements and the use of certain pesticides. The feebler the light the richer the plants are in nitrate (crop under grass, particularly out of season, shady locations or overcrowding, short days and weak sunshine) (Gao et al.,1989). If leaves are insufficiently developed or damaged by a parasitic attack the nitrate content in the roots is raised. Lack of minerals slows protein synthesis which affect the accumulation of nitrates. All factors which obstruct normal biochemical processes in the soil or in the plant can slow down protein synthesis and therefore accelerate the accumulation of nitrates. For example: lack of water, inadequate drainage and inadequately cultivated soil, use of 2-3 D herbicide etc. (Aubert, 1983).

It is not nitrate that is toxic, but the compounds derived from it - nitrites and nitrosamines. The nitrites and nitrosamines are formed in microbiological process which takes place immediately after harvest whenever vegetable - medium reach in nitrate - is placed in anaerobic conditions at ambient temperature. The very same processes will take place in cooking vegetable if it is stored at ambient temperature. There are two reasons why nitrates derived compounds are toxic:

- a) nitrates oxidize the ferrous ion of the haemoglobin into ferric, methemoglobin is formed and transport of oxygen is troubled (methaemoglobinaemia - blue baby syndrome) and
- b) nitrosamines are carcinogenic (Aubert, 1983).

The aim of the research was to study water-fertilizer production functions for crops and vegetables, to establish the influence of different water and nitrogen application levels on yield and to find out ranges of water and fertilizer applications where the production functions have maxima, this means quantitative evaluation, as well as qualitative evaluation of crops: to observe the influence of different application levels of water and mineral nitrogen on the content of nitrate and nitrite in crops and vegetables and establish the optimal interaction of water management in the soil (drainage, irrigation) with nitrogen application for environmentally sound and antitoxic food production.

2. METHODOLOGY

Experiments were conducted at Agrohydrological Station of Biotechnical Faculty, Ljubljana in 1985-1993 (Matičič et al., 1991, 1992, 1993). Plants were growing in 72 weighing lysimeters. Plants were treated with six different water application levels ($I_1, I_2, I_3, I_4, I_5, I_6$) and four different nitrogen application levels (N_1, N_2, N_3, N_4). Therefore we got 24 different combinations each repeated three times to give 72 independent yield values. Automatic movable canvas was stretched above the weighing lysimeters to prevent rainfall. Water application rates were so absolutely controlled.

2.1 Fertilization

All lysimeters were fertilized with 150 kg of P_2O_5/ha and 300 kg K_2O/ha. Crops were treated with four different nitrogen application levels: 0 up to 400 kg of nitrogen/ha, which were divided into two to three parts. Nitrogen fertilizer was KAN (calcium-amon nitrate) with 27% of nitrogen; 50% of this is in form of nitrate, another 50% is in ammonium form.

2.2 Application of water

Plants in the lysimeters were watered through the dripping system on the soil surface. Daily amount of water was determined by the water application level and the potential evapotranspiration measured by weighing lysimeter as well as calculated by Penman's method for the previous day.

Water application levels were changed according to the following criteria: phenologic state of plants, change of soil moisture in lysimeters and change of lysimeter weight. Lysimeters were weighted twice a week (daily and weekly data were so obtained) while soil moisture in lysimeters was determined often gravimetrically.

2.3 Chemical analysis of nitrate, nitrite and ammonium

Immediately after the harvest the samples of fresh plants were grinned with mixer and analysis of nitrate (NO_3) and nitrite (NO_2) was done using Technicon autoanalyzer.

3. RESULTS AND DISCUSSION

3.1 Water production functions

Our study of water production functions is based on the assumption that the yield-irrigation-fertilization relationship can be expressed with a function of irrigation (I) and fertilization (N):

$$Y(I,N) = f(I)g(N) \qquad (1)$$

where Y is yield, I is irrigation and N is fertilization with nitrogen.

We expressed a function Y as a polynomial approximation. Regression plane for yield (Y) was determined considering the average measured values of three plants for each replication. Measured yield values were approximated with square polynomial using the least square method:

$$Y = b_0 + b_1 I + b_2 n + b_3 I^2 + b_4 N^2 + b_5 IN \qquad (2)$$

where Y = crop yield (g/lysimeter)
  I = irrigation (mm of water/vegetation period)
  N = fertilization with nitrogen (kg of N_2/ha)

As example two water-fertilizer production functions for cabbage and chicory are presented in Fig. 1a and 1b.

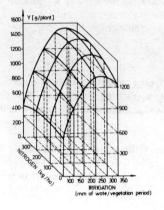 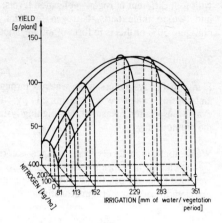

Figure 1: Water-Fertilizer Production Functions: a) Cabbage, b) Chicory

3.2 Yield

Maximal and minimal average weight of crops achieved, together with irrigation-fertilization combination are shown in Tab. 1.

Table 1: Average weight of crops achieved

CROP	MAXIMUM YIELD t/ha	COMBINATION	MINIMUM YIELD t/ha	COMBINATION
cabbage	56.2	I5 x N3	4.9	I1 x N4
celery	21.0	I6 x N4	2.9	I1 x N1
red beet	15.5	I3 x N4	5.1	I5 x N2
chicory	13.4	I6 x N3	2.2	I1 x N1
lettuce	25.4	I4 x N4	9.9	I1 x N1
sugar beet 91	95.0	I5 x N2	38.5	I1 x N1
sugar beet 92	130.1	I3 x N2	51.0	I1 x N1
potatoes	96.1	I6 x N2	43.3	I1 x N1

3.3 Content of nitrate and nitrite in the crop

As the example, NO_3 and NO_2 contents are presented here for some crops. Trends for all crops are similar.

3.3.1 <u>Nitrate - NO_3</u>. In fresh samples the amount of nitrate is varying in cabbage from 695 to 1320 mg/kg in red beet roots from 624 to 3427 mg/kg and in celery from 0 to 236 mg/kg in bulbs and from 1 to 1014 mg/kg in leaves. The amount of nitrate is the lowest in samples which were not treated with nitrogen, while the highest is in samples which were treated with the highest nitrogen doses (see Tabs 2,3 and 4).

Table 2: NO_3 content in cabbage.

FERTILIZATION			IRRIGATION (mm)			
kg N/ha	72	121	166	220	280	324
			NO_3 in mg/kg			
0	1124	1043	694	693	739	839
120	1059	1072	843	825	882	736
260	1319	1222	1113	990	797	1099
360	1183	1058	1095	941	992	1067

Table 3: NO_3 content in celery bulbs.

FERTILIZATION			IRRIGATION (mm)			
kg N/ha	203	305	458	630	789	1022
			NO_3 in mg/kg			
0	42.16	82.55	3.41	0.31	0.13	0.89
100	62.44	11.16	1.20	0.31	0.75	0.27
200	94.42	78.16	20.99	0.89	0.49	0.13
400	236.18	236.66	178.47	11.69	5.40	22.41

Table 4: NO_3 content in celery leaves.

FERTILIZATION			IRRIGATION (mm)			
kg N/ha	203	305	458	630	789	1022
			NO_3 v mg/kg			
0	19.62	121.65	8.86	0.97	23.91	0.80
100	311.11	55.89	24.14	0.71	1.73	2.26
200	778.81	259.65	154.65	2.57	0.75	1.11
400	1013.75	954.45	442.59	127.77	14.70	101.90

The increase of water application levels caused decrease of NO_3 content in cabbage until the water application reached the optimal level (in our experiment 280 mm of water/vegetation period), but further increase of water application also increased NO_3 content. Influence of irrigation on the content of NO_3 is greater than the influence of fertilization with nitrogen. Increasing water application levels (in our experiment up to 789 mm of water) caused decrease in amount of NO_3 in celery. Further increase of water application doses also increases NO_3 content (see Figs 2a and 2b). The extreme amounts of water: minimal, when plants are suffering from drought and maximal, when plants have too much water for normal growth, increase the value of NO_3 in celery as well as in other crops.

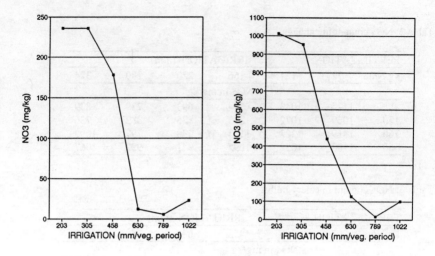

Figure 2: Nitrate Content in Celery at Different Water Levels and 400 kg/ha of Nitrogen:
a) Bulbs   b) Leaves

3.3.2 <u>Nitrite - NO_2</u>. The analysis of NO_2 content in experimental plants show high differences: in cabbage for example, nitrite contents is varying from 0.02 to 0.32 mg/kg, in red beet from 4.47 to 11.99 mg/kg, in celery from 0.08 to 0.40 mg/kg in bulbs and from 0.04 to 1.56 mg/kg in leaves. Nitrate content in red beet is extremely high. On average the lowest NO_2 content (0.03 mg/kg) in cabbage is found in samples which were treated with the maximal water level (see tables 5,6 and 7).

Table 5: NO_2 content in cabbage.

FERTILIZATION	IRRIGATION (mm)					
kg N/ha	72	121	166	220	280	324
	NO_3 in mg/kg					
0	0.022	0.025	0.026	0.012	0.018	0.009
120	0.030	0.026	0.022	0.096	0.029	0.008
260	0.018	0.035	0.069	0.024	0.006	0.018
360	0.026	0.012	0.058	0.042	0.038	0.022

Table 6: NO_2 content in celery bulbs.

FERTILIZATION	IRRIGATION (mm)					
kg N/ha	203	305	458	630	789	1022
	NO_3 in mg/kg					
0	0.09	0.11	0.10	0.10	0.06	0.06
100	0.40	0.10	0.14	0.10	0.09	0.11
200	0.09	0.08	0.11	0.09	0.08	0.10
400	0.12	0.11	0.11	0.11	0.11	0.13

Table 7: NO_2 content in celery leaves.

FERTILIZATION	IRRIGATION (mm)					
kg N/ha	203	305	458	630	789	1022
		NO_3 in mg/kg				
0	0.04	0.15	0.05	0.04	0.05	0.05
100	0.30	0.07	0.23	0.03	0.04	0.11
200	0.41	0.56	0.15	0.04	0.05	0.05
400	0.54	1.56	0.54	0.18	0.09	0.40

The amount of NO_2 in celery leaves is higher than in bulbs. The highest NO_2 content in leaves is in samples which were treated with the maximal nitrogen level (400 kg N/ha) and in bulbs which were treated with 100 kg N/ha. The lowest values of NO_2 are in samples which were not fertilized with nitrogen.

3.4 Approximation polynomials for nitrate content in crops

For the approximation of nitrate and nitrite content in crops similar polynomial approximation as for yield were used:

$$N = b0 + b_1*N + b_2*I + b_3*N^2 + b_4*I^2 + b_5*I*N \qquad (3)$$

where N is the amount of NO_3 (or NO_2) in sample.
The coefficients for the NO_3 content as examples in cabbage, celery and red beet are shown in table 8:

Table 8: Polynomial coefficients for NO_3 content in celery, cabbage and red beet.

	celery	cabbage	red beet
b_0	114.1	1439	2564
b_1	-0.3586	-5.449	-14.57
b_2	0.2512	0.773	2.298
b_3	0.00027	0.0102	0.0238
b_4	0.00092	-0.0012	-0.0029
b_5	-0.00068	0.0015	0.0025

The correlation coefficient for cabbage is $R^2=0.75$, for celery is $R^2=0.86$ and for red beet is $R^2=0.81$. Figure 3 shows the three-dimensional approximation polynomials for the NO_3 content in cabbage and in celery and Fig. 4 the same approximation for red beet. The minimal content of NO_3 in cabbage was reached only in relation to irrigation at 268 mm water/vegetation period which corresponds to the irrigation level where maximal yield was achieved (at 274 mm water/vegetation period).

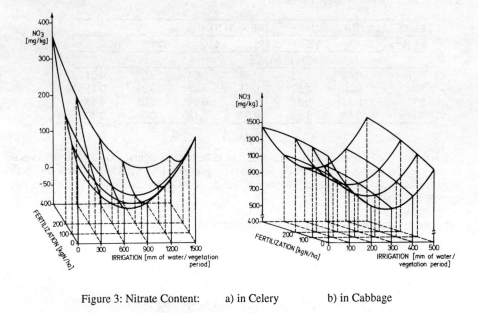

Figure 3: Nitrate Content: a) in Celery   b) in Cabbage

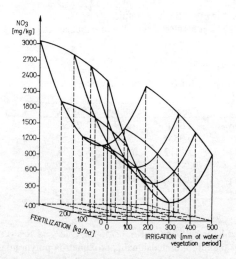

Figure 4: Nitrate Content in Red Beet

The nitrate content increases faster when irrigation is high then when irrigation is low, while absolute values are greater at low irrigation levels.

The minimal nitrate content in celery bulbs was reached at 196 kg N/ha and 899 mm water/vegetation period. The NO$_3$ minimum is moving from lower irrigation levels at low fertilization towards greater irrigation levels at high fertilization. For celery which was not treated with nitrogen the minimal concentration of NO_3 is at 650 mm water/vegetation period while for the maximal nitrogen application level the minimum lies at 1200 mm of applied water. The NO_3 content increases rapidly with fertilization at minimal irrigation while with increasing water application level the slope decreases. The same behaviour of NO_3 content can be seen in leaves.

The minimal content of NO_3 in red beet was reached at irrigation 333 mm water/vegetation period. Trends for all crops are similar.

5. CONCLUSIONS

The study of water production functions for crops and vegetables conducted in the period 1985-90 gave the following results of crop yield quantity/quality influenced by different water and nitrogen application doses: Influence of irrigation on yield was higher than the influence of fertilization with nitrogen.

Maximal estimated yield of celery bulbs (19.44 t/ha) was reached at I6 x N4. Maximal red beet yield was 15.5 t/ha at I3 x N4, maximal yield of chicory was 13.4 t/ha at I6 x N3 and maximal yield of lettuce was 25.4 t/ha at I4 x N4. Maximal estimated yield of sugar beet was 130.1 at I3 x N2 combination and potato yield was maximal at I6 x N2 combination and yielded 91.6 t/ha.

Analysis of NO_3 and NO_2 content in experimental plants which were treated with different water and nitrogen doses indicates that irrigation and fertilization influenced the nitrate and nitrite levels considerably, depending on the type of crop and vegetable and cultivar. The extreme amounts of water (minimal, maximal), when the crop is under water stress enlarge the value of NO_3 content enormously.

According to our experiment we can conclude:
1. Fertilization with nitrogen increases the nitrate level in crops and vegetable. The increase on N doses from 0 to 400 kg N/ha caused more than fifteen times higher NO_3 content in celery leaves and more than ten times higher nitrate content in celery bulbs as an example.

2. The increase of water application levels caused decrease of NO_3 content in crops and vegetable until the water application reached the certain limit (optimal water treatment), further increase of water application also increased NO_3 content. Low water content and high water content in the soil when plants are in water stress increases NO_3 content in plants considerably.

The analysis of approximation polynomials shows that the minimum of NO_3 content was reached in relation to irrigation for all crops with optimal water treatment.

3. Content of nitrite in plants was low (average 0.03 mg/kg). Plant samples which got optimal water doses and were not treated with nitrogen have the lowest level of nitrite.

Important reduction of nitrate and nitrite in crops and vegetable was observed with optimal water supply and with small doses of nitrogen fertilizers. The results achieved in lysimeters have relative values but they are an excellent indication for our agricultural technology behaviour in future: among all other agricultural technological measures a proper soil water management (drainage, irrigation) is an obligation for environmentally sound and antitoxic food production.

6. LITERATURE

1. Claude Aubert, M.: Nitrates in Vegetable: Some Possible Toxic Effects, Nutrition and Health. Vol. 2, No. 2 (1983), pp 77-84

2. Boddy, P. L., Baker, J. L.: Conservation Tillage Effects on nitrate and atrazine leaching, 1990 International Winter Meeting sponsored by The American Society of Agricultural Engineers, Chicago, Illinois, 1990

3. Eysinga, J.P.N.L.R.; Meijs, M.Q.: Effect of nitrogen nutrition and global radiation on yield and nitrate content of lettuce grown under glass, Communications in Soil Science and Plant Analysis (1985) 16. 1293-1300, Naaldwijk, Netherlands

4. Gao,Z.M.; Li,S.J.; So,C.J.; Hu,Y.B.; Kong,X.G.: Effect of NO_3N and Mo on nitrate accumulation in leaf vegetables under different light intensities, Acta Horticultural Sinica (1987) 14 192-196, Department of Horticulture Nonjing Agricultural University, China

5. Magette, W.L., Wood, J.D., Ifft, T.H.: Nitrate in shallow grownd water, 1990 International Winter Meeting sponsored by The American Society of Agricultural Engineers, Hyatt Regency Chicago, Illinois

6. Matičič, B. et al. (1991): Agrohydrology - Soil Water Management for Higher Crop Production, Final Technical Report, . 1991, Project No.: YO-ARS-91,JF-PP-619, University of Ljubljana

7. Matičič, B. et al. (1992): Agrohydrology - Soil Water Management for Higher Crop Production, Technical Report, 1992, University of Ljubljana

8. Matičič, B. et al. (1993): Agrohydrology - Soil Water Management for Higher Crop Production, Technical Report, 1993, University of Ljubljana

9. Spalding, R. F., Exner, M. E., Eaton, C. W.: Investigation of Sources of Groundwater Nitrate Contamination in the Burbank- Wallula Area of Washington, U.S.A., Journal of Hydrology, Vol. 58, No.3/4, (1982) pp 307-325.

ENERGY AND IRRIGATION IN SOUTHEAST AGRICULTURE UNDER CLIMATE CHANGE

Robert M. Peart, R. Bruce Curry, J. W. Jones, K. J. Boote, L. H. Allen, Jr.[*]

ABSTRACT

The effects of three predicted scenarios of climate change under effectively doubled carbon dioxide has been studied for yield effects, but this paper shows the possible effects on energy efficiency, and irrigation water demand under these conditions. Forty years of daily weather data for many weather stations has been assembled into the IBSNAT format for running crop simulation models under a user interface named DSSAT, Decision Support System for Agrotechnology Transfer, developed in the Soils Dept. at the University of Hawaii, in cooperation with Dr. J. W. Jones, at the University of Florida. The three General Circulation Models of the global climate provide monthly parameters to adjust these historical data to form three new data sets to represent weather under doubled carbon dioxide conditions, predicted to occur around the year 2040.

The models of soybean, peanut and maize give maturity dates and yields and irrigation water usage for irrigated and for non-irrigated crops. With these data, we have analyzed the differences in energy inputs and outputs under irrigated and rainfed conditions for historical data and for climate change. In general, climate change would result in a greater demand for irrigation water and the energy efficiency of production would be somewhat less.

BACKGROUND

The current general consensus among scientists and engineers is that the long term effect of increasing CO_2 levels in the atmosphere will produce a significant climate change. The amount and timing of these changes are still subject to a great deal of uncertainty. A recent study by the Council for Agricultural Science and Technology (Waggoner, 1992) points out the importance of this problem in the future of U.S. agriculture. To evaluate the potential impact of climate change on crop agriculture, we have used crop simulation models, historic climate data and the results of General Circulation Models (GCM's) predicting climate changes. This work has been partially supported by the U.S. EPA and currently by the Southeast Regional Climate Center, Columbia, SC.

These simulations show climate change effects on both rainfed and irrigated yields, but also on the amount of irrigation required. With these results, we have calculated the effects on energy inputs into crop production and on the output yields expressed in energy units.

---

[*]Robert M. Peart, Graduate Research Professor, Agr. Eng. Dept., R. Bruce Curry, Visiting Professor, Agr. Eng. Dept., J. W. Jones, Professor, Agr. Eng. Dept., K. J. Boote, Professor, Agronomy Dept., and L. H. Allen, Jr., USDA-ARS, and Adjunct Professor, Agronomy Dept., University of Florida, Gainesville, FL 32611, USA.

METHODOLOGY

The tools used in this work were readily available systems from a variety of sources that have been previously developed for other purposes. The climate change parameters used in developing the scenarios came from General Circulation Models (GCM's) developed to study world weather dynamics. The soybean crop model, SOYGRO, was developed to study the soybean production system (Jones, et al., 1988). The peanut model, PNUTGRO, was developed by the same group (Boote, et al., 1989). The maize model, CERES/Maize, was used for the maize simulations (Jones and Kiniry, 1986). They are physiological-based crop simulators that include processes for photosynthesis, respiration, partitioning, phenological development, soil water balance, etc. The models include factors for the change in photosynthetic rate and in evapotranspiration caused by carbon dioxide concentration.

Simulation runs were made for each year of the weather data base for each scenario at each location for both rainfed and irrigated cases. These runs were managed by the DSSAT system the Decision Support System for Agricultural techology Transfer, an executive program developed to assist in agricultural decision making (IBSNAT, 1989, and Jones, et al., 1990).

An energy conversion spreadsheet was developed using data from the Midwest (Doering and Peart, 1977) for factors that would be typical for the Southeast, also, and from a recent study on energy inputs into Florida agriculture (Fluck, et al., 1992).

Energy equivalents of the seed yields (output) and the inputs, including fuel, fertilizers, and pesticides are shown in Table 1. We used data for conventional tillage, fertilization and pesticide application practices from the Midwest, which would be typical for the Southeast, also, except that peanut pesticide application was doubled over that used for soybean. In this study, we included a variation in fuel for harvesting proportional to yield and a similar factor relating fertilizer to yield. The following expressions specify the functions used in the spreadsheet.

Fuel inputs for plowing, disking and planting were 1478, 349, and 349 MJ/ha, respectively, for soybean and peanut. Plowing for maize was 1337 MJ/ha, with disking and planting the same as for soybean and peanut. Since maize and soybean are usually alternated, plowing for soybean is on harvested maize land and requires slightly more energy than the reverse.

For harvesting, 441, 561, and 1121 MJ/ha were assumed for soybean, maize and peanut, respectively, for yields of 2.35, 9.42 and 2.5 t/ha, then adjusted according to actual yield.

Fertilizer for soybean and peanut was estimated as 352 MJ/ha plus 352 MJ/ha adjusted in proportion to the actual yield divided by 2.35 t/ha for soybean or 2.5 t/ha for peanut. Maize requires much more fertilizer, especially nitrogen, and it was calculated as 7821 MJ/ha plus 4419 MJ/ha multiplied by actual yield divided by 9.42 t/ha.

Pesticide energy was estimated at 1251 MJ/ha for soybean, 1681 MJ/ha for maize and 3363 MJ/ha for peanut. Drying of maize and peanut was calculated on the basis of 5044 MJ/ha for a normal maize crop of 9.42 t/ha, and the actual energy was computed as proportional to the ratio of the actual yield to 9.42 t/ha.

Irrigation energy was calculated assuming a pumping depth of 30.48 m (100 ft.), a head pressure of 2,758 kPa (40 psi), and an overall engine/pump efficiency of 20%. Electric pumping was assumed to have the same efficiency, including the energy losses in generating the electricity. These agreed well with the extensive recent study on energy inputs into Florida agriculture (Fluck, et al., 1992). However, since we are projecting possible future practices, we should note that it is feasible to improve on this energy efficiency by using minimum-tillage practices, lower-pressure irrigation, and more efficient pumps and power units.

Weather data for some 27 locations in the Southeast 30 years from 1951-80 was provided by NOAA/NCDC, Ashville, NC, and NCAR, Boulder, CO. The Southeast Regional Climate Center supplied the data for 1981-90 for these locations to make a 40-yr record.

Climate change scenarios used in the study were developed using 40-yr standard historic daily weather data bases for each location plus parameters supplied by the 3 GCM's. These parameters included monthly temperature change in degrees, a monthly change ratio for precipitation and for solar radiation for each location. The historic weather included temperature and precipitation, and WGEN, a synthetic weather generating program, was used to generate solar radiation based on temperature and precipitation. The change parameters from the GCM's were applied to the standard historic data base to provide an additional 3 climate change scenarios, thereby giving 4 weather data sets. The precipitation ratio for each month was simply applied to each period of rainfall in the historic data, so the amount of rainfall was changed, but the length and the frequency of the precipitation events was not changed.

Table 1. Energy Equivalents for Crop Production Inputs and Outputs

ITEM	English Units	SI Units
Nitrogen Fertilizer, per unit of N	25,000 Btu/lb.	58.09 MJ/kg
Phosphate Fertilizer, per unit of P_2O_4	3,000 Btu/lb.	6.97 MJ/kg
Potassium Fertilizer, per unit of KO_2	2,000 Btu/lb.	4.65 MJ/kg
Diesel Fuel	135,250 Btu/gal.	35.91 MJ/Liter
LP Gas	93,500 Btu/gal.	24.82 MJ/Liter
Electricity, input/unit of output	3 * 3,413 Btu/kwh	3.0 Kwh/Kwh
Pesticides, per unit active ingred.	120,000 Btu/lb.	278.85 MJ/kg
Soybean Seed Yield	7,239 Btu/lb.	16.80 MJ/kg
Peanut Seed Yield	9,000 Btu/lb.	20.91 MJ/kg
Maize Seed Yield	6,270 Btu/lb.	14.57 MJ/kg

The GCM's used in this study were: 1) the GISS model developed by the Goddard Institute for Space Studies, located at Columbia University, New York, NY, (Hansen, et al., 1988), 2) The GFDL model developed by the Geophysical Fluid Dynamics Laboratory, located at Princeton, Princeton, NJ, (Manabe and Wetherald, 1987), and 3) the UKMO model developed by the United Kingdom Meteorological Office in the UK (Wilson and Mitchell, 1987).

Assumptions: 1) Precipitation change from the GCM's would affect the amount of each rainfall event, not the number of events. 2) CO_2 level for equivalent doubling of atmospheric CO_2 was assumed to be 555 umol mol^{-1} compared to the current ambient level of 330, and the other greenhouse gases would account for the effect of doubled CO_2 alone. 3) Soybean cultivars were chosen that are in current use in the area around each location. 4) Varieties would not change with changed climate. 5) Soil property data were used from generic soils. 6) Planting dates are the same as used currently around each location. 7) For irrigation cases, water supply for irrigation was non-limiting.

RESULTS

Here we present results of simulations in two states in the Southeast, Charlotte and Raleigh, NC, and Memphis and Nashville, TN. (Raleigh and Nashville were more typical of peanut production areas.) These results in Tables 2-7 are averaged over 40 years of historical weather data for each location. The simulations were done for each weather year, day-by-day, and results were averaged. This gives a much more realistic result than running one simulation of an "average" weather year. Then for each of the 3 "doubled" CO_2 climate change scenarios, these weather data were adjusted by the appropriate parameters and the simulations run again. These climate change results in the tables are the averages of the three different scenarios.

The actual CO_2 concentration used was 555 ppm rather than double the 330 ppm, since the effect of doubling CO_2 is estimated to occur at the 555 ppm level because of the added effects of other greenhouse gases, mainly methane, which are also increasing. Runs were made for non-irrigated (rainfed) and irrigated management. Irrigation was applied by the model when the low soil moisture produced moisture stress in the crop and enough water was added to bring the top 30 cm of soil up to its field capacity. Improved irrigation management for humid areas such as these could probably improve the efficiency of this irrigation strategy.

Table 2. Energy Output/Input, Soybean, Memphis, TN, 40-year ave.

FACTOR	Historic Rainfed	Climate Change, Rainfed	Historic Irrig.	Climate Change, Irrig.
Seed Yield, t/ha	2.08	1.11	4.01	3.55
ENERGY OUT, MJ/ha	35,023	18,690	67,520	59,775
Field Fuel, MJ/ha	2566	2384	2927	2841
Fertilizer, MJ/ha	663	518	951	883
Pesticide, MJ/ha	1251	1251	1251	1251
Irr. Water, mm	0	0	194.9	311.5
Irr.Fuel, MJ/ha	0	0	5599	8949
ENERGY IN, MJ/ha	4480	4154	10,729	13,924
OUTPUT/INPUT	7.82	4.50	6.29	4.29

Tables 2-7 show yields and the equivalent energy value of this yield, based on the constants given in Table 1. Each input factor is shown in energy value, and the amount of irrigation water used is listed. The values in the "Climate Change" columns reflect the effects of the higher concentration of CO_2 and the increased temperature and changed monthly distribution of rainfall, according to the location.

These tables show the "bottom line" as the ratio of output to input, always greater than 1.0 because of the huge amount of solar energy used by the photosynthesis process, which is not counted along with the non-renewable inputs. Also, "implicit" energy inputs such as the energy used for manufacturing the equipment are not counted.

Rainfed yields of soybean (Tables 2 and 3) are greatly reduced, almost 50% less in Memphis, under climate change. This somewhat reduced input totals, but the ratio of output to input was also greatly reduced. Irrigated yields were about the same in Charlotte, and about 10% less for climate change in Memphis. However, due to the greatly increased requirement for irrigation water under climate change, total inputs were much greater and the output/input ratio much less.

Table 3. Energy Output/Input, Soybean, Charlotte, NC, 40-year ave.

FACTOR	Historic Rainfed	Climate Change, Rainfed	Historic Irrig.	Climate Change, Irrig.
Seed Yield, t/ha	2.35	1.79	4.02	4.07
ENERGY OUT, MJ/ha	39,569	30,196	67,689	68,475
Field Fuel, MJ/ha	2617	2512	2929	2938
Fertilizer, MJ/ha	703	620	953	960
Pesticide, MJ/ha	1251	1251	1251	1251
Irr. Water, mm	0	0	162.3	262.5
Irr.Fuel, MJ/ha	0	0	4663	7541
ENERGY IN, MJ/ha	4571	4384	9796	12,690
OUTPUT/INPUT	8.66	6.89	6.91	5.40

The results for peanut in Nashville and Raleigh follow the pattern of soybean in Memphis and Charlotte. Rainfed yields are reduced more in Raleigh, but are reduced some 25% in Nashville, and the output/input inputs are reduced accordingly for climate change. Drastic increases in irrigation water are required for peanut in both locations, but yields are still reduced by climate change, making for greatly reduced efficiency of use of the irrigation water.

A significant factor for both soybean and peanut is the marginal increase in yield due to irrigation. These ratios or percentages, (increase in yield due to irrigation)/(rainfed yield), are much higher for climate change than for historic weather, about double in 3 out of 4 cases. This means that the economic incentive for a grower to add irrigation will be much greater. For example, in Charlotte, the increase due to irrigation is 2.3 t/ha, and at a current price of $7/bu., this amounts to $586/ha ($237/acre), a strong incentive to irrigate.

Table 4. Energy Output/Input, Peanut, Nashville, TN, 40-year ave.

FACTOR	Historic Rainfed	Climate Change, Rainfed	Historic Irrig.	Climate Change, Irrig.
Seed Yield, t/ha	2.85	2.12	5.29	4.61
ENERGY OUT, MJ/ha	59,662	44,380	110,741	96,576
Field Fuel, MJ/ha	3314	2987	4408	4105
Fertilizer, MJ/ha	753	650	1096	1001
Pesticide, MJ/ha	3363	3363	3363	3363
Irr. Water, mm	0	0	185.0	243.1
Irr.&Dry, MJ/ha	1529	1138	8154	9460
ENERGY IN, MJ/ha	4480	4154	10,730	13,925
OUTPUT/INPUT	6.66	5.45	6.51	5.39

Table 5. Energy Output/Input, Peanut, Raleigh, NC, 40-year ave.

FACTOR	Historic Rainfed	Climate Change, Rainfed	Historic Irrig.	Climate Change, Irrig.
Seed Yield, t/ha	3.89	2.58	5.54	4.88
ENERGY OUT, MJ/ha	81,433	54,080	115,974	102,158
Field Fuel, MJ/ha	3780	3194	4520	4224
Fertilizer, MJ/ha	900	715	1132	1039
Pesticide, MJ/ha	3363	3363	3363	3363
Irr. Water, mm	0	0	140.1	211.9
Irr.&Dry, MJ/ha	2088	1386	6998	8706
ENERGY OUT, MJ/ha	10,130	8,659	16,013	17,332
OUTPUT/INPUT	8.04	6.25	7.24	5.89

Maize does not follow the pattern of the other two crops because it is a C_4 plant, while soybean and peanut have C_3 photosynthetic pathways. Maize responds only about one third as much as soybean or peanut to increased CO_2 concentration. In addition, maize in the Southeast is planted earlier than soybean or peanut, and therefore the weather pattern for maize at the same location will be different.

Table 6. Energy Output/Input, Maize, Memphis, TN, 40-year ave.

FACTOR	Historic Rainfed	Climate Change, Rainfed	Historic Irrig.	Climate Change, Irrig.
Seed Yield, t/ha	8.92	8.69	10.88	9.54
ENERGY OUT, MJ/ha	150,195	146,322	183,197	160,635
Field Fuel, MJ/ha	2567	2553	2684	2604
Fertilizer, MJ/ha	12,006	11,898	12,926	12,297
Pesticide, MJ/ha	1681	1681	1681	1681
Irr. Water, mm	0	0	109.0	82.7
Irr.& Dry, MJ/ha	4778	4655	8959	7485
ENERGY IN, MJ/ha	21,033	20,788	26,250	24,067
OUTPUT/INPUT	7.14	7.04	6.98	6.67

Interestingly, irrigated maize yields were reduced more by climate change than were rainfed yields. Also, irrigation water requirements were about 30% less at both locations. However, the yield increase due to irrigation under climate change (about 10%) was about half that for historic weather. Thus, irrigation efficiency was less under climate change, even though actual water requirements were also less. However, with maize the increase due to irrigation is less under climate change, so in the Southeast, climate change should not increase the demand for water for irrigating maize.

Table 7. Energy Output/Input, Maize, Charlotte, NC, 40-year ave.

FACTOR	Historic Rainfed	Climate Change, Rainfed	Historic Irrig.	Climate Change, Irrig.
Seed Yield, t/ha	9.56	9.18	11.75	10.02
ENERGY OUT, MJ/ha	160,971	154,537	197,847	168,661
Field Fuel, MJ/ha	2605	2582	2735	2632
Fertilizer, MJ/ha	12,307	12,128	13,334	12,521
Pesticide, MJ/ha	1681	1681	1681	1681
Irr. Water, mm	0	0	110.3	78.9
Irr.&Dry, MJ/ha	5121	4917	9463	7632
ENERGY IN, MJ/ha	21,714	21,309	27,214	24,467
OUTPUT/INPUT	7.41	7.25	7.27	6.89

REFERENCES

Boote, K. J., J. W. Jones, G. Hoogenboom, G. G. Wilkerson and S. S. Jagtap. 1989. PNUTGRO V1.02, IBSNAT version, User's Guide. Univ. of Florida, Gainesville, FL.

Doering, O. C. and R. M. Peart. 1977. Evaluating alternative energy technologies in agriculture. NSF/RA-770124, Agr. Exp. Station, Purdue Univ., W. Lafayette, IN 47907.

Fluck, R. C., B. S. Panesar and C. D. Baird. 1992. Florida Agricultural Energy Consumption Model, Final Report, Agr. Eng. Dept., Univ. of Florida, Gainesville, FL 32611.

Hansen, J., I. Fung, A. Lacis, S. Lebedeff, D. Rind, R. Ruedy, G. Russell, and P. Stone. 1988. Global climate changes as forecast by the GISS 3-D model. J. Geophys. Res.

IBSNAT. 1989. Decision Support System for Agrotechnology Transfer (DSSAT v2.1), User's Guide. IBSNAT Project, Dept. of Agronomy and Soil Science, University of Hawaii, Honolulu, HI 96822.

Jones, C. A. and J. R. Kiniry, (eds.) 1986. CERES-Maize: A Simulation Model of Maize Growth and Development, Texas A & M University Press, College Station, TX.

Jones, J. W., J. W., S. S. Jagtap, G. Hoogenboom, and G. Y. Tsuji. 1990. The structure and function of DSSAT. pp. 1-14. In: *Proc., IBSNAT Symposium: Decision Support System for Agrotechnology Transfer,* 16-18 Oct., 1989. Part I: Symp. Proc., Dept. of Agronomy and Soil Science, Univ. of Hawaii, Honolulu, HI 96822.

Jones, J. W., K. J. Boote, S. S. Jagtap, G. Hoogenboom, and G. G. Wilkerson. 1988. SOYGRO V5.41: Soybean crop growth simulation model. User's Guide. Florida Agr. Exp. Sta. Journal Series No. 8304, IFAS. Univ. of Florida. Gainesville. 53 pp.

Manabe, S. and R. T. Wetherald. 1987. Large-scale changes of soil wetness induced by an increase in atmospheric carbon dioxide. J. Atmos. Sci., 44:1211-1235.

Waggoner, Paul E. 1992 (ed.) Preparing U.S. agriculture for global climate change. CAST Task Force Report No. 119, Council for Agr. Science and Technology, Ames, IA 50010.

Wilson, C.A. and J.F.B. Mitchell. 1987. A Doubled CO_2 Climate Sensitivity Experiment with a Global model including a Simple Ocean. J. Geophys. Res., 92:13315-13343.

Effects of Companion Seeded Berseem Clover on Oat Grain Yield,
Biomass Production, and the Succeeding Corn Crop

M.Ghaffarzadeh and R. M. Cruse[1]

ABSTRACT

Sustainability of Iowa agriculture depends on changes in Iowa's predominant corn-soybean cropping systems. These two annual crop's rotation result in inadequate ground cover for approximately six months of each year. Alternative crops are vital for providing temporal diversity and enhancing sustainability. Small grains create the potential to establish a three crop rotation and also enable the inclusion of a forage legume as a companion crop. However, inclusion of small grains in crop rotations has been unpopular, largely because of their low grain market value and the high year to year variability in grain yield. A field study was established in 1991 on Kenyon (fine-loamy, mixed, mesic Typic Hapludoll) soil to evaluate the economic and biological benefits of an oat (*Avena sativa* L.) crop underseeded with berseem clover (*Trifolium alexandrinum* L.). Three rotation treatments were compared: 1) corn/soybean; 2) corn/soybean/oat; and 3) corn/soybean/oat+berseem clover. In Treatment 2, hairy vetch was seeded after oat grain and straw harvest. In Treatment 3, berseem clover was seeded with oats in spring. Oat grain yields were 22 and 26% lower, and hay/straw yield was 35 and 48% higher from oat/berseem plots compared to the sole oat plots in 1991 and 1992 respectively. In 1993, oat grain yields were not significantly different, although 3.6 Mg/ha more hay was produced from the plots containing berseem clover. Additionally, berseem clover produced one to two cuttings of hay after oat harvest, with late season regrowth remaining as ground cover. Corn grain yields in 1992 were 7.1, 9.4, and 9.5 Mg/ha, in Treatment 1, 2, and 3 respectively. Contributions of fixed N by berseem clover in 1993 with excessive precipitation caused a significant increase in grain yield for the succeeding corn crop. Corn grain yields were 3.3, 3.6, and 6.2 Mg/ha in Treatment 1, 2, and 3 respectively.

INTRODUCTION

Increased corn (*Zea mays* L.) and soybean [*Glycine max* (L.) Merr.] production throughout the Midwest has decreased landscape diversity, contributed to significant environmental problems, and concurrently resulted in a farm economy reliant on government subsidies. In Iowa approximately six out of the ten million harvested hectares are managed with an annual corn/soybean rotation. This rotation of two annual crops results in lack of live ground cover for approximately six months of each year. Particularly, low plant residue levels from soybeans result in exposure of the soil to water and wind erosion, increased potential for leaching of nitrate nitrogen remaining in the soil, and reduce N availability for the subsequent crop rooting zone. Inclusion of an alternative small grain as a third crop can significantly improve soil productivity. Research in Minnesota (Crookston et al., 1991) indicated that total production increased when a small grain/legume is added to a corn-soybean rotation. However, the inclusion of small grains in crop rotations in Iowa has been unpopular, largely because of the low grain market value and the high variability in grain yield. Small grains permit the inclusion of a forage legume as a companion crop or cover

---

[1]The authors are Mohammadreza Ghaffarzadeh, Assistant Scientist, Agronomy Department, Iowa State University, Ames; Richard M. Cruse, Professor, Agronomy Department, Iowa State University, Ames, IA.

crop after small grain harvest. Oat (*Avena sativa* L.) is the most common small grain used in Iowa. Over 340 thousand hectares of oats are planted each year, but only 140 thousand hectares are harvested for grain (Iowa Agricultural Statistics, 1993). The rest are either planted on highly erodible Conservation Reserve Program (CRP) acres, Set Aside Program acres, or are harvested as hay or silage. The use of cover crops in conservation tillage is increasing due to benefits such as reduced soil erosion (Ebelhar et al., 1984; Scott et al., 1987).

Legume cover crops may also contribute nitrogen (N) to the subsequent crop, reducing the N fertilizer requirement by 121 Kg N/acre or more (Hargrove, 1986; McVay et al., 1989; Holderbaum et al., 1990). In northern climates, approximately 100 Kg N/ha is contributed by underseeded or interseeded legumes to the succeeding corn crop (Bruulesema and Christie, 1987). Conventional small grain/legume combinations in Iowa include oats underseeded with Alfalfa (*Medicago sativa* L.), or Mammoth red clover (*Trifolium pratense* L.). Rye (*Secale cereale* L.) or oat mixture with hairy vetch (*Vicia villosa* Roth.) and seeded after oat harvest is also a common practice. In some cases, a cover crop may deplete soil moisture necessary for the subsequent crop grain production (Ebelhar et al., 1984; Frye et al., 1988; Badaruddin and Meyer, 1989). Chemical or mechanical elimination of these cover crops in the fall, or prior to the germination of the succeeding crop in spring, is also a major management consideration. In selecting a legume cover crop, characteristics and associations of a legume with a nurse crop, management practices, and environmental factors must be considered. Berseem clover (*Trifolium alexandrinum*) seeded as a companion crop, in contrast to traditional legumes, grows rapidly after seeding, responds very well to multiple cutting schedules (Baldridge et al., 1992) and winter kills in Iowa. Forage quality is high (Brink and Fairbrother, 1992; Singh et al., 1989) and biomass production in the seeding year may also be high. Research (Singh et al., 1989) indicates that the oat/berseem clover combination may produce 10 t/ha of high quality biomass, which creates more harvesting options (i.e. forage, grain, straw, or grazing). Grazing as a harvest option has been investigated, and no bloat by ruminant animals has been reported (Baldridge et al., 1991; Sims et al., 1991). Investigators also compared berseem clover to other clovers and found significantly lower levels of toxins in the berseem clover forage (Ayalon and Lindner, 1976).

The magnitude of legume N contribution under different environmental, cropping system, and soil conditions is unclear. Most evaluations of legume N contributions have been conducted specifically with the legume as a green manure, or within a rotation following the complete life cycle of the legume species. However, companion legume cover crop contributions on an annual basis remain unknown and confounded with other factors. Tillage practices also influence availability of the contributed N to the succeeding crop (Heichel, 1987); plow-down or incorporation of biomass is limited with no-till and ridge tillage. Studies in Nebraska showed that in no-till, a hairy vetch cover crop increased soil nitrate, but not until 50 to 78 days after the succeeding corn crop was planted (Brown et al., 1993). This increase in soil nitrate concentration occurred after corn silking, which has insignificant effects on corn grain yield (Hanway, 1963; Chevalier and Schrader, 1977). The objectives of this study were to evaluate the economic and biological benefits of an oat crop underseeded with berseem clover in a ridge-till system.

MATERIAL AND METHODS

A field experiment was conducted for three years (1991 to 1993) at the Northeast Iowa Research Farm on Kenyon (fine-loamy, mixed, mesic Typic Hapludoll) soil near Nashua, Iowa. A randomized, complete block, split-plot design with four replication was used. Whole-plots were three rotation treatments. Fertilizer (N) rates were sub-plots. Prior to the experiment, the site had been in ridge-till soybean production for one year. Three rotation treatments, whole-plots, were compared: 1) corn/soybean; 2) corn/soybean/oat; 3) corn/soybean/oat+berseem clover. Corn, 60,000 seed/ha and soybean 258,000 seed/ha were planted each year in 6-76 cm rows. Oats were planted in 15 cm rows at a seeding rate of 135 kg/ha. In Treatment 2, a mixture of oats and hairy vetch was seeded approximately one month after oat grain and straw harvest. Seeding rates were approximately 100 kg of oat and 30 kg/ha of hairy vetch. In Treatment 3, berseem clover was seeded with oats in spring. Seeding rates were 135 kg oat and 18 kg/ha berseem clover. In each of the treatments, four nitrogen rates (0, 56, 112, and 168 Kg N/ha) as ammonium nitrate were applied by hand to the corn plots prior to first cultivation. Whole-plot size was 4.6 by 60.4 m. Sub-plots were 4.6 by 15.1 m. Ridge rebuilding was the only other field operation in corn and soybean plots. Phosphorus and potassium were not applied, because of the high soil test levels for these nutrients. Preemergence weed control was obtain by banding granular alachlor (2-chloro-N-(2,6-diethylphenyl)-N-(methoxymethyl)acetamide) at planting time for corn and soybeans. Hairy vetch regrowth was terminated one week after corn planting by applying 2,4-D ((2,4-dichlorophenoxy)acetic acid). No herbicide was used on the oat/berseem clover plots.

Corn and soybean grain yields were determined by harvesting the two center rows using a small plot combine. Grain yields were adjusted to the basis of 15.5% and 13% grain moisture for corn and soybean respectively. Oat grain and straw yields were determined by hand-harvesting 1 m^2 sections from the center of each sub-plot. In Treatment 3 berseem clover biomass was separated from straw and measured at oat harvesting time. Berseem regrowth dry matter production was measured prior to killing frost. Plant material was dried, and ground to 1 mm with a Cyclone mill. Nitrogen concentration was determined by micro-Kjeldahl digestion (Bremner, 1969).

In 1993, soil samples were collected, eight composite soil core, 1.9 cm to a depth of 30 cm every other week from each corn subplot, starting early spring till fertilizer application in June. Soil NO_3-N was extracted with a solution containing 0.025 moles/L aluminum sulfate and 0.02 moles/L boric acid (Mills, 1980). Filtered extracts were analyzed using a Fisher[2] Scientific Accumet pH meter equipped with a Hach[2] nitrate electrode.

Statistical analyses were conducted using SAS (SAS Institute, Inc., 1985) analysis of variance procedures. Means comparison were made using the least significant difference (LSD 0.05 level) procedure. Iowa market values for corn, soybean, oat, straw and hay were used for economic evaluation.

RESULTS AND DISCUSSION

Oat grain yield in 1993 was not significantly different between the sole-seeded oat plots and the plots containing berseem clover. But, grain yield was 23% lower in the plots containing berseem in 1991 and 25% lower in 1992. Hay and straw production, in contrast, was 35 and 46% greater in plots containing berseem in the same years. Iowa's small grain production suffered in 1993, however, berseem clover hay production underseeded in oat was 5.6 Mg/ha in the first cutting after oat grain harvest (Fig. 1). High year to year variability of oat grain yield is a consideration of farmers in selecting crops. This variability is also experienced in our field data. However, straw and especially hay production in three years of this experiment had low variability. The economic loss of oat grain can be recovered from hay production. Total production cost and income from the three rotation treatments indicate that a three crop rotation with small grain underseeded with berseem is more profitable than the most popular rotation of corn and soybean (without considering the government programs. Data also signifies the economic value of companionship of oat and berseem clover. In 1991 and 1993 underseeding oat with berseem resulted in greater profit than sole-seeded oat. In 1992, one of the greatest year for small grain production in Iowa,

---

[2] Mention of companies or trademarks, are for the information and convenience of the readers and does not constitute an official recommendation or endorsement by Iowa State University.

no significant income differences were calculated between oat and oat underseeded with berseem. These economic comparisons are based on production only up to oat grain harvesting, the possible profits of a second or third cut of hay are not considered. Regrowth of berseem was left as cover crop. In the last three years, berseem has shown favorable companionship with oat and appears to be well adapted to climatic conditions in Iowa.

In three crop rotation treatments, including berseem clover with oat had no effect on soybean yield. However, corn grain yield showed significant differences caused by three crop rotation treatments (Fig. 2). Corn grain yield was significantly higher for the corn/soybean/oat+berseem treatment than for the corn/soybean or corn/soybean/oat rotation in both years (Fig. 2).

Economic evaluation of the three treatments are summarized in Table one. Including oat in the rotation both in 1992 and 1993 improved economic return. Additionaly, underseeding oat with berseem enhanced this positive effect. Opportunity costs (direct and indirect cost associated with cover crop) of underseeding berseem is limited to seed cost, extra baling and labor cost, and loss of oat grain yield. Additional hay production not only compensate for opportunity cost, as a consequence, profits were higher due to the underseeding. Most winter legume cover crops used in Iowa either delay planting or compete with succeeding crop, however, berseem is an annual clover and winter kills in this climate presenting no management obstacle the following spring.

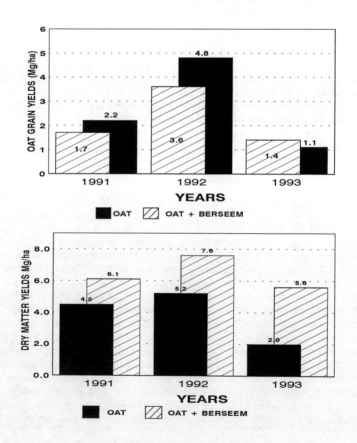

Figure 1. Oat grain yield and biomass (straw and hay) production as influenced by underseeding berseem clover in 1991, 1992, and 1993.

Table 1. Economic Comparison of the three rotation treatments in 1992 and 1993 at Northeast Iowa Research Center.

	Corn/Soybean/Oat		Corn/Soybean		Corn/Soybean/Oat +berseem		Market Value¶ $/Mg		Total§ Production Cost $/ha
	1992	1993	1992	1993	1992	1993	1992	1993	
Corn Yield Mg/ha Net $/ha	7.1 -$92	3.6 -$268	7.6 -$54	3.2 -$309	9.3 $75	6.4 $14	$76	$101	$632
Soybean Yield Mg/ha Net $/ha	2.7 -$23	2.9 $117	2.4 -$80	2.5 $27	2.4 -$80	2.9 $117	$192	$227	$541
Oat Yield Mg/ha Net $/ha	4.8 $48	1.2 -$225	---- ----	---- ----	3.6 -$122	1.4 -$300	$91	$107	$353‡
+ Straw Yield Mg/ha Net $/ha	5.2 $257	2.0 $110	---- ----	---- ----	---- ----	---- ----	$49	$55	
+ Hay Yield Mg/ha Net $/ha	---- ----	---- ----	---- ----	---- ----	7.6 $418	5.6 $339	$55	$65	$450†
Net/ha/treatment	$75	-$89	-$67	-$141	$97	$57			
Two year Avg.	-$7		-$104		$77				

¶ Market value at harvesting time source (reported by local news)
§ Source; Estimated cost of crop production in Iowa 1993. Iowa State University, Cooperative Extension Service November 1992.
‡ Production cost of oat and straw included.
† Production cost of oat and Hay included.

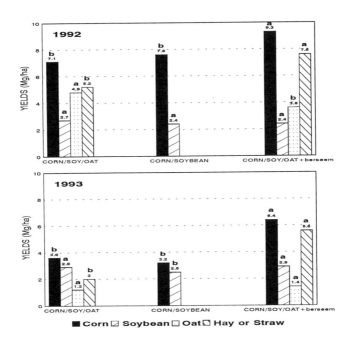

Yields with same letter for a given crop were not significantly different (0.05 level)

Figure 2. Yields of three rotations in 1992 and 1993.

In 1992 no corn yield response to applied nitrogen was observed beyond the 112 Kg N/ha rate. Above normal precipitation and flooding in 1993 resulted in a linear corn grain yield response to applied nitrogen, however corn grain yield in both years was significantly greater in the corn/soybean/oat+berseem rotation than in the other two rotations (Table 2).

Table 2. Rotation treatment and four nitrogen fertilizer rate effect on corn grain yield in 1992 and 1993 at the Northeast Iowa Research Center.

Treatment	N Rate (Kg/ha)	1992[‡]	1993	Means
		---	Corn yield (Mg/ha)	---
Corn/Soybean	0	6.0	1.5	3.7
	56	8.0	2.7	5.3
	112	8.4	3.5	5.9
	168	8.0	4.9	6.4
	Mean	7.6[b]	3.2[b]	**5.3[B]**
Corn/Soybean/Oat	0	5.1	2.5	3.8
	56	6.9	2.9	4.9
	112	8.2	4.6	6.4
	168	8.3	4.4	6.3
	Mean	7.1[b]	3.6[b]	**5.3[B]**
Corn/Soybean/Oat+berseem	0	7.3	4.6	5.9
	56	9.8	6.1	7.9
	112	10.3	7.0	8.6
	168	10.0	8.1	9.1
	Mean	9.3[a]	6.4[a]	**7.8[A]**

[‡] 1991 yield not included due to confounded effect of the past rotation.
Common letters within a column indicate no significant differences (0.05 level)

Higher corn yields following oat underseeded with berseem may be due to the N contribution of the legume cover. At the end of the growing season in 1991 and 1992, respectively, 2.2 Mg/ha and 2.6 Mg/ha berseem clover biomass remained on the soil surface as a cover crop. The nitrogen concentration of this dry matter was 2.3%. However, determining N contribution by calculating N content of the cover crop is misleading. Many studies (McVay et al., 1989) use an indirect measurement of the N contribution by comparing yield response to N fertilizer with and without legume cover. Regression equations for corn grain yield as a function of fertilizer N applied were developed for both 1992 and 1993 (Fig 3). Estimations of N fertilizer replacement values were 71 and 188 Kg/ha for 1992 and 1993 respectively. However 1993 was a very unusual year and N application were not sufficient to optimize corn grain yield. Considering this, 188 Kg N/ha may be unrealistic.

Rotation treatments and sub-plots shown no significant impact on soil NO_3-N in 1993 samples. Propose of monitoring soil nitrate was to determine early spring status and progress of N contribution by cover crop. In 1993, lack of conclusive conclusion may have been caused by excessive rainfall or insufficient monitoring technique.

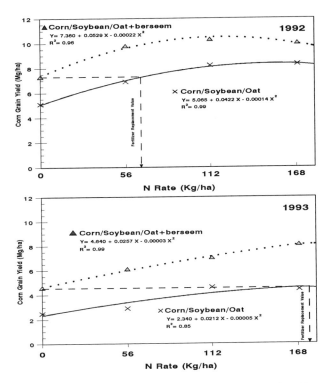

Figure 3. Corn grain yield in 1992 and 1993 as a function of fertilizer N response with and without a berseem cover crop at the Northeast Iowa Research Center.

The effect of a berseem cover crop on soil productivity and soil physical properties was not part of this research objectives, however, soil surface with this cover crop had 95% cover over winter. Visual inspection of the plots with berseem clover as a cover crop showed lower weed population and more uniform germination of the corn crop seedlings than other plots. Forage harvesting potential of oat underseeded with berseem may help integrate crop and livestock in Iowa agriculture.

REFERENCES

Badaruddin, M., and D.W. Meyer. 1989. Water use by legumes and its effect on soil water status. Crop Sci. 29:1212-1216.

Baldridge, D., R. Dunn, R. Ditterline, J.Sims, L. Welty, D.Wichman, M. Westcott, and G. Stalkecht. 1992. Berseem Clover: A Potential hay and green manure crop for Montana. MT 9201 (AG). Montana Ag. Extension Services.

Bremner, J.M. 1965. Organic forms os soil nitrogen. *In* C.A. Black et al. (ed.) Methods of soil analysis. Part 2. Agronomy 9:1238-1255.

Brink, G.E. and T.E. Fairbrother. 1992. Forage quality and morphological components of diverse clovers during primary spring growth. Crop Sci. 32:1043-1048.

Brown, R.E., G.E. Varvel, and C.A. Shapiro. 1993. Residual effects of interseeded hairy vetch on soil nitrate-nitrogen levels. Soil Sci. Soc. Am. J. 57:121-124.

Bruulesema, T.W., and B.R. Christie. 1987. Nitrogen contribution to succeeding corn from alfalfa and red clover. Agron. J. 79:96-100.

Chevalier, P., and L.E. Schrader. 1977. Genotype differences in nitrate adsorption and translocation in corn genotypes following silking. Agron. J. 68:418-422.

Crookston, R.K., J.E. Kurle, P.J. Copeland, J.H. Ford and W.E. Lueschen. 1991. Rotational cropping sequence affects yield of corn and soybean. Agron. J. 83:108-113.

Ebelhar, S.A., W.W. Frye, and R.L. Blevins. 1984. Nitrogen form legume cover crops for no-till corn. Agron. J. 76:51-55.

Frye, W.W., R.L. Blevins, M.S. Corak, and J.J. Varco. 1988. Role of legume cover crops in efficient use of water and nitrogen. p. 129-154. *In* W.L. Hargrove (ed.) Cropping strategies for efficient use of water and nitrogen. ASA Spec. Publ. 51. ASA, CSSA, and SSSA, Madison, WI.

Hanway, J.J. 1963. Growth stages of corn (*Zea mays* L.). Agro. J. 55:487-492.

Hargrove, W.L. 1986. Winter legumes as a nitrogen source for no-till grain sorghum. Agron. J. 78:70-74.

Heichel, G.H. 1987. Legume as a source of nitrogen in conservation tillage systems. pp. 29-34. *In* J.F. Power (ed.) The role of legumes in conservation tillage systems. Soil Conservation Society of America, Ankeny, IA.

Holderbaum, J.F., A.M. Decker, J.J. Meisinger, F.R. Mulford, and L.R. Vough. 1990. Fall-seeded legume cover crops for no-tillage corn in the humid East. Agron. J. 82:117-124.

Iowa Crop Report, 1993. Iowa crop report. Vol. 93-9. Iowa Agricultural Statistics, Des Miones, IA.

McVay, K.A., D.E. Radcliffe, and W.L. Hargrove. 1989. Winter legume effects on soil properties and nitrogen fertilizer requirements. Soil Sci. Soc. Am. J. 53:1856-1862.

Mills, H.A. 1980. Nitrogen specific ion electrodes for soil, plant, and water analyses. J. Assoc. Off. Anal. Chem. 63:797-801.

SAS Institute. 1985. SAS user's guide: Statistics. SAS Institute, Inc., Cary, NC.

Scott, T.W., J. Mt. Pleasant, R.F. Burt, and D.J. Otis. 1987. Contributions of ground cover, dry matter, and nitrogen from intercrops and cover crops in a corn polyculture system. Agron. J. 79:792-798.

Sims, J.R. et al,. 1991. Yield and bloat hazard of berseem clover and other forage legumes in Montana. Montana Ag. Research, Vol.8, 1:4-10.

Singh, V., Y.P. Joshi, and S.S. Verma. 1989. Studies on the production of Egyptian clover and oats under intercropping. Expl. Agric. 25:541-544. U.K.

ROLE OF MYCORRHIZAE IN SUSTAINABLE AGRICULTURE

D.M. Sylvia[1]

ABSTRACT

The organisms found in soil are a critical resource for sustaining productive agriculture. A dominate component of the microbial community in soil are the mycorrhizal fungi which form beneficial symbioses with the fine roots of plants. Mycorrhiza contribute to sustainable agriculture by enhancing the efficiency of nutrient uptake by plants and by channelling carbon into soil, thereby improving soil aggregation. When propagules of mycorrhizal fungi in soil are few or ineffective, application of selected inocula may significantly improve the survival, growth and yield of a crop. Three examples from research in Florida are described to illustrate the potential importance and use of arbuscular mycorrhiza in agricultural practice. These include revegetation of beaches, establishment of forage legumes, and drought resistance of maize. Technologies for the efficient production and use of arbuscular mycorrhizal inocula are also discussed. Scaling up nutrient-film and aeroponic-culture technologies, along with sheared-root processing of colonized roots produced in these systems, should provide readily available, low-cost, high-quality inocula for nursery inoculation and field testing.

INTRODUCTION

For agricultural practice to be sustainable it must produce adequate amounts of high-quality food or fiber while protecting, or even repairing previous damage to, the land resource (Reganold et al., 1990). In other words, it has to be both profitable and environmentally sound. To achieve this goal, those who practice sustainable agriculture must make use of natural, beneficial processes and renewable resources available on the farm. It is now generally appreciated that the soil itself is the critical resource for sustaining productive agriculture. As Reganold et al. (1990) state, "Soil is...a complex, living, fragile medium that must be protected and nurtured to ensure long-term productivity and stability."

The living component of the soil, made up of plant roots and soil organisms (both macro and micro), plays a vital role in the development and maintenance of soil structure, in nutrient cycling, and plant health. Management technologies that conserve the biodiversity of organisms in the soil may provide the greatest benefits for long-term sustainability (Lee and Pankhurst, 1992). Microbial communities in the soil are composed of bacteria, actinomycetes, fungi, algae, protozoa, nematodes, and microarthropods such as colembola and mites. The density of these organisms is usually much greater in the vicinity of roots (rhizosphere) than in the bulk soil. This is because roots provide a rich pool of nutrients and growth factors that serve to stimulate microbial activity (Bolton et al., 1993).

A dominate component of the rhizosphere microbial community are the mycorrhizal fungi. These fungi form beneficial symbioses with the fine, ephemeral roots of plants (Harley and Smith, 1983) and have unique functions relative to nutrient uptake by plants and carbon flow into the soil. Mycorrhiza benefit plant growth by (i) greatly increasing the absorbing surface area (ASA) of plants, thereby providing uptake of nutrients beyond the nutrient-depletion zone of the root (ii) having narrow diameter ASA that allow more nutrient to be removed from the soil solution, and (iii) producing enzymes for enhanced mineralization of organic nutrients. Mycorrhizal fungi

---

[1] D.M. Sylvia, Professor, Soil and Water Science Department, University of Florida, Gainesville, FL 32605-0290

also channel a significant amount of carbon into the soil, thereby contributing to soil aggregation and stability (Finlay and Soderstrom, 1992; Miller and Jastrow, 1993).

The mycorrhiza condition is the rule in nature, rather than the exception (Newman and Reddell, 1987). Some plant families (e.g. Dipterocarpaceae, Fagaceae, Pinaceae) predominately form ectomycorrhiza (hyphae in the root grow between cortical cells) while other plant families (e.g. Gramineae, Leguminosae, Compositae) predominately form arbuscular mycorrhiza (hyphae penetrate root cortical cells with the formation of the highly branched arbuscule within the cell). Even though the morphology of these two forms of mycorrhiza are very different, they are surprising similar in function (Tinker et al., 1992).

Various cultural practices associated with conventional agriculture may reduce the distribution and effectiveness of the mycorrhizal symbiosis. For example, high levels of inorganic fertilizer or certain pesticides can drastically reduce the distribution and function of mycorrhiza in the field. Recently, Douds et al. (1993) reported that a conventional farming system had a lower level of arbuscular mycorrhizal fungi (and presumably less potential to benefit from the symbiosis) compared to low-input sustainable agriculture with cover crops planted between cash crops. Further considerations of arbuscular mycorrhiza in sustainable agriculture are discussed in Mosse (1986) and Bethlenfalvay and Linderman (1992).

CASE STUDIES OF MYCORRHIZAL APPLICATIONS

When propagules of arbuscular mycorrhizal fungi are few or ineffective, application of selected inocula may significantly improve the survival, growth and yield of a crop. Three examples from research in Florida are described below to illustrate the potential importance and use of arbuscular mycorrhiza in agricultural practice.

Revegetation of beaches

Coastal sand dunes of the southeastern U.S. are stabilized naturally by perennial grasses such as sea oats. Roots of sea oats in established dunes are colonized extensively by arbuscular mycorrhizal fungi; however, in many coastal areas building construction has destroyed this source of inoculum for replants. Sand placed on the beach from an offshore site is devoid of arbuscular mycorrhizal fungi and beach grasses transplanted into it are colonized very slowly. Sea oats are usually produced in container nurseries, providing an opportunity for economical inoculation of a large number of plants. We conducted inoculation experiments under standard cultural practice at a commercial nursery (Sylvia, 1989). Inoculation with an arbuscular mycorrhizal fungus resulted in 28% colonization of the roots after 8 weeks, even with routine applications of soluble fertilizer and fungicides. These colonized plants were outplanted at Miami Beach. The low-level of colonization established in the nursery resulted in well-colonized plants in the nourishment sand. Twenty months after outplanting, sea oats colonized with arbuscular mycorrhizal fungi in the nursery had 219% greater biomass compared to sea oats outplanted without arbuscular mycorrhizal fungi. The conclusion is that sea oats colonized with arbuscular mycorrhizal fungi are better adapted for growth in restored beach sand than are noncolonized plants.

Establishment of forage legumes

We studied the effect of inoculation with two selected isolates of arbuscular mycorrhizal fungi on the growth and nutrient content of two tropical forage legumes (*Macroptilium atropurpureum* Urb. cv. 'Siratro' and *Aeschynomene americana* L.), at various phosphorus (P) levels under field conditions (Medina et al., 1990). At all P levels and for all harvests, shoot dry mass of both legumes was greater for the plants inoculated with the arbuscular mycorrhizal fungi than control plants. Differences between inoculated and control plants were most marked between 30 to 90 kg ha^{-1} of applied P and diminished at 120 kg ha^{-1} (Fig. 1). Fungal inoculation resulted in at least a 30% savings in the amount of P fertilizer required for maximum yield. Plants inoculated with mycorrhizal fungi had greater tissue concentration and total content of P and N than control plants at low and intermediate levels of applied P. The conclusion is that under amended (limed and fertilized) soil conditions inoculation with selected arbuscular mycorrhizal fungi can improve the establishment and growth of forage legumes in fields that contain ineffective populations of native mycorrhizal fungi.

Fig. 1. Effect of P application on shoot dry mass of field-grown *Aeschynomene americana* inoculated with the arbuscular mycorrhizal fungi, *Glomus etunicatum* (ETU), *G. intraradices* (INT), or the control (CON).

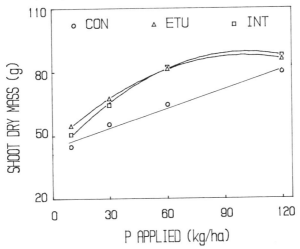

Drought resistance of maize

We conducted a three-year field study to test the effects of mycorrhiza and water management on the growth and grain yield of maize (Sylvia et al., 1993). In each year, two inoculation treatments (inoculated or not with an arbuscular mycorrhizal fungus) and three water-management treatments (fully irrigated, moderate stress, and severe stress) were applied to fumigated and fertilized Millhoppper fine sand. Inoculum was placed in a furrow 10-cm deep at an average rate of 1500 propagules m^{-1} of row. Six to seven weeks after planting, colonization ranged from 0 to 6% of total root length on noninoculated plants and from 10 to 61% on inoculated plants. Twelve to thirteen weeks after planting, colonization ranged from 2 to 30% on noninoculated plants and from 21 to 56% on inoculated plants. Water stress had little effect on root colonization. By 52 days after planting, one more leaf had appeared and one additional leaf had formed a collar on inoculated plants. Inoculation increased the concentrations of P and Cu in both shoots and grain on all measurement dates. Overall, total above-ground biomass (458) and grain yields (306 kg ha^{-1} cm^{-1} of water) increased linearly with irrigation (Fig. 2). A significant positive response to mycorrhizal inoculation was constant across irrigation levels (1,170 for biomass and 802 kg ha^{-1} for grain). Due to the smaller size of water-stressed plants, but a consistent growth response to inoculation across water treatments, the proportional response of maize to inoculation with the mycorrhizal fungus increased with increasing water stress.

Fig. 2. Effect of irrigation and mycorrhizal inoculation on maize biomass and grain yield. Solid lines represent inoculated plants, dotted lines represent noninoculated plants.

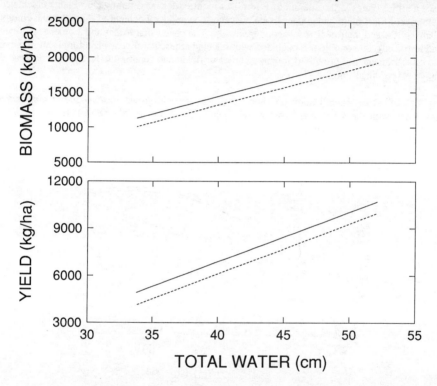

TECHNOLOGY FOR USING ARBUSCULAR MYCORRHIZA

<u>Inoculum production</u>

Culture of arbuscular mycorrhizal fungi on host plants in soil (pot cultures) has been a limiting factor in the production of these fungi (Jarstfer and Sylvia, 1992). However, as more efficient technologies become available, inoculum production should progress towards commercial application. Soilless culture and adaptations of hydroponics can produce high quality inoculum with propagule numbers many times greater than the pot culture inoculum used in the past. The lower cost of inoculum and ease of production are making these technologies applicable in less-developed agricultural areas as well as in highly-industrialized agricultural systems which currently use phosphate fertilizers.

Culturing arbuscular mycorrhizal fungi in soilless media avoids the need for soil sterilization and allows better control over the physical and chemical characteristics of the growth medium (Jarstfer and Sylvia, 1992). Soilless media are more uniform in composition, weigh less, and facilitate aeration better than does soil. The ideal soilless mixture should hold sufficient water for plant growth and also allow good aeration. Bark, calcined clay, expanded clay and perlite provide good aeration while peat and vermiculite hold more water. The availability of P in most soilless media requires prudent management of nutrients. Frequent addition of dilute, soluble nutrient solutions, incorporation of time-release fertilizer or the use of less-available forms of P are strategies for nutrient management in soilless media.

Inoculum of arbuscular mycorrhizal fungi may be produced in hydroponic or aeroponic culture. Plants are usually inoculated in sand or vermiculite before they are transferred to these culture systems. The key to successful colonization is maintenance of low P concentrations (3 to 24 µmol). Inoculum can be produced using the nutrient film technique (NFT) by growing precolonized plants in a defined nutrient solution which flows over host roots (Mosse and Thompson, 1984). A culture system which applies a fine mist of defined nutrient solution to the roots of the host plant is termed aeroponic culture and has been adapted for growing arbuscular mycorrhizal fungi (Hung and Sylvia, 1988, Fig 3). Because the colonized-root inoculum produced in this system is free of any substrate, it can be sheared, resulting in very high propagule numbers (Sylvia and Jarstfer, 1992).

Fig 3. Illustration of an aeroponic chamber. The nutrient mist may be provided by a misting impeller or centrifugal pump system. Drawn by A.G. Jarstfer.

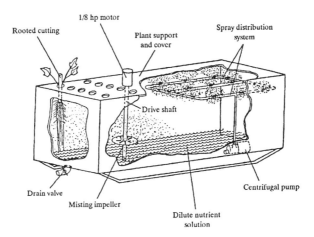

Use of inoculum

An inoculum must be infective and effective in the agricultural system for which it is intended. Infectiveness is the ability of a fungus to penetrate and spread in roots (Abbott and Robson, 1981). Many of the benefits of arbuscular mycorrhizal fungi begin early, often within the first 3 wk in the plant growing cycle. Effectiveness is the ability of arbuscular mycorrhizal fungi to enhance growth or stress tolerance of the host. Effectiveness can vary dramatically among arbuscular mycorrhizal isolates so screening studies should be undertaken before large quantities of inoculum are produced for field use (Hung et al., 1990). Effectiveness should be tested under crop production conditions since indigenous arbuscular mycorrhizal fungi, pathogens, fertility levels, and edaphic factors will influence effectiveness.

An inoculum should be small in dose and large in response (Jarstfer and Sylvia, 1992). The application of 20 to 30 metric tons of inoculum per hectare has produced growth responses but these rates are impractical on a field scale (Powell, 1984). The production of concentrated inoculum should yield a product that is less costly to store, transport, and apply. For example, with aeroponic culture of roots it is possible to produce inocula with hundreds of thousands of propagules per dry gram of root (Sylvia and Jarstfer, 1992).

Colonized roots from NFT culture have been used with success for field inoculation (Elmes et al., 1984). Colonized roots from aeroponic cultures could also be used in a similar manner or sheared for more efficient application (Sylvia and Jarstfer, 1992). Expanded-clay pellets colonized by arbuscular mycorrhizal fungi have been used for field application, either directly or air-dried before use (Dehne and Backhaus, 1986).

Any method by which viable propagules of arbuscular mycorrhizal fungi are delivered efficiently to the rhizosphere can produce colonized plants. The most common application method is to place the inoculum below the seed or seedling prior to planting (Hayman, 1987). For example, expanded clay pellets have been applied through the same tube used to drop maize seeds from an air-powered seed drill (Baltruschat, 1987). Seed coating (Hattingh and Gerdemann, 1975) and placing inoculum in hydrogels used with somatic embryos (Strullu et al., 1989) are additional methods that should be investigated for distributing inoculum. Sheared-root inocula from aeroponic cultures have been suspended in a hydrogel for application (Sylvia and Jarstfer, 1992). Applications will be most efficient when existing machines and practices are used or only slightly modified for incorporating arbuscular mycorrhizal fungi.

CONCLUSION

In response to increased concern for environmental quality, sustainable technologies need to be incorporated into agricultural systems. Management of arbuscular mycorrhizal fungi is an important aspect of such an approach (Bethlenfalvay and Linderman, 1992). However, the obligate, symbiotic nature of these fungi and limited inoculum supplies continue to impede research aimed at managing these beneficial fungi. High production costs, as well as problems associated with soil-based pot cultures, have reduced enthusiasm for their commercial development. Nonetheless, the inability to use fermentation processes to produce arbuscular mycorrhizal inocula should not discourage further attempts to develop innovative technologies to mass produce these fungi.

Several areas of inoculum production and application technology merit further investigation. Scaling up nutrient-film and aeroponic-culture technologies, along with sheared-root processing of colonized roots produced in these systems, should provide readily available, low-cost, high-quality inocula for nursery inoculation and field testing. Little is known about the storage of arbuscular mycorrhizal inocula. Further knowledge of the storage properties of arbuscular mycorrhizal inocula would encourage commercial production and the development of new formulations. These new and innovative inoculation technologies need to be developed and implemented in order to provide the most efficient application of inocula in a wide array of crop production systems.

REFERENCES

Abbott L.K. and A.D. Robson. 1981. Infectivity and effectiveness of five endomycorrhizal fungi: competition with indigenous fungi in field soils. Aust. J. Agric. Res. 32:621-630.

Baltruschat H. 1987. Field inoculation of maize with vesicular-arbuscular mycorrhizal fungi by using expanded clay as carrier material for mycorrhiza. Zeitschrift fur Pflanzenkrankheiten und Pflanzenschutz 94:419-430.

Bethlenfalvay G.J. and R.G. Linderman. 1992. Mycorrhizae in Sustainable Agriculture. ASA Special Publication no. 54, Madison, WI. 124 pp.

Bolton, H., J.K. Fredrickson, and L.F. Elliott. 1993. Microbial ecology of the rhizosphere. p. 27-64. *In* B. Metting (ed.) Soil microbial ecology. Marcel Dekker, NY.

Dehne H.W. and G.F. Backhaus. 1986. The use of vesicular-arbuscular mycorrhizal fungi in plant production. I. Inoculum production. Zeitschrift fur Pflanzenkrankheiten und Pflanzenschutz 93:415-424.

Douds D.D.,Jr., R.R. Janke, and S.E. Peters. 1993. VAM fungus spore populations and colonization of roots of maize and soybean under conventional and low-input sustainable agriculture. Agriculture, Ecosystems and Environment 43:325-335.

Elmes R.P., C.M. Hepper, D.S. Hayman, and J. O'Shea. 1984. The use of vesicular-arbuscular mycorrhizal roots by the nutrient film technique as inoculum for field sites. Ann. Appl. Biol. 104:437-441.

Finlay, R. and B. Soderstrom. 1992. Mycorrhiza and carbon flow to the soil. p. 134-160. *In* M.F. Allen (ed.) Mycorrhizal Functioning. Chapman and Hall, NY.

Harley, J.L. and S.E. Smith. 1983. Mycorrhizal Symbiosis. Academic Press, NY.

Hattingh M.J. and J.W. Gerdemann. 1975. Inoculation of Brazilian sour orange seed with an endomycorrhizal fungus. Phytopathology 65:1013-1016.

Hayman, D.S. 1987. VA mycorrhizas in field crop systems. p. 171-192. *In* G. Safir (ed.) Ecophysiology of VA Mycorrhizal Plants. CRC Press, Inc., Boca Raton, FL.

Hung L.L., D.M. Sylvia, and D.M. O'Keefe. 1990. Isolate selection and phosphorus interaction of vesicular-arbuscular mycorrhizal fungi in biomass crops. Soil Sci. Soc. Amer. J. 54:762-768.

Hung L.L. and D.M. Sylvia. 1988. Production of vesicular-arbuscular mycorrhizal fungus inoculum in aeroponic culture. Appl. Environ. Microbiol. 54:353-357.

Jarstfer, A.G. and D.M. Sylvia. 1992. Inoculum production and inoculation strategies for vesicular-arbuscular mycorrhizal fungi. p. 349-377. *In* B. Metting (ed.) Soil Microbial ecology:

Applications in Agriculture and Environmental Management. Marcel Dekker, NY.

Lee K.E. and C.E. Pankhurst. 1992. Soil organisms and sustainable productivity. Aust. J. Soil Res. 30:855-892.

Medina O.A., A.E. Kretschmer, and D.M. Sylvia. 1990. Growth response of field-grown Siratro (*Macroptilium atropurpureum* Urb.) and *Aeschynomene americana* L. to inoculation with selected vesicular-arbuscular mycorrhizal fungi. Biol. Fertil. Soils 9:54-60.

Miller, R.M. and J.D. Jastrow. 1993. The role of mycorrhizal fungi in soil conservation. p. 29-44 *In* G. Bethlenfalvay and R.G. Linderman (ed.) Mycorrhizae in sustainable agriculture. ASA Special Publication Number 54, Madison, WI.

Mosse, B. 1986. Mycorrhiza in a sustainable agriculture. p. 105-123. *In* J.M. Lopez-Real and R.D. Hodges (ed.) The role of microorganisms in a sustainable agriculture. A.B. Academic Publishers, Great Britian.

Mosse B. and J.P. Thompson. 1984. Vesicular-arbuscular endomycorrhizal inoculum production. I. Exploratory experiments with beans (*Phaseolus vulgaris*) in nutrient flow culture. Can. J. Bot. 62:1523-1530.

Newman E.I. and P. Reddell. 1987. The distribution of mycorrhizas among families of vascular plants. New Phytol. 106:745-752.

Powell, C.Ll. 1984. Field inoculation with VA mycorrhizal fungi. p. 205-222. *In* C.Ll. Powell and D.J. Bagyaraj (ed.) VA Mycorrhiza. CRC Press, Inc., Boca Raton, FL.

Reganold J.P., R.I. Papendick, and J.F. Parr. 1990. Sustainable agriculture. Sci. Amer. 262:112-120.

Strullu D.G., C. Romand, P. Callac, E. Teoule, and Y. Demarly. 1989. Mycorrhizal synthesis *in vitro* between *Glomus* spp. and artificial seeds of alfalfa. New Phytol. 113:545-548.

Sylvia D.M. 1989. Nursery inoculation of sea oats with vesicular-arbuscular mycorrhizal fungi and outplanting performance of Florida beaches. J. Coastal Res. 5:747-754.

Sylvia D.M., L.C. Hammond, J.M. Bennett, J.H. Haas, and S.B. Linda. 1993. Field response of maize to a VAM fungus and water management. Agron. J. 85:193-198.

Sylvia D.M. and A.G. Jarstfer. 1992. Sheared-root inocula of vesicular-arbuscular mycorrizal fungi. Appl. Environ. Microbiol. 58:229-232.

Tinker, P.B., M.D. Jones, and D.M. Durall. 1992. A functional comparison of ecto- and endomycorrhizs. p. 303-310. *In* D.J. Read, D.H. Lewis, A.H. Fitter and I.J. Alexander (ed.) Mycorrhizas in Ecosystems. CAB International, Wallingford, UK.

DESIGNS FOR WINDBREAKS AND VEGETATIVE FILTERSTRIPS THAT INCREASE WILDLIFE HABITAT AND PROVIDE INCOME

B.K. Miller, B.C. Moser, K.D. Johnson, and R.K. Swihart*

ABSTRACT

Windbreaks and vegetative filterstrips are often encouraged in agricultural systems to reduce soil erosion (Ticknor, 1988) and improve water quality by reducing nutrient and chemical run-off (Karr and Schlosser, 1977). In spite of their environmental advantages and government cost-share and incentive payments offered (e.g., Conservation Reserve Program [CRP], Agricultural Conservation Program [ACP], and Stewardship Incentive Program [SIP]), widespread adoption of these structures has not occurred. Lost income resulting from retiring this acreage from row crop production is often expressed by landowners as a concern.

A long-term demonstration/research project was funded by and located at Purdue University's Center for Alternative Agricultural Systems in Spring 1990 to evaluate planting designs which met environmental objectives (conserve soil, improve water quality, provide wildlife habitat) and provided income to landowners which equaled or exceeded row crop production on the same acreage. Windbreaks and filterstrips were designed which did not require labor inputs during the planting and harvest season for row crops, provided better wildlife habitat, and were more aesthetically pleasing than conventional designs.

Tree and shrub species providing ornamental cut branches utilized by the florist trade (Jenkins, 1991) are capable of giving the environmental benefits desired in these plantings and are compatible with most row crop operations as most labor inputs occur between December and April. Pussy willow (*Salix discolor*) branches were harvested as early as two years post-planting. Preliminary harvest studies indicate gross returns in excess of $13,590/ha are possible.

KEYWORDS. Windbreaks, Vegetative filterstrips, Income.

PROJECT OVERVIEW

Soil erosion continues to be a major concern and is the target of several state and federal programs designed to control it. Windbreaks and vegetative filterstrips play important roles in soil erosion control (Black and Aase, 1988; Karr and Schlosser, 1977). However, past governmental programs and proven crop yield increases from windbreaks (Kort, 1988) have not been sufficient incentives to get these systems established on a large scale (Sturrock, 1988). With the increased interest in low input sustainable agriculture and alternative crops, it seems logical to pursue some of these ventures in windbreak and filterstrip systems in order to receive benefits from both and to increase the economic return obtained from these areas taken out of row crop production to a level that is equal to or exceeds previous production of this acreage. A farmstead windbreak, vegetative filterstrip, and field windbreak were established at the Throckmorton- Purdue Agricultural Center, Romney, IN in Spring 1990 to demonstrate the potential these structures have in providing valuable wildlife habitat, reduce soil erosion, improve water quality, and provide an additional source of income. Ongoing research is designed to evaluate the environmental and economic advantages of these systems.

---

*Brian K. Miller, Extension Wildlife Specialist and Robert K. Swihart, Assistant Professor of Wildlife Ecology, Department of Forestry and Natural Resources; Bruno C. Moser, Professor and Head, Department of Horticulture; and Keith D. Johnson, Professor, Department of Agronomy; Purdue University, West Lafayette, Indiana 47907.

Farmstead Windbreak

A windbreak is a planting of trees, shrubs and/or grasses designed to reduce the speed and energy of the wind to protect a specific field, farmstead or other defined area. The primary purpose and advantages of a farmstead windbreak is to reduce home heating and cooling costs (DeWalle and Heisler, 1988), reduce snow drifting (Shaw, 1988), and to protect livestock (Dronen, 1988). Conventional designs usually consists of a minimum of two to three rows of conifer trees (Wight, 1988). The improved design demonstrated in this project (Table 1) improves the aesthetics of the windbreak, improves its value to wildlife by providing a diversity of nesting structures and an additional food source, and provides additional products for family use (i.e., fruit and nut production, Christmas trees). Depending on the scale of production, these products can be used strictly for family use, or the surplus can be sold. In addition, hardwood species can be added which increase the windbreak density and can provide firewood, fenceposts, and timber production.

Table 1. Woody Plant Species Providing Edible Fruit Planted in a Farmstead Windbreak on the Throckmorton-Purdue Agricultural Center.

Species	Spacing Between rows (m)	Spacing Within rows (m)
Persimmon (*Diospyros virginiana*)	6.7	4.6
Chinese Chestnut (*Castanea mollissima*)	6.7	4.6
Apples (*Malus cultivars*)	6.7	4.6
Pears (*Pyrus cultivars*)	6.7	4.6
Grapes (*Vitis cultivars*)	3.7	1.8

Vegetative Filterstrip

Vegetative filterstrips are bands of vegetation located adjacent to wetlands, streams, and watercourses. When properly designed, filterstrips have proven effective in removing nutrients (Dillaha et al., 1989; Osborne and Kovacic, 1992) sediment and other suspended solids from run-off produced by agricultural fields and feedlots (Dillaha et al., 1989). In addition, filterstrips may influence aquatic systems by altering water temperature (Gray and Eddington, 1969) which in turn reduces rates of phosphorous disassociation from sediments (Sommers et al., 1975) and increases oxygen carrying capacity.

The accepted width of filterstrips for CRP enrollment is 20.1 - 30.2 m and both woody and vegetative material is accepted. In addition, some herbicide labels now require a setback from watercourses. When adapted perennial forage species are planted along streams and drainage ditches, they reduce soil erosion along the bank and intercept nutrients and soil particles before reaching water. Typical filterstrips are narrow, mowed frequently thus providing poor wildlife habitat, and the only compensation for removing this acreage from crop production comes from government incentive programs.

Forages are essential for farming operations including ruminant livestock and horses. When concentrating the forage along watercourses or in strips between vegetable crops, the forage serves as a filterstrip or windbreak, respectively, (Black and Aase, 1988) while providing a return to the producer. Opportunities also exist for cash cropping hay. Forages planted between rows of trees and shrubs can be used as wildlife habitat, or livestock feed. If distance between rows is great enough, the forage can be mechanically harvested as hay or silage. If small trees are protected with tubes, grazing, between narrow spacings is possible. When properly managed, these forages provide excellent nesting, feeding, and roosting habitat for wildlife (Hull and Robel, 1992).

The improved filterstrip design demonstrated in this project uses both shrub and forage species which enhances the aesthetic appearance of the filterstrip, improves its value to wildlife by providing a diversity of nesting habitat, escape and winter cover, and an additional food source. The shrub species used (Table 2) in this project have branches with unique stem form, stem color, and flowers which are of economic importance to the florist trade (Jenkins, 1991). Branches were harvested as early as two years post-planting during the winter months (December - March). Shrubs resprout quickly and continue to perform their environmental function while providing a short-term economic return to the landowner (Yoder and Moser, 1993). Fine hardwood tree species were planted in a filterstrip 30.2 m wide on one portion of the drainage ditch. This planting provides a long-term economic return to the landowner and is most suitable when consistent with other ownership objectives.

Table 2. Shrub Species Planted in a Vegetative Filterstrip at the Throckmorton-Purdue Agricultural Center Which Provide Income From the Florist Trade.

Species	Characteristics			Spacing (m)	
	Stem Color	Stem Form	Flowers	in row	between row
Corkscrew Willow (*Salix matsudana 'Tortuosa'*)		X		2.4	6.1
Pussy Willow (*Salix discolor*)			X	1.8	3.0
Yellow Twig Dogwood (*Cornus sericea 'Flaviramea'*)	X			1.5	3.0
Red Osier Dogwood (*Cornus sericea*)	X			1.5	3.0
Bailey's Red Osier Dogwood (*Cornus sericea 'Bailey'*)	X			1.5	3.0

Field Windbreak

The primary purpose of a standard field windbreak is to reduce wind erosion by slowing the speed and energy of the wind (Heisler and DeWalle, 1988). In addition, less drying winds during hot summer conditions often reduces transpirational moisture loss and results in increased crop yields (Kort, 1988; Davis and Norman, 1988). A conventional design usually consists of a single row of trees. Row crops are planted as close to the trees as possible leaving very little herbaceous vegetation as a buffer. The improved design demonstrated in this project consists of two-row windbreaks that have a rich herbaceous vegetative component between rows and on each side of the windbreak. This improves the aesthetics of the windbreak, improves its value to wildlife by providing a diversity of nesting structures, improved roosting habitat, escape and winter cover and an additional food source (Johnson and Beck, 1988). The trees selected for these plantings (Table 3) can provide additional products for sale or farm use (fruit and nut production, Christmas trees, firewood and fenceposts, balled and burlapped landscape stock, and timber production). The shrub species used (Table 3) provide edible fruit or nuts, can be sold as balled and burlapped landscape stock, or have branches with unique stem form, foliage, flowers, or ornamental fruit which is of economic importance to the florist trade. Multiple row windbreaks and replanting are required to accommodate practices resulting in plant removal (i.e., landscape stock, firewood, fenceposts, timber, or Christmas trees). Timing of planting and removal has to be staggered to assure that an effective windbreak remains in place. The addition of some hardwood species can provide timber, firewood, or fenceposts. However, the selection of these species is a more long-term venture, as a return may not be realized for 15-60 years. Practices requiring branch removal for harvest do not harm the existing plants and therefore the cost of repeated planting is not necessary. This practice actually thickens the windbreak due to resulting regrowth.

Table 3. Shrub Species Planted in a Field Windbreak at the Throckmorton-Purdue Agricultural Center Which Provide Income.

Species	Fruit Orn.	Edible	Stem Form	Flowers	Unique Foliage	Spacing within rows (m)
Hazelnut (*Corylus americana*)		X				1.8
Elderberry (*Sambucus pubens*)	X					1.8
Nanking Cherry (*Prunus tomentosa*)		X		X		1.8
Meadowlark Forsythis (*Forsythia 'Meadowlark'*)				X		1.8
Redbud (*Cercis canadensis*)			X	X		2.4
Sea Buckthorn (*Hippophae rhamnoides*)	X		X		X	1.8
Red Chokeberry (*Aronia arbutifolia 'Brilliantissima'*)	X					1.8
Vernal Witch Hazel (*Hamamelis vernalis*)				X	X	1.8

MATERIALS AND METHODS

Four designs which met three income objectives (1) short term -- six forage mixes providing annual forage, 2) midterm -- four grass/legume mixes with noted wildlife value planted with strips of shrubs (13 species) whose branches can be harvested within 2-3 years post-planting and used by the nation's seven billion dollar per year florist industry 3) midterm -- edible fruits which can be raised for home use or sale and 4) long term -- hardwood species which will provide timber, firewood, or fenceposts were planted in spring 1990.

<u>Planting Design</u>

Farmstead Windbreak

Two rows of conifers spaced 4.6 m apart extend along the north and west sides of all buildings. Several combinations of the following conifer species were used: white fir, white pine, white cedar, red pine, blue spruce, and Norway spruce and were spaced 1.8 - 3.7 m (depending on species) within the row.

Various plant species that provide additional benefits to the farm family were planted on the leeward side of the windbreak (Table 1). Horticultural species were selected which contribute to the shelterbelt effect, while at the same time, providing fruits suitable for personal consumption or sales at roadside market or the other fresh market distributors.

Six varieties of newly introduced scab immune apples from the Purdue University apple breeding program have been planted as part of the shelterbelt. These apples do not require spray for the major fungus disease attacking apples, apple scab, and can be managed with minimal spray programs compared to popular varieties. These apples represent the results of 35 years of apple breeding research at Purdue University and should contribute significantly to low input apple production in the future. Varieties include Coop 27, Coop 28, Coop 29, Enterprise, Dayton, and Williams' Pride. The range of selections will provide apples maturing throughout the normal apple season with colors ranging from dark red to yellow, depending on variety.

Four varieties of grapes have been planted to provide primarily fresh fruit for the farm family. The grape arbor could be expanded to include quantities for roadside marketing, etc. Varieties include Catawba, Niagara, Concord, and Fredonia which are older varieties requiring minimal management strategies to ensure reasonable crop production.

Additional plantings of Chinese chestnut trees, fireblight resistant pear trees from the Purdue pear breeding program ('Honeysweet'), and persimmons were added from 1991-1994.

Vegetative Filterstrips -- forages

Various forages have been planted throughout the filterstrip to reduce soil erosion and nutrient run-off into the water course and to increase wildlife habitat. No trees or shrubs were planted on one side of the ditch to facilitate ditch access for maintenance. The 20.1 m wide grassy filterstrip also provides an area for field access, shrub maintenance and harvesting, and equipment turning.

Six planting treatments, (reed canarygrass with or without alsike clover, low-endophyte tall fescue with or without red and ladino clovers, and high-endophyte tall fescue with or without red and ladino clovers) were planted in plots approximately 50 m long and replicated three times. Size of grass plots are adequate to evaluate habitat preference of small mammal species.

Vegetative Filterstrips -- trees

Three fine hardwood species (green ash, black walnut, and northern red oak) were planted 2.4 m apart in nine rows (three rows of each species) spaced 3.4 m. This planting fits nicely in a filterstrip 30.1 m wide (the maximum size cost-sharable). Since the planting of fine hardwoods is a long term venture, the selection of these species is most appropriate in areas that will be taken permanently out of crop production. Two forage species, ladino clover, orchardgrass, and native weeds (a legume, cool-season grass, and control treatment respectively) were planted in plots approximately 50 m long. Three replications of these three treatments were randomly located along the length of the hardwood planting creating a 3 x 3 latin square design.

Vegetative Filterstrips -- shrubs

Several species of shrubs suitable for filterstrip conditions were planted in spring 1990. Species were selected for their potential for growing in wet sites along waterways and streams and for potential production of cut branches suitable for the florist trade during winter months. Shrubs selected are those which grow with multiple stems from the ground, spreading over time to control erosion and run-off (Table 2). Branches harvested in January through March are suitable for florist trade use because of their ornamental flower buds, unique bark color or contorted growth habit. Due to the rapid growth of these species, an economic return is possible two-three years post planting. Shrubs were planted in two-three rows and spacings varied by species (Table 2).

Field Windbreak

A two row windbreak in which trees are not removed for harvest (just trimmed) was planted. The northern windbreak is composed of 1 row of conifers (1/2 white pine and 1/2 Norway spruce spaced 3.7 m and 3.0 m, respectively) and 1 row of shrubs with income potential. The western windbreak is composed of one row of deciduous trees (2/3 pin oak and 1/3 bald cypress, custom planted according to micro site characteristics and spaced 2.4 m apart) and one row of shrubs with income potential. The spacing between shrubs ranged between 1.8 - 2.4 m. Rows were spaced 3.7 m apart.

A number of species used by the florist trade and are a source of fruits and nuts are suitable contributors to windbreak plantings, particularly those of a shrub-like nature that grow 1.8 - 3.0 m tall and provide low growing density to compliment trees and evergreens included in the windbreak. Species selected were those which have potential for attracting wildlife because of their fruit and growth structure while providing fruits and cut branches for both edible and florist associated uses (Table 3).

Three forage mixes (orchardgrass-birdsfoot trefoil; redtop-timothy-red clover-ladino clover; and native warm-season grasses, big bluestem-Indiangrass-little bluestem-sideoats grama and an adapted wildflower mix) were planted with conifers and income-producing shrubs in this field windbreak design. These mixes were selected because of their accepted value to wildlife, aesthetic qualities, and low maintenance required. These three treatments were randomly placed in plots approximately 70 m long and replicated three times. Total windbreak width including forages was approximately 7 m. The vast differences in forage growth characteristics will permit evaluation of tree or shrub-forage competition.

Monitoring

Wildlife

A long-term monitoring project has been established to evaluate the changes in bird and small mammal use as these plantings mature. Small mammal species are censused every three years using standard mark-release-recapture techniques for mammals (American Society of Mammalogists 1987) and birds are sampled using point transects (Buckland et al. 1993). Comparison of species density and abundance will reflect the importance of these plants to various wildlife species as they mature.

Tree-forage interaction

Tree heights and caliper at ground level from a subsample of each experimental unit in the vegetative filterstrip fine hardwood plantings have been measured between 1990-1994. Future monitoring and analysis will reveal if the associated grass or legume cover has any significant affect on tree growth.

Branch Production Potential and Harvest Strategies

Four harvest strategies (annual and biannual selective harvest and annual and biannual non-selective harvest) are being evaluated on pussy willow plantings in the vegetative filterstrip (Yoder and Moser 1993). A similar evaluation was initiated on the other filterstrip shrub species in January 1994 and will be initiated in windbreak plantings when shrubs reach the appropriate stage of maturity.

Future Research

A research project to evaluate the effects of various vegetative filterstrip designs on water quality is being initiated in spring 1994.

RESULTS AND DISCUSSION

Small mammals

In 1992, trapping before (September, 820 trap nights) and after (November, 812 trap nights) harvest of adjacent fields resulted in capture of 134 individuals of six species, including short-tailed shrews (*Blarina brevicauda*), white-footed mice (*Peromyscus leucopus*), deer mice (*P. maniculatus*), house mice (*Mus musculus*), prairie voles (*Microtus ochrogaster*), and thirteen-lined ground squirrels (*Spermophilus tridecemlineatus*). Species richness was greater in areas consisting of clover and grass ($\bar{x} = 4.8$) than in grass monocultures ($\bar{x} = 2.5$).

Short-tailed shrews were the dominant small mammal present (76% of individuals captured). Our trapping data revealed abundances of shrews in filter strips (5-6 per 100 trap nights [TN]) that far exceed abundances reported for shrews in other preferred habitats (0.2-3.2/100TN based on a review of the literature). Thus, the filter strip appears to provide remarkably good habitat for short-tailed shrews.

Birds

Standard data on bird use of the filter strip was obtained in June-August 1991 in the filter strip as well as 150 m away. Peak numbers and species diversity occurred in June in both the filter strip (11 species) and the fields (10 species). Over the course of the summer, monthly diversity averaged 9.3 species in the filter strip and 6.0 species in the fields. Abundance was 3.4-6 times greater in the filter strip than in the fields on a monthly basis, with average values of 69 birds and 17 birds in the two habitats, respectively. Red-winged blackbirds (*Agelaius phoeniceus*) were the most common species occurring at the site in both habitat types.

Branch Production

The branch production potential and harvest strategies of shrubs used by the florist trade are being evaluated (Yoder and Moser, 1993). A selective harvest technique beginning two growing seasons post-planting yielded an average of 84 marketable pussy willow branches (>60 cm) per plant. At a conservative market wholesale price of $.10 per branch and minimum plant density of 1631 plants/ha, a minimum gross return in excess of $13,590/ha seems possible.

The greenhouse and nursery industry is one of the fastest growing sectors in U.S. agriculture, representing 10 percent of all farm crop cash receipts. Ongoing research will assess the potential market for these products, evaluate the demand and profitability for value added strategies, and assess the best distribution system for producers. An ongoing study of post-harvest care and storage of these branches will provide cost and input assessments needed in this analysis.

CONCLUSION

This demonstration/research project is a long term venture designed to evaluate income options in windbreak and filterstrip planting. Ongoing and future long-term research will evaluate the effects of these plantings on water quality, soil erosion, and wildlife habitat. Many species and design options which provide income equal to or exceeding row crop production are possible in these plantings. This diversity makes this design concept flexible enough to be integrated into most farming operations. It is important to consider the landowner's farm management objectives, existing operation, income goals, and geographic location when designing a planting.

Establishing windbreaks and filterstrips which meet all environmental objectives (water quality improvement, erosion control, and wildlife habitat) throughout the entire agricultural landscape is imperative if we are to practice environmentally sound agriculture in the future that is in balance with the surrounding ecosystem. The use of these systems on a large scale in agricultural landscapes would enhance biodiversity within the landscape, provide critical habitat areas, and provide travel corridors between isolated habitats for species that otherwise could not traverse agricultural land. The concept of designing these systems in a way that they pay for themselves will ensure their maintenance and retention long term. The existing model of relying on government incentive programs is too erratic to provide environmental stability within the agricultural ecosystem.

REFERENCES

1. American Society of Mammalogists. 1987. Acceptable field methods in mammalogy: preliminary guidelines approved by the American Society of Mammalogists. J. Mammalogy 68:supplement. 18 pp.

2. Black, A.L. and J.K. Aase. 1988. The use of perennial herbaceous barriers for water conservation and the protection of soils and crops. Agric. Ecosystems Environ. 22/23:135-148.

3. Buckland, S. T., D. R. Anderson, K. P. Burnham, and J. L. Laake. 1993. Distance sampling: estimating abundance of biological populations. Chapman and Hall, London. 445 pp.

4. Davis, J.E. and J.M. Norman. 1988. Effects of shelter on plant water use. Agric. Ecosystems Environ. 22/23:393-402.

5. DeWalle, D.R. and G.M. Heisler. Use of windbreaks for home energy conservation. Agric. Ecosystems Environ. 22/23:243-260.

6. Dillaha, T.A., R.B. Reneau, S. Mostaghimi, and D. Lee. 1989. Vegetative filterstrips for agricultural non-point source pollution control. Trans. Amer. Soc. Agric. Eng. 32:513-519.

7. Dronen, S.I. 1988. Layout and design criteria for livestock windbreaks. Agric. Ecosystems Environ. 22/23:231-240.

8. Gray, J.R. and J.M. Eddington. 1969. Effect of woodland clearance on stream temperature. J. Fish. Res. Bd. Can. 26:399-403.

9. Heisler, G.M. and D.R. DeWalle. Effects of windbreak structure on wind flow. Agric. Ecosystems Environ. 22/23:41-69.

10. Hull, S.D. and R.J. Robel. 1992. Avian relative abundance and productivity in Conservation Reserve Program and rowcrop fields in the Flint Hills of Kansas. Tech. Paper No. 239. Midwest Fish and Wildlife Conf. 54:239.

11. Jenkins, D.F. 1991. Woody plants as cut flowers. Proc. Assoc. of Specialty Cut Flower Growers, Inc., Oberlin, OH. p. 68-74.

12. Johnson, R.J. and M.M. Beck. 1988. Influences of shelterbelts on wildlife management and biology. Agric. Ecosystems Environ. 22/23:301-335.

13. Karr, J.R. and I.J. Schlosser. 1977. Impact of Nearstream vegetation and stream morphology on water quality and stream biota. USEPA, Athens, Georgia. Project Report EPA/600/3-77-097.

14. Kort, J. 1988. Benefits of windbreaks to field and forage crops. Agric. Ecosystems Environ. 22/23:165-190.

15. Osborne, L.L. and D.A. Kovacic. 1992. Effectiveness of riparian vegetated buffer strips for reducing non-point source pollution of streams. Tech. Paper No. 105. Midwest Fish and Wildlife Conf. 54:185.

16. Shaw, D. 1988. The design and use of living snow fences in North America. Agric. Ecosystems Environ. 22/23:351-362.

17. Sommers, L.E., D.W. Nelson, E.J. Monke, D. Beasley, A.D. Bottcher, and D. Kaminsky. 1975. Nutrients are concentrated in subsurface flow. EPA-905/9-75-006. U.S. Environmental Protection Agency, Chicago, Illinois. pp. 63-154.

18. Sturrock, J.W. 1988. Shelter: Its Management and Promotion. Agric., Ecosystems and Environ. 22/23:1-13.

19. Ticknor, K.A. 1988. Design and use of field windbreaks in wind erosion control systems. Agric. Ecosystems Environ. 22/23:122-132.

20. Wight, B. 1988. Farmstead windbreaks. Agric. Ecosystems Environ. 22/23:261-280.

21. Yoder, K.S. and B.C. Moser. 1993. Pussy willow branches - A new crop for sustainable agriculture. Hort. Science 28(5):191.

APPLICATION OF MSF PROGRAM FOR SIMULATION OF ECOLOGICAL SYSTEMS

W.Wojcik and M.Wojcik [*]

ABSTRACT

For the analyses of the ecological systems the systems analysis proposed by H.T.Odum can be applied (Odum, 1983). Starting point of this method is a diagram and model of a system with all its components, outside sources of energy, and the paths of flow of mass and energy. The diagram visualizes the functioning of a systems and helps to understand relationship between the components. Then the mathematical equations could be connected with each path for the simulation of a model. The Mass Flow Simulation Program has been developed for simulation of such models. The MSF has been written in a Turbo Pascal 6.0 Object Oriented Programming and can be used by the persons not experienced with the programming. A simulated system is based on practically unlimited number of the sources, storages, and flows. The user can select the types of energy sources, storages, and simulation steps. The mathematical equations can be optional from proportional up to differential equations.
KEY WORDS: ecological systems, systems analysis, ecological modelling

INTRODUCTION

In analysis of the ecological systems a necessity exists to make a systems analysis and a simulation of a mass and energy flow. For such purpose a method can be proposed which apply a systems analysis developed by H.T.Odum. Several excellent textbooks and papers describe this method (Odum, 1983, Odum,1988, Odum,1989), therefore only the basic principles of the method will be described here. The analysis begins from preparing a diagram of an ecological system with all the external sources of energy, components and different kinds of connections describing the flow of mass and energy. Most convenient way of drawing such a diagram is to apply energy symbols proposed by T.T.Odum (Odum, 1983). This phase of research helps to understand how the system is functioning and what are the connections between the components of the system (Fig.2). In the next step of the analysis, mathematical equations should be designated for each pathway of mass and energy flow. These equations describes the correlation between the flow and the components of the system. To simulate the performance of the system one has to set the initial values of the energy received from the external source and of the mass in the so called storages as well as coefficients of flow equations for all given connections. Also, a time step and a number of steps has to be set follow the changes in the time of the analyzed parameters of the ecological system.

PROGRAM DESCRIPTION

This program can be used in simulation of many kinds of ecological systems, although original version of the program was prepared mainly for the simulation of the wetland systems.
The program was written as an Object Oriented Program in Turbo Pascal 6.0 using Turbo Vision Communication. All components of the model of the ecological system are the dynamic objects in the program. The objects (components of the model) of the same categories (as for example sources or storages) create a corresponding dynamic list, i.e. list of sources or storages. This list is easily accessible using the extended pull-down menu (Fig.1).

---

[*]University of Mining of Metallurgy, al. Mickiewicza 30, 30-059 Kraków, Poland

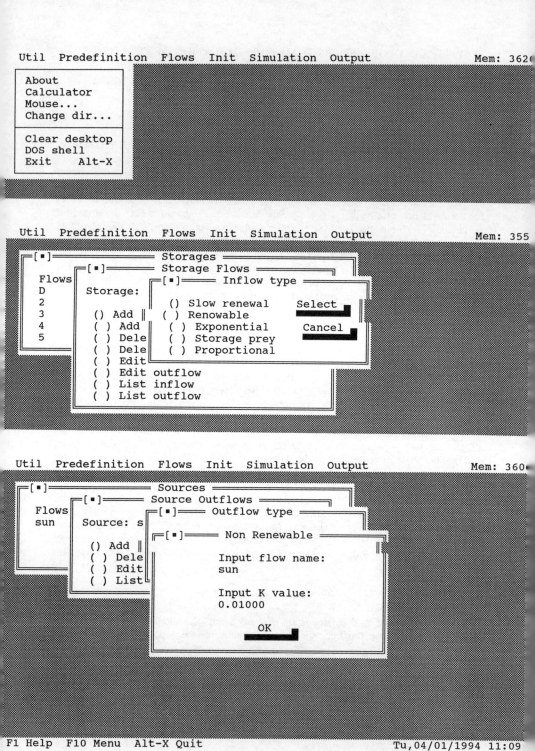

Figure 1. The Menu Screen printout of the MFS Program

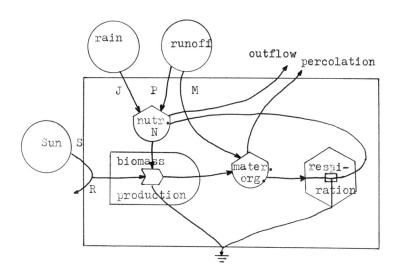

Figure 2. Enegry language diagram of a pond (after Odum, 1988)

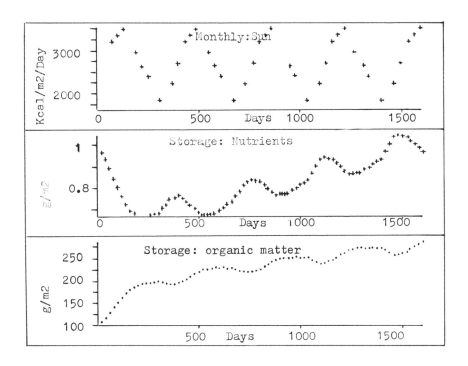

Figure 3. Results of biomass simulation with application of the MFS Program.

The sources contain a dynamic list of the outflows, and the storages the lists of both the inflows and the outflows. All these objects and their lists have corresponding virtual methods, which in a moment of a simulation made sequential calculations for the particular elements in the system in the sequantial time steps.

So the user makes in a very beginning the definition (Menu: Predefinition) on how many and what types of sources and storages of his model will consists of. The elements of the model receive their names, and physical units are defined in which the values representing the elements will be described. Also the number and length of the time steps for the simulation process, and the time units should be defined (Menu: Predefinition). Next, in Menu: Flows, a character of
of inter-connections between the elements should be defined, as well as a rate of the changes by the proper flow coefficients. The program itself shows declared objects and possible types of connections between the elements. In Men: Unit, the initial values (for a time-step equal 0) are defined for all initiated elements on the lists of sources and storages.

Next the computation can be made (Menu: Simulation) and later the results will be shown on the graphs (Menu: Output) for chosen component of the system.

EXAMPLE

For an example one ecological system described by H.T.Odum (Odum, 1989) has been chosen. This was a pond with the external sources of energy such as sun, rainfall, and runoff (Fig. 2). The diagram has been drawn with application of the symbols to distinguish outside energy sources, biomass production, storage of nutrients, storage of organic matter, and respiration as a consumers unit which the organic matter. These symbols are so called energy language symbols (Odum, 1983).

In the described system the respiration is equal to photosynthesis. The sun and the nutrients are the basis for biomass production, which is accumulated in the organic matter storage. Respiration makes the cycle of nutrients closed. There is also dissipation of energy drawn in the bottom of the diagram.

Some results of the simulation are shown on Figure 3. At the beginning a decrease in quantity of nutrients can be observed. Later the nutrients accumulation is recorded as a cyclic process due to changing character of supply of sun energy. The organic matter however is steadily increasing its quantity just from the beginning.

Described application of the program to the simple system does not show all of its features, and advantages. This program was also used to simulate more complicated ecological systems with consideration of the interaction if the environment with the contaminants.

ACKNOWLEDGEMENTS

The MFS Program has been written as part of the Sendzimir Family Project in wich Prof. H. T. Odum of the University of Florida was a principal investigator. Also we would like to acknowledge Prof. Odum for teaching us his energy analysis method.

REFERENCES

1. Odum, H.T.1983. Systems ecology. Wiley & Son

2. Odum, H.T. and E.C.Odum. 1989. Computer Minimodels and simulation Exercises. University of Florida.

3. Odum, H.T. Microcomputer simulation models for introducing principles of ecological and economic systems. University of Florida.